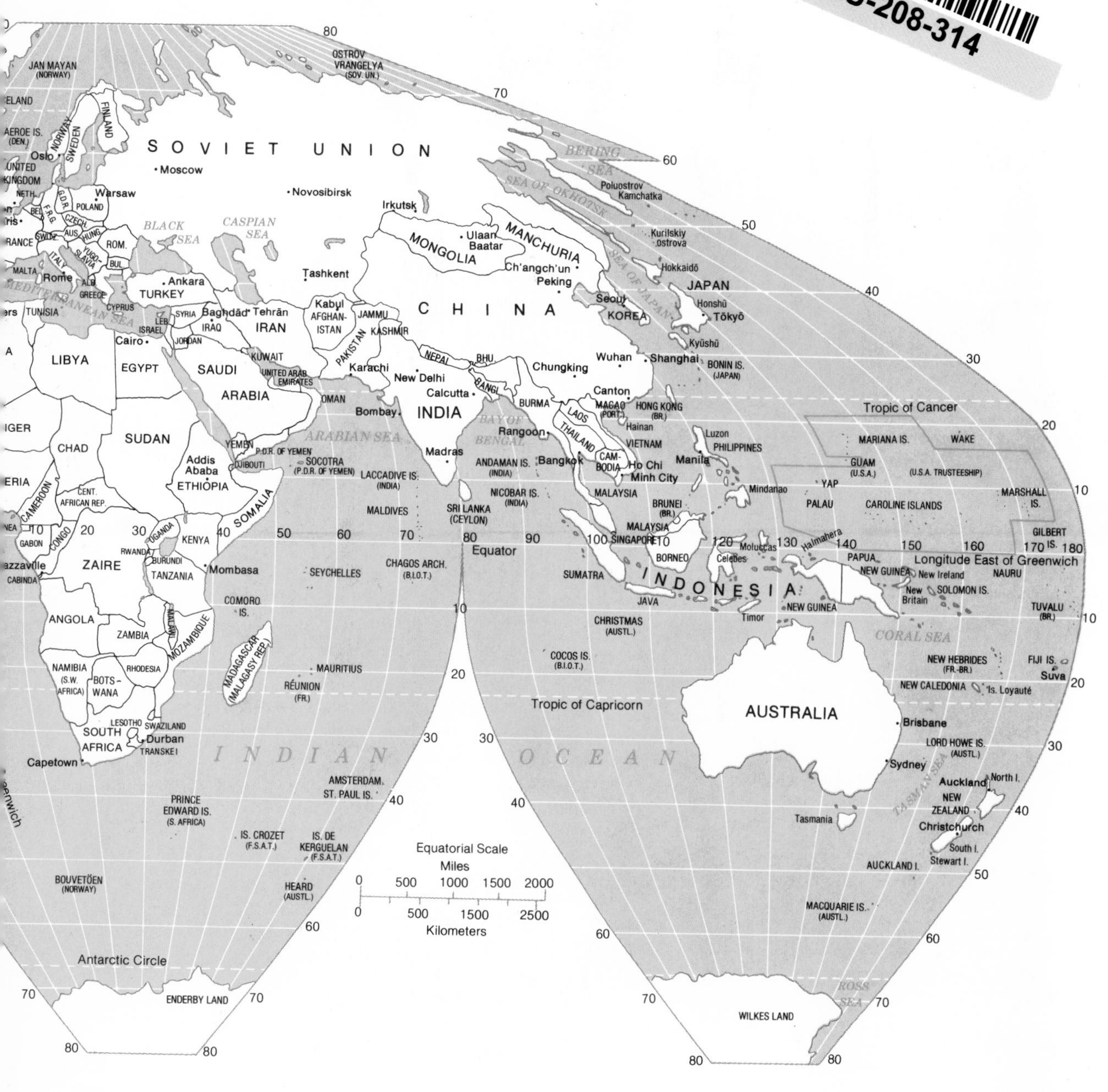
JAN MAYAN
(NORWAY)
OSTROV
VRANGELYA
(SOV. UN.)
FINLAND
NORWAY
SWEDEN
Oslo
SOVIET UNION
Moscow
Novosibirsk
Irkutsk
BERING
SEA
SEA OF OKHOTSK
Poluostrov
Kamchatka
Warsaw
POLAND
G.D.R.
F.R.G.
CZECH.
AUS.
HUNG.
SWITZ.
YUGO-
SLAVIA
ROM.
BUL.
ITALY
ALB.
GREECE
MALTA
Rome
BLACK
SEA
CASPIAN
SEA
MONGOLIA
Ulaan
Baatar
MANCHURIA
Kurilskiy
ostrova
Hokkaidō
SEA OF JAPAN
JAPAN
Honshū
Tōkyō
Kyūshū
BONIN IS.
(JAPAN)
Ch'angch'un
Peking
Seoul
KOREA
Tashkent
Ankara
TURKEY
CYPRUS
MEDITERRANEAN SEA
TUNISIA
LEB.
SYRIA
ISRAEL
JORDAN
Baghdād
IRAQ
Tehrān
IRAN
Kabul
AFGHAN-
ISTAN
JAMMU
KASHMIR
PAKISTAN
CHINA
Cairo
LIBYA
EGYPT
KUWAIT
SAUDI
ARABIA
UNITED ARAB
EMIRATES
OMAN
Karachi
New Delhi
NEPAL
BHU.
BANGL.
Calcutta
Chungking
Wuhan
Shanghai
Canton
MACAO
(PORT.)
HONG KONG
(BR.)
Hainan
BURMA
LAOS
THAILAND
Rangoon
Bombay
INDIA
BAY OF
BENGAL
ARABIAN SEA
VIETNAM
Bangkok
CAM-
BODIA
Ho Chi
Minh City
Luzon
PHILIPPINES
Manila
Tropic of Cancer
MARIANA IS.
WAKE
GUAM
(U.S.A.)
(U.S.A. TRUSTEESHIP)
YAP
PALAU
CAROLINE ISLANDS
MARSHALL
IS.
Mindanao
IGER
CHAD
SUDAN
YEMEN
P.D.R. OF YEMEN
DJIBOUTI
SOCOTRA
(P.D.R. OF YEMEN)
Addis
Ababa
ETHIOPIA
SOMALIA
Madras
ANDAMAN IS.
(INDIA)
LACCADIVE IS.
(INDIA)
NICOBAR IS.
(INDIA)
MALAYSIA
BRUNEI
(BR.)
ERIA
CAMEROON
CENT.
AFRICAN REP.
MALDIVES
SRI LANKA
(CEYLON)
MALAYSIA
SINGAPORE
GILBERT
IS.
GABON
CONGO
UGANDA
KENYA
RWANDA
BURUNDI
ZAIRE
TANZANIA
Mombasa
SEYCHELLES
CHAGOS ARCH.
(B.I.O.T.)
Equator
SUMATRA
BORNEO
Celebes
Moluccas
Halmahera
PAPUA
NEW GUINEA
New Ireland
NAURU
Longitude East of Greenwich
INDONESIA
JAVA
New
Britain
SOLOMON IS.
NEW GUINEA
Timor
TUVALU
(BR.)
CABINDA
ANGOLA
ZAMBIA
MALAWI
MOZAMBIQUE
COMORO
IS.
CHRISTMAS
(AUSTL.)
CORAL SEA
COCOS IS.
(B.I.O.T.)
NEW HEBRIDES
(FR.-BR.)
FIJI IS.
Suva
NAMIBIA
(S.W.
AFRICA)
BOTS-
WANA
RHODESIA
MADAGASCAR
(MALAGASY REP.)
MAURITIUS
RÉUNION
(FR.)
NEW CALEDONIA
Is. Loyauté
Tropic of Capricorn
AUSTRALIA
Brisbane
LESOTHO
SWAZILAND
SOUTH
AFRICA
Durban
TRANSKEI
Capetown
INDIAN OCEAN
LORD HOWE IS.
(AUSTL.)
Sydney
TASMAN SEA
AMSTERDAM.
ST. PAUL IS.
Auckland
North I.
NEW
ZEALAND
Christchurch
South I.
Stewart I.
Tasmania
PRINCE
EDWARD IS.
(S. AFRICA)
IS. CROZET
(F.S.A.T.)
IS. DE
KERGUELAN
(F.S.A.T.)
Equatorial Scale
Miles
0 500 1000 1500 2000
0 500 1500 2500
Kilometers
AUCKLAND I.
BOUVETØEN
(NORWAY)
HEARD
(AUSTL.)
MACQUARIE IS.
(AUSTL.)
Antarctic Circle
ENDERBY LAND
ROSS
SEA
WILKES LAND

The Spatial Order

The Spatial Order
An Introduction to Modern Geography

Richard L. Morrill
University of Washington

Jacqueline M. Dormitzer

Maps and Diagrams by
Helen Sherman

Duxbury Press
North Scituate, Massachusetts

The Spatial Order: An Introduction to Modern Geography was edited and prepared for composition by **Beverly Harrison Miller.** Interior design was provided by **Oliver Kline and Duxbury Press.** The cover was designed by **Joseph Landry.**

Duxbury Press
A Division of Wadsworth, Inc.

Library of Congress Cataloging in Publication Data

Morrill, Richard L.
The spatial order.

Includes bibliographies and index.
1. Geography—Text-books—1945-
I. Dormitzer, Jacqueline M., 1939- joint author. II. Title.
G128.M67 910 78-32061
ISBN 0-87872-180-0

Printed in the United States of America
1 2 3 4 5 6 7 8 9 — 83 82 81 80 79

To Joann and Ralph

Contents

Preface

To many people who were first introduced to geography in elementary school, geography consisted only of descriptions of the world's places, peoples, and customs. But modern geography is much more than simple description, and *The Spatial Order: An Introduction to Modern Geography* is dedicated to a new kind of exploration: to present an understanding of how human societies organize their space and why certain patterns emerge.

Our landscape is the product of many different forces and influences, physical and human. In this text we limit our treatment to the human use and modification of the landscape. As the discipline of geography comes of age, what has the geographer's way of looking at the world led us to discover? Human occupation of the landscape is the result of choice and purpose, creating somewhat predictable patterns and uses.

The Spatial Order explores how and why human societies organize themselves on the earth's surface. Using a simplified location theory, it reveals the basic spatial order that underlies our often confusing world. Readers will become aware of the patterns of land use, trade, and movement that result from our attempts to survive and prosper. They will discover that the landscape around us makes sense when viewed as the outcome of human decisions to use land rationally in the context of a particular culture, level of technology, and physical environment.

The current generation of introductory texts in geography stresses the broad differences in world cultures as revealed in descriptions of language, race, religion, and political economy. Although this traditional approach to human or cultural geography is interesting, most introductory texts deemphasize the spatial organization of technically advanced societies. The theory and methods of analysis that constitute modern geography have been derived from work on technically advanced societies, and for this reason we give considerable attention to these societies. We frequently draw comparisons between the organization of societies at various stages of development in our approach.

The Spatial Order describes and analyzes patterns of location and interaction primarily in high technology market economies, in which most of our readers participate. But it also considers spatial patterns in centrally planned and less developed societies. The first three chapters provide an overview of the human landscape and its evolution. Chapter 1 describes world patterns of population, economic development, and culture; chapter 2 examines factors influencing location, interaction, and development and outlines the principles of location theory; and chapter 3 gives a brief summary of the evolution of the landscape, concentrating on the human forces that modify the face of the earth.

The next nine chapters deal with specific patterns of location and interaction. Chapter 4 discusses patterns formed in low technology societies; the other chapters focus on advanced societies and their patterns of agricultural location, urban settlement, central places, industrial location, urban activities, and trade, transportation, and human movement, including the diffusion of ideas and innovations. Chapter 12 reviews these patterns and the principles of human location and explores regional organization.

Chapter 13 points out some differences between theoretical and actual spatial order. It briefly considers the causes and effects of society's problems in organizing satisfactory landscapes. Chapter 14 weighs the relative importance of environment and space in shaping the human landscape. And the final chapter, 15, looks to the landscape of the future.

Geographic concepts are explained in language that is as nontechnical as possible, and key terms, printed in boldface, are defined in the glossary. Maps, diagrams, photographs, and examples drawn from real-world situations will help readers understand the concepts and relate theoretical principles to their own experience of the world.

The Spatial Order: An Introduction to Modern Geography is a successor to *The Spatial Organization of Society,* which stressed the role of economic decisions. This new book is more broadly human, with greater attention given to the influence of social, cultural, and political factors and to the environmental consequences of location decisions. It should provide a useful introduction for students taking only one course in the field and also serve as a foundation to the range of courses—industrial, agricultural, urban, and social—offered in most geography programs.

We gratefully acknowledge the comments of colleagues and students who used *The Spatial Organization of Society* and thereby helped bring about the present book. We particularly want to express our appreciation to the following people who provided reviews of the manuscript: Ronald Abler, John Baxevanis, Terence Burke, Roland E. Chardon, Robert E. Huke, John C. Lowe, Robert Mayfield, Louis D. Rosenthal, Robert Stoddard, and Michael Sullivan. We also wish to thank Helen Sherman for her superb cartography and John Sherman for his rendering of the world maps, as well as Jane MacQuarrie, our photo researcher, Beverly Miller, our copyeditor, and the production staff at Duxbury Press for their interest and help. Finally our deepest gratitude is owed to our editors, Alex Kugushev, who gave freely of his time and expertise in developing this book, and Jerry Lyons, who guided the book to completion.

The Spatial Order

Introduction

Perspective of the Book

Basic World Patterns

Spatial Experience and Level of Technology

1 Introduction

In September 1673, two French explorers, Louis Jolliet and Père Jacques Marquette, stepped out of their canoe onto a low ridge between two rivers, a branch of the Mississippi and a small river flowing into Lake Michigan. Surrounded by prairies and marshland and the scent of wild onion, they had discovered the place Indians called *Checagou* (a name referring to the wild onion).

This area had been shaped by natural processes long before humans entered the wilderness of North America. Glaciers had leveled most of the elevations and had scoured the basin of Lake Michigan. Mounds of glacial drift deposited by the last glacier, some 13,500 years ago, paralleled the flat lake plain at the southwestern tip of Lake Michigan. Above the poorly drained lake plain rose a few sand bars, spits, and ridges that marked the receding shoreline of an earlier, postglacial lake. Forests of beech and maple, watered by 100 to 150 centimeters (40 to 60 inches) of rain a year, lined the lakeshore and riverbanks. In the drier climate west of the forests and marshes, tall-grass prairies blanketed the fertile soil formed over thousands of years.

As the last glacier receded, exposing large areas for human habitation, Indian tribes began to appreciate the site's convenient location in the midcontinent, where prairie, lake, and river converged. Here at the tip of Lake Michigan, the Mississippi River system joined the Great Lakes, providing an immense interconnected waterway from the Gulf of Mexico in the south to the St. Lawrence River in Canada and to the Atlantic Ocean. The chain was broken only at a sandy ridge known as the Checagou portage.

For several hundred years there was little change in the area. Indians may have cleared some land for campsites or set fires to trap game, but they had neither the technology nor apparently the desire to alter the landscape. Then Europeans arrived in the New World. It was three hundred years ago that Marquette and Jolliet penetrated the interior of North America through the Great Lakes–Mississippi waterway. Jolliet observed in his journal that the linkage to the ocean could be completed by cutting a canal through the prairies at the Checagou portage. European interest in this strategically located region began to grow.

At first, the Europeans simply traded with Indians near Checagou or trapped beaver for the profitable overseas fur trade (beaver tophats were the rage in Europe). In 1803, Fort Dearborn was built to safeguard the passage between the Great Lakes and the Mississippi Valley and to protect the few European settlers from hostile Indians. But European settlement did not really begin until the 1830s, after the Indians were relocated west of the Mississippi and the Erie Canal was opened through New York State (1825), creating a new pattern of westward migration from older, more established centers in the Northeast.

In 1833, the Fort Dearborn settlement officially became the town of Chicago. Over the next twenty years, its population leaped from 350 to 30,000. By 1890, Chicago had more than a million inhabitants. What caused this astounding growth?

Many settlers were drawn to the area because they perceived that their lives might improve there. Farmers saw the advantages for agriculture—the fertile soil, level terrain, adequate rainfall and growing season, and good access to northeastern urban markets. European immigrants were drawn by the open land, chances for a better life, and freedom from political or religious oppression in their own countries. Entrepreneurs accompanied the growing supply of cheap labor. Merchants liked Chicago's central location and transport position. Speculators were quick to note the potential for development and rising land values. And other people were attracted by the beauty of the prairies and the lake.

Within a century, settlers transformed Checagou, the "place of wild onion," to Chicago, one of the largest metropolises in the world (figure 1.1). They cleared much of the forest, filled the marshes, widened and straightened the Chicago River, and eventually (in 1900) reversed the river's course. The canal Jolliet had envisioned was completed in 1848. That same year, Chicago's first railroad, the Galena and Chicago Union, was built. By 1856, Chicago was the largest railroad center in the world. Products from midwestern farms were processed and packed in Chicago and distributed to the rest of the nation. Logs from Michigan and Wisconsin were milled in Chicago and shipped to communities in the treeless prairies and plains. Furniture, clothing, and farm equipment were manufactured in Chicago and sent to farmers throughout the Midwest. Iron ore from northern Michigan and Wisconsin, coal from Pennsylvania, Ohio, and southern Illinois, and limestone from Michigan were fed to Chicago's steel mills. By 1875, Chicago was the nation's largest producer of steel.

Thus Chicago was first a major commercial and distribution center and then one of the nation's leading industrial cities. It was the hub of a vast national and international trade and transportation network. In the words of the poet Carl Sandburg, Chicago had become the "Hog Butcher for the World,/Tool Maker, Stacker of Wheat,/Player with Railroads and the Nation's Freight Handler;/Stormy, husky, brawling,/City of the Big Shoulders." * More and more people became aware of Chicago's favorable environment, central location, and access to agricultural and industrial raw materials (figure 1.2). Many were willing and able to invest in the city's future. Their ambition, capital, and technology spurred Chicago's development into a modern metropolis after 1840.

Today as in the past, the Chicago landscape is changing as a result of human activity. Taller skyscrapers are rising in the city center, population is spreading into the suburbs and beyond, new transport facilities are emerg-

A

B

C

1.1. The evolution of Chicago. By comparing these views of Chicago, you can see how the original landscape has been transformed. Photograph A shows an artist's conception of how Chicago looked in 1820, when it was still a small prairie outpost and fur-trading station. By 1892 (B), it had become a bustling trade and transportation center. Now one of the world's largest metropolises (C), Chicago has grown not only outward from the lake but upward, forming an impressive skyline. (Photos A and B courtesy Chicago Historical Society; photo C from World Wide Photos.)

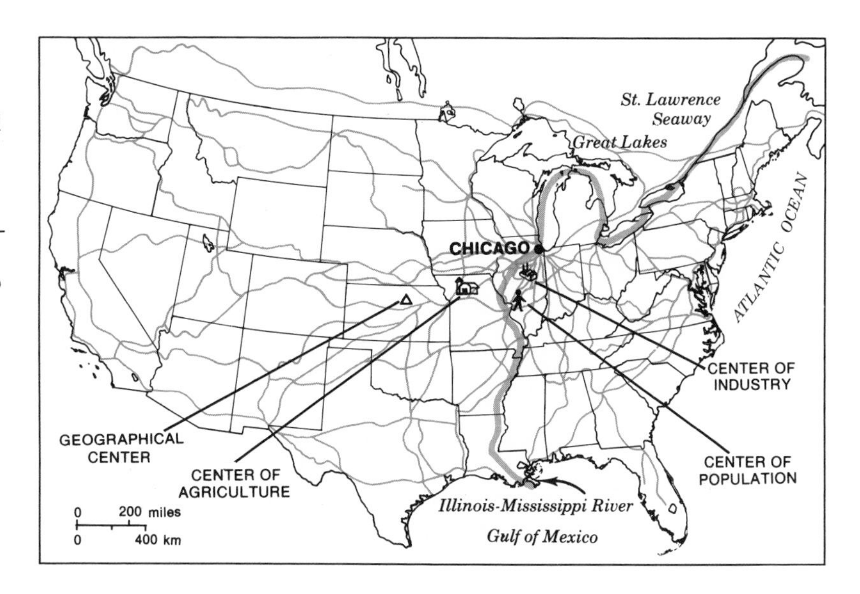

1.2. Chicago's centrality. Chicago's location in the midcontinent, at the junction of the Great Lakes and Illinois-Mississippi waterway, helped it become a center of trade and transportation. Chicago has benefited not only from its geographic centrality but also from its location near the centers of population, industry, and agriculture. In turn, the convergence of transport networks in Chicago contributed to the growth of population, industry, and agriculture in the region.

ing, and patterns of land use are shifting. With the changes have come problems: heavier loads of waste and pollution; greater loss of open space and productive farmland at the urban fringes; higher levels of traffic congestion, urban blight, racial unrest, and crime; and increasing political fragmentation. Such problems are common to large cities nearly everywhere. As the world becomes more and more urbanized, industrialized, and interdependent, the ability of cities to solve their problems and plan their development intelligently and equitably will have important implications for the future of cities in many other countries.

Chicago is unique in several respects. But the environmental, spatial, and cultural forces that shaped its settlement and development help us understand the location of human activities and the evolution of the landscape. In the next two chapters, we will explore the dynamics of human location and the drama of the changing landscape. Because Chicago so clearly illustrates major contemporary geographic concepts, we will refer to it often as a familiar, small-scale example of the larger human landscape.

PERSPECTIVE OF THE BOOK

The mark that human societies leave on the land is the object of human geographic study. This **human landscape** can be understood only as the product of the relationship between a society, defined by a particular social, economic, and political structure, and its environment.* Generally, three perspectives dominate the study of the human landscape. One is primarily concerned with how society adapts to and modifies the physical **environ-**

* Words printed in **boldface** are defined in the glossary at the back of this book.

ment; another with how environment and **culture** work together to give unique character to various regions of the earth; and the third with how the more abstract qualities of **space** influence society's use of the land, and with general principles that help explain the society-environment system. This concern with general principles has resulted in a set of theories, known as **location theory,** explaining and predicting patterns of land use and movement.

For a complete understanding of the human landscape, no single perspective is truly sufficient. The landscape should be examined in the context of physical environment, cultural elements, and spatial relationships. Most geographers, however, focus on either environment, culture, or space. Our perspective is basically spatial; we stress the role of the geometric properties of space—area, shape, relative location, distance separating places—rather than concentrate on the influence of physical or cultural characteristics of particular territories. We point out, for example, that the location of Chicago relative to the urban-industrial Northeast and to the Mississippi River system contributed to the city's success as much as, if not more than, did its climate and terrain or the particular ethnic groups that settled the region.

We believe that the human landscape can be understood as the outcome of human decisions to use land in ways that will best serve human interests or maximize human well-being, and that these decisions are greatly influenced by such spatial relationships as distance to markets and resources. Taking a spatial viewpoint, we demonstrate that individuals and societies tend to organize space to make the best possible use of the land, satisfying basic human needs for food and water, shelter, security, and community.

Peoples of widely differing technical abilities and value systems in extremely varied physical environments pursue this goal. The environment limits or constrains the ways in which a society can feed, house, and protect itself, and it influences the nature of human settlements. A society's way of life also affects its perception and use of land in its best interests. Thus Indians in the Chicago region perceived and used land in ways that differed from those of European settlers. Culture itself is influenced by both spatial and environmental factors. Distinctive languages and customs, for example, developed over a long period when most societies were spatially separated and had little contact with others; and various ways of life—hunting, herding, agriculture—were influenced by terrain, climate, and other environmental conditions.

Highlighting the role of space in human geography, our thesis is that society organizes the space it occupies. We will demonstrate the spatial order that exists in a complex world. In Chicago, the European settlers organized their space by creating a **transport network** that linked them to places with which they traded and to which they traveled, and by building residential, commercial, and industrial structures in convenient sites within the city. We will use location theory to explain the process and evolution of such organization. The theory stresses the economic and social forces that govern **location** (places where human activities occur) and **interaction,** (movements of goods, people, and ideas), assuming that society attempts to use space efficiently and rationally. We have already seen that Chicago was an efficient, rational location for trade, commerce, and industry and that human needs could well be satisfied there.

Decisions on location and interaction may also reflect goals other than those suggested by location theory. Entrepreneurs may locate their businesses in certain areas for the sake of prestige, and people who work in the city may choose homes far in the countryside because they prefer the rural environment. We will therefore discuss the human landscape in terms of a modified location theory, taking into account the environmental, cultural, behavioral, political, and historical factors that also influence society's use and organization of space.

To define the human landscape, we will first consider basic world patterns of population, development, and culture. We will also compare the ways in which space is used and organized to satisfy human needs in low and high **technology** societies. Low technology societies tend to be more self-sufficient, in that most goods and services required by the people are produced locally. Indians in the Chicago region, for example, survived mainly by hunting and gathering and engaged in only limited trade with other tribes. Their patterns of location and interaction were fairly simple, and the landscape changed little as a result of these activities. In contrast, higher levels of technology permit societies to specialize in the activities favored by local conditions and to exchange their specialized products for other needed goods and services from other regions. The more advanced agricultural technology of the European settlers in the Chicago region enabled them to take advantage of favorable soil and climate for farming. They produced corn on a large scale and sold it to regions that lacked good farmland or that needed more corn than could be grown locally. Later industrial development promoted specialized manufacturing in Chicago, which led to further trade with other regions and to greater changes in the landscape. Location theory helps us understand the resultant patterns of location and interaction.

Many of the factors that influence human location, interaction, and development will be analyzed, and the theoretical principles that govern location and interaction outlined, in chapter 2. Chapter 3 shows how these factors and principles have operated over time to create a continually evolving human landscape.

Because a society's level of technology strongly affects its use of land and its interaction with other peoples and places, we will first examine spatial patterns in low technology societies (chapter 4) and then proceed to study the more complex patterns of high technology societies. Chapter 5 describes and analyzes the rural landscape in high technology societies, focusing on the organization of agriculture.

Reflecting the importance of cities in a world that is becoming highly urbanized and industrialized, the next four chapters are devoted to the urban-industrial landscape. Chapter 6 reveals the general spatial order underlying the urban landscape. Chapter 7 specifically analyzes central place, or service center, patterns, and chapter 8 concentrates on industrial patterns. The internal order of the city is examined in chapter 9.

The specialized activities that have developed in high technology societies depend on trade flows and efficient transportation. We will see that patterns of trade and transportation tend to mirror the pattern of specialized locations, but they may sometimes lead to changes in location (chapter 10). Change also results from the spread of people and ideas (chapter 11).

In chapter 12 we review the spatial patterns explained and predicted by location theory; we also show how the human landscape can be divided into regions on the basis of distinguishing characteristics or relationships.

Throughout the book we point out the differences between the spatial patterns that would result if people made location decisions according to the principles of location theory and the actual patterns that prevail in the human landscape. Chapter 13 focuses on the problems that arise when poor decisions are made, for a variety of reasons, generating inefficient and inequitable spatial patterns.

Although we emphasize the role of spatial factors in influencing patterns of location and interaction, we do not ignore the importance of the physical environment. The relative significance of both sets of factors is evaluated in chapter 14. In the last chapter, we project the consequences of society's present use and organization of space into the future, suggesting a range of potential levels of human well-being.

To sum up, this book is structured around the ways in which societies organize space to serve their interests and needs best. This effort creates predictable patterns of land use and movements of goods and people. Though our perspective is spatial, we will often remind you how differing cultural and physical environments affect human landscapes. We will also point out that these landscapes, like our sample, Chicago, are continually evolving as individuals and societies cope with problems of food, shelter, security, and human development.

BASIC WORLD PATTERNS

To begin our study, we will consider three basic geographic patterns that help define the human landscape: the distribution of population, economic development, and cultural variation. These patterns, in turn, raise three important questions: Why do people locate in some places and not in others? Why are some places prosperous and others poor? Why do ways of life vary in different parts of the world, and how do these variations influence society's use and organization of space? Some of the answers will be suggested here and in the following chapter. The rest of the book will develop these themes.

The Distribution of Population

World maps reveal that the distribution of population, economic development, and cultural variation is highly irregular. The distribution of population is the key pattern that human geography seeks to understand because it is people who organize and modify the landscape. From figure 1.3, you can see that Japan, China, India, and Europe appear to teem with people. In contrast, North and South America, the Soviet Union, Australia, New Zealand, and most of Africa are more thinly populated, except for a few dense concentrations. And Greenland, north central Africa, and central Australia are virtually unpopulated. Some very small nations, such as the Netherlands, Belgium, and Mauritius, are very densely populated, and some large nations, such as the United States and the Soviet Union, are only moderately dense, although their total populations are quite large. The basic pattern of concentration reflects the evolution of human settlement over thousands of years. Indeed China, India, and Europe were already quite populous a thousand years ago. These were areas where superior agricultural technology permitted gradual but major rises in population.

Before the industrial revolution, when machine technology transformed society, both birthrates and death rates were high, and war, famine, and disease held population growth in check. But after the industrial revolution, rising standards of living and better health care reduced death rates. Population began to grow very rapidly and continues to do so in many less developed societies, which are in the so-called population explosion stage. The world's population has increased from 1 billion persons in 1800 to over 4 billion today.

As societies become highly developed, birthrates tend to fall and the population gradually stabilizes. There are at least three reasons for this trend (known as the *demographic transition*): the emancipation of women and their entry into the labor force, greater literacy and acceptance of contraception, and urban conditions, such as crowding and job opportunities, which dissuade people from wanting more children. It is important to observe the pattern of birthrates and death rates—infant mortality rates are considered an especially good indicator of the level of economic and social development (figure 1.4). Generally Europe, North America, Australia, and Japan have very low birthrates and infant mortality rates and low rates of natural increase (less than 1 percent a year). Several countries are stabilized or are even losing population. Some developing countries, such as China, Sri Lanka (Ceylon), and parts of India, have sharply declining birthrates, largely because of the rising literacy and status of women. Other moderately developed countries, particularly in Latin America and the Middle East, still have rapid growth rates (reflecting in part the influence of Roman Catholicism and Islam), as have most of the least developed countries of Africa and Asia. Some measures of development in major world areas are summarized in table 1.1.

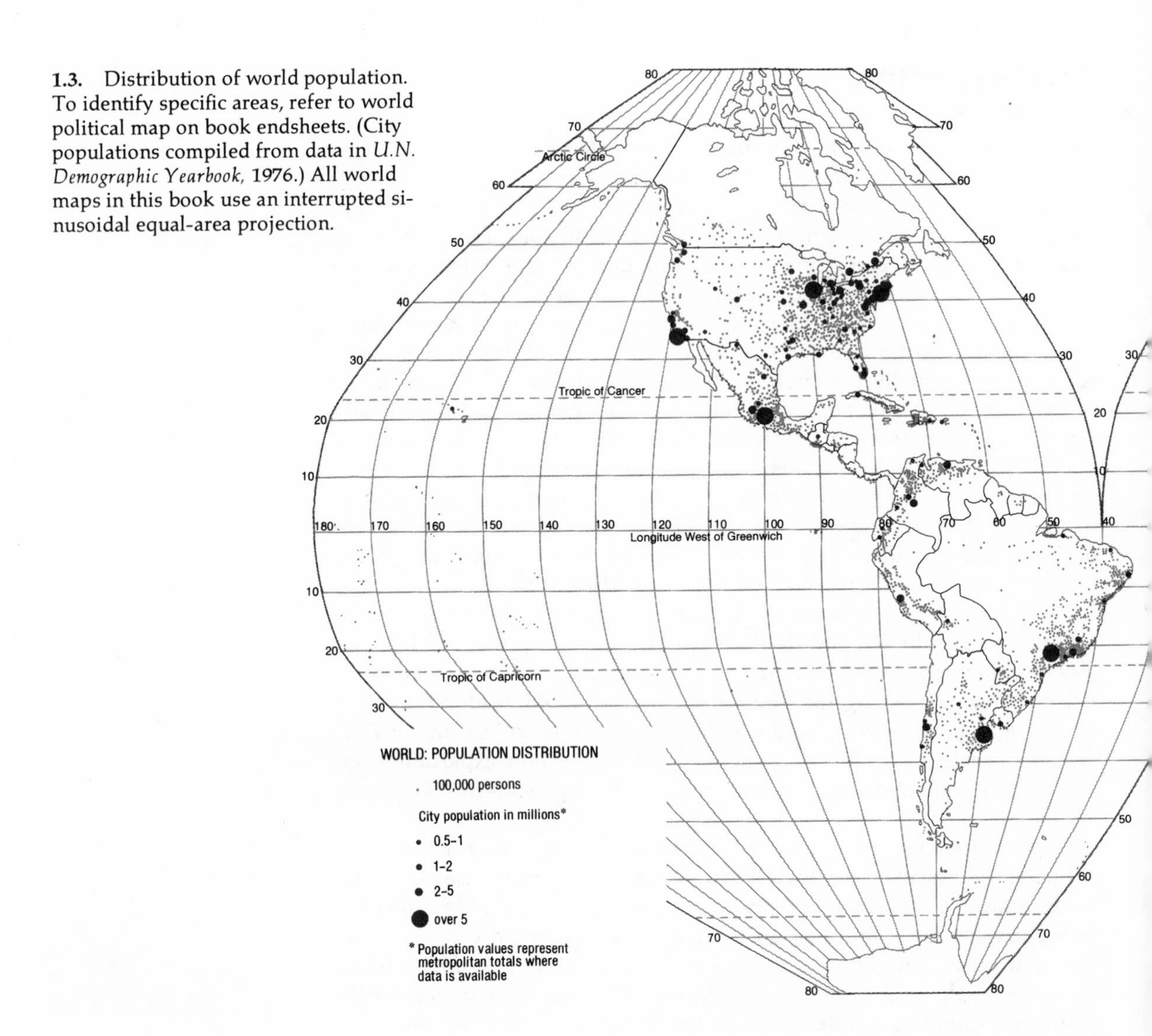

1.3. Distribution of world population. To identify specific areas, refer to world political map on book endsheets. (City populations compiled from data in *U.N. Demographic Yearbook*, 1976.) All world maps in this book use an interrupted sinusoidal equal-area projection.

The growth of world population to at least 8 billion persons by the year 2075 is almost inevitable, even if birthrates fall rapidly, because a large proportion of the world's population is under the age of fifteen, considered to be the lower limit of the child-bearing years. Such growth will create serious problems, but the idea of overpopulation is an elusive one. Very populous or very dense nations, such as Japan, are not overpopulated if they can provide adequately for their populations, whereas small, low-density nations like Mali may be overpopulated, given their present technology and organization.

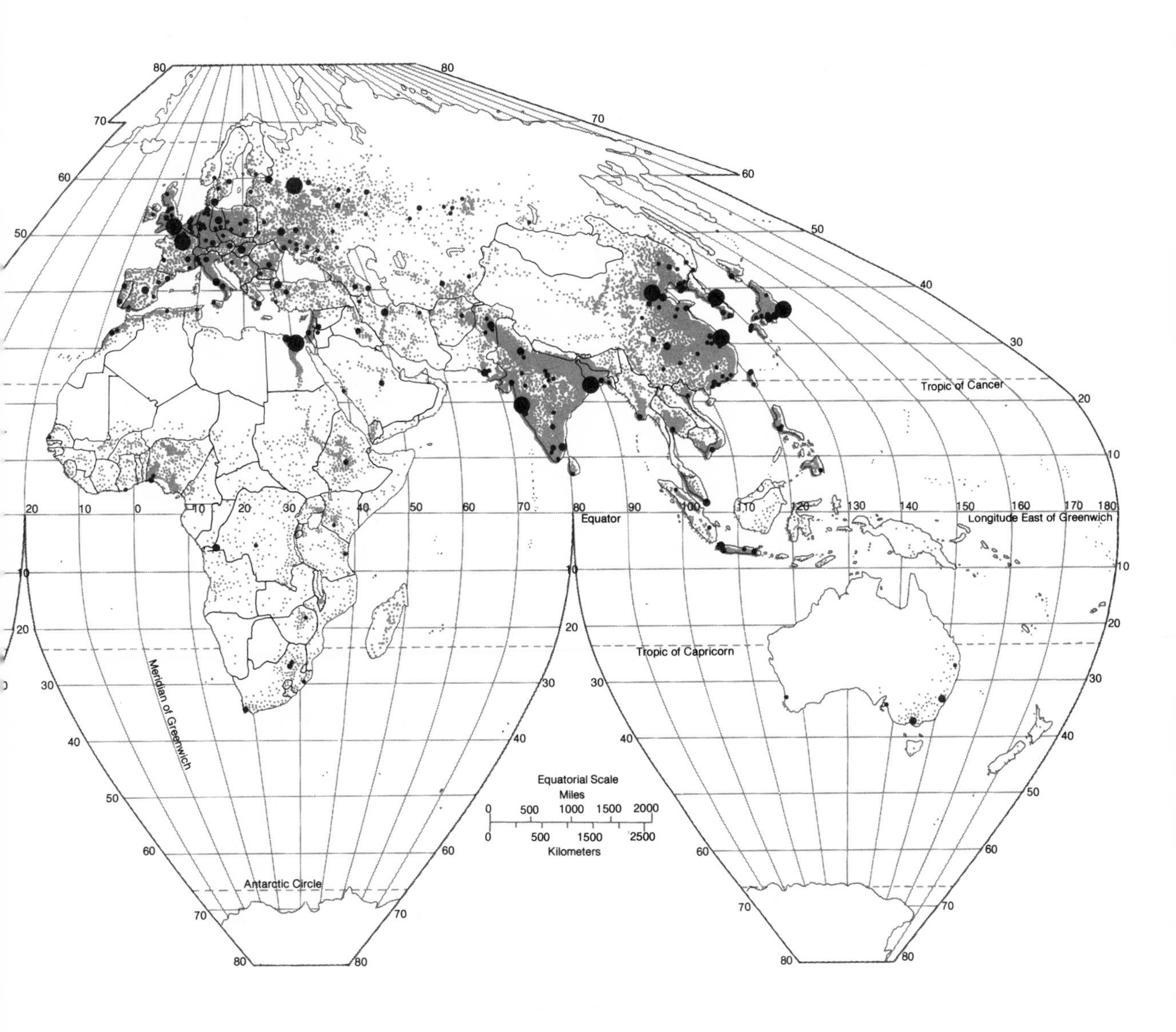

The Distribution of Economic Development

The pattern of economic development is as irregular as that of population (figure 1.5). The highest standards of living, which can be defined by the quantity and quality of goods and services, are found in North America, Venezuela, Argentina, Western Europe, Japan, Singapore, Australia, New Zealand, and Israel. The lowest standards of living are found chiefly in India, China, parts of Southeast Asia and South America, and most of Africa.

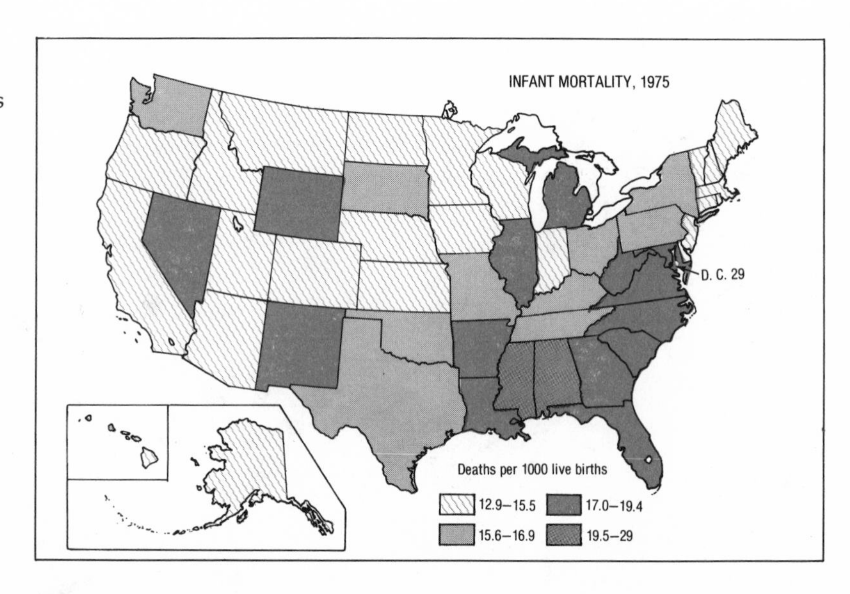

1.4. Infant mortality in the United States. Even within one country, there is substantial variation in mortality rates, which reflect variation in access to health care and in economic development. Such differences would be equally striking at the regional or local (within cities) scale.

Table 1.1. Some Measures of Development for Selected Areas

Area	*Population 1977 (millions)*	*Percentage Increase per Year*	*Infant Mortality, 1975*	*Percentage Urban*	*Per Capita GNP, 1975*	*Per Capita Energy, 1975[a]*
World	4,100	1.6	103	38	$ 1,530	2,030
Africa	425	2.3	154	24	400	400
North Africa	100	2.5	130	39	570	450
Egypt	39	2.1	116	45	310	405
Libya	3	3.5	130	30	5,100	1,300
West Africa	125	2.3	175	19	300	120
Ghana	10	2.4	156	31	460	182
Nigeria	67	2.4	180	18	310	90
Mali	6	2.3	188	13	100	25
Ivory Coast	7	2.3	164	20	500	366
East Africa	120	2.4	151	12	220	100
Ethiopia	29	2.2	181	12	100	29
Kenya	14	3.0	119	10	220	174
Tanzania	16	2.3	162	7	170	70
Rhodesia	7	2.3	122	20	540	764
Zambia	5	2.8	160	36	540	504
Central Africa	49	2.1	165	24	270	110
Zaire	26	2.2	160	26	150	78
Angola	6	2.0	203	18	680[b]	174
South Africa	30	2.2	119	44	1,220	2,960
Asia	2,330	1.8	116	26	530	700
Southwest Asia	90	2.5	114	44	1,370	1,055
Iraq	12	3.0	99	64	1,280	713
Israel	4	1.8	22	82	3,600	2,810
Kuwait	1	3.5	44	56	11,500	8,720
Saudi Arabia	8	2.6	152	21	3,010	1,400
Turkey	42	2.5	119	43	860	630

Area	Population 1977 (millions)	Percentage Increase per Year	Infant Mortality, 1975	Percentage Urban	Per Capita GNP, 1975	Per Capita Energy, 1975[a]
South Asia	865	2.1	125	20	200	230
Bangladesh	83	2.4	132	9	110	28
India	625	1.9	122	21	150	221
Iran	35	2.5	139	44	1,440	1,350
Pakistan	75	2.6	121	26	140	183
Southeast Asia	332	2.2	116	21	260	300
Burma	32	2.2	126	22	110	51
Indonesia	137	2.2	131	18	180	178
Philippines	44	2.5	74	32	370	326
Singapore	2	2.0	14	100	2,510	251
Thailand	44	2.2	89	13	350	284
Vietnam	47	2.0	—[b]	22	160[b]	260
East Asia	1,050	1.4	22	31	820	1,000
China	850[b]	1.5	—[b]	24	350	700
Japan	114	1.0	10	72	4,500	3,625
Korea	36	1.5	47	48	550	1,040
Taiwan	17	1.6	26	63	900	1,500
Australia	14	0.9	16	86	5,650	6,490
North America	240	0.6	16	74	7,020	11,000
Canada	24	0.8	15	76	6,650	9,900
United States	216	0.6	16	74	7,060	11,200
Latin America	340	2.5	78	59	1,030	1,170
Middle America	85	3.2	70	56	1,060	1,100
Mexico	64	3.3	66	62	1,190	1,220
Caribbean	28	2.0	75	45	970	1,000
Tropical South America	183	2.6	84	59	960	1,000
Brazil	112	2.6	82	59	1,010	670
Colombia	25	2.3	97	64	550	671
Venezuela	13	2.9	49	74	2,220	2,640
Temperate South America	40	1.3	63	79	1,340	1,200
Europe	480	0.4	22	65	4,100	4,040
Northern Europe	82	0.2	14	74	4,600	5,000
Sweden	8	0.2	8	81	7,900	6,180
United Kingdom	56	0.1	16	78	3,900	5,265
Western Europe	152	0.1	15	77	6,150	4,900
France	53	0.4	12	70	5,800	4,300
West Germany	61	−0.2	20	88	6,600	5,400
Eastern Europe	110	0.7	26	57	2,800	3,700
Czechoslovakia	15	0.8	21	67	3,750	7,150
Poland	35	1.0	25	54	2,910	3,805
Southern Europe	136	0.8	28	52	2,500	2,500
Italy	57	0.5	21	53	2,950	3,012
Spain	37	1.0	14	61	2,700	2,150
U.S.S.R.	259	0.8	28	60	2,700	5,550

Sources: World Bank, *World Data Sheet*, 1978, and *United Nations Statistical Yearbook*, 1976.

[a] Energy in kilograms of coal equivalent.

[b] Exact figures unavailable.

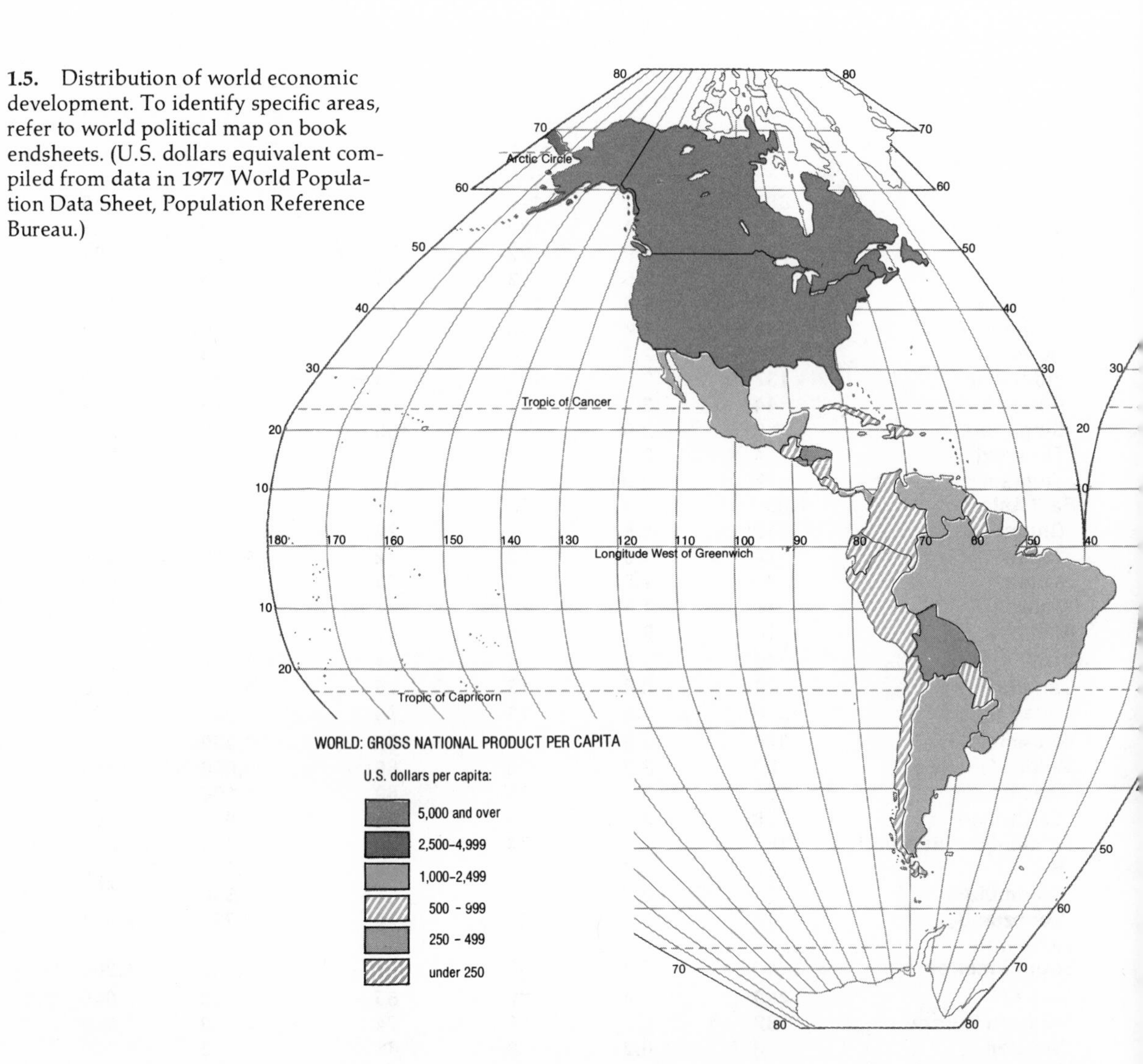

1.5. Distribution of world economic development. To identify specific areas, refer to world political map on book endsheets. (U.S. dollars equivalent compiled from data in 1977 World Population Data Sheet, Population Reference Bureau.)

Of course, per capita income, measured in terms of gross national product (GNP), is derived only from an economy's market or commercial sector and is not a true indicator of well-being. Much of the effort involved in providing for food and shelter in subsistence societies is not shown by the data. But per capita income is closely related to a society's technological level and the energy needed to maintain it. The richest countries have the most highly developed technology, are the most urbanized and industrialized, and use a disproportionate amount of energy, with some exceptions. The oil-rich countries of the Middle East, for example, have very high per capita income but have not yet reached comparable levels of urbanization and industrialization. And some countries in Latin America have high levels of urbanization, while their industry, technology, and income lag. (See table 1.1 for measures of development in some of these areas.)

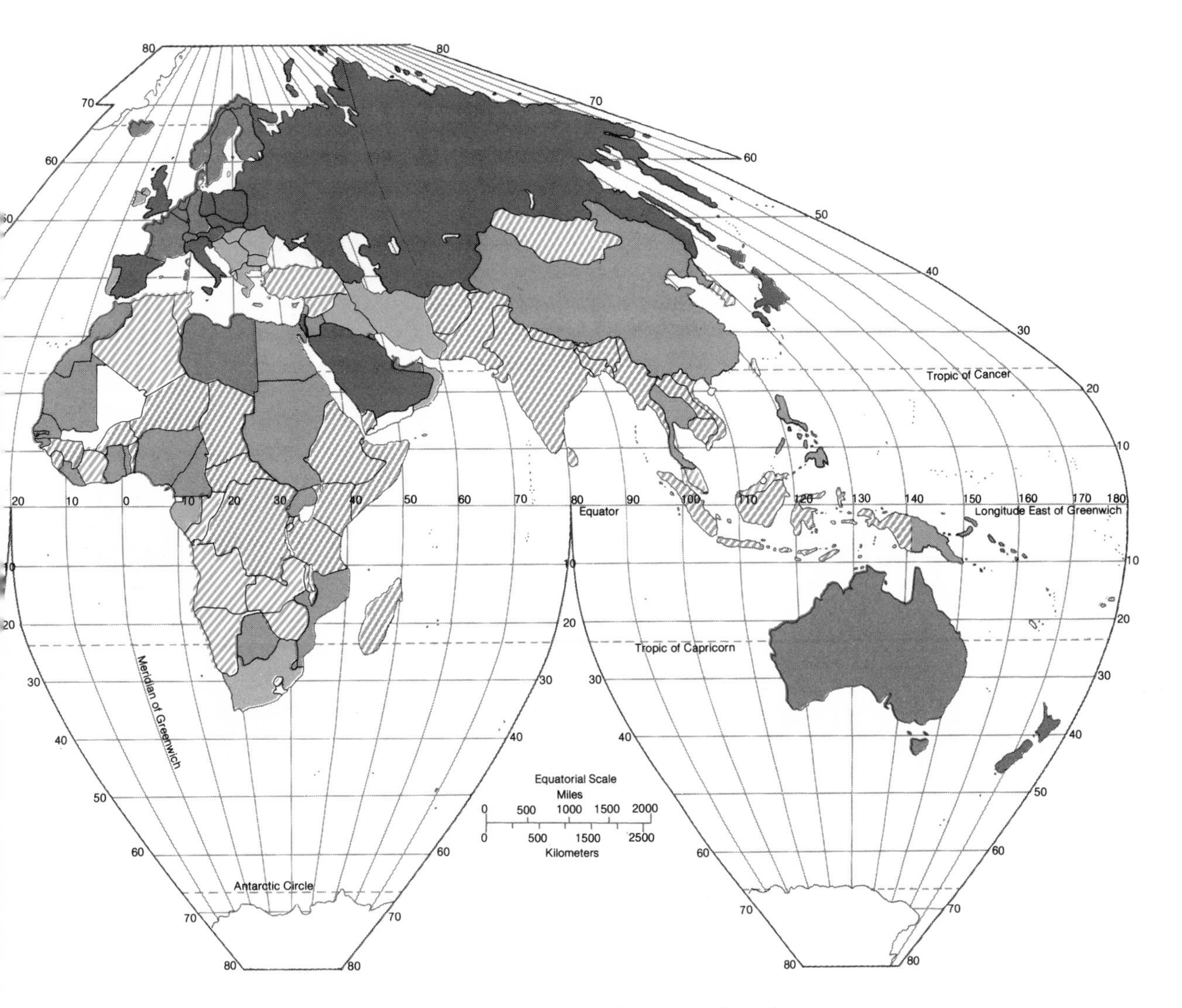

Even within a single country—the United States, for example—there are significant differences in levels of urbanization, industrialization, birthrates and death rates, and average income. The northeastern United States is very densely populated, industrialized, and urbanized and has a low rate of natural increase and high levels of income. The South and plains states have moderate density, are less urbanized, and have higher rates of natural increase and lower levels of income. The West, except for coastal California, has lower density, is highly urban but not so industrial, and has a lower rate of natural increase, except for Mormon Utah and Idaho, and higher income levels. These patterns will be discussed frequently in other chapters.

The distribution of population and economic development form a complex socioeconomic landscape, which has evolved from gradual changes in human location and development. Variations in the process of economic

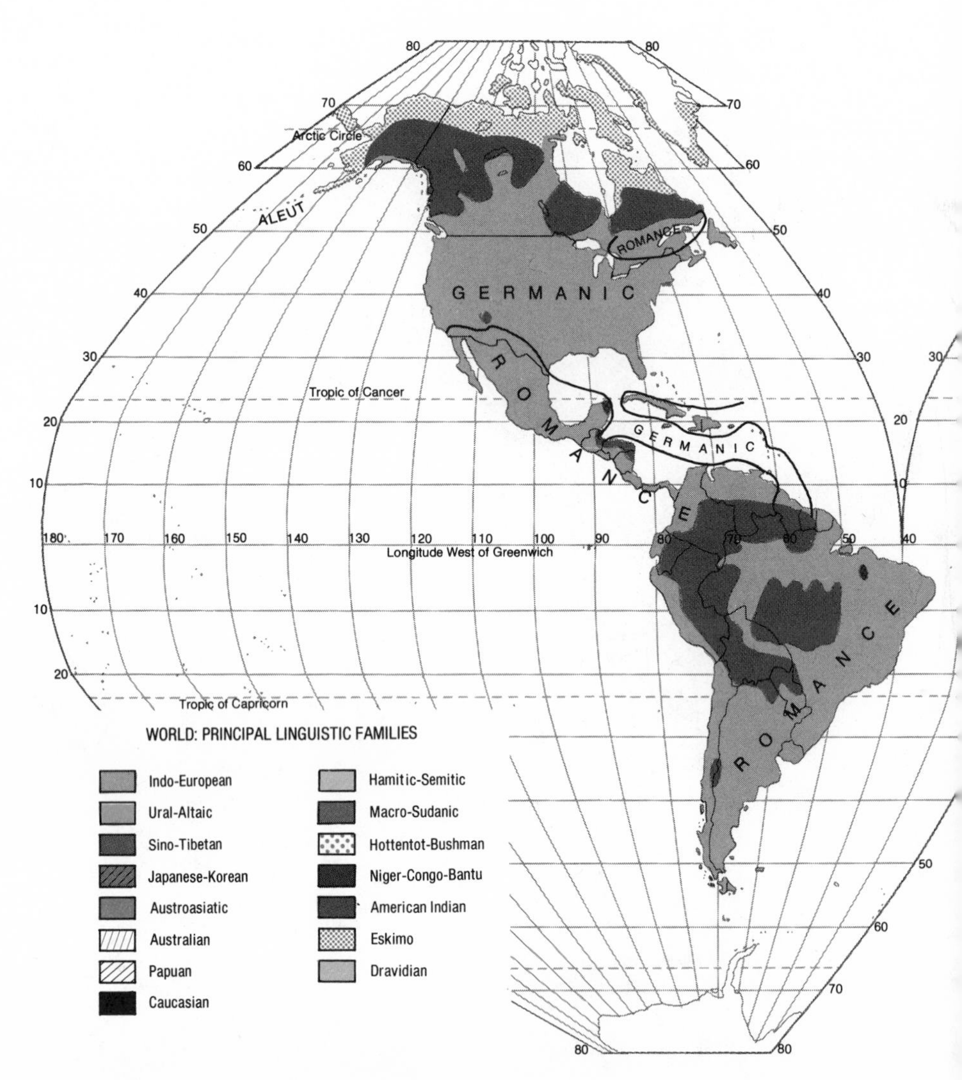

1.6. Distribution of major world languages. As a vehicle of communication, language influences the social, political, and economic structure of human societies. The principal linguistic families can be subdivided into smaller groups. The Indo-European family, for example, includes Romance, Germanic, Slavic, Celtic, Iranic, and Indic languages. Each of these may be further subdivided, such as Romance languages into French, Spanish, Italian, Portuguese, and Romanian.

Source: Redrawn with permission from J. O. M. Broek and J. W. Webb, *A Geography of Mankind* (New York: McGraw-Hill, 1973).

develoment are essential in explaining the distribution of population. But variations in cultural traits also help explain the human landscape. It is to this distribution that we now turn.

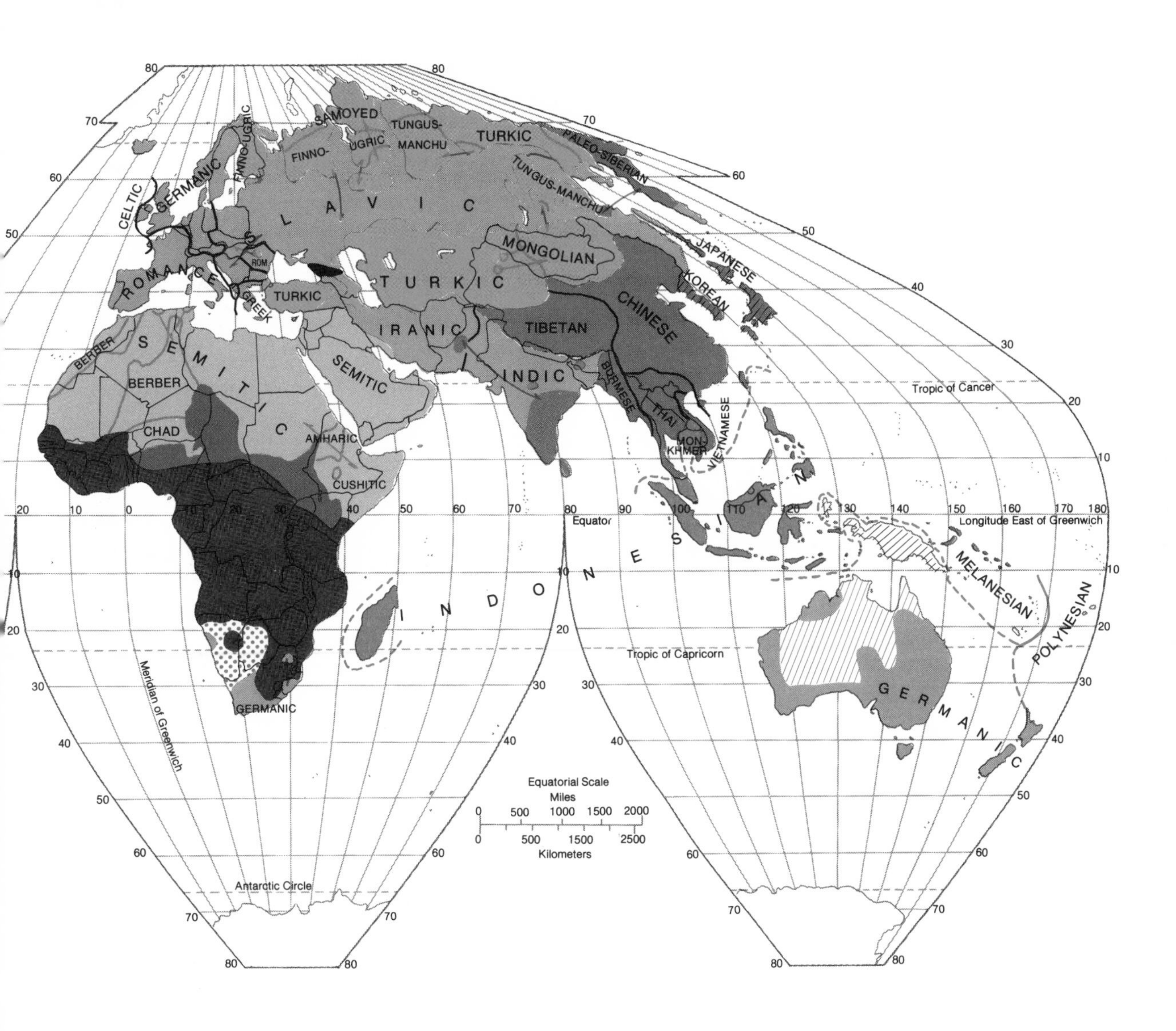

The Distribution of Cultural Variation

Distinctive cultural traits may give unity to a social group and separate it from other peoples, affecting patterns of location and interaction. The role of cultural variation in shaping the human landscape will be discussed in the next chapter. For now, we will briefly describe the general distribution of race, language, and religion around the world.

World Cultural Variation

Race and language evolved together over the millennia when human societies developed in relative isolation, leading to real and perceived cultural

Table 1.2. Regional Structure of Racial and Linguistic Groups

Dominant Race	Language Group	Region
Caucasoid (with some Negroid admixture)	Indo-European	Europe
		North America
		Latin America
		Australia–New Zealand
		South Africa (in part)
		South Asia India, Pakistan Iran, Iraq
	Hamitic-Semitic	Southwest Asia, Arabia
		North Africa
	Dravidian	South India
Mongoloid (with some Caucasoid peoples and admixtures)	Ural-Altaic	Siberia, Central Asia, Turkey
		Hungary, Finland, Estonia
	Amerindian	North and Latin America (in part)
	Sino-Tibetan, Japanese	East Asia
	Malayo-Polynesian	Indonesia, Polynesia
Negroid	Niger-Congo-Bantu	Africa (south of Sahara)
		North and South America [a]
Australoid	Papuan-Australian	Australia, Papua

Note: In many regions intermingling occurs among races and languages.
[a] Does not include language group.

variations and conflict. The distribution of languages is shown in figure 1.6, and the general structure of both race and language is presented in table 1.2. The peoples of Europe and North America and their direct colonies (Australia, New Zealand, and, in part, South Africa) form one interactive racial-linguistic group. It is racially Caucasoid and linguistically Germanic-Romance-Slavic and some Celtic, although indigenous, precolonial races and languages survive. The Hamitic-Semitic–speaking peoples of North Africa and the Middle East form a fairly distinctive Caucasian group, as do the peoples of South Asia who speak Persian (an Iranic language), Hindi (an Indic language), and other Asian languages. The peoples of India developed a way of life differing significantly from that of the Europeans partly because of India's relative isolation from Europe and because of the population's intermixture with other groups. Negroid peoples forming many distinct language groups dominate sub–Saharan Africa and represent a significant racial element in the United States, the Caribbean, and Brazil. Mongoloid peoples can be divided into three major language groups: Sino-Tibetan-Japanese, Malayo-Polynesian, and Ural-Altaic. They are dominant in much of East Asia and also constitute a major racial element in North and South America (Amerindians).

Despite cultural biases toward the preservation of differences, much racial and linguistic intermingling has occurred over many thousands of years. Mixed racial and linguistic groups are found along the southern rim of the Sahara and in India, Europe (notably in Hungary and Finland), and Latin America. Mexico and Brazil have fairly distinct Amerindian, European, and African groups but also contain complex racial and linguistic mixtures.

Some religions—for example, Confucianism and Taoism in China, Shinto in Japan, Hinduism in India, and local tribal religions in many other areas—developed simultaneously with the language of a people and with their sense of being a distinct group or nation. Such religions were exported only insofar as the peoples themselves migrated. Other religions, notably Christianity and Islam, spread far beyond their areas of origin, often at the cost of conflict and bloodshed. East Asia contains a strong Buddhist element. The disciples of Buddha, an Indian, carried his teachings to Tibet, Ceylon, Burma, and Thailand. The religion also diffused to China, Japan, and Korea, but India itself remained largely Hindu.

Both Christianity and Islam have a common origin in Judaism, and both developed in a small area in the Middle East. They diffused widely, conflicting with each other and with indigenous religions. Christians, Moslems, and, to a lesser degree, Jews have fought and persecuted each other for centuries in the Mediterranean basin and the Middle East (even today, as in Lebanon). Similarly Moslems and Hindus have clashed in India, leading to the partition of India and Pakistan. The Islamic faith, carried by Mohammed and his disciples into many countries, provides a cultural bond among peoples of different languages and races in Africa, the Middle East, the South Asian subcontinent, Pakistan, Malaysia, parts of the Philippines, and Indonesia. Christianity diffused throughout Europe and, accompanying colonization, spread throughout the New World, Australia, New Zealand, and, on a smaller scale, parts of Africa and Asia.

U.S. Cultural Variation

Because we will often cite examples from the United States to illustrate geographic concepts, it is appropriate here to describe major U.S. cultural variations.

The United States has traditionally opened its doors to many different peoples, although until recently U.S. immigration policy favored those of European origin. Compared to many other countries, the United States has been highly tolerant of individual and group diversity; for example, persecuted religious groups found refuge here and established scores of isolated colonies. But despite the relative openness, much of the immigrants' culture was suppressed or, particularly in the case of Africans, destroyed, and a recognizably American national culture developed. The early cultural influence and political control of the English and Scots, who colonized America in the seventeenth and eighteenth centuries, enabled them to impose the English language on all later groups. They were also able to control land, resources, capital, social relations—in short, wealth and power—and to set overall norms and customs. Nevertheless the United States still contains a rich cultural diversity based on the contributions of many immigrant groups.

The immigrant cultures of the United States could not have survived without the preservation of native languages in colonies and ghettos. Ethnicity was preserved for a generation or so in the rural Midwest and the metropolitan Northeast. Inevitably, however, immigrants from Germany, France, Scandinavia, Italy, Poland, and Russia—all white Europeans—were absorbed into the majority culture, which was of white Anglo-Saxon Protestant origin. Yet some large concentrations of distinctive nationalities have retained their own identity, expressed in architectural style, regional customs, religious traditions, and voting patterns (figure 1.7).

Scandinavian influences are still significant in Minnesota and the Northwest; German in Wisconsin, Missouri, Pennsylvania, Ohio, and Texas; and Italian and Polish in metropolitan centers like Chicago, Detroit, and New York. To the north in Canada, a distinctive French culture developed within the dominant English culture and sometimes conflicted with it; serious differences have culminated in the separatist movement of French-speaking people in Quebec. Elements of French tradition have extended into the northeastern United States and survive on a small scale in Louisiana. Spanish and then Mexican control over what is now the U.S. Southwest gave rise to a territorially separate base of Spanish-speaking peoples, whose cultural identity is maintained by continuing in-migration from Mexico.

Racially at least 11 percent of the American population is of African ancestry. While sharing the language and culture of the white majority, especially that of the South, blacks have developed a modified form of American culture, expressed in part in music, dress, and slang. Black culture stems from the spatial and social isolation imposed by slavery and discrimination and from the group identity fostered by more recent ghetto isolation. No longer tied to the rural South, black culture has spread to metropolitan areas throughout the United States. Blacks began to migrate to northern cities after World War I. In Chicago, the proportion of blacks rose from 2 percent of the population in 1919 to 32 percent in 1970. Although some black families have moved to the suburbs, where they still tend to be isolated within communities, most remain in Chicago's South Side ghetto.

To a lesser degree than blacks experienced, racial differences also impeded the entry of the Chinese, Japanese, and others of Asian ancestry into the majority culture. Ghettoization enabled Asians to preserve their identity for several generations. But only small enclaves of Asian culture survive to-

A

B

1.7. Cultural variation in the United States. Photograph A shows a Pennsylvania Dutch farm; decorative hex signs characterize barns and houses of Amish farmers, whose ancestors migrated from Germany and Switzerland in the seventeenth and eighteenth centuries. (Pennsylvania "Dutch" is derived from *Deutsch,* meaning "German.") Photograph B shows adobe ranch home in Taos, New Mexico; hispanic architectural styles diffused from Latin America to the U.S. Southwest. (Photo A by Elliott Erwitt from Magnum Photos, Inc.; photo B by Josef Muench.)

day, and the trend is toward rapid assimilation. Another minority, Jewish immigrants from Germany and Eastern Europe, has been set apart on the basis of religion. On the whole, however, their essentially European cultural character has enabled Jewish immigrants to become part of and to help shape the American majority culture.

In the United States, religion, both as a heritage of the immigrant and as an indigenous development, is a significant but less obvious cultural element than is race. Protestant groups have predominated in most parts of the United States. Exceptions include the metropolitan centers of the northeastern seaboard, where Catholic and Jewish immigration was concentrated, and the Southwest, where early Spanish Catholic influence was strong. In most rural areas, particularly in the South, socially more conservative Protestant sects flourished and contributed to the development of other social attitudes. A socially more liberal Protestantism developed in the North and in most other urban areas, where levels of education were higher and change was more readily accepted. To the west, frontier society weakened organized religion and fostered the rise of independent and often rather unusual sects, such as the Snake Handlers and the Holy Rollers. The opportunity for isolated development in Utah was an important factor in the survival and growth of one of the more established independent sects, the Church of Jesus Christ of Latter-Day Saints (the Mormons). We will discuss U.S. and world regional patterns based on cultural variation in chapter 12.

Classification of World Countries

Viewed as a whole, the patterns of population, economic development, and culture are complex. At the risk of oversimplification, however, we will classify the countries of the world according to this conventional fourfold division based on economic development:

1. Already affluent and technologically advanced countries (such as the United States, Canada, and the countries of Western Europe);
2. Increasingly affluent, urban-industrial or resource-rich countries (such as the Soviet Union, the countries of Eastern Europe, and the Republic of South Africa);
3. Third world, or developing, countries, still in the early stages of industrialization and evolving from a predominantly subsistence economy to a commercial one (such as China, India, Peru, and Nigeria);
4. Fourth world, or poorest, countries, still largely rural and subsistent and unable or unwilling to modernize (including many countries in Asia and Africa).

As we will see, these levels of development tend to correspond to distinct patterns of the human landscape.

Perhaps the most meaningful distinctions regarding population can be made between the populous countries with tens or hundreds of millions of people, such as China, India, the Soviet Union, and the United States, and the ministates with fewer than a million people, such as Botswana and Qatar; and between countries with a high ratio of land and **resources** to people (low population density), such as the United States and Australia, and countries with a low ratio of land and resources to people (high population density), such as the United Kingdom and Japan.

Culturally the most important distinctions are based on economic and political systems and on cultural history. Some countries, such as the United States and West Germany, are characterized by market economies, in which producers decide what they will produce according to consumer demand and prices secured in the market. Other countries, such as the Soviet Union, China, and East Germany, are characterized by centrally planned economies, in which government commissions determine what will be produced, how the goods will be distributed, and what the prices will be. (These distinctions are not clear-cut in reality; modern economies feature some aspects of both market and centrally planned systems.) Still another set of countries—mostly in Africa and Asia—is characterized by peasant or tribal subsistence economies, in which most of the goods produced are consumed directly by the producer rather than sold at market, although there are some cash crops. Political systems may be roughly divided into authoritarian, in which power is concentrated in the hands of an elite, and democratic, in which power is diffuse and rests ultimately in the hands of the citizens. And finally, countries may be classified according to their cultural history—Western, African, Islamic, or Eastern.

Almost all of the affluent countries have market economies, belong to the Western cultural realm, and are democratic. East Germany, however, has a centrally planned economy and a more authoritarian political structure; Japan belongs to the Eastern cultural realm; and some Middle Eastern countries are traditionally more authoritarian in political structure and are Islamic in culture. Some of these countries are populous but large and therefore have low population densities, such as the United States and Canada, while others are small but densely populated, such as Belgium, Denmark, and the Netherlands.

The countries that are becoming increasingly urban and industrial include many of those in Europe and Latin America with market economies and belonging to the Western cultural realm. Some of these countries are more democratic than authoritarian in political structure (Portugal and Greece); others are more authoritarian than democratic (Argentina and Chile). This second set of countries also comprises most of the centrally planned economies, notably the Soviet Union and Eastern Europe; several market economies within the Eastern cultural realm (Singapore, Korea, and Taiwan are examples); and that very complex country, the Republic of South Africa. These countries also range from very large (the Soviet Union) to very small (Singapore, Hong Kong).

Many third world countries have market economies often coexisting with peasant subsistence economies. Examples are Peru and Colombia in Latin America; the Philippines, India, and Thailand in Asia; and Kenya, Nigeria, and the Ivory Coast in Africa. Other third world countries have centrally planned economies. China is the most notable example, along with Tanzania, Angola, Mozambique, and Vietnam. Some of these countries—primarily those in Latin America—are of Western cultural heritage, but most are African or Eastern (Asian). A minority are democratic, notably India. While a few are very large (India, Pakistan, China, Indonesia), many are extremely small, particularly the most recently independent states in Africa, Oceania, and the Caribbean.

The fourth world countries still have largely peasant or tribal subsistence economies. Most are in Africa (Mauritania, Mali, Niger, Chad, and Ethiopia) or Asia (Bangladesh and Afghanistan). A few of the countries, such as Bangladesh, are large, but most are small.

We want to discover what forces account for the basic patterns of world population, economic development, and cultural variation. Why have people chosen to concentrate in some areas and not in others? One significant factor is the quality of the physical environment. The Sahara Desert cannot readily support abundant human life, nor can the North and South Poles. Climate and other physical conditions influence population patterns, but they are not the whole explanation. After all, some sparsely settled areas lie well within the range of human tolerance, while many densely populated sites have become so crowded and polluted that they can hardly be considered attractive habitats today. Human location is affected by a variety of factors, the most important of which will be discussed in the next chapter. The pattern of population can perhaps best be understood by examining the spread of agricultural technology and sedentary (settled) culture over thousands of years (see chapter 3).

The second question is, why are some regions highly developed and prosperous and others backward and impoverished? As we will see, the imbalance of development and wealth stems from the economic advantages or disadvantages of a region, the scarcity or abundance of resources, and the technical skills applied to increase productivity. Development is also strongly affected by the more subtle influences of culture and human behavior. The pattern of development is closely related to the spread of industrial technology and urban culture from Europe during just the past few centuries (chapter 3).

The final question is, why do people in different countries, or within the same country, speak a variety of languages or dialects (regional variations in language), have different racial characteristics, practice different religions, follow different traditions and customs, and live under different socioeconomic systems? Cultural variation is chiefly due to the separate development of social groups over long periods of time. And the pattern of the contemporary cultural landscape is a result of the spread, isolation, and interaction of various **cultural groups** over the millennia.

SPATIAL EXPERIENCE AND LEVEL OF TECHNOLOGY

The level of a given cultural group's interaction with other people and places is related to its **spatial experience,** its awareness of and contact with other areas. Some cultural groups, or societies, have few connections with the outside world. Their movements are restricted by poor transportation, and trade and communication are relatively limited. Other societies, in contrast, are highly interconnected with other areas. The difference in spatial experience reflects variations in level of technology.

Low Technology Societies

In low technology societies, the individual's awareness of distant places and peoples is limited to direct experience. A small unit of organization, such as a tribal village, may be fairly self-sufficient. Most basic necessities are produced locally; other goods can be obtained from nearby villages and itinerant merchants. Hence spatial experience is usually limited to the village and its lands and to neighboring villages. Much effort may be devoted to interacting with others, but inadequate transportation limits the spatial extent. Villagers know their local area intimately and can differentiate it in detail. Their relationships with other members of the village may be dictated by tradition. Tradition dominates everyday life, and changes occur only gradually (see chapter 4).

With the development of more efficient technologies, societies become capable of producing greater surpluses and engaging in higher levels of trade and other interaction. If the local region favors a certain kind of production, the society may specialize, sell the surplus, and purchase other needed or desired goods from other specialized regions. In this way regional **specialization** and **interdependence** develop.

The republic of Tanzania, Africa, provides examples of different levels of spatial experience. Some Tanzanian tribes still practice subsistence farming, using primitive tools and producing only enough food to support themselves. Other tribes have specialized their production to some extent and have become more interdependent. For instance, the Chagga tribesmen dwelling at the foot of Africa's highest mountain, Kilimanjaro, farm for themselves but also raise enough surplus coffee and cotton for export. More specialized are the sugar-cane producers on plantations along Tanzania's coastal plain, bordering the Indian Ocean. They use modern equipment and grow sugar cane for export.

High Technology Societies

As the technological and economic development of a society progresses, certain kinds of economic activity become concentrated by region. The traditional cotton belt of the U.S. South was a specialized *agricultural region* (figure 1.8). Specialization in cotton as a major cash crop developed as a re-

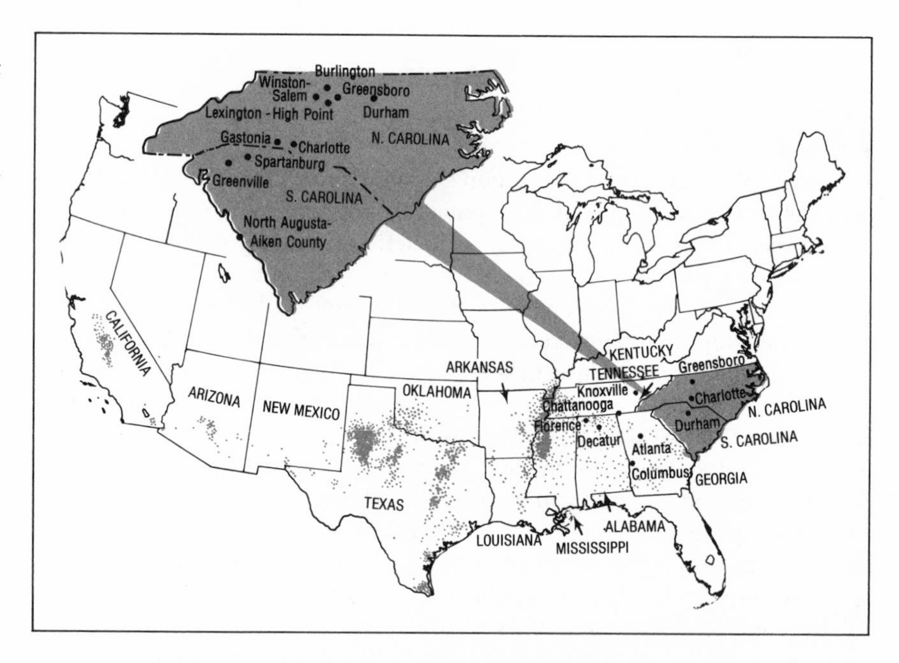

1.8. Cotton production in the United States. (Each dot represents 25,000 bales.) Actually a fragmented rather than continuous band of specialized production, the cotton belt spans fourteen states. Cotton production spread from the South to the West mainly after World War II. The largest number of major textile centers are located in the Carolinas, with convenient access to cotton supplies. Yet today Texas outproduces the other cotton belt states, and California's San Joaquin Valley produces nearly as much cotton as does the Mississippi River Valley.

sult of favorable climate and soil conditions, plentiful slave and, later, sharecropper labor, technological **innovations,** and high economic demand first from Europe and later from within the United States. Early American colonists found that cotton flourished in the fertile, well-drained soils and subtropical climate of the South. But harvesting and processing the cotton was a time-consuming chore requiring the help of many hands (hence the attraction of slave labor); production was therefore limited.

The shift to large-scale production was made possible by the invention of the cotton gin in 1793, which mechanically separated fiber from seed and enabled farmers to bale cotton rapidly. European demand for cotton provided a ready market during the early decades of the nineteenth century. The strength of the cotton market and the available technology gave cotton production a long-term advantage over other specialties such as rice, indigo, and tobacco.

Although the old cotton belt is now fragmented—farmers in western Texas, New Mexico, Arizona, and California have been able to produce cotton more intensively and cheaply on their large, irrigated, and more mechanized farms—the South is still a major producer of cotton, especially in the rich delta areas of the Mississippi and other rivers.

The Ruhr district in West Germany is an example of a specialized *industrial region* (figure 1.9). Geographic location, transport position on a network of rivers and railroads, and an abundant supply of coal were keys to the evolution of the Ruhr into one of the world's most highly industrialized regions. Its origins, however, were rural and agrarian. During the Middle Ages some iron and steel was produced with local supplies of iron ore and charcoal, and small quantities of linen were made in households with locally grown flax. But farming remained the dominant activity.

Industrialization began in the Ruhr in the 1850s, when steam-powered textile mills were introduced to the region, turning the cottage textile industry into a large-scale enterprise. Significant changes occurred after 1871 when the separate states of Germany were politically unified. Coal production boomed as mining technology became more sophisticated and a new technique was developed for processing steel with coke made from coal. Thousands of rural people were employed in the mines to meet the rising demand for coal. Steel mills emerged near the coalfields, and small mining villages mushroomed into urban-industrial centers.

When local supplies of iron ore proved inadequate, the steel companies imported iron from Sweden, Spain, and Lorraine (in northeast France). Coal, iron, and steel were transferred by rail or barge along the Ruhr and Rhine rivers to the Dutch ports of Amsterdam and Rotterdam and to Belgium, Luxembourg, and France. The Ruhr district became highly interconnected with other parts of the world through trade. While its economy has become more diversified in recent decades, it continues to be the leading industrial region of Germany and central Europe.

Thus regional **self-sufficiency** eventually gives way to specialization and interdependence. Both the old cotton belt and the Ruhr district were at one time relatively self-sufficient. Each region became increasingly specialized as its natural advantages were developed through a combination of technical advances and growing total and per capita demand. Now these regions concentrate in the production of just a few major commodities, and they import other needed or desired goods from other regions.

Other differences in spatial experience can be observed in advanced and less advanced nations. People in advanced nations are more likely to be

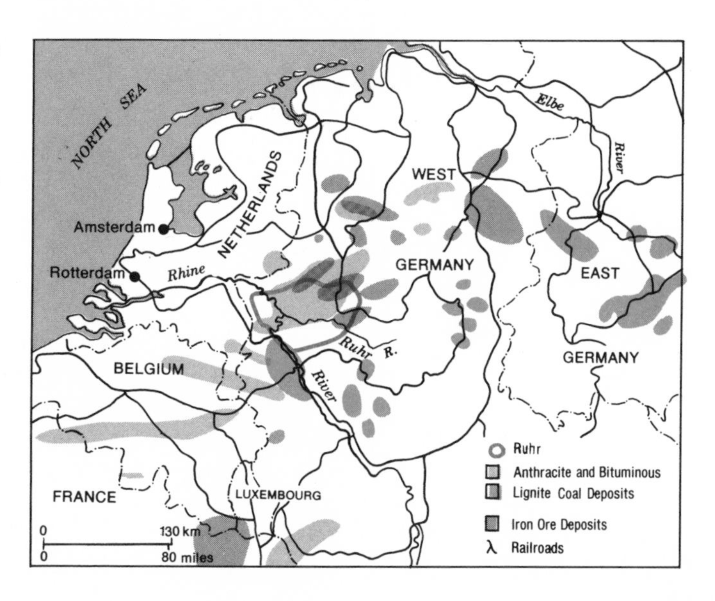

1.9. The Ruhr industrial district, West Germany. Central location, high accessibility, abundant coal, and advanced industrial technology helped transform this formerly agrarian region into one of the world's most intensive producers of iron and steel and other manufactures. Rivers and railroads provide access to the Atlantic Ocean and to other European countries.

aware of the distant events that affect their lives. They realize that their very survival depends on the goods and services produced in other places. Americans, for instance, are coming to know that much of their oil is produced in the Middle East and that the Organization of Petroleum Exporting Countries (OPEC) has, in recent years, raised the price of oil. Villagers in India have also come to use imported oil, and many have found that they can no longer afford the higher price of petroleum-based fertilizer. Yet most Indian villagers are probably not aware of the chain of events that have resulted in the higher fertilizer price. (On the other hand, they may be able to return to organic fertilizer more readily than could their American counterparts.)

Indeed the very perception of distance is different for city dwellers in advanced societies and villagers in subsistence societies. Although most individual experience in both kinds of society occurs at a local scale, "local" typically implies a radius of 80 kilometers (50 miles) for city dwellers in contrast to 20 kilometers (14 miles) for villagers.

Communications instantly link people in advanced societies with the rest of their nation and the world. Radio, television, and daily newspapers inform them of events and innovations around the globe. If a villager in Saudi Arabia contracts a serious disease such as cholera, the Western world may learn of it as soon as his neighbors do. Individuals in advanced societies are constantly confronted with new information, new ideas, and new products. Enjoying a greater freedom of choice, higher income, and superior transportation, their spatial experience is broadened and their mobility enhanced. They are more likely to change residence and travel frequently, covering greater distances in less time. In contrast, most of the world's population is spatially more restricted; freedom of choice is limited, and subsistence must be won from the resources of the local environment.

CONCLUSION

We have described some characteristics of the human landscape on a global scale, giving you a general idea of the pattern of population, economic development, and culture and introducing the concept of spatial experience. Our purpose is to help you become aware of the complexity of the human landscape—and of the fundamental spatial order that underlies this landscape.

If you think about your local area or neighborhood, you may be able to relate some of these ideas to your own experience. What are the characteristics of the area you live in? Is it densely or sparsely populated? Are most of the people relatively prosperous, or are they poor? What do they do for a living? Do many of them share the same values, attitudes toward the physical environment, food preferences, and other traits that have implications for the use of land? Do most of your neighbors work close to home, or do they commute some distance to their jobs? Is the housing old or new? Once you have defined your local human landscape, you can begin to think about the reasons why so many or so few people have settled there, have prospered or struggled to survive, and have made certain choices about the way they use and organize the space they live in. You will also come to understand the social, economic, and environmental consequences of these choices. The next chapter will provide some of the background you need.

SUGGESTED READINGS

Abler, Ronald; Adams, John; and Gould, Peter. *Spatial Organization: The Geographers' View of the World.* Englewood Cliffs, N.J.: Prentice-Hall, 1971.

Berry, B.J.L., and Conkling, E. C. *The Geography of Economic Systems.* Englewood Cliffs, N.J.: Prentice-Hall, 1976.

Broek, Jan. *A Geography of Mankind.* New York: McGraw-Hill, 1973.

Cox, Kevin. *Man, Location and Behavior: An Introduction to Human Geography.* New York: Wiley, 1972.

De Blij, Harm. *Human Geography: Culture, Society, and Space.* New York: Wiley, 1977.

Haggett, Peter. *Geography: A Modern Synthesis.* New York: Harper & Row, 1975.

Jakle, John; Brunn, Stanley; and Roseman, Curtis. *Human Spatial Behavior.* North Scituate, Mass.: Duxbury Press, 1976.

Jones, Emrys. *Human Geography.* London: Hutchinson University Library, 1964.

Jordan, Terry, and Rowntree, Lester. *The Human Mosaic.* San Francisco: Canfield Press, 1976.

Lloyd, P. E., and Dicken, P. *Location in Space.* New York: Harper & Row, 1977.

Murphey, Rhoads. *An Introduction to Geography.* Chicago: Rand McNally, 1971.

Taafe, Edward, ed. *Geography.* Englewood Cliffs, N.J.: Prentice-Hall, 1970.

Thoman, Richard, and Corbin, Peter. *The Geography of Economic Activity.* New York: McGraw-Hill, 1974.

Wagner, Philip. *The Human Use of the Earth.* Glencoe, Ill.: Free Press, 1960.

2

Factors Shaping the Human Landscape

Factors Influencing Location, Interaction, and Development

Location Theory and Human Spatial Behavior

2 Factors Shaping the Human Landscape

The areas occupied and used by people reflect their perception, judgment, and choice of locations suitable for activities. The location decision is central to our understanding of the human landscape. Indeed the patterns of population and development presented earlier are shaped by the collective decision making of various cultural groups over many generations.

Location decisions are also an individual matter. Suppose you were considering moving to California. Any one of several factors, or a combination, might convince you to locate there. You might have learned of interesting job opportunities in the San Diego shipyards. Or perhaps you are a surfer or skindiver and want access to southern California beaches. You might enjoy the open social climate or lively intellectual atmosphere of Berkeley and other university communities. You might be enticed by the cosmopolitan flavor of San Francisco or the tranquil beauty of the San Joaquin valley.

Because it is difficult to draw meaningful global or even regional conclusions from individual decisions, location theory gathers individual data regarding location and unifies them into a comprehensive whole. In this way, certain broad principles emerge and enable us to make better sense of the landscape. These principles, which will be outlined at the end of the chapter, are basically economic, and they assume that people try to use space as efficiently as possible. But it is clear that noneconomic forces motivate human **spatial behavior** as well. The choice of a location for human activities, the level of interaction between locations, and a location's potential for development are influenced by a variety of factors—historical, economic, spatial, environmental, cultural, and political.

FACTORS INFLUENCING LOCATION, INTERACTION, AND DEVELOPMENT

The Historical Factor

Perhaps the most important influence on patterns of location, interaction, and development is the force of history. Past patterns set a precedent for the future. Because most people rely on previous experience when making a location decision, past patterns are strongly reinforced (figure 2.1). Large economic investments made in the physical structures (towns, farm or factory buildings, cathedrals, stores, warehouses, transport facilities) and the

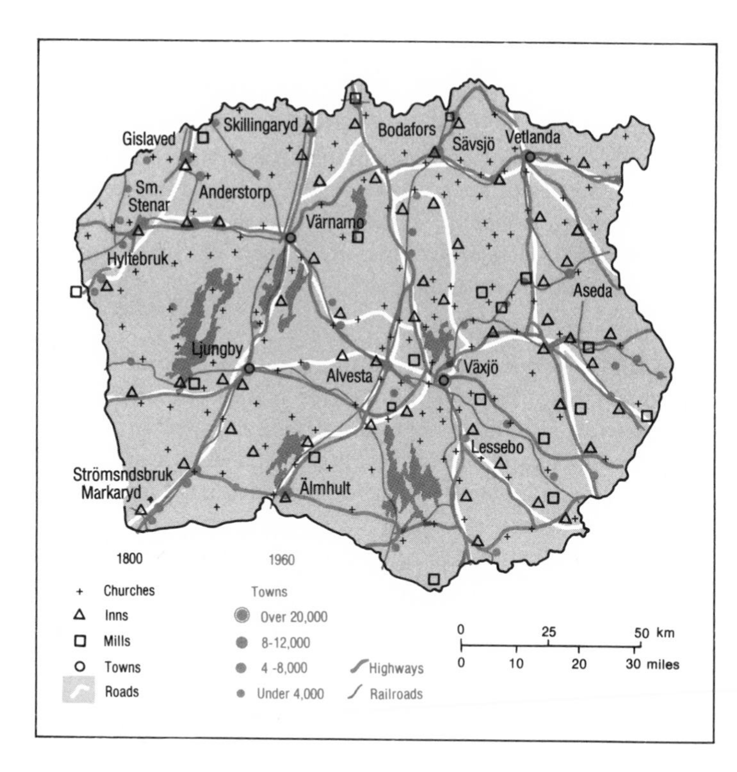

2.1. The influence of past patterns of development. Location decisions made during the nineteenth century in a sample area of Sweden have affected twentieth-century settlement patterns. Small settlements have developed near nineteenth-century churches and mills, larger towns and cities have evolved from nineteenth-century towns, highways have been built over many of the older roads, and railroads now parallel major nineteenth-century roads.

human resources (employees, customers, local suppliers) of existing locations militate against radical change in location. Traditional ways of farming, methods of construction, styles of housing, and procedures of buying and selling all tend to perpetuate themselves. The psychological investments channeled into homes, neighborhoods, and personal associations tend to make individuals immobile. (This is true particularly in countries other than the United States. Since World War II, Americans generally have become highly mobile.) Most people would prefer to improve existing structures than to begin again somewhere else. Hence new investment is concentrated in existing locations. For example, a greater proportion of investment is devoted to modifying existing plants than to constructing new ones.

Explaining the present on the basis of past patterns, however, begs the question of why these patterns came about in the first place. And while past patterns do condition the future, the landscape is constantly changing as people find it more profitable or satisfying to move.

Economic Factors

A fundamental aim of individuals, families, and societies is to satisfy the human need for food and shelter. This goal underlies the organization of economies—systems of resource use; production, consumption, and exchange of goods and services; and allocation of value. (These systems range from private enterprise to centrally planned. Later in the chapter we will see how economic systems affect location decisions and the landscape.) The basic economic factor governing the location of activities is the desire to meet essential material needs, that is, to satisfy demands for goods and services. Ideally people locate activities in such a way as to produce the maximum output of goods and services at least cost, thereby realizing the greatest net gain (profit) or satisfaction. The geographic expression of this economic goal is discussed at the end of the chapter.

Prices of the Factors of Production

Producers of goods and services seek the combination of factors that will best satisfy demand. Factors of production include land, both as a natural resource and as an area or site for the location of activities; labor; capital; and transportation. Costs of land, labor, resources, and even capital vary greatly over space, giving some places a comparative advantage over others for certain activities. (The concept of comparative advantage will be discussed later.) The abundant and relatively cheap hydroelectric resources of the U.S. Pacific Northwest, for example, make that area a good place to locate electricity-consuming aluminum plants. Producers can substitute to some extent among the factors of production. When labor is expensive, for instance, more machines (capital) can be used, and vice versa. As we will see in chapters 5 and 8, substitution has a significant impact on location. Location decisions are also influenced by transport costs, the economic measure of distance, or spatial separation. To minimize transport costs, related economic activities often find it desirable to cluster at certain points. This principle, known as agglomeration, is discussed further below.

Economies of Scale

The goal of producing the maximum output of goods and services at least cost is reached not only through substitution among factors of production and through agglomeration with related activities, but also by increasing the scale of operations ranging in size from farms to factories, business firms, and nations. The cost advantages of large-scale production—economies of scale—and the role played by investments in better transport systems to make these economies possible will be treated in detail in chapters 8 and 10.

Spatial Factors

Traditionally geographers emphasized the role of the physical environment in determining location and influencing human spatial behavior. They studied space chiefly for its environmental content. But today geographers consider space to be important in itself, regardless of environmental variation, because it is required for all human activities. The total area needed for agriculture, forestry, mining, commerce, settlements, recreation, and other pursuits involves most of the earth's land surface.

Because activities are spread out over vast areas, distance inevitably separates suppliers from producers, producers from markets, homes from jobs, friends from each other, and so on. The cost of overcoming the distance between specific locations plays an important role in planning the most efficient spatial pattern for a given set of activities. (This concept will be developed further in chapters 5 through 11.) *Space,* as an area to be used and organized for human purposes, and *distance,* as a spatial barrier to be overcome, are two of the most important geographic factors we will be studying.

The amount of space required by activities and the distances they can tolerate vary greatly. To operate successfully, a tavern may require only a few hundred square feet of space, but a grain-hog farm requires hundreds of acres. Although the tavern needs to be located fairly close to its patrons, the farm can tolerate distances of several hundred miles between itself and food-processing and packing plants. Because of these different requirements, space is differentiated into a structure consisting of areas of various sizes and shapes, specific points (locations), and interconnecting lines (transport routes and lines of communication). Each location can be characterized by its relation to the whole spatial structure—for example, by its proximity to other locations. Spatial characteristics tend to determine a location's potential for development as much as physical characteristics do. Chicago's high **accessibility** (a spatial quality) to the farmlands of the Midwest and the industrial centers of the Northeast contributed to its development perhaps even more than did its temperate climate and level terrain.

The qualities of space are abstract. They are defined in terms of distance, size, shape, relative location, and accessibility. Each spatial quality affects human location, interaction, and development.

Distance

Distance—spatial separation—can be measured physically in miles (or kilometers) or by the time, effort, and cost of overcoming it. Distance has always been viewed as a barrier to communication, movement, and trade. The cost of overcoming a given physical distance may vary considerably. In one direction the cost may be so high that it completely discourages interaction, while in another it may be low enough to invite interaction. The difference can often be attributed to the presence or absence of political, physical, psychological, or economic barriers. For example, West Germany can ship goods tariff free—and therefore at lower cost—to its trade partners to the north, south, and west than it can to other nations the same distance away to the east—for political rather than physical distance reasons. Similarly the cost and time of shipping goods are higher for the same distance if there are mountain barriers to overcome (southward across the Alps from Germany to Italy) or major waterways to cross (northward to Sweden or Norway). Psychological barriers, such as fear or anxiety in unfamiliar surroundings, may distort an individual's perception of distance. The distance across streets considered dangerous or stressful, for example, may seem greater than it actually is (figure 2.2).

The economic cost of distance is usually measured in monetary terms.

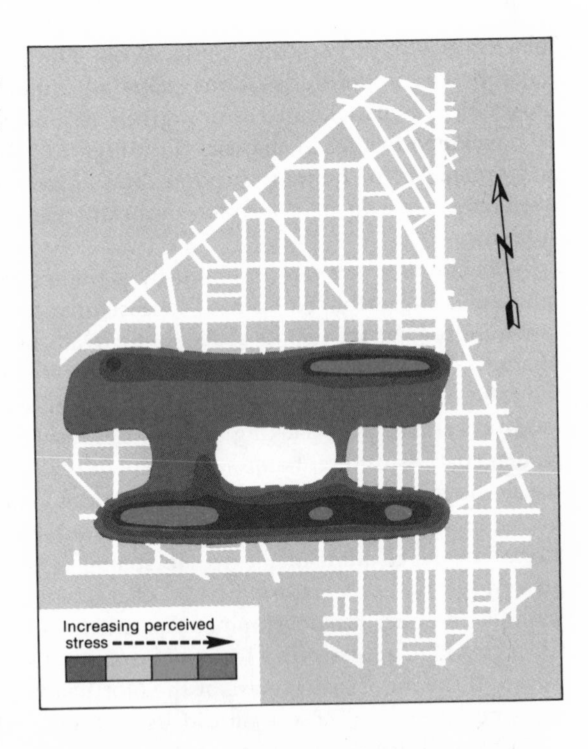

2.2. Perceived environmental stress surface, Philadelphia. Emotional factors such as fear of street gangs or fast-moving traffic influence inner-city residents' perception of certain streets, which in turn affects their behavior. Residents are afraid to cross the lines of high stress, so they take circuitous routes around the stressful areas.

Source: D. F. Ley, *The Black Inner City as Frontier Outpost* (Washington, D.C.: Association of American Geographers, Monograph Series No. 7, 1974), Fig. 36, p. 221.

Transport costs vary directly with distance and are influenced as well by physical qualities, such as degree of slope and amount of curve; by the quality of the transport link (since time is highly valued, the faster the mode of transport, the higher the price); by the cost of facilities, such as bridges, tunnels, causeways, and canals, that reduce the time it takes to overcome distance; and by governmental decisions on tolls, tariffs, subsidies, and the like.

These economic costs prevent people and production from becoming more concentrated than they already are. Beyond a certain range, transport becomes too expensive, so similar combinations of activities must be repeated again and again over space. There are many markets for grain or livestock and many cities catering to the exchange of goods. Although Chicago is a major market and wholesale-retail trade center, other cities such as Minneapolis–St. Paul, Fargo, and Sioux Falls also compete in the Midwest (see figure 7.13).

Much of economic history concerns the gradual surmounting of distance barriers through the development of faster transportation. Society has invested huge sums of money and devoted much human talent to the grading and construction of roads, the building of canals, tunnels, bridges, railroads, and harbors, and the design and manufacture of even faster and larger vehicles and carriers. However, the cost and effort of overcoming distance still limits the exchange of goods and information and personal movement. But modern transportation enables us to ship things and to travel farther with less effort and at less cost than ever before, and it permits spatial relationships once considered impossible (witness the many American students hitchhiking through Europe at very little cost).

In the distant past, the oceans were an effective barrier to trade and other forms of interaction. But with the invention of the sailing ship about four thousand years ago, goods could be exchanged across water at less cost per unit of distance than over land. In the 1840s, for instance, it was cheaper to transport goods from New York to San Francisco by ship around Cape Horn at the tip of South America than by wagon overland. When trains and trucks came along later in the nineteenth and twentieth centuries, respectively, land transport regained importance, and economic development again proceeded rapidly in the interior of countries. Today air travel has revolutionized passenger transportation and has virtually dissolved the time needed to overcome distance. The Mayflower trip took sixty-five days; today the same distance can be covered in three and a half hours by plane—and very probably at a cost that is lower relative to income.

Distance is often evaluated by the time required to cover it. Thus in less developed, spatially restricted societies, the lack of vehicles or all-weather roads limits how far villagers can go to their fields and still have time to work, or how far and how frequently they can visit markets or friends and relatives. In technically advanced societies, the distances traveled are greater but are still evaluated in terms of how much time can be afforded for the journey to work or for shopping and other kinds of trips.

Size

The size (and complexity) of units of social, economic, and political organization ranges from tribe to nation, and nations range in size from tiny Monaco, which is smaller than New York City's Central Park, to the Soviet Union, which occupies one-sixth of the earth's total land surface. Although conflict among various ethnic and interest groups may prevent large nations from developing evenly and acting as a unit, large size is usually an advantage. A variety of raw materials is more likely to be found in a large nation than in a small one, given the sporadic distribution of natural resources. The Soviet Union has abundant supplies of lumber, minerals, fossil fuels, and other important resources; Monaco, with few natural resources, has had to create its own—gambling casinos, luxury hotels, nightclubs, and the Grand Prix sportscar races. Specialization and efficient production develop more readily if a nation contains most of the resources it needs for an activity—for example, coal (to fuel furnaces), iron ore, and limestone for the manufacture of steel. In addition, large internal markets and labor pools permit industries to produce on a larger scale. This reduces the costs of producing each unit and yields greater revenues than could be obtained with the same effort from a smaller operation (figure 2.3). For these reasons, it is much easier for large nations to be reasonably self-sufficient. We will see in later chapters that large size also allows greater specialization and more efficient patterns of location.

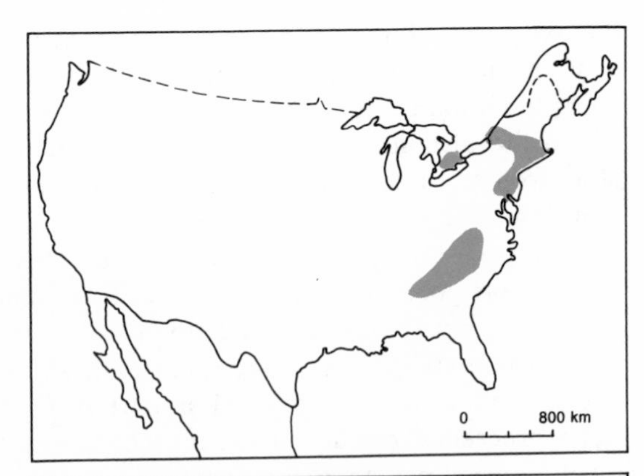

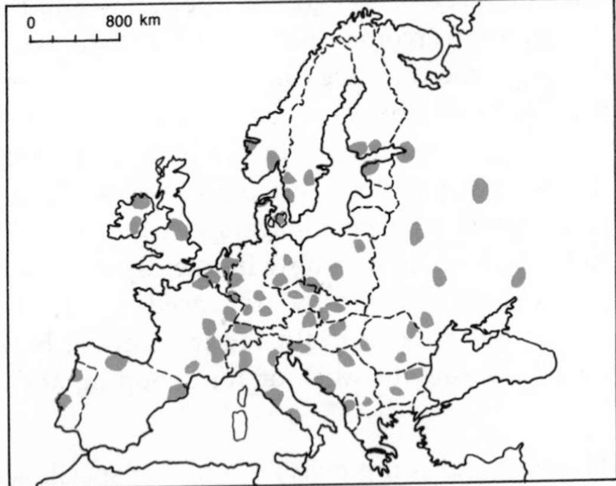

2.3. The effect of size and political unity on production. The large size of the United States and the absence of tariffs and other internal restrictions on movement have fostered a concentration of textile production. In contrast, the small countries and many separate national economies of Europe have given rise to many local zones of textile production.

Source: Redrawn from F. E. Ian Hamilton, "Models of Industrial Location," in R. J. Chorley and P. Haggett, eds., *Models in Geography*, Metheun & Co. Ltd., London, 1967. Reproduced by permission.

Shape

The shape of the space occupied by a tribe, nation, or other organizational unit affects control, transportation, and communication. A compact shape often encourages interaction because of the relative ease and low cost of movement. For example, the major cities of France—a compact nation—are linked by a network of highways, railroads, rivers, and canals. There is much trade and travel between regions; and Paris, the capital, dominates the nation politically and culturally. In contrast, an irregular shape, such as that of Chile and the island nations of Indonesia and the Philippines, may hinder control, prevent cultural unity, and discourage interaction among regions (figure 2.4).

Relative Location

Relative location defines a site in terms of its position with respect to other places—its proximity to natural passageways, transport networks, urban centers, and so on. Gibraltar and Singapore, for example, developed largely because of their location on natural passageways (figure 2.5). Gibraltar, a small peninsula on the southern coast of Spain, partly controls access to the Mediterranean Sea from the Atlantic Ocean and has served as a strategic military base for many centuries. The island of Singapore commands the

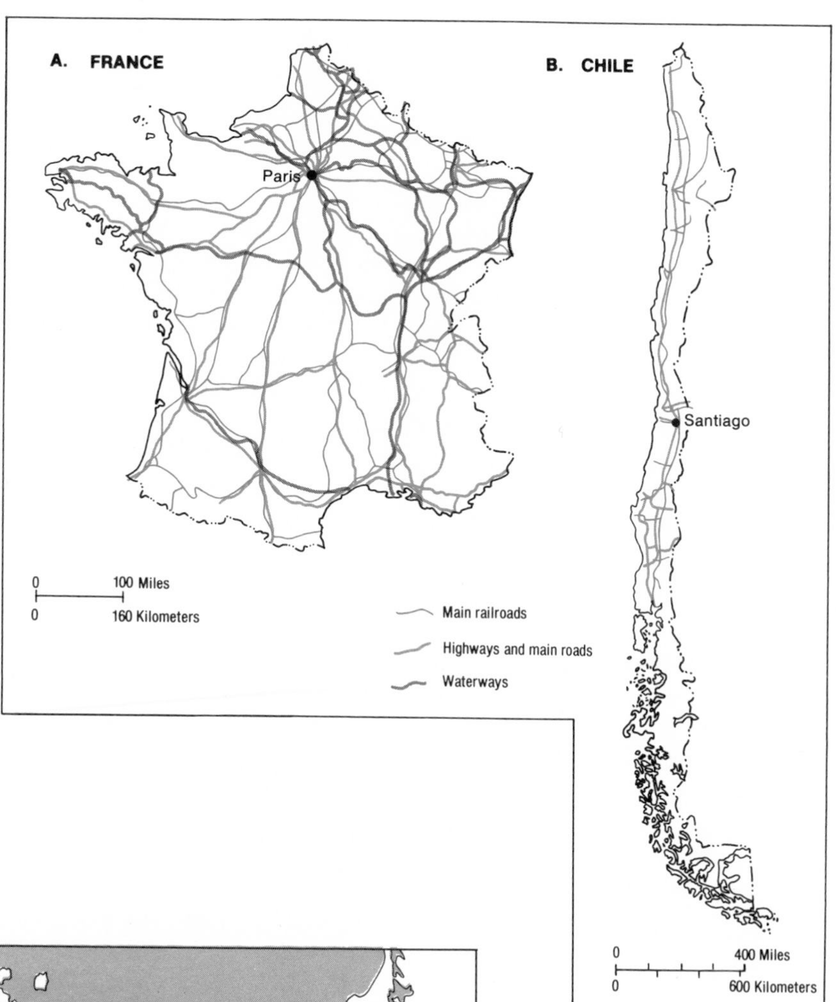

2.4. The effect of shape on interaction. Both Paris and Santiago occupy strategic central locations, but France's compact shape gives Paris easier and faster access to its territory.

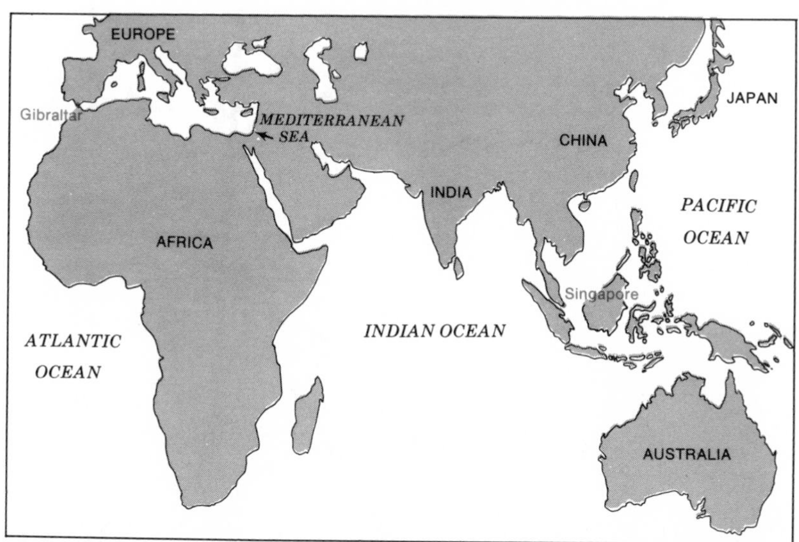

2.5. The advantage of location on natural passageways. The favorable relative location of Singapore and Gibraltar, each positioned at strategic points between large bodies of water, has been a major factor in their development.

principal water route between India and the Far East and is one of the world's busiest ports, as well as a major banking and commercial center.

Location on a transport network has a very favorable impact on development, even if the area lacks resources and has a poor physical environment. Rotterdam became one of Europe's leading commercial centers not because it had a superior environment (the city is below sea level, and poor drainage has always been a problem), but because it was located on the North Sea and had access to Germany, Switzerland, and France via the Rhine River. Conversely there are places that have splendid environments but support fewer economic activities partly because of inaccessibility or poor location relative to existing transport routes. Uganda, which lacks direct access to the sea, and Chile, which is rather remote from world markets, are examples.

Finally the potential of any site may depend as much on its location relative to already developed, well-populated areas as on its inherent physical qualities. The industrial satellites around Chicago, Detroit, and Pittsburgh have benefited from their proximity to the markets, labor pools, banking facilities, advertising agencies, and other resources of the larger cities.

Accessibility

Accessibility is a measure of relative location. It refers to the relative ease with which a location can be reached from other points. Since human beings are social animals and many human activities entail gathering for the

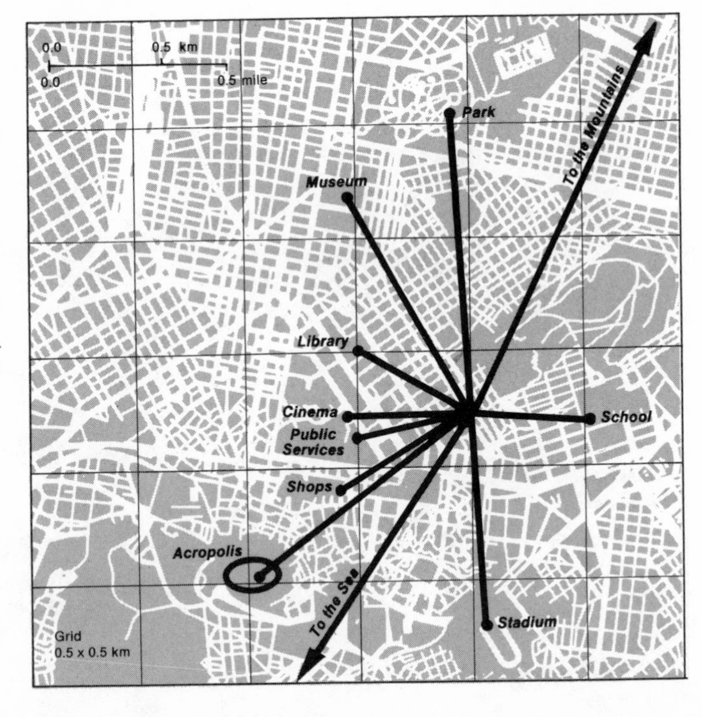

2.6. The advantage of accessibility. This ekistics map shows the accessibility of a given site in Athens to a selection of activities. ("Ekistics" refers to the study of human settlements, a discipline developed by Greek city planner Constantinos A. Doxiadis.) A person living in such a central location could satisfy a variety of needs and activities, including a visit to the Acropolis, at relatively little travel time, cost, and effort.

Source: From *Ekistics* by Constantinos A. Doxiadis. Copyright © 1968 by Constantinos A. Doxiadis. Reprinted by permission of the publisher, Hutchinson Publishing Group, Ltd., London, and the U.S. publisher, Oxford University Press, Inc.

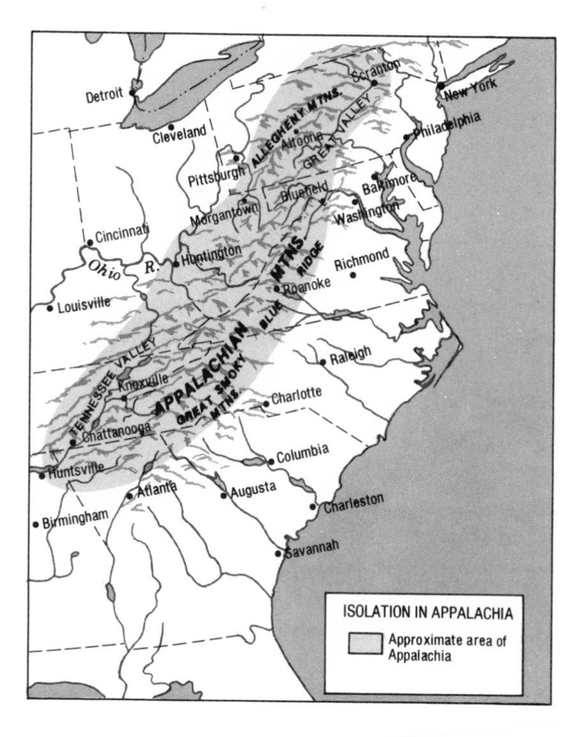

2.7. The disadvantage of isolation. On a national scale Appalachia seems close to major urban regions. But its rugged local topography and lack of all-weather roads have resulted in a high degree of isolation as measured, for example, by the time it takes to reach a hospital or major shopping area. Recent highway construction has somewhat reduced this isolation.

Source: Data based on percentages of isolation from S. S. Birdsall and J. W. Florin, *Regional Landscape of the United States* (New York: Wiley, 1978).

exchange of goods or for common protection, an area that is highly accessible, or central, has a great advantage. A location may be central because of its position in a physically defined basin; but more generally it is central with respect to the distribution of people or activities to which it is related (figure 2.6). We have already noted the role of accessibility in Chicago's development.

The concept of accessibility, or **centrality,** implies its opposites, isolation and location on the periphery. The development of fringe areas has traditionally lagged. Many isolated hamlets in Appalachia in the United States, for example, are backward and impoverished: unemployment is a problem, schools are poor, and access to medical facilities is limited (figure 2.7). In contrast to the cultural centers of a nation, peripheral areas are often sparsely settled and culturally and materially deprived. Noncentrality has prompted some governments to give inducements to people to settle in marginal regions and develop them economically. Appalachia, Nova Scotia, Siberia, and western Brazil have been targets of such aid.

Accessibility is important not only to regions but to nations. Nations with access to the ocean can engage in international trade without depending on the goodwill of neighbors. This freedom is vital to the development of nations. The desire for direct access helped generate Russian expansionism and the development of St. Petersburg (now Leningrad), ice-free Murmansk, and Vladivostok on the Pacific. Today Bolivia is trying to obtain a thin corridor to the sea through a territorial treaty with Chile (wars having failed).

In sum, the *spatial* quality of a location may be even more important than its environmental quality. As an exercise, think of how spatial factors affect your personal life. Does the friction of distance limit the range of your activities? Was accessibility an important factor in your choice of residence at college? Did proximity to recreational or cultural resources influence your choice of college? How close or far away are most of your friends? Spatial considerations are often paramount in the decision of where to live. As we will see, location theory is based almost entirely on such factors.

The Principle of Agglomeration

The spatial factors described above, particularly the qualities of relative location and accessibility, provide the basis for defining an important geographic and economic principle of human location: **agglomeration.** Agglomeration refers to the grouping of people or activities in one area for mutual benefit. This principle of human spatial behavior is expressed in the clustering of huts in a tribal village, as well as in the concentration of people and activities in a modern metropolis. Agglomeration minimizes the cost and effort of overcoming distance and maximizes productivity and interaction. People who live in villages, towns, or cities rather than in isolated homesteads enjoy both closer contact with their neighbors and the convenience of nearby shopping, cultural, and recreational facilities. A person living in Chicago, for example, can easily shop at Water Tower Place, visit the Art Institute, and meet a friend at Orchestra Hall—in the same day. The agglomeration of people in one area satisfies human social needs and makes it easier to exchange goods and pass along information useful in making location decisions.

Agglomeration also increases the **efficiency** of production and distribution. That is, it helps firms produce goods and distribute them to customers at least cost or effort. For example, when small subcontractors locate near large automotive and aircraft complexes, they can supply parts quickly, and transport costs are reduced. When buyers and sellers get together at fairs and in market towns and shopping centers, sellers save on delivery costs and customers on travel costs. An agglomeration of stores not only reduces the total distance that customers must travel; it also enables them to do several errands with relatively little additional effort (figure 2.8).

The principle of agglomeration is perhaps best illustrated by the structure of a city, in particular, its central business district, which contains banks, retail stores, wholesale facilities, restaurants, specialty shops, nightclubs, and theaters. But the fact that the city may also contain competing business districts indicates that the degree of agglomeration is limited.

Agglomeration results from the human desire to maximize productivity

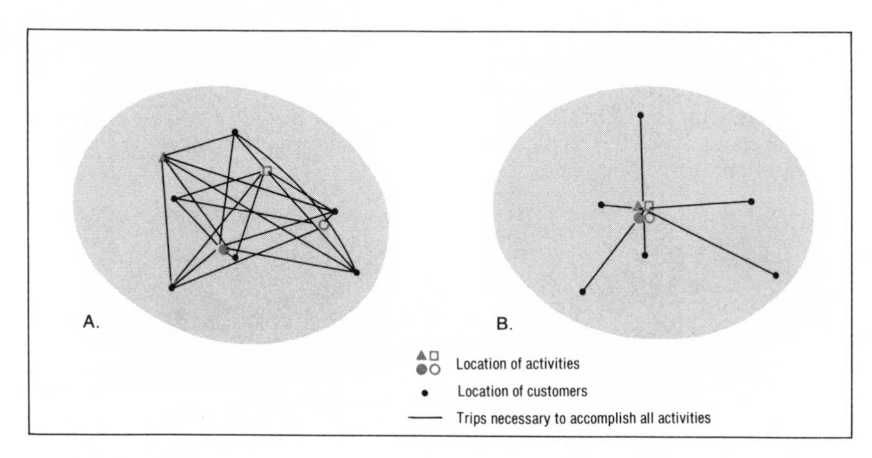

2.8. The advantage of agglomeration. Activities are dispersed in A and agglomerated in B. Customers are in fixed locations. If customers wish to visit all activities, they will have to travel farther in A. Travel time, cost, and effort are reduced when activities are agglomerated (B).

and interaction for both social and economic reasons. It is precisely this attempt at concentration, continually warring against the reality of the dispersion of people and activities across the earth's surface, that gives rise to predictable patterns in the human landscape.

Environmental Factors

The qualities of the physical environment, including landforms, water features, climate, soils, vegetation, and natural resources, also shape the human landscape. These environmental factors have been basic to human survival, and they alone may appear to determine a site's potential for development. Yet we must guard against overstating the role of nature. After all, technology has permitted human beings to prosper in rather harsh environments, as in Iceland and Saudi Arabia. Conversely apparently excellent environments, as in East Africa, have as yet proved no guarantee of development.

The physical environment affects development insofar as it either facilitates or hinders a given economic activity in a given location for a given cultural group. Society often evaluates the environment in terms of the costs it might impose on human activities. In this sense, the influence of physical features on location, interaction, and development can be measured. Sometimes nature must be modified at great cost before economic activities can take place. Amsterdam's favorable relative location justified development in spite of rather poor physical conditions. Much of the arable land around Amsterdam has been reclaimed from the sea by means of expensive drainage systems and protected from flooding by dikes (these reclaimed areas are known as *polders;* see figure 14.4). In the United States, irrigation has transformed California's Imperial Valley from a barren desert into one of the nation's richest, most productive farmlands (figure 2.9A and B). In these examples, human modifications have clearly enhanced the pro-

2.9. Human modification of the physical environment. Irrigation has made the desert near the Salton Sea in southern California (A) bloom into lush, productive farmland (B). But environmental values are not always respected, as in the large-scale strip mining of coal in Montana (C). (Photo A by Alan Pitcairn from Grant Heilman; photo B by J. C. Dahilig from U.S. Bureau of Reclamation; photo C from U.S. Bureau of Land Management.)

A

B

C

ductivity of the land. Sometimes, however, the physical environment invites development but is itself severely damaged in the process. Low-sulfur coal, for instance, may be found close to the surface of a scenic area, inducing exploitation by massive open-pit mining (figure 2.9C).

Just as we analyzed the role of spatial factors, let us now see how elements of the physical environment influence human location, interaction, and development.

Landforms

Landforms refer to the topographical features of the earth's surface, such as plateaus, plains, valleys, hills, mountains, chasms, small islands, and swamps. Although the latter five have discouraged many activities, especially farming, they have sometimes attracted settlement for reasons of defense. Athens, Edinburgh, and Salzburg were built around fortresses located on hills; the city of Rotterdam was protected by marshland; and Paris was founded on a small island in the middle of the Seine River. Such landforms were a source of refuge and protection for human beings throughout much of history and helped to preserve minority cultures. Some Gaelic-speaking Scots, for example, still survive in the islands of the Hebrides off the northwest coast of Scotland. On the whole, landforms that are most conducive to development include plains, valleys, and harbors.

The slope and ruggedness of the terrain often determine whether a given location can be used for economic activities. Certainly farming can be practiced far more readily and at less cost per unit of land on relatively level terrain; soil erosion is less of a problem, and labor and machine costs are lower. Moreover mechanized agriculture requires fairly level land. Since the development of modern farm equipment, agriculture has been abandoned in many rugged areas. Today most of the world's farmland is found in plains and valleys.

Landforms also affect the development of transport networks. In rugged areas, highways and railroads are usually concentrated in passes of least grade. As slope and curvature increase, construction, maintenance, and operation costs rise too. Overcoming landform barriers often entails the costly construction of tunnels, canals, and causeways. Mountains may act as barriers between lowland regions, fostering cultural and economic isolation, as in Appalachia, the Himalayas (Nepal, Bhutan, Tibet), and the Andes (Chile, Peru, Bolivia).

Urban and industrial development in rugged terrain is prohibitively expensive unless high-value resources can be exploited. Copper mining permitted the development of Butte, Montana, at an altitude of 1,730 meters (5,765 feet) in the Rocky Mountains, and tin mining helped the development of La Paz, Bolivia, one of the world's highest cities (3,900 meters or 13,000 feet). In recent years, skiing has become an important industry in many mountainous areas, including Mont Tremblant, Quebec, Sun Valley, Idaho, Stowe, Vermont, and Aspen, Colorado.

Hydrography (Water Features)

Certain water features have long attracted human settlement. Some early groups of humans were lake dwellers, and the earliest civilizations were based on irrigated farming along rivers. Even today, about 75 percent of the U.S. population lives within fifty miles of a major water body. In fact, many cities in Europe, Japan, North and South America, and Africa originated as seaports or lakeports. When land transportation was poor, waterways served to unite people and activities. They still provide fairly cheap transportation, particularly for bulk commodities, such as grain, iron ore, and steel. Major waterways have helped cities and regions develop. For example, the Great Lakes–St. Lawrence Seaway, completed in 1959, has stimulated both the agricultural development of the Midwest and the industrial development of such port cities as Chicago, Detroit, Cleveland, Buffalo, and Montreal (see figure 2.10).

River junctions, where tributaries flow into the main stem of the river, and river mouths are convenient sites for trade and processing. Raw materials can easily be brought in by barge and processed in local factories. The finished goods can be distributed to markets by barge, train, or truck (railroads and highways often follow level river valleys). St. Louis, located on the Mississippi River just south of the Missouri River junction, is an important center of transportation, trade, and industry (see figure 2.10). New Orleans, at the mouth of the Mississippi River, also benefited from its location on the river and its direct access to the Gulf of Mexico to become a national and world trade center (see figure 2.10). Many examples of comparable port cities may be found in other parts of the world. Alexandria, Egypt, located on the Mediterranean coast at the mouth of the Nile, was an ancient center

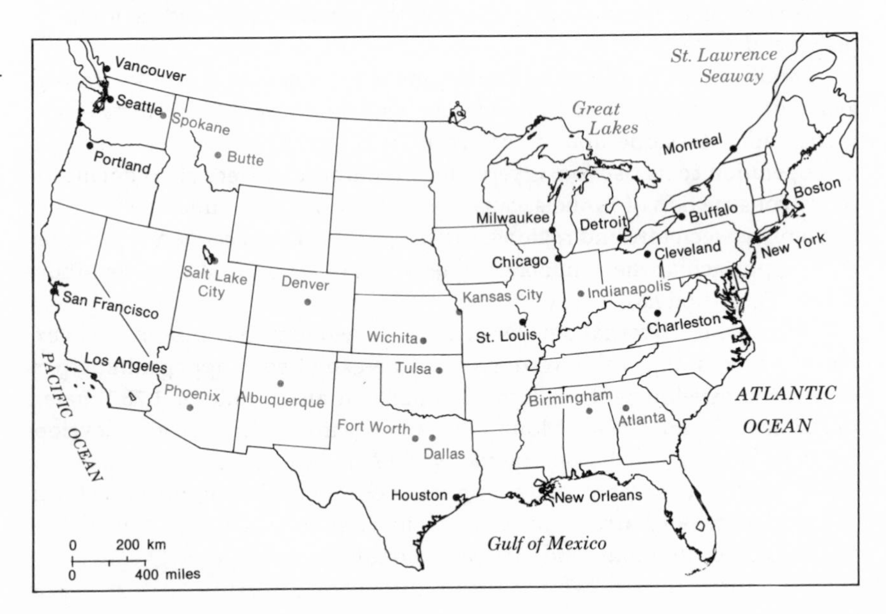

2.10. Major cities in the United States. Cities not located on navigable rivers are shown in orange.

of Greek culture and trade and is now an important seaport and the second largest city in Egypt. Calcutta, India, a few miles upstream from the mouth of the Hooghly River near the Bay of Bengal, is a commercial, manufacturing, and cultural center and one of the busiest ports in the world. And Bangkok, on the Chao Praya River near the Gulf of Siam, is Thailand's capital and leading trade center.

Many cities and towns have developed at points where waterways intersect land routes and can be crossed at relatively little cost. Such sites include shallow fords, narrow bridgeable canyons or the first bridgeable point after an estuary. (Names like Ox*ford* and Cam*bridge* reveal the practical origins of these renowned British university towns; and Frank*furt* in Germany was the ford of the Franks, an early Germanic tribe that lived in the Rhine region.) Because highways and railroads tend to converge at such crossings, adjacent towns often become transshipment points, where goods are transferred from one kind of transport to another.

Given the crucial importance of transportation to the urban economy, location on a navigable river is especially advantageous for larger cities and towns. It is said that Chicago is "a city that had to be" because of its ideal location at the convergence of the Chicago River and Lake Michigan. But successful development does not depend on location. Some major river junctions have been completely ignored for settlement, and some major cities are riverless or are located on nonnavigable rivers (figure 2.10).

Climate

Location and development are often attributed to favorable climate, but some temperate, relatively humid areas are sparsely settled (parts of Africa and South America are examples), and some arid or frigid areas support fairly dense populations—with the help of modern technology (among them are Phoenix, Arizona, and Norilsk, Siberia). Because of pressure from large populations, land is used much more intensively in Japan and Western Europe than in areas of comparable climate in New Zealand and North America.

Farming is generally confined to sufficiently warm and humid climates. Yet agriculture has been somewhat freed from the constraints of climate by irrigation, drainage, erosion control, and the development of drought-resistant plant species. These technological victories have often released immense latent productivity at relatively low cost. At times, however, nature has been modified at high cost under difficult conditions. The irrigation of California's Imperial Valley and the polder development in the Netherlands are notable examples. There still exist many seemingly unfavorable environments that need only a small human investment in irrigation, drainage, or disease control to become productive. This is particularly true of parts of Africa, such as the Sudan, where present low levels of technology prevent the realization of very high agricultural potential.

The understandable desire of certain cultural groups to remain in their familiar territories despite a harsh and unresponsive environment may entail costly efforts to transform nature. Desert nomads, for example, may attempt to desalinize brackish water in their own homelands rather than move to a more hospitable environment. Extensions of crop cultivation into semiarid regions and the overgrazing of arid regions have often resulted in severe erosion as well as crop failure. Evidently the world's deserts are encroaching into major grassland regions at least partly because people are trying to extend crop farming into areas that lack sufficient rainfall.

Climate influences industrial location in more limited ways. Warmer areas may be preferred for the sake of lower heating costs and greater recreation opportunities. But hot and humid conditions may also lower human productivity and increase spoilage problems. The technologies of air-conditioning and heating are about equally costly. They can free people and industries from climatic constraints but only at higher cost. Air-conditioning has been a significant factor in the rapid industrial development of the U.S. South—but it has contributed, among other factors, to the very high rate of energy consumption in the country.

Finally climate affects human location through its control over water supplies. Water shortages limit agricultural, mineral, and urban development in arid regions. Settlement in such regions closely reflects either the location of oases (the islands of population in the Sahara Desert and in the Tarim basin of Sinkiang province, China) or the course of large rivers (the concentration of population along the Nile, the Amu Darya and Syr Darya in the Soviet Union, the Niger in Africa, the Euphrates in Syria, and the Indus in Pakistan). Where water is scarce, ocean water must be desalinized, as in Kuwait, or fresh water must be brought in from great distances, as in Los Angeles and San Diego (figure 2.11). Even in humid areas, local water supplies often cannot keep up with the demands of large concentrations of people, and additional water must be brought in at high cost.

Soils

Soils vary not only in depth, structure, texture, and composition but also in their productivity for agriculture and their stability for supporting manmade structures. In many metropolitan fringe areas, the influence of soil is critical in terms of its implications for septic systems; such considerations affect zoning decisions, investment decisions, and hence location decisions. Areas that are poorly drained because of uneven glacial deposition or location on river or tidal floodplains have often discouraged human settlement. As a rule, such marginal areas are occupied only when population pressures are great or when society is willing to pay for costly water control systems or to endure disastrous periodic floods. Yet many riverbanks, such as those of the Nile, invited agricultural use because of the rich silt deposited during periodic flooding.

Permafrost (permanently frozen subsoil) increases the difficulty and cost of development in Canada, Siberia, and Alaska, as demonstrated by the construction of the Alaska pipeline. Frozen soil is rock hard and thus difficult to use in construction projects. Moreover the ice-bound topsoil above the permafrost may thaw during the short summer season and lose its

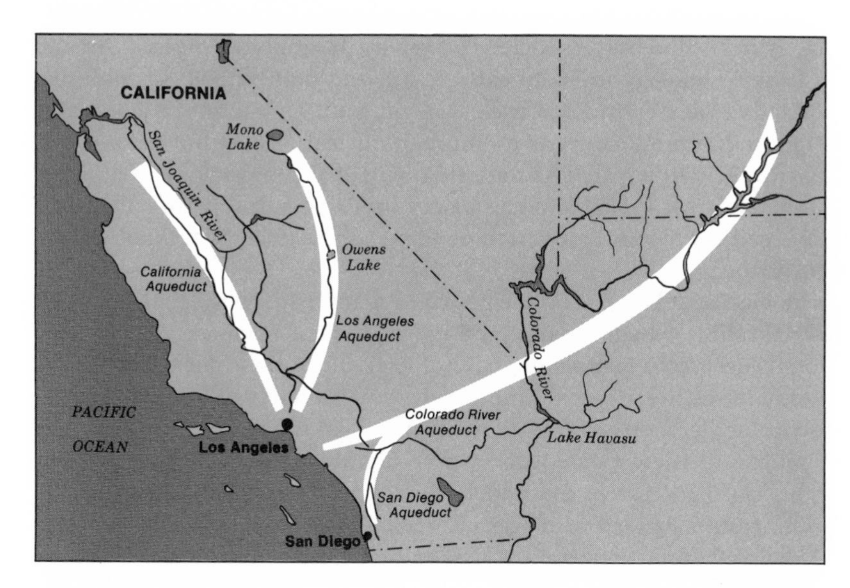

2.11. Water supply systems serving Los Angeles and San Diego. Note the extremely long distances needed to supply water to metropolitan southern California. Much of this development was subsidized by the federal government.

strength as a foundation; in winter the topsoil refreezes, causing frost heaves, or displacement of the soil. To prevent damage from frost heaves and thawing, portions of the Alaska pipeline had to be elevated above the ground on steel piles.

Vegetation

The four principal kinds of vegetation are grasslands, forests, desert shrubs, and tundra. Primitive agricultural technology limited early human societies to the use of existent vegetation: grassland for grazing herds; forests for hunting game, cutting timber, and gathering firewood; and easily cultivated river valleys for planting crops. The vegetation found in a region was often a barrier to intensive agricultural settlement. The thick sod underlying grasslands was too unyielding for tillage, and forests were difficult to clear. With the development of stronger plows for working the sod and more efficient techniques for cutting trees, large areas were opened to agricultural settlement. It was only after the invention of the steel plow in 1837 that the rich but tough midwestern soils in the United States could be easily worked for farming. Dense forests in rugged terrain and forests and tundra in poorly drained regions still impede settlement and transportation in much of Alaska, northern Canada, and western and northern Siberia. However grasslands in many areas continue to be valued as a source of forage for animals and forests as a source of game, firewood, wood for charcoal, and timber for houses and ships.

Natural Resources

Natural resources are the elements of nature that society uses to help satisfy human needs and desires. What constitutes a natural resource is a matter of cultural judgment. A raw diamond, for example, is in itself a neutral substance. It has value only if society perceives its utility or desirability and only if the technology exists to extract, cut, and polish it. Many resources valued today were considered useless in the past. Uranium is a high-value mineral used in producing energy through nuclear fission. But before 1940 it had relatively little value because no one had the technology to utilize it for atomic energy. Probably some places on earth still contain substances whose significance will be realized at some future time, as technology develops further.

The high standard of living enjoyed in many nations today is based on industrialization, which requires the use of natural resources, particularly the fossil fuels (coal, petroleum, natural gas). Since these fuels came to be used only in relatively recent times, their location tends to be random with respect to settlement patterns formed in the preindustrial era, when the chief sources of energy were wood, water, and the power of animal and human muscle. Particularly in northwestern Europe, high-quality coal and other minerals coincided with agricultural lowlands near densely populated sites. Hence these fields were the first to be exploited. Gradually more remote coalfields and, to a lesser extent, other resource areas attracted industry and settlement (see figure 3.4). Population shifted locally to these places from agriculturally more productive ones. For example, in England during the industrial revolution, large numbers of people moved away from Norfolk, an agricultural area northeast of London, to the coal-rich Midlands near Birmingham.

Natural resources are distributed unevenly over the world and are often concentrated in small areas (see figure 14.2). Within most nations, local concentrations of natural resources have given rise to clusters of small settlements, such as mining or fishing villages and mill towns. Once the resource is depleted, the settlement may be abandoned. Ghost towns in California and Nevada are relics of the gold-mining era. Yet even if local resources run out, modern technology allows some industries in settled areas to continue production. The steel industry, for example, can use scrap iron rather than raw ore to manufacture steel. Manufacturing facilities are often sustained by importing the needed raw materials to the location and exporting the finished goods.

Resource exploitation is a major source of growth in regions remote from the center of an economy. As accessible sources of raw material become depleted, capital is invested in developing more distant supply areas. In the Soviet Union, for example, population and industry are concentrated in the western third of the country. Yet over three-fourths of the Soviet Union's energy resources are located in Siberia, the eastern two-thirds of the country—a vast, rugged region extending from the Ural Mountains eastward to the Pacific Ocean. In the past Siberia's harsh environment deterred agricultural and industrial development. But to meet rising energy needs of the densely populated industrialized portion of the nation, investment will increasingly be channeled into the exploitation of Siberian resources. The Soviet Union will construct the necessary housing, services, railroads, and utilities for settlement near Siberian coal, oil, and natural gas fields. Soviet investment will be supplemented by foreign capital. Resource-poor Japan is especially interested in tapping Siberian coal, oil, and timber resources for its own use. (And, incidentally, Japan illustrates that a country need not be richly endowed with natural resources to enjoy a high level of development; but it also demonstrates that a country dependent on external resources is vulnerable to potential price rises and embargos.)

The environmental factors influencing location, interaction, and development shift in importance over time. Arable land and moderate climate determined location and intensity of settlement for agricultural societies, but these factors are less critical for advanced industrial societies. Today arid or rugged regions can successfully be settled and developed. Nonetheless hospitable environments are once again inducing human settlement—not so much for economic reasons as for personal and aesthetic ones. In the United States, for example, many people have left the older northeastern industrial centers for the warmer, more scenic, and unspoiled environments of the sun belt areas of the South and Southwest.

Comparative Advantage

We have seen that spatial and environmental factors make some locations more attractive than others for certain activities, subject to the technological level and values of the cultural group. A location is considered particularly advantageous for a given economic activity when net productivity, or excess of revenues over costs, is greater than it would be in other locations. Chicago's centrality, large markets, access to low-cost water transportation, and proximity to supplies of iron ore, coal, and other raw materials have, for the present at least, given it a **comparative advantage** over many other areas for steel production.

The spatial and environmental requirements of different kinds of activities vary so markedly that places with a comparative advantage for steel production, for instance, may be totally unsatisfactory for another activity, such as cranberry production. Steel can be produced relatively cheaply in Chicago and cranberries in Cape Cod, Massachusetts, but not vice versa. If there is a demand for cranberries in Chicago and for steel in Cape Cod, the two regions can benefit each other through trade. The potential for such interaction is known as **complementarity.**

Some places have a comparative advantage for many kinds of activities, while others are relatively unproductive for most or even all activities. Such differences may result in strong competition for the favorable locations and extremely intensive use of the land. Less profitable activities are crowded out. Marginal locations (those with inferior physical environments—low rainfall, poor soils, and the like), if they are to develop at all, must be satisfied with activities that can at least survive, even though the same activities might be more profitable elsewhere. For example, cattle grazing would be more productive on the lush prairies near midwestern urban markets than on the western high plains. But grazing survives on the plains and is squeezed out from the prairies because more profitable activities, such as corn and hog production, preempt the more advantageous locations.

Comparative advantage may result from human decisions to group activities into a mutually beneficial system—an agglomeration. Even though an enterprise may not in itself seem profitable, its fortunes are likely to improve if it is located near related activities. A corner hamburger stand in a small town may be largely ignored; but if it is moved near an amusement park, business will probably boom.

Comparative advantage is a relative quality, not an absolute one. Relative price levels and land costs must be taken into consideration. For example, the central business district of a city has a comparative advantage for retail trade and other services because of its high accessibility; but a suburban competitor may succeed because of lower land costs and easier parking. To take another example, a farmer in the Great Plains may shift from wheat production to feed-grain and cattle production, depending on the relative price of wheat and beef. A higher market price for beef gives the region a comparative advantage for beef rather than wheat production.

The comparative advantage of a location is never static (see figure 6.10). Changes occur constantly in the relative influence of land, climate, resources, transport position, and concentration of development, and the demands of society change too. Until the mid-1800s, Boston and other Massachusetts seaports were important centers of trade in the United States. When population began to migrate westward, the commercial advantage shifted to the ports of New York, Baltimore, and New Orleans, which had better access to the interior. The Berkshire Hills and the greater distance to the frontier created a barrier between Massachusetts and the rest of the country. Nonetheless Massachusetts adjusted to the change by shifting from mercantilism to manufacturing. Some former shipbuilding towns became mill towns, and some fishing villages became shoemaking centers.

History is filled with examples of formerly profitable locations rendered obsolete by change, and poor areas invigorated by new technologies. The experience of Sweetwater County, Wyoming, illustrates both swings of fortune. Sweetwater County lies in a rather desolate, windblown corner of southwestern Wyoming. Its abundant coal deposits induced the Union Pacific Railroad, a major consumer of coal, to run its line through the county in the mid-1800s. But the steam locomotive era came to an end, and with it, Sweetwater's comparative advantage for coal mining. Now, after a long slump, Sweetwater has recovered. Not only has the energy shortage renewed the demand for coal, but other important minerals—soda ash, oil, oil shale, and natural gas—have been discovered under the sand and the sagebrush. A large power plant and several chemical companies have moved into the region. Attracted by job opportunities, construction workers, miners, and power plant employees have flowed steadily into the area, setting off a local population explosion and reviving the region's economy. The rapid development of large-scale mining and industry has drawn national attention of more than one kind, however. Sweetwater County is now a central concern of environmentalists, who are seeking to protect it from potential air pollution, surface derangement, erosion, and other kinds of environmental damage.

Cultural Factors

Comparative advantage and past patterns of development do not by themselves explain the complexity of the contemporary landscape. Society's use and organization of space is also influenced by *cultural variation*—differences in ethnic composition, religion, literacy, health, attitudes, customs, technology, and socioeconomic systems. Cultural diversity reflects the partially isolated nature of the development of human societies. It also furnishes the complex behavioral context within which the overriding goals of providing food, shelter, and security are carried out and expressed in the landscape.

Ethnic Composition (Race and Language)

The ethnic composition of a population is based primarily on the attributes of race and language. These traits—particularly language—are more important than economic factors in determining the system of nation-states within whose borders and under whose laws most location and interaction decisions must be made. Hostility and competition between different racial and linguistic groups have been at the root of innumerable disputes and wars within and between nations. Often the result has been the destruction of existing settlement patterns and their replacement with new ones. For example, in 1521 Cortes and his Spanish soldiers conquered the Aztec island capital of Tenochtitlán in Mexico, razed it, and built a new city on top of the ruins. The government buildings, churches, and plazas of Mexico City now stand in the place of the Aztec dwellings, pyramids, and temples of Tenochtitlán.

Language differences constitute an effective barrier to interaction, even among groups that live near each other. In Pakistan, for instance, there are at least four major cultural groups, each with its own language. But linguistic differences also have a positive side: they help preserve the values and customs of distinct cultures. These qualities are transmitted from generation to generation in a society's sayings, proverbs, expressions, and oral history.

Cultural groups tend to defend their racial and linguistic integrity from dilution by outside influences by developing a somewhat independent set of institutions (school systems), sources of information (newspapers), and service centers (distinctive shopping areas or markets). Ethnic culture is often preserved in enclaves within cities. In Chicago, for example, German, Scandinavian, Irish, and Polish immigrants formed their own communities in the mid-1800s. Although most minority groups eventually disperse and become assimilated into the larger society, many cities still have a Chinatown, Little Italy, Germantown, and the like.

Language boundaries between groups within a single nation or between nations are often reflected in visible differences in housing styles, village organization, agricultural methods, level of development, and intensity of interaction. Obvious differences may be seen between French- and Flemish-speaking regions in Belgium, between French- and English-speaking regions in Canada, and between English- and Spanish-speaking regions near the U.S.-Mexico border.

Religion

Religious differences are often expressed in the landscape through distinctive styles of architecture (mosques, temples, cathedrals, churches), through the duplication of facilities (many countries, for example, have both Protestant and Catholic school systems), and through local segregation (such as the separation of Moslem and Christian areas in Lebanon). (See figure 2.12.)

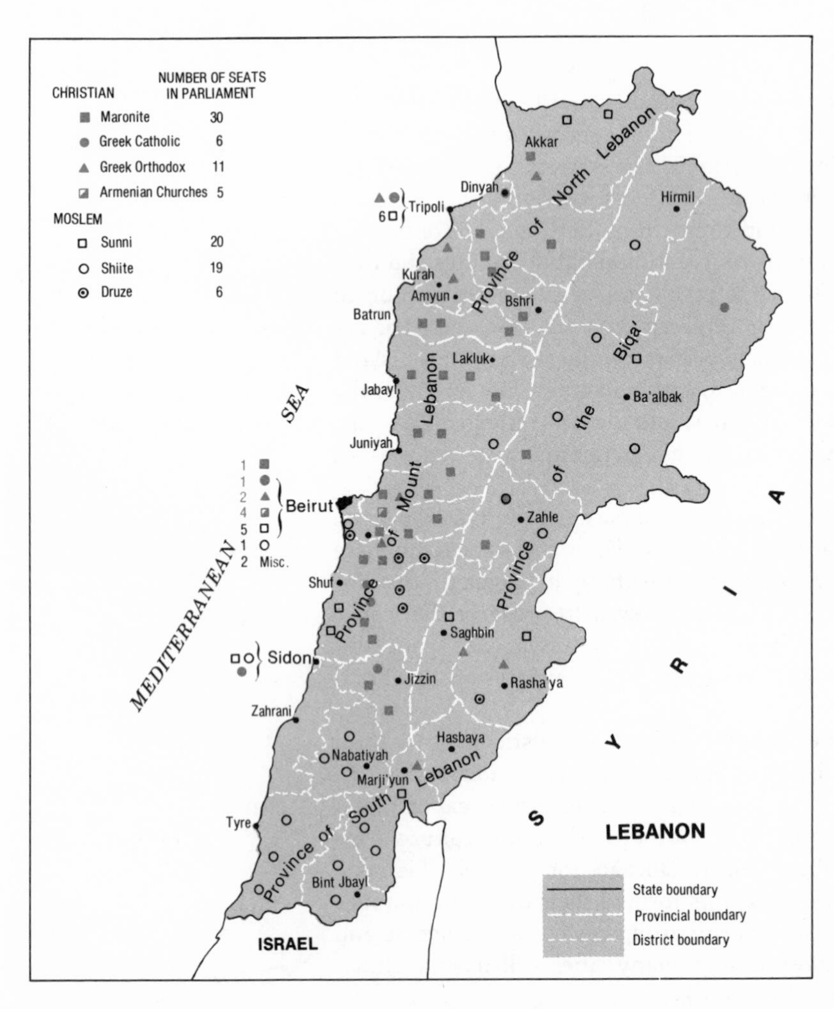

2.12. Moslem and Christian areas in Lebanon, as reflected in the number of seats each group holds in parliament. Although Christians are dominant in the north and Moslems in the east and south, the intermingling is sufficient for continuing conflict.

More indirectly, religious differences are reflected in birthrates (the Mormons in Utah, for example, have a higher birthrate than other religious groups in the state) and in levels of social interaction and intermarriage. Some Hindus and Moslems in central India, for example, practice *purdah*, the seclusion of women from society, particularly from males outside the family; and marriages in this area are often still arranged by parents of the same religion and caste.

Historically the role of religion was even greater. In Europe, for example, the very pattern of villages and cities was in part based on the location of churches and cathedrals, which was determined by the church. In addition, religious wars between Protestants and Catholics and between Christians and Moslems dramatically altered the pattern of settlement and the nature of economic development. Protestant Europe, for example, became more receptive to industrialization and urbanization than was the more conservative Catholic population (perhaps partly because the Calvinist stress on hard work and thrift encouraged the accumulation of capital, which could be most profitably invested in industrial and urban activities). And under the Ottoman (Turkish Moslem) Empire, southeastern Europe lagged far behind the rest of Europe in social and economic development, particularly after the empire began to decay militarily and administratively during the seventeenth century.

Attitudes and Customs

Changes in patterns of location and development often reflect a society's attitude toward migration and change of occupation. Americans in particular are noted for their willingness to migrate and try something new, resulting in a high level of geographic and social mobility. During and after World War II, for example, many blacks migrated from the South to Chicago and other northern industrial cities in response to job opportunities and in the hope of a better life. This trend influenced the rise of black population and development in the North and their decline in the South. Today that trend is reversing itself. More and more people and businesses are moving to the rapidly industrializing South, in part because the attitude of Americans toward the South has become much more positive during the past decade.

Less developed societies tend to be far less mobile and more tradition bound. During the long drought that ravaged sub-Saharan Africa between 1968 and 1974, for instance, several nomadic tribes chose to face starvation with their herds in their traditional homelands rather than try to relocate in fertile river valleys and shift to farming. When migration does occur, the choice of destination may be dictated by the location of relatives or friends from the same villages.

As we will see in chapters 4 and 15, many less advanced societies are adopting Western technology and forms of social and economic development. Yet there still prevails a wide variety of traditional attitudes and customs bearing on all aspects of life, including relations between people, attitudes toward children and childrearing, modes of farming and construction, dietary and housing preferences, and attitudes toward the young, the old, and the feeble, toward nature, and toward change. The willingness or ability of a society to practice birth control, for example, is influenced by the attitude of men and women toward each other, by the perceived need or desire for children, by widely held religious beliefs regarding abortion and birth control, by literacy levels, and by the effectiveness of social networks.

Attitudes toward eating certain kinds of food may affect a society's use of space for food production. Japan harvests the ocean much more than the United States does partly because the Japanese enjoy eating a greater variety of seafood. Conversely the Hindu sanction against eating beef has prevented India from more fully exploiting its enormous cattle population (the largest in the world) or from using the land occupied and grazed by cattle for more productive activities. And the preference of Americans and Argentinians for meat accounts for the high proportion of acreage in those countries planted in feed crops rather than food crops.

Racist attitudes often generate spatial patterns that are inequitable or inefficient, or both. Residential segregation, for example, is enforced by the white-dominated government of South Africa through its apartheid policy, which mandates the separate development of whites, black Africans (the majority population), Asians, and coloreds, those of mixed race. Special territories have been set aside for the blacks; the Asians and coloreds also live in segregated quarters. Besides isolating certain groups and preventing their full development, discrimination gives rise to unnecessary duplication of facilities and inefficient travel patterns, as from black townships to city jobs. Racial discrimination has had a long tradition in the United States as well, beginning with the institution of slavery and currently reflected in the isolation of black ghettos in large cities (see chapters 9 and 13).

Attitudes toward women, the handicapped, children, and the elderly are also reflected in the human landscape and in a society's level of economic development. In the United States and Europe, for example, the elderly usually live apart from their children, often in special housing, and even tend to retire to separate regions, while in many other parts of the world, extended families, where several generations live together, are typical.

A society's attitude toward nature influences the manner and degree to which it modifies the physical environment. The Western concept of human mastery over nature has encouraged an exploitation of natural resources and modification of the physical environment to serve human needs. In contrast, many tribal societies (to some extent illustrated by the American Indian) tend to live within the limits set by nature. You recall that the Chicago landscape remained relatively intact until Europeans settled the area.

Finally preferences for certain residential patterns are reflected in the human landscape. The English, Scandinavian, and American tradition of living in isolated homesteads is expressed in the suburban single-family home and garden.

Technology

We have already noted the critical role technology plays in shaping patterns of location, interaction, and development. Advances in technology and in society's ability to harness energy have in many ways raised the quality of human life. Mechanized agriculture has increased the earth's ability to support humankind. Superior transportation and communication technologies have enabled advanced societies to overcome the barrier of distance and to develop worldwide patterns of location and interaction. And the technologies of irrigation, flood control, and temperature control have at least partially liberated human society from environmental constraints.

One of the most dramatic recent examples of a society transformed by modern technology is provided by Kuwait, an oil-rich sheikdom on the Persian Gulf. With the help of oil revenues, its capital, Kuwait City, has been converted from a small, traditional, mud-walled desert town into the world's most air-conditioned city, complete with gleaming glass and concrete buildings, modern supermarkets, luxurious homes, water distillation plants, and illumined highways.

Modern technology has given human society material abundance and comfort. It is unquestionably the key to progress for developing nations. Yet technologically induced change has exacted a heavy toll in many instances. The drain on sources of energy is creating serious shortages, which may have severe implications for the future (see chapter 15). Crop and animal hybridization is reducing genetic diversity. Industrial pollution has fouled the environment and endangered human life. More unsettling still, the long-term consequences of some forms of pollution, such as chemical contamination of lakes and rivers, are yet unknown; some ecologists fear that they may be catastrophic.

Socioeconomic Systems

Another important cultural influence on patterns of location, interaction, and development is a society's socioeconomic system. (Indeed classical location theory is based on the economic competition and profit motivation of private enterprise or market economies.)

The least complex system is found in *tribal societies,* which are usually small in size, somewhat communal in character, and demonstrate comparatively simple levels of technology, economy, and spatial organization. The location of activities follows repetitive patterns; decision making is relatively unaffected by outside forces; and economic activities tend to be local, traditional, and small scale.

Feudal systems (although fairly rapidly disappearing) still exist in the Middle East, Asia, Africa, and Latin America. They may be found, for example, in parts of Egypt, Burma, Ethiopia, Saudi Arabia, Bolivia, Paraguay, and Peru. Under a feudal socioeconomic system, a large number of peasants usually support a small ruling class. Incentive is lacking for growth or change. As long as control is maintained and the expectations of the people remain low, there is little reason to increase productivity. Changes in spatial structure are minimal, consisting mainly of further land subdivisions as the population expands. Feudal systems usually break down when the need arises for the state to defend itself, requiring greater total wealth, or when a middle class emerges (often as a result of the exchange of surpluses across system boundaries) and begins to demand more goods, greater power, and broader freedom. Industrialization and urbanization are key processes in such change. The transition was evolutionary in most of Europe and Japan and revolutionary in Russia and China. (The spatial order of tribal and feudal societies will be examined in greater detail in chapter 4.)

Highly differentiated patterns of location and interaction have evolved in technologically advanced capitalist societies such as the United States, Canada, Western Europe, and Japan (see chapters 5–11). Most theories of **spatial organization** stress such goals as the short-term maximization of profit, which are most easily realized in capitalist societies. Location decisions in a capitalist system are made by individuals under conditions of **risk** and **uncertainty.** Investment tends to reinforce and only gradually extend (by virtue of the greater risk takers) existing patterns of location and interaction. Typically some regions are much more highly developed than others, despite government inducements to stimulate economically depressed regions (see figure 13.7). Market economies are characterized by large numbers and varieties of economic units, such as farms and businesses, and by relatively rapid change (such as changes in occupation or the demise of some businesses and the birth of others). These enterprises tend to be highly responsive to changes in demand and are not sentimental about the displacement of workers as a result of changes in demand or technology or about the alteration of scenic or historic landscapes.

In socialist societies such as the Soviet Union, Eastern Europe, and China, investment and location decisions are centrally planned and controlled. Since decisions are made by a few government agents or committees who have access to fairly complete information and can view the system as a whole, patterns of location and interaction in these areas should theoretically be more efficient than those in a private-enterprise society. On the other hand, a society that lacks a *market system* (in which resources are allocated according to how much the consumer is willing to pay) often uses resources unwisely. It may, for example, tie up too much capital in long-term construction. No system is immune from error. A poor decision under central planning can have far-reaching and ruinous consequences. The 1958 Great Leap Forward in China, a five-year plan to double industrial production, resulted in waste of scarce capital and inefficient use of raw materials; and the extension of continuous grain cropping into the "virgin lands" of Soviet Kazakhstan in the 1960s resulted in severe land erosion and repeated crop failures. However since the state is better able and perhaps more willing than individual investors to sustain losses for a longer period, it should be possible to achieve a more rational and balanced pattern of development with a central system. In the long run, regional differences in a socialist system probably will come to reflect the actual potential of a region more than its economic history, and regional differences are usually less pronounced.

Political Factors

Finally location and interaction decisions are made in a political context. Relevant political factors include systems of government, national boundaries, and the impact of colonialism, imperialism, and nationalism.

Government Systems

The concentration of authority in a *centralized government* promotes greater regional economic and cultural uniformity than does a *federalized system* in which authority is divided. States with a strong central government, such as the United Kingdom, France, China, and Poland, can impose nationwide policies—uniform wage rates, educational standards, and the like. National economic and cultural interests are likely to supersede regional interests, as reflected in more balanced patterns of resource exploitation and the insistence on a single national language. But even strong governments have proved unable to create cultural uniformity over long periods; Welsh and Scottish nationalism, for instance, still persist in the United Kingdom.

States that have evolved from several smaller units or that contain radically diverse cultural elements, such as West Germany, the United States, Canada, and, in theory, the Soviet Union, tend to form federal systems. Economic power is divided, and regional and national interests often compete with one another. Federalism helps preserve regional economic and cultural variation, which has a visible impact on the appearance of housing, business facilities, farmsteads, and other structures.

Several nations with traditionally centralized governments, but also with great cultural diversity, are moving toward federal systems. Examples include Belgium, with its Walloons (French-speaking people) and Flemings; Spain, with its Basque, Catalonian, Galician, and Castilian peoples; Italy, with its numerous cultural groups, each with its own dialect (historically a result of the isolation of various regions separated by mountain barriers); and even the United Kingdom, which is considering the possibility of forming separate Welsh and Scottish assemblies.

One might think that a predominantly authoritarian regime would be more effective than a predominantly democratic one in generating faster development. Indeed the charge of economic chaos is often the justification used for imposing authoritarian rule, as in Chile, Brazil, and Argentina. But the evidence of recent economic history suggests that economic as well as social stagnation is more likely under an authoritarian system and that decentralization of decision making encourages innovation and productivity. These relationships have been demonstrated by the recent progress of Greece and Portugal and by the stagnation of Paraguay, Burma, Argentina, and Uruguay.

Boundaries

National boundaries usually signify a sharp break in political and economic authority; many also reflect linguistic divisions, particularly in Europe and Asia. Some boundaries follow physical barriers, such as mountains and rivers, while others arbitrarily separate areas that would more logically be united physically, culturally, and economically. The boundaries of many former European colonies in Africa, for example, cut across natural, linguistic, and tribal regions. Closer to home, the United States–Canada border imposes a political barrier between nearly identical regions, severing transportation and settlement patterns (figure 2.13) and weakening the economies of the immediately contiguous areas. And the linguistic and cultural boundary between Mexico and the United States has been an even greater deterrent to interaction.

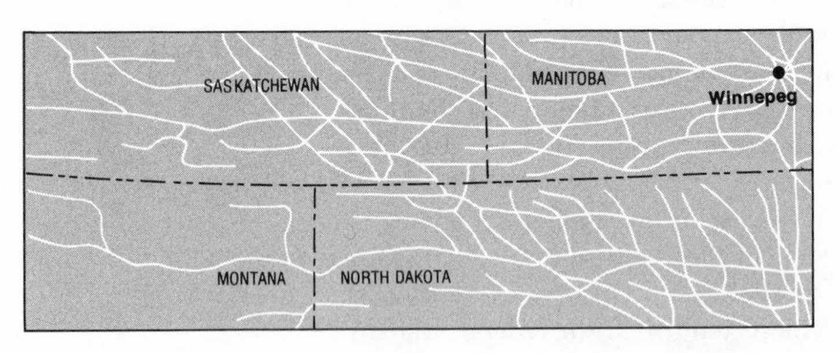

2.13. Broken railroad network at the Canada–United States boundary. Political boundaries often interrupt connections and reduce interaction between contiguous areas.

On the other hand, boundaries may stimulate development. Military posts and major ports have emerged as national responses to the independence symbolized by political boundaries. Examples are the separate but nearby ports of Rotterdam in the Netherlands and Antwerp in Belgium. Political boundaries have also permitted undeveloped nations to evolve beyond being mere suppliers of raw materials for the developed nations. In the early nineteenth century, for example, an independent United States could support its own industry and protect it by imposing high tariffs on imports. Today the petroleum-exporting countries require that some portion of oil be refined within their own borders.

Colonialism and Imperialism

For thousands of years powerful nations have created empires. The empires have come and gone, but many have left a long-term imprint on the land. The Roman Empire, for example, ended almost two thousand years ago, but it remains visible in the pattern of settlements and roads throughout its former realm, as well as in many aspects of Western culture, including language, law, and religion. More recent systems of colonies, most notably those of Britain, France, and Portugal, continue to have an impact on the landscape and on patterns of interaction. They are responsible for the existence of mines, estates, and other foreign-dominated investments and for traditional exchanges of raw materials for manufactured goods. Current political-military alliances continue to affect patterns of development and trade, as exemplified by the U.S. trade embargo against neighboring Cuba and the enduring effects of the division of Germany after World War II.

Nationalism

Nationalism, the pursuit of goals perceived to be in the best interests of the nation-state, has an extremely powerful impact on location, interaction, and development. It may stimulate economic and social development by fostering national identity or by inspiring creativity and productivity. Nationalism has been a factor in the construction of Brazil's Trans-Amazon Highway, a symbol of the country's emerging national unity. But because nationalism may also foster an exaggerated "we-against-others" mentality, it is often a destructive force. Many conflicts and wars over territory and resources are incited by nationalism and result in the destruction of people and the man-made and physical environment. In recent years, nationalistic fervor has destroyed communities and bloodied the soil of the Middle East, Northern Ireland, and Somalia-Ethiopia.

Strong nationalistic feelings may tempt small but politically independent nations, such as the new nations of Africa, to try to become self-sufficient. Their size, however, prevents them from succeeding readily as independent economic units. Nonetheless they may impose such protective measures as tariffs and subsidies for internal industries and maintain large military forces. These actions often hinder regional or national specialization and interdependence. The result is that space is not used as productively as it might be, international ties are weakened, and already low standards of living may decline even more.

LOCATION THEORY AND HUMAN SPATIAL BEHAVIOR

A complex world landscape has evolved from the interlocking of the factors discussed above. Some degree of underlying order to this landscape exists because society tries to utilize space rationally and organize it efficiently, despite the frequency with which other, noneconomic motivations—political and cultural chauvinism and protectionism, for example—intervene. We will conclude our overview of the human landscape by defining and discussing the theoretical principles that govern human spatial behavior and result in predictable spatial patterns. We will also consider some of the behavioral constraints that prevent the realization of ideal spatial order.

Principles of Human Spatial Behavior

According to *location theory,* three general principles, or goals, govern human spatial behavior:

1. To maximize the net utility of places at least cost;
2. To maximize spatial interaction at least cost;
3. To minimize the distance separating related activities.

The first principle concerns the use of space. It implies that people try to use space as efficiently as possible, subject to their level of technology, to get the greatest satisfaction or economic return for the least expenditure of time, money, or effort. To illustrate, suppose that a farmer has bought a parcel of land. Theoretically he or she would want to use that space as productively and profitably as possible. The farmer's choice of potential activities might range from raising sheep to growing tomatoes. After considering all the variables—physical conditions (soil, terrain, climate); the size of the parcel; the costs of seed, fertilizer, equipment, and fuel; distance from market; the cost of transporting various crops; and the current market prices—the farmer would presumably choose the activity that would net the greatest *profit,* or excess of returns over cost, at the location.

People can maximize the profitability of their activities by taking advantage of the potential for regional specialization. Specialization, as we have seen, involves interaction—the exchange of goods between regions. For example, it is often desirable for farmers located in the U.S. upper Midwest to specialize in dairying because of the combination of terrain, climate, and relative location. And it is desirable for certain heavy-machinery producers and food processors to locate near Chicago for reasons of raw-material, energy, and labor supply as well as market demand. Each specialized region produces something the other needs. It is in their mutual interest to maximize their trade interaction at the least possible cost, fulfilling the second goal of efficient spatial behavior. Similarly people tend to organize their settlements in such a way as to minimize the distance to work, to shopping places, and to friends' homes; their goal is to maintain a high level of interaction with others at least effort.

The first two principles (maximizing the utility of places and spatial interaction at least cost), if adhered to by society, should lead to an economically optimal pattern of land use, degree of specialization, and volume of trade. The third principle—minimizing the distance between related activities—is a corollary of the first two and can logically be inferred from them. The concept of agglomeration is based on this principle: related activities are more profitable when grouped close together in a mutually beneficial system. This defines the very notion of a **market.**

The extent to which a society's spatial behavior is governed by these principles might be determined by answering the following questions:

1. Do enterprises successfully produce demanded goods and services at maximum but not excess profit? That is, is location optimal and are **economies of scale** realized?
2. Is the allocation (apportionment) of value fair? (a) Are wages, salaries, rent, royalties, interest, and profits allocated to people according to their level of productivity, and are they adequate to provide all with acceptable levels of food, shelter, and security? (b) Are natural resources, including land, utilized as needed but not wasted?

3 Does the pattern of interaction reflect in a simple way the structure of economic specialization and social differentiation, avoiding unnecessary cross-movements? And does the level of interaction satisfy the needs and desires of the population?
4 Are regional and intraurban differences in income and wealth relatively insignificant?
5 Are all costs of production accounted for and paid by the producer, thereby reducing pollution, blight, and other social costs?

The actual spatial organization of the world will be evaluated against these theoretical criteria in later chapters, particularly chapter 13.

Behavioral Constraints

Human spatial behavior in fact departs greatly from the theoretically ideal model. Individuals, groups, businesses, and governments often lack the information, resources, ability, or motivation needed to make an optimal location decision. More important, the majority of people are satisfied with less than maximum long-term profitability (figure 2.14). As long as the business remains profitable at a given location, it is easier not to move. As long as the home location provides enough social and environmental satisfaction, it may be worth traveling farther to work than is really necessary.

2.14. Actual and potential labor productivity. Indexes of actual farm labor productivity in a sample area of Sweden are shown in A, in units of ten kroner per hour. Maximum potential productivity is shown in B. A comparison between the two maps reveals that actual labor productivity falls far below potential productivity. For example, farmers in the northernmost portion could earn over forty kroner an hour, but most earn only twenty-eight to thirty-four or less. This is due partly to the farmers' lack of information about more efficient techniques, but primarily to their unwillingness to take the risk or make the effort required for greater productivity. Most people are satisfied with less than maximum potential productivity and profits.

Source: Redrawn by permission from the *Annals* of the Association of American Geographers, Volume 54, 1964, J. Wolpert.

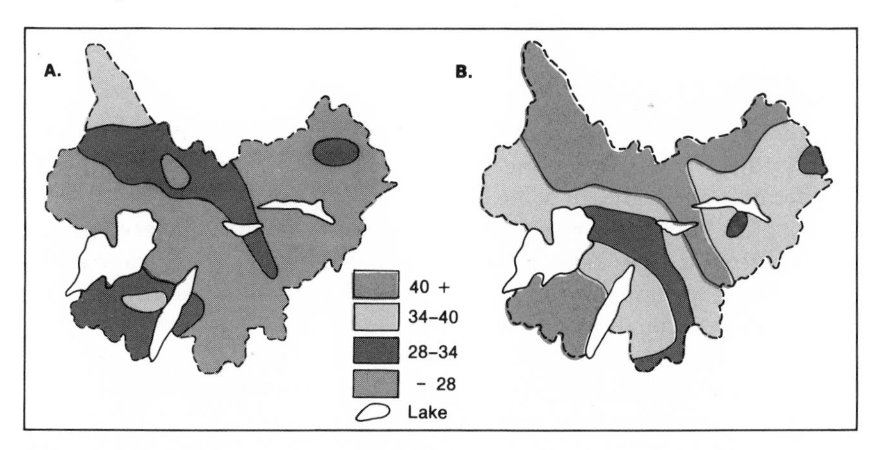

Individuals are often unwilling to forgo immediate satisfaction for the realization of possible long-term benefits. It may be more profitable to move to a region that is experiencing rapid economic growth, but most families would prefer to stay in a familiar location.

Natural disasters (floods, hurricanes, earthquakes) and irrational, self-destructive human behavior (racism, fear, war) may destroy nearly optimal patterns of location and interaction or prevent them from being achieved. More commonly, however, most location decisions are made in the context of a specific time frame and set of circumstances, and their value may erode as technology improves and dramatic changes occur (the lessening of racial discrimination in the United States is an example). Yet a substantial investment has usually been made in the physical structures and human resources of a location. As long as the investment continues to pay for itself, the individual or firm may feel it is better to operate at less than peak efficiency than to abandon the location altogether. Though nearly all locations are less than perfect, only the extremely inefficient ones will fail. It is also possible for a poorly located activity to adjust profitably to its location through the efforts and resourcefulness of its operators.

Some locations are inefficient to begin with. Typically decision makers have acted on insufficient or inaccurate **information** and have made errors in the location, scope, and size of investments. Then, too, a nonoptimal location that is still profitable or satisfactory may deliberately be selected. A farmer may prefer to raise cattle on land suitable for wheat even if more money could be made by growing wheat. Many investors, unsure of long-term possibilities, are content to achieve satisfactory rather than maximum profitability, or they may have overriding social or personal reasons for choosing a location or for using it in a less than optimal way.

The factors noted above inhibit the realization of ideal patterns based on the stated economic principles of location theory. But in addition, location is governed by major social principles as well: to maintain the social status of the individual and to preserve the social integrity of the group. Particularly in the context of the geography of the city, people often value these goals as much as, if not more than, the economic ones.

CONCLUSION

Location theory simplifies reality by focusing on spatial considerations, particularly the cost of overcoming distance. It provides fairly simple models that permit us to highlight some essential principles and factors of human location. The real world, of course, does not correspond very closely to the patterns projected by location theory because the human landscape is the complex product of many different forces—historical, physical, cultural, political, and behavioral, as well as economic (spatial). These noneconomic factors distort theoretical spatial patterns and prompt us to modify the theory. In the next chapter we will consider the role these various influences play in the evolution of our world's landscape.

SUGGESTED READINGS

Bunge, William. *Theoretical Geography.* Lund Studies in Geography, Series C, no. 1. 1966. Lund University, Sweden.

Chapin, F. S., and Weiss, S. *Factors Influencing Land Development.* Chapel Hill, N.C.: University of North Carolina Press, 1962.

Ehrlich, Paul, and Ehrlich, Anne. *Population, Resources and Environment.* San Francisco: W. H. Freeman, 1970.

Firey, W. *Man, Mind and Land: A Theory of Resource Use.* Glencoe, Ill.: Free Press, 1960.

Haggett, Peter; Cliff, A. D.; and Frey, A. *Locational Analysis in Human Geography.* London: Arnold, 1977.

Kasperson, R. E., and Minghi, J. V. *The Structure of Political Geography.* London: University of London Press, 1970.

Manners, G. *The Geography of Energy.* Chicago: Aldine, 1964.

Nystuen, John D. "Identification of Some Fundamental Social Concepts." *Papers of Michigan Academy of Sciences, Arts and Letters* 48 (1963): 377–384.

Pred, Allan. *Behavior and Location.* Lund Studies in Geography, Series B, no. 27, 1967; 28, 1969. Lund University, Sweden.

Semple, E. C. *Influences of Geographic Environment.* New York: Holt, 1911.

Sopher, D. E. *Geography of Religions.* Englewood Cliffs, N.J.: Prentice-Hall, 1967.

Webber, M. *The Impact of Uncertainty on Location.* Cambridge, Mass.: MIT Press, 1972.

Zipf, G. R. *Human Behavior and the Principle of Least Effort.* Cambridge, Mass.: Harvard University Press, 1949.

3

The Evolution of the Landscape

3 The Evolution of the Landscape

Ancient myths, chronicles, and maps reveal that the world's landscape has changed profoundly over many thousands of years (figure 3.1). Even now, wind and water erode and transfer enormous loads of material, often in highly visible ways. Blinding sandstorms and dust storms sweep across deserts and plains, erasing familiar contours and forming new ones. Cascading rivers carve trenches through solid rock. Meandering rivers slowly destroy old stream courses and create new ones. River branches change their courses, leaving ports incongruously stranded on dry land. Coastal lagoons emerge, then disappear. Promontories erode into the sea. Glaciers slowly advance and retreat, sculpturing mountains, scouring valleys, and flecking the earth with new lakebeds. Volcanoes give spectacular birth to new land areas, bury old landforms, and modify the climate in ways we do not fully understand. Volcanoes, earthquakes, floods, and fire have ravaged human settlements and structures, demolishing the long-term efforts of society time and again.

Yet during just the past two thousand years, human society has modified the landscape even more dramatically than have the forces of nature. Humans have terraced slopes, dug canals, dammed rivers, cultivated crops, cut and planted forests, and built walls, roads, and cities. They have also modified the impact of natural forces. For example, they have at times strengthened the impact of wind and water on land by clearing forests and overgrazing livestock. As a result, their removal of topsoil has more rapidly turned forests into grasslands and grasslands into deserts. At other times and in other places, humans have slowed this erosion and increased the productivity of the land by irrigating deserts and draining marshes, by planting trees and building dikes.

Much of the natural landscape has evolved into a human landscape through the gradual development and modification of settled areas and through the spread, or diffusion, of people and ideas across the earth. In this chapter, we will see how the dual processes of **development** and **diffusion** have contributed to the evolution of the landscape.

DEVELOPMENT

Development refers to changes that improve human well-being and often transform human spatial order. The term includes the idea of individual growth, but we will stress its more conventional meaning of cumulative change in a society's economy and culture. Development may be measured by gains in real per capita income, literacy, and life expectancy. This kind of change stems from the innovation and application of new technologies, the

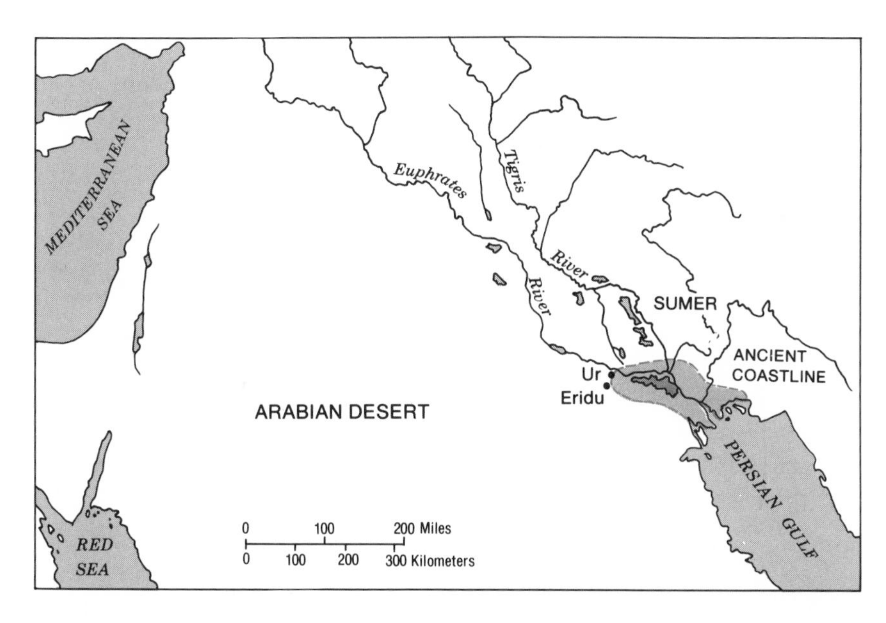

3.1. Natural changes in the landscape over time. An inscription written about four thousand years ago in Mesopotamia states that "Shulgi, the son of Ur-Nummu, devoted special care to the city of Eridu, which lay on the seacoast." If these words accurately describe the location of Eridu (modern Abu Sharain in Iraq), then the shore of the Persian Gulf must have extended farther northwest in ancient times. Sedimentation from the Tigris and Euphrates rivers gradually changed the shoreline.

development of new customs and values, and the accumulation of knowledge and wealth. Some forms of development can and indeed must be generated within a society. Others are borrowed from or imposed by different societies. *Diffusion* refers to the spread over space and time of people, goods, technology, and ideas, usually from just a few places of origin.

The Impact of Agricultural Technology and Sedentary Culture

Preagricultural Societies

Humanity left its first mark on the landscape nearly a million years ago. Armed with fire, clubs, and spears, human beings began to migrate—evidently from **hearth areas** (places of origin) in Africa and perhaps parts of Southeast Asia. They were successful hunters, capable of exploiting a great variety of animals. They hunted some species even to extinction, modifying natural patterns of evolution, including their own. As tribes roamed from one initially rich environment to another, they altered patterns of vegetation as well. In burning vast areas to capture their prey, they extended the range of grasslands, as in North Africa, and speeded natural rates of erosion.

The Agricultural Revolution

For hundreds of thousands of years, the impact of human beings was limited to the effects of hunting. Then about ten thousand years ago, a breakthrough occurred in human development: some tribes learned to cultivate plants and to domesticate animals—to control their own food supply, at least partially, rather than depend on hunting. This *agricultural revolution* led to new forms of human spatial order and to changes in the landscape. It still continues to transform large portions of the earth. Agricultural societies have settled nearly all habitable parts of the world. About 15 to 20 percent of the earth's land area, excluding Antarctica, has been plowed and cultivated. Much of this area was originally forested. Agriculture made possible permanent settlement and the aggregation of people into villages. About another 20 percent has been modified, though less obviously, through the grazing of domesticated animals.

By 3000 B.C. farmers in Egypt, Mesopotamia, Persia, northern India, and China had developed an innovation, irrigated agriculture, that made possible the rise of civilization—and cities—under an authoritarian social structure. Irrigation raised productivity by ensuring a more reliable water supply and by permitting the cultivation of drier lands. Farmers were able to produce enough surplus to support people who specialized in nonfarming activities: priests, administrators, warriors, merchants, and artisans. More and more people gathered in villages, some of which grew into cities (see chapters 6 and 7). Irrigation spurred the development of cities, transportation to bring surplus crops to market, legal systems to regulate land ownership and water rights, and specialization of labor. It also led to widening social gaps as some people acquired more wealth and power than others.

Changes in the landscape were no less dramatic. Fields, canals, roads, and villages spread across the landscape. Well-organized cities emerged as centers of commerce and control. The earliest cities probably arose in the Middle East: Jericho, a preagricultural oasis in Israel; Susa, Ur, and Kish in Mesopotamia, between the Tigris and Euphrates rivers; and Mohenjo-Daro, on the Indus River in what is now Pakistan.

Perhaps the greatest long-term effect of the agricultural revolution was a sharp and continuing rise in population. China, a hearth area of intensive agriculture, is estimated to have reached a population of 60 million by A.D. 900. The growth of population, in turn, led to larger, denser settlements and further modification of the landscape.

A simpler, nonirrigated agriculture penetrated into the more humid regions of Europe and East Asia and gradually diffused into most areas capable of supporting crop agriculture. (Some scholars believe that agriculture in Western Europe dates back to 4000–3000 B.C.) Regions too dry or rugged for crops were used for grazing sheep, goats, and other livestock.

By the second century after Christ, a fairly productive agriculture thrived in parts of Europe, the Mediterranean, the Middle East, and throughout the river valleys of China, India, and Japan and the lowlands of Southeast Asia. It also supported the centers of Mayan culture in Middle America, which flourished between A.D. 250 and 900. (Whether this was one of the origins of the agricultural revolution or the practices diffused to Middle America from Asia or Africa is still not known.) Huge expanses of forest were probably destroyed in the quest for crop and grazing land, particularly in China and North Africa. Here, too, the development of agriculture was accompanied by the construction of villages, roads, defensive walls, and cities.

Beyond the centers of early civilization, a less productive *shifting cultivation* was commonly practiced. This more primitive form of agriculture, which occurred throughout northern and western Europe, central Africa, Southeast Asia, and Central and South America, entailed cutting and burning trees and brush to clear patches of land for crops. After the nutrients provided by the ashes were depleted, the cultivators cleared new areas and either abandoned the old ones or left them fallow (idle) for several years (see chapter 4). Society's impact was visible in the huge strips of deforested land that mottled the earth.

The Development and Spread of Sedentary Culture

As we have seen, irrigated agriculture fostered the growth of cities and the rise of civilization. While European civilization slowly developed and then emerged as the most vigorous and expansionist culture (250 B.C.–A.D. 1750), other civilizations—Arabic, Indian, Chinese, African—were also growing, declining, and changing. We will focus on the development of Europe, beginning with the Roman Empire, but parallels may be drawn among various cultures. (We have selected Europe because of its widespread influence, particularly during the period of colonization, and because the industrial revolution started there.)

By the third century B.C., the landscape near such great centers of civilization as Rome and Peking began to acquire an unmistakably human, ordered appearance. Both cities were located in agriculturally productive valleys that also had advantages for defense and administration. These centers controlled long supply lines, which enabled them to expand their earlier patterns of growth. The Roman Empire developed an economically interdependent network over a wide area. To balance the supply and demand of production in different regions, the Romans imposed on the landscape a system of high-quality roads, canals, and port facilities (figure 3.2). The Chinese empire similarly depended on rivers, canals, and roads to redistribute grain and to develop and maintain the capital's power and splendor.

Meanwhile nomadic and seminomadic tribes roamed the fringes of early civilizations. As the tribes grew larger, straining their primitive hunting, herding, and agricultural economies, some would periodically invade the frontier and even the core of the more civilized areas. Although many of these outsiders, or barbarians, were eventually absorbed into civilization, the empires themselves were ultimately transformed. Weakened by repeated invasions and internal problems, the Roman Empire collapsed in the fifth century. Trade and travel declined. The once highly interdependent empire shattered into many small, poorly connected, more self-sufficient kingdoms and principalities that continually fought each other. Roman civilization was succeeded in part by an Islamic one, in which cities and science

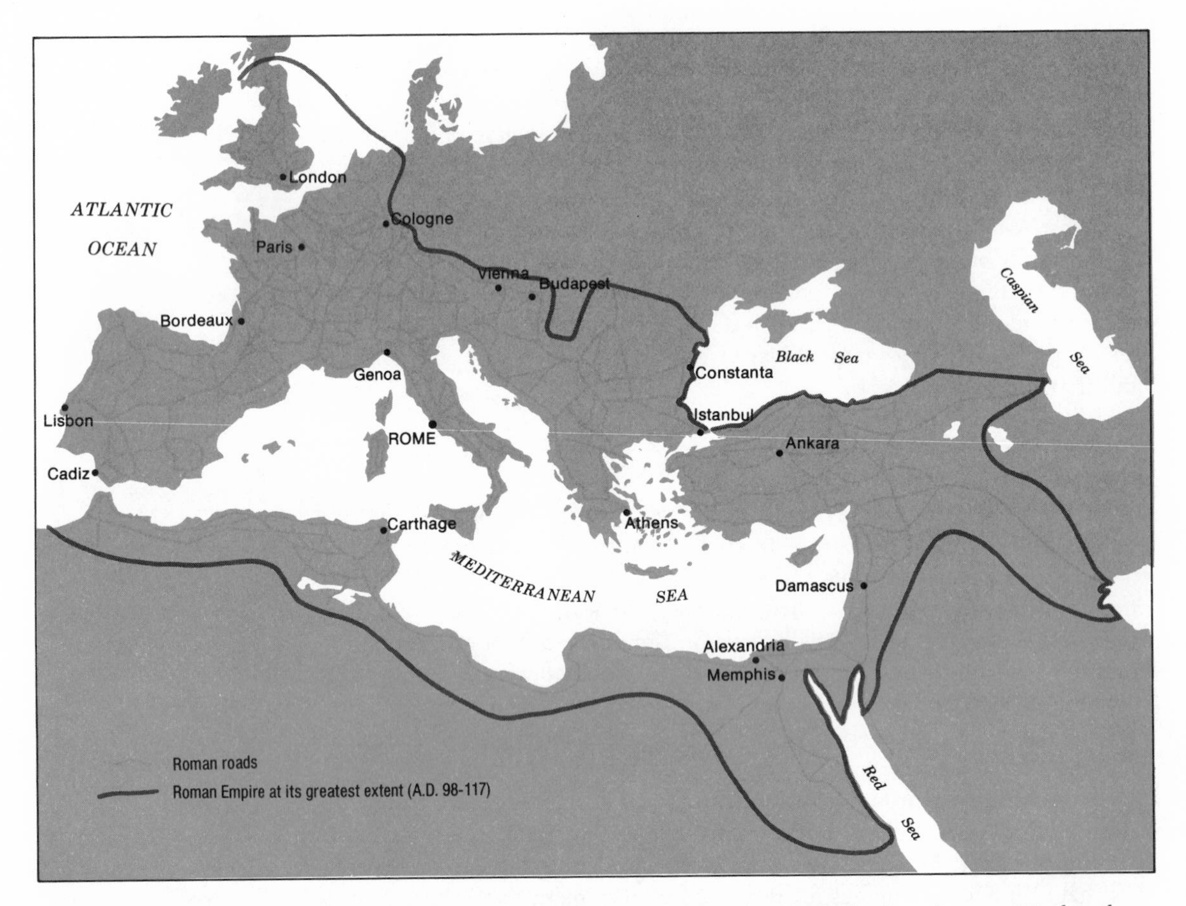

3.2. Major roads of the Roman Empire. This network of roads was critical to the empire's political unity and military defense.

flourished from the sixth to the twelfth centuries. (An alternating pattern of strong empires that encouraged cultural development and trade and of warring ministates had also appeared in China.)

Once again the landscape was modified. Roads became poor and dangerous. Formerly prosperous cities and farms declined with the breakdown of trade and commerce. For the next thousand years, a system of feudal agriculture, similar to that of the Eastern civilizations, prevailed in Europe. Poor and uneducated peasants labored as serfs on the estates of a small aristocracy. The nobility, in turn, protected the serfs from brigands and invaders.

The feudal landscape consisted primarily of many small peasant villages surrounded by small cultivated fields. The lord's castle or manor dominated and protected the villages in its estate. The higher nobility and prelates lived in walled cities, many of which contained no more than three hundred or four hundred people. Because of the decline in trade, local areas were forced to become more self-sufficient. Farmers developed systems of rotation and fallowing that permitted some self-sufficiency in the production of grain, meat, and fibers. The villages of an estate were able to make many handicraft products for themselves. A few of the largest cities, notably Milan, Rome, Paris, and London, maintained a limited commerce and nourished the beginnings of national cultures.

For nearly a thousand years after the fall of the Roman Empire, the population of Europe (and of China and India) rose and fell with famines, plagues, and war. But gradually improvements in agricultural technology, medicine, and science led to a steady increase in population. In the thirteenth and fourteenth centuries, trade, commerce, and cities revived, partly as an aftermath of the Crusades, Christian expeditions to recapture the Holy Land from the Moslems. Among the most prominent trading centers were the Hanseatic cities of northern Europe, such as Hamburg, Lübeck, Copenhagen, and London, and the city-states of Italy, such as Venice and Genoa. Portugal and Spain were in the forefront of exploration and colonization in Africa and Asia, as well as in the New World. The Renaissance, Europe's great intellectual and cultural reawakening, rose to its height in Italy during the fourteenth and fifteenth centuries. The late Renaissance, a period of vigorous exploration, commerce, and religious reformation, peaked in northern Europe—Holland, Germany, England—in the sixteenth century.

The voyages of discovery and exploration were launched partly by the interruption of Europe's trade routes to the East following the rise of Islam. These ventures were aided by the development of the compass (a Chinese invention) and other nautical instruments and by continuing improvements in the design and construction of sailing ships. As new lands and resources were discovered, the landscape changed further. Trade flourished, and European port cities expanded. By the seventeenth century, population growth had begun to press on the existing farm technology and the limits of the land. Nearly all areas in Europe suitable for farming had been cleared of forests and occupied. Europeans aggressively colonized the new territories in Africa, India, Indonesia, the Caribbean, and the Americas and exploited their resources. The flow of exotic foods and minerals—and slaves—helped to reestablish specialization and interdependence.

The impetus for the European settlement of North and South America, Africa, and other colonial territories sprang from religious, commercial, and military zeal and from a desire to escape the poverty resulting from apparent overpopulation. Inevitably conflict erupted between the European and native populations. The Europeans, with their superior application of military technology—horses, armor, steel-bladed swords, and firearms—had a decided advantage, which they readily employed to displace or subdue the natives. Western forms of agriculture and settlement were transplanted and imposed over large areas (see chapter 4). As in Europe, a new landscape was created by **deforestation.** This cutting of trees made possible extensive crop cultivation and the construction of roads and towns. By the end of the eighteenth century, the more developed forms of the agricultural revolution—whether they had spread from Europe, the Middle East, China, or India—nearly encircled the globe.

Contrary to common opinion, the agricultural revolution has not yet run full course. In Latin America and Africa, primitive agricultural technology still prevails in lands capable of more intensive and productive agriculture. In another thirty years, the agricultural revolution will probably be complete, though we may expect to see continuing technological change.

The Impact of Industrial Technology and Urban Culture

The Industrial Revolution: Early Stages

The next major landscape transformation began in Great Britain in the eighteenth century. The force behind this new stage was the industrial revolution, which is only now taking place, two centuries later, in at least half the world. It was and continues to be an energy revolution, a mechanical revolution, a production-consumption revolution, and an urban revolution.

The key to this revolution was the harnessing of *inanimate energy*—first in waterfalls, then in coal, and now in petroleum, natural gas, and nuclear materials. Inanimate energy replaced animal and human muscle power in driving the machines of agriculture, manufacturing, transportation, and construction. New machines and tools raised the productivity of labor. The astounding rise in per capita energy and productivity encouraged specialization and mass production. As economies of scale reduced the cost of each unit produced, many goods could be afforded by even poorly paid factory workers. New patterns of production and consumption thus emerged. The mass of people no longer made all their own clothing, household articles, tools, and so on; they depended instead on factory-made goods.

The exploitation of colonial resources permitted both the rapid industrialization of Western Europe and the United States and the accumulation of wealth in the hands of a prosperous middle class. But even more important was the continual flow of surplus labor from the countryside to the factory and the city. The gradual commercialization and mechanization of agriculture left thousands of farm workers unemployed. For them, fifteen-hour days and low-paying jobs in coal mines and textile mills were preferable to poverty in the countryside. Some of the more fortunate and ambitious rose into the middle classes.

Cities became extremely crowded, despite the development of lower-density housing along streetcar tracks radiating out from the central city. Nineteenth-century tenements, not to mention the mines and mills, were unhealthy and unnatural environments. Low wage levels and the need to be within walking distance of their jobs forced many factory workers and their families into urban hovels. Diseases spread rapidly, and death rates in the city exceeded even those in the countryside. Not until the twentieth century—when industrialization was fairly complete, when unions won recognition, and when the propertyless gained the right to vote—did the material comforts, new educational opportunities, improved sanitation, and other benefits of the industrial revolution reach the lower classes. Gradually the balance of population shifted from the countryside to the city, and economic power passed from the landed gentry to the industrialists and financiers.

At first the industrial revolution did not change the landscape very much. Small mills and mill towns sprang up near waterfalls on streams and rivers, and canal networks were extended to transport raw materials to mills and finished goods—iron, glass, textiles—to markets and ports (figure 3.3). Local forests were cut to provide charcoal for small steel mills and lumber

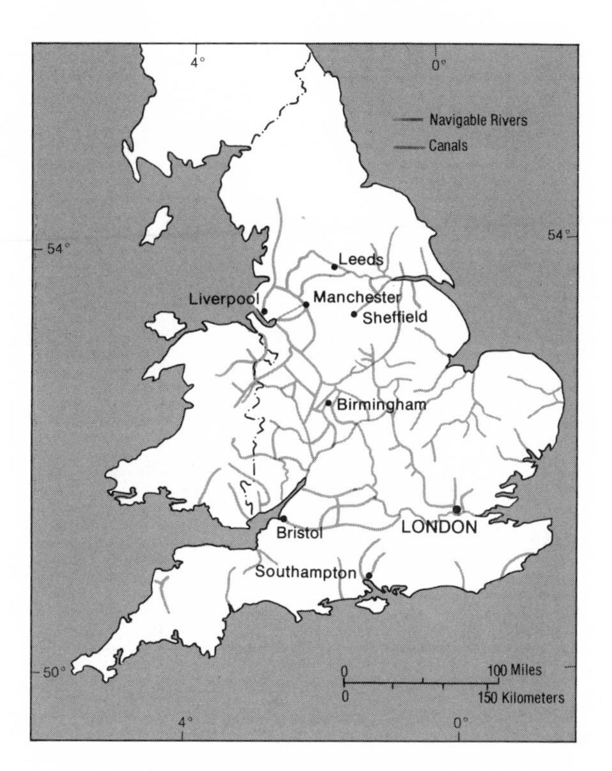

3.3. Navigable rivers and canals in England. Location on waterways was vital to early industrialization and urbanization. These waterways, which brought supplies to mill towns and linked industrial centers to markets and ports, were the most efficient transport routes until the development of the railroad.

for shipyards. Most of these changes were a matter of degree, not of kind. Canal building and deforestation had been going on for years, though not necessarily in the same places.

The Industrial Revolution: Later Stages

Nineteenth-century developments were more impressive. This was the age of the coal-fired steam engine and the railroad, of smoke- and soot-blackened cities, and of railroad and mining towns. New spatial patterns developed as commercial centers expanded to serve the growing population. Small rural centers declined, and industrial towns and cities grew up near coalfields (figure 3.4). Many farmlands turned into cityscapes as worker settlements were strung along railways and clustered around factories, and middle-class residences sprawled comfortably beyond the cities.

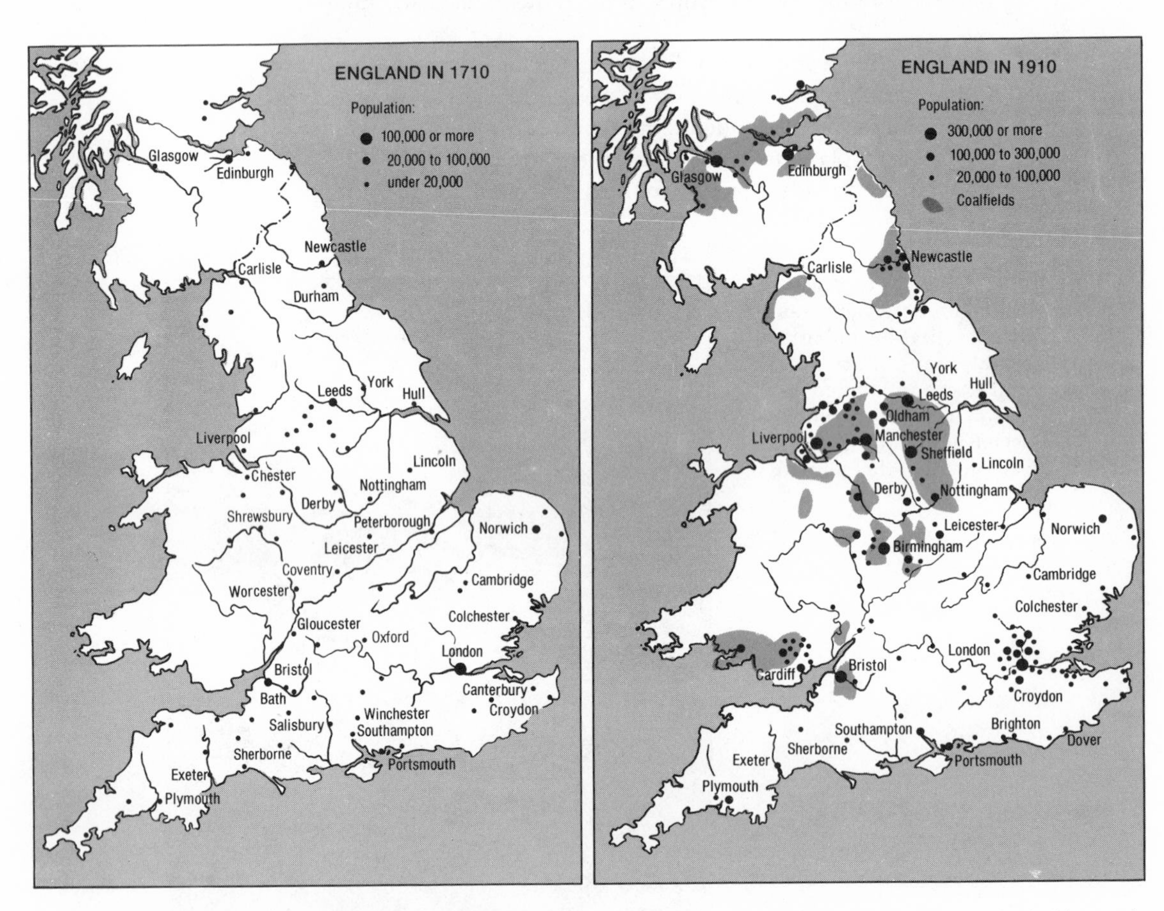

3.4. Rise of industrial cities and towns in England. Industrialization fueled by coal led to population shifts toward the resource.

While coal transformed many rural landscapes into urban ones, the railroad opened new areas to settlement and made possible new patterns of trade. Railroads, station settlements, and track-side factories spread across the landscape of industrializing nations. The rail system penetrated formerly inaccessible areas, particularly in the interior of the United States, Canada, Russia, and central and eastern Europe. It provided a cheap and efficient means of transporting lumber, coal, and other resources and was a lifeline for sparsely settled areas.

The industrial revolution also transformed the agricultural landscape. Grazing lands were fenced, and small farms were consolidated into larger fields and holdings. New machines mechanized agriculture, and urban markets helped commercialize it. Farmers' sons and daughters left for the city or immigrated to the New World. The whole pattern of agricultural production changed too. The railroad permitted farmers to market their produce over long distances. Thus grain production shifted westward in the United States and eastward in Europe, away from population centers. Distant, marginal lands were planted or used for livestock. Farms became larger, more specialized, and responsive to the demands of metropolitan markets.

The Industrial Revolution Today

The twentieth century has brought still other changes. Rather than depending on coal as an energy resource, industrial nations now rely as well on petroleum and natural gas. This shift was caused by the development of new kinds of engines—internal combustion, diesel, jet—and the increasing use of electrical energy. Particularly in the United States, migration, urban development, and industrial growth have spread to Texas, California, and other areas rich in petroleum and natural gas. These new fuels, however, are easier to transport than coal, and their production is *capital intensive* rather than *labor intensive;* the machines replace much of the human labor. Thus the redistribution of population is no longer as much to the location of available energy as it had been during the coal era, but to the location of energy use.

The modification of the twentieth-century landscape has stemmed chiefly from the development of the automobile, truck, tractor, and airplane. Automobiles and trucks have created a denser, more visible pattern of roads and highways on the landscape. The tractor has speeded the mechanization of agriculture and the depopulation of farming areas. And the airplane has drawn distant places closer together.

The contemporary U.S. landscape reveals another legacy of the industrial revolution: an emphasis on consumption as a driving force for greater development. Across the country there is evidence of ever greater consumption and replacement of goods and services: new homes, new cars (and junkyards), new boats, new golf courses and shopping centers and abandoned old ones, and demolition and new construction of streets, apartment buildings, and office towers.

Perhaps the most striking phenomenon of recent times has been the concentration of people and activities in large cities and the spread of these cities into the surrounding countryside. In expanding to accommodate low-density housing and businesses, many cities have obliterated productive farmland at the *rural-urban fringe* (figure 3.5). Although they cover only a small portion of the earth's total land surface, cities have become large enough to show clearly in satellite photographs of the earth—primarily because of the concentrated output of pollutants in and around the cities. They may also be capable of disrupting local weather and drainage patterns and of damaging local vegetation.

Not only cities but farms, factories, vehicles (such as jumbo jets and supertankers), corporations, and governments have tended to become larger and larger. This rise in scale has come about to meet the economic goal of

3.5. Suburban sprawl. Residential development is consuming more and more farmland at the rural-urban fringe, as in this area west of Lancaster, Pennsylvania. (Photo by Grant Heilman.)

maximizing output. As we will see in chapter 15, the trend toward further growth is now being questioned.

The long process of human development, at least in advanced nations, has led to more intensive exploitation of land, water, minerals, and other resources. It has also resulted in the concentration of people in ever larger cities interconnected by an increasingly sophisticated network of transport and communication. The industrial revolution is far from complete in most of the world. But because of both growing concern about resource and energy shortages and some disenchantment with prevailing models of development (Western capitalist or Soviet socialist), patterns of further industrialization and urbanization may change. (Various scenarios of the future will be presented in chapter 15.)

DIFFUSION

New forms of development leading to changes in the landscape may be generated within a society, or they may be borrowed from other societies through diffusion. The industrial revolution, for example, originated in northwestern Europe and diffused to the Americas, Australia, New Zealand, and parts of Africa and Asia. The theory of diffusion will be examined in chapter 11. Here our concern is with its practical role in the evolution of the landscape.

The principal mechanisms by which diffusion is accomplished include **migration** (itself a diffusion process); exploration and colonization; military conquest; and travel, trade, and communication. Exploration, colonization, and military conquest may be considered forms of migration and will be treated as such in this chapter.

The successful diffusion of innovations such as new agricultural methods depends on several factors: the quality and social or economic value of the innovation; the nature and strength of competitive ways of doing things; the nature and strength of other forms of resistance, including the cost of overcoming distance; the quality of information about the innovation; the determination and numbers of those carrying the innovation; and the quality of communication and transportation. The diffusion may promote development in the sense of greater social complexity and diversity, as well as material and cultural enrichment—for example, the spread of the Roman Empire and of agricultural technology. But diffusion may also destroy an existing culture (American Indian culture in the United States), enslave a society (the Spanish conquest of the Aztecs), or supplant simpler ways of life with more specialized ones (the decline of general farming and the rise of commercial agriculture).

The Impact of Migration

Migration, the movement of population to a new area, is an important vehicle of diffusion. Though perhaps less evident in settled agricultural civilizations, migration is basic to human experience; people believe that their lives may improve in a different location.

Primitive hunters and shifting cultivators migrated because they had to. They ultimately covered most of the world searching for game and more productive land. Tribes would often exchange primitive technologies, which sometimes led to changes in the landscape and in forms of spatial organization. Shifting cultivators might show hunters how to grow crops or build better shelters. By adopting new techniques and customs and by coming into contact with other societies with different skills, tribes developed and reorganized their use of space.

Once human society had mastered the technology of a more productive agriculture, migration assumed both centripetal and centrifugal forms (figure 3.6). In *centripetal* migration (movement from a low-density periphery to a high-density core), less developed peoples on the fringes of civilization

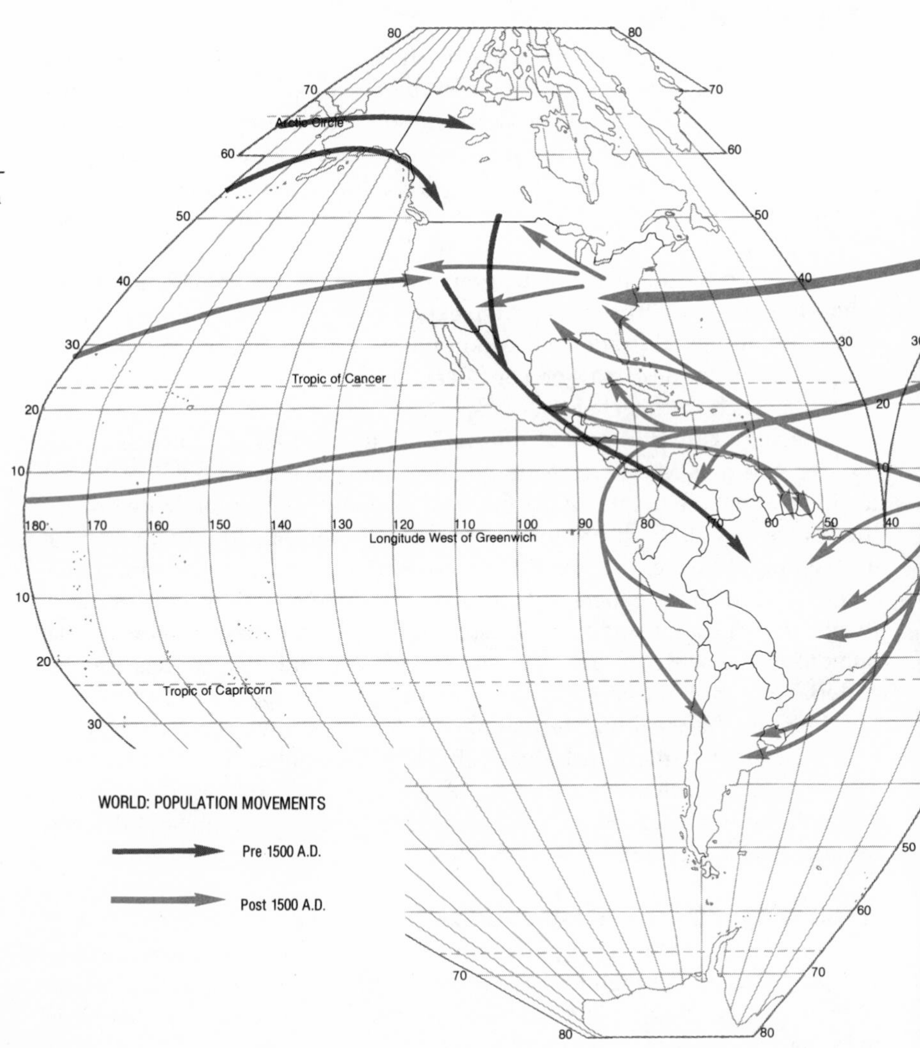

3.6. World population movements. Movements stemming from colonization and the slave trade dominate the post–1500 flows; in modern times political movements, as from India and Germany, have been important.

were attracted to more developed centers of intensive agriculture. Many raiding tribes became assimilated into the more advanced culture. But they often infused it with their own ideas and innovations too. The horse, for example, was introduced into civilized society by the nomadic tribesmen of central Asia.

These early centripetal migrations must have been relatively very large, particularly those originating from the heartland of central Asia and diffusing into China, India, and Europe over long periods. The extensive migrations of the Vikings (Scandinavian peoples) from the fifth through the tenth centuries to what are now parts of England, France, Italy, and the Soviet Union illustrate how a less civilized but vigorous group interacted with lo-

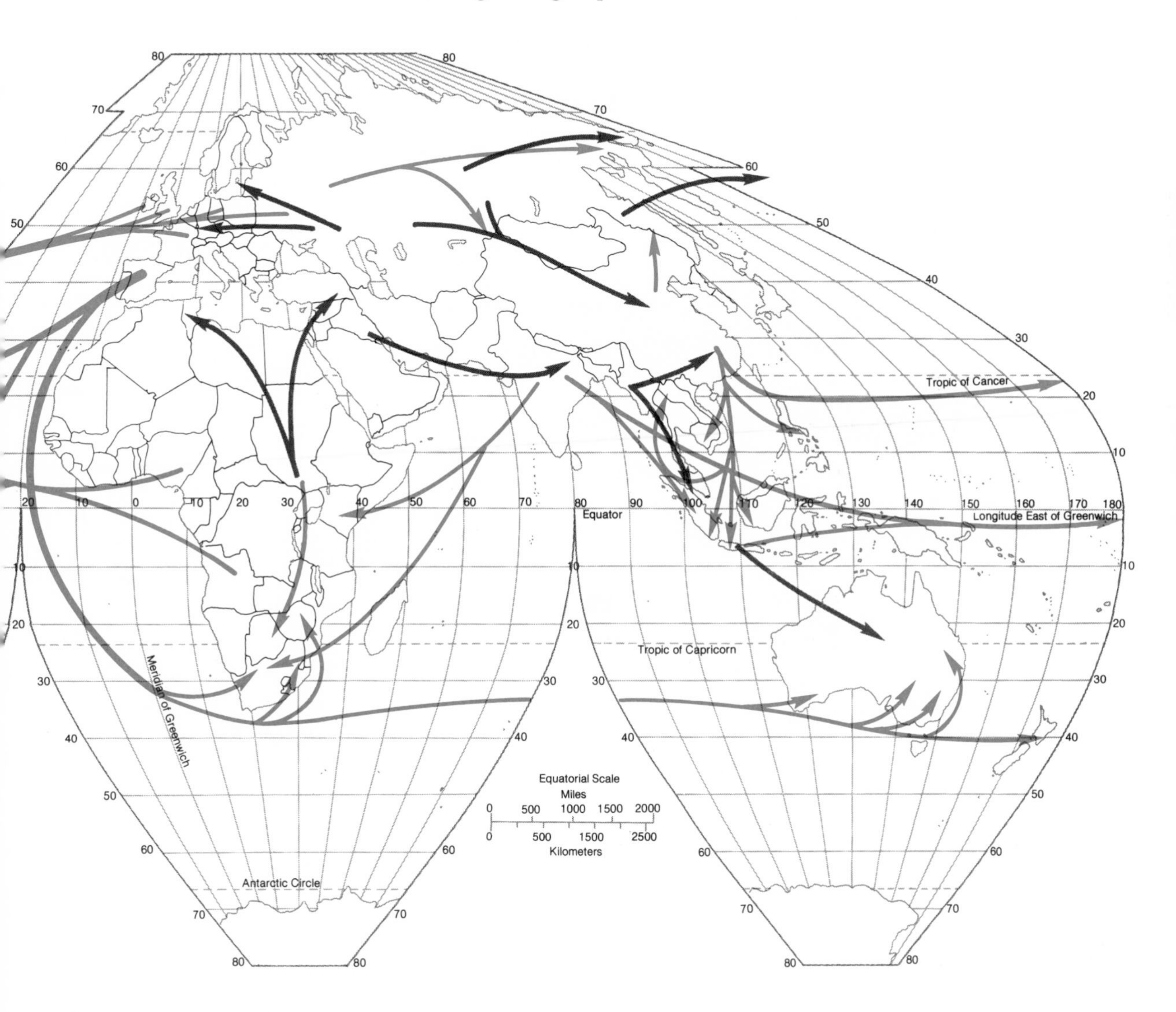

cal peoples and institutions to create new and vital societies, such as the French Normans and the early Russian state.

At other times, dominant societies in a *centrifugal* movement expanded their territory by settlement or colonization and, if necessary, by conquest over vast areas. Examples include the early Egyptian, Greek, Macedonian,

Indian, Chinese, and Roman empires. One of the greatest and longest migrations in world history was that of the Bantu-speaking peoples of Africa. During the first century after Christ, they left what is now eastern Nigeria and slowly migrated south and southeast. By the sixteenth century they had introduced agriculture and metal working to hunting-gathering societies throughout central and southern Africa and began to conflict with Dutch and British settlers as their southward movement continued.

During the seventeenth century, the Russians pushed eastward through Siberia to the Pacific coast (figure 3.7), and Europeans by the tens of millions migrated to the New World. Again the conquering or colonizing society not only imposed its own culture on the native peoples but also

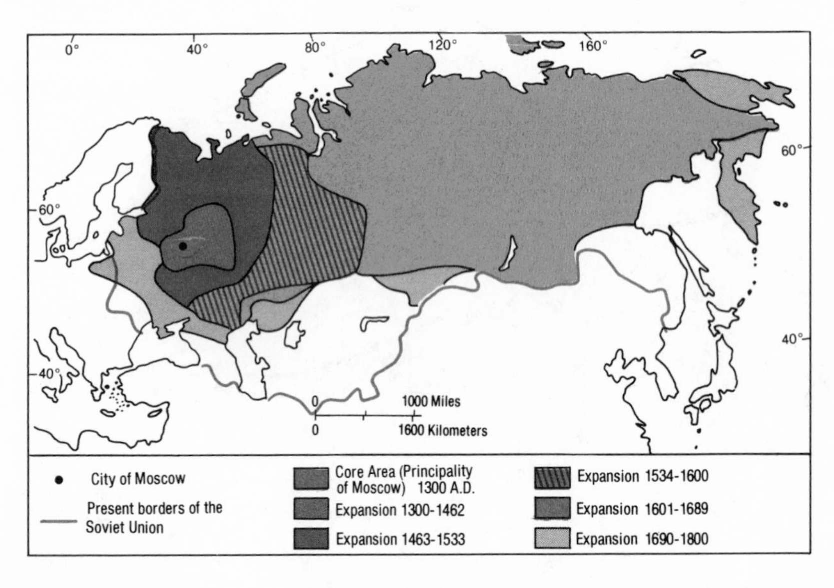

3.7. Expansion of the Russian state. Desire to secure access to the Pacific Ocean was an important factor in the great westward push through Siberia between 1601 and 1689.

Source: After Fig. 4–3 "Map" (p. 151) in *The Human Mosaic: A Thematic Introduction to Cultural Geography* by Terry G. Jordan and Lester Rowntree. Copyright © 1976 by Terry G. Jordan and Lester Rowntree. By permission of Harper & Row, Publishers, Inc.

learned from them. Thus innovations such as tobacco, potatoes, and chocolate diffused from the New World to Europe.

This latter-day colonization, like that of the Greeks in the eastern Mediterranean two thousand years earlier, was motivated by many incentives. One was the need to relieve population pressure on older settlements by opening new land resources. Others included the desire to convert the heathen, civilize the "savages," and exploit forest and mineral resources. Once the industrial revolution was well underway, Europeans and North Americans felt that their prosperity depended on the partial colonization of the rest of the world. The colonies were a welcomed and indeed essential source of food and raw materials. They also provided controlled markets for manufacturers from the mother country.

The settlement of the United States shows how several waves of migration can change the landscape (figure 3.8). In the sixteenth century, a few European explorers, hunters, trappers, and miners came to North America.

Their impact was limited to the exploitation of resources and the building of trading posts. Some of these posts later developed into small towns. In the seventeenth century, settlers from Europe began to occupy the Atlantic coast. They created a fairly dense pattern of farms, hamlets, and roads in the East and a more expansive pattern of plantations in the South.

The interior of the country was first occupied by semisettled livestockmen. During the nineteenth century, the railroad enabled commercial agriculture, accompanied by new villages and towns, to extend west of the Ohio River (see figure 10.3). The settlers fenced most of the open range and divided it into farms and ranches. Other newcomers exploited mineral resources that were far from the older population centers. And industry (initially based on technologies from northwestern Europe) and urbanization diffused from the hearth areas of the Northeast into the Great Plains, the South, and the West.

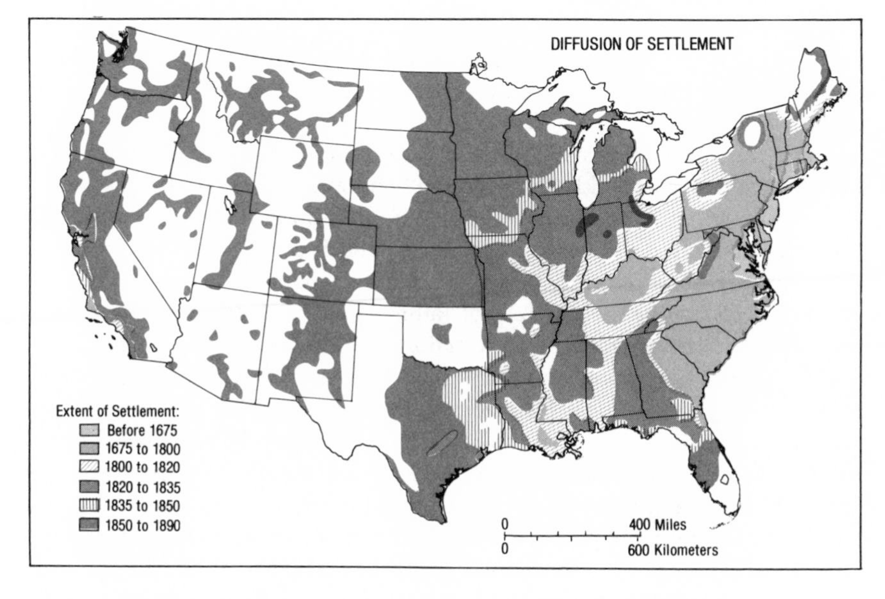

3.8. Diffusion of settlement in the United States. The settlement of the United States was accompanied by the development of villages, towns, and cities. While the diffusion process was rapid, it nonetheless extended over two hundred years.

In the twentieth century, a more recent wave of migration has flowed over the continent to the South and Southwest from the North and East. The large numbers of migrants have stimulated the development of these areas. At a local scale within regions, the diffusion of development into the suburbs has also transferred millions of people and further extended the urban landscape.

Behind each wave of migration was the desire for a better life. The migrants believed that they would find new opportunities in their new location. Families were and still are willing to take risks, to sever old ties and abandon familiar surroundings to attain what they perceive to be a more satisfying way of life.

The Impact of Trade, Travel, and Communication

The diffusion of new forms of development through migration is one way in which the landscape evolves as a result of forces generated outside a society. Perhaps equally effective today is the diffusion of new ideas and technology through trade, travel, and communication. Particularly since the adoption of printing in the sixteenth century and the explosion of technical inventions in the eighteenth century, new forms of development have spread rapidly without migration or colonization. Such innovations include artistic modes, political and economic ideas, and medical, industrial, agricultural, and transportation technologies, as well as the capital investment needed for their adoption.

Some of the fastest and most successful diffusions have been in the fields of medicine and public health, which have contributed to the population boom of the past two centuries. Vaccines against smallpox, measles, and polio were fairly enthusiastically accepted. But innovations in agriculture and transportation, such as mechanization and all-weather roads, have not always been so quickly or successfully adopted. This has been demonstrated by periodic crises in the supply and distribution of food in less developed areas, such as sub-Saharan Africa and India. Sometimes these innovations are accepted only slowly because of inadequate information or education, lack of capital, or fear of failure, but many other subtle factors also influence the acceptance or rejection of innovations.

Through trade, travel, and communication, every variety of innovation, from soft drinks, chemical fertilizer, and transistor radios to clothing fads and hard drugs, is introduced into a society. Trade directly transmits innovations through the exchange of technology and goods. New products are introduced into a society. People's tastes and preferences change. And patterns of production and consumption are modified. Trade raises demand for raw materials from less developed countries. It gradually fosters a cash economy (rather than the barter of goods) in the exporting nation and increases the economic, social, and physical mobility of the population.

These conditions stimulate the development of cities and industry in less advanced countries. Urban and industrial development set off other spatial processes: the migration of surplus rural population to cities; the growth of trade, travel, and communication among ever more specialized centers of production and consumption; and the further diffusion of new ideas and technologies.

Some landscapes have evolved largely through the selective acceptance and adaptation of innovations borrowed from other societies. Japan is an outstanding example. When the industrial revolution was sweeping through Europe, Japan was still a feudal society. In a deliberate attempt to preserve traditional ways of life, Japanese rulers of the Tokugawa dynasty (1603–1867) severely restricted trade and travel. Development virtually ceased for more than two and a half centuries. But in 1867, when Emperor

Mutsuhito came to power, a development strategy was pursued just as deliberately and successfully. Through a highly disciplined, hard-driving effort to modernize, beginning in 1867, Japan leaped from feudalism to industrialism. Japanese businessmen studied Western technology and adapted it. Factories sprouted up along the coast. As in other industrializing nations, hordes of people left their farms for the city. Today Japan is one of the most urbanized and industrialized nations of the world. Tokyo, its capital, is the world's largest city. Its metropolitan area, including suburbs, contains 23 million people—over 5 million more than New York City.

Many people in both advanced and less advanced nations feel that the less developed ones should be spared the pollution, blight, and other problems resulting from industrialization and urbanization. In their view, undeveloped environments should be preserved and simpler ways of life left intact. But a realistic look at conditions in such societies reveals not the prevalence of idyllic contentment but hunger, disease, and illiteracy. Given the need and demand for higher standards of living and the tremendous power inherent in the spread of new ideas, nothing short of a natural or man-made holocaust is likely to stem the eventual worldwide diffusion of industrialization and urbanization.

INTERNAL DEVELOPMENT VERSUS DIFFUSION

It is difficult to determine whether a landscape-changing innovation, such as irrigated agriculture and the rise of cities, stems from independent development or from the diffusion of ideas from outside sources. Authorities have argued, for example, that irrigated agriculture began independently in both China and Southwest Asia. Evidence for the independent origin of many early innovations is inconclusive, but the development of any one region may be the result of changes generated both internally and externally.

Change within a region first comes about through the natural stages of the human life cycle: the growth, aging, death, and renewal of a population. Adulthood, family formation, entering and leaving the labor force, physical decline—all require decisions about the location and nature of work and residence. Of course, net population growth creates a need to expand production. But the driving force behind development is the human desire to achieve greater security, a higher level of living, and personal fulfillment and recognition. This force, though more strongly institutionalized in Western capitalist societies, is present in all societies.

Change within a region occurs through both internal and external forces. A combination of internal development and the diffusion of ideas from outside sources is often evident in the intensification and modernization of agriculture; growth in the variety and volume of mineral exploitation and industrial production; urbanization; the growth of services; the development of education and the arts; and changes in social, economic, and political relationships and in government structure. Modern China, for example, is adopting many agricultural and industrial processes from abroad (diffusion), but it has depended mostly on its own resources and innovations (internal development) and is looked upon by many third world nations as a source of alternative or less capital- and energy-intensive technologies.

A fairly advanced level of organization is necessary to bring about the forms of development mentioned above. Sufficient time and talent must be available to a society for the training and education of its youth. More important, and in fact a key notion for our understanding of the landscape's evolution, the society must produce enough surplus to make investment possible, for only investment in plant, equipment, education, and ideas can actually bring about any significant change in the organization of society.

THE ENVIRONMENTAL IMPACT OF DEVELOPMENT

The preceding discussion has shown how society has changed the landscape through the development of agriculture, human settlement, transportation, industry, and urbanization. Development often requires overcoming physical constraints, such as inadequate water or rugged terrain. But our survival ultimately depends on the capacity of the physical environment to support life. How has human development affected this physical base?

First, let us consider the various impacts of agriculture. Livestock—particularly goats—have at times stripped protective vegetation from the soil, increasing the rate and severity of erosion. The risk of erosion and flooding has also been greater where forests have been cleared for farmland. As agriculture has expanded, the range of wildlife and vegetation has diminished. Runoff from fertilizer, insecticides, and herbicides has polluted streams and poisoned some plants and animals. And the overextension of crops into marginally dry lands, common in densely populated regions, has often resulted in severe erosion. Yet as a whole, agricultural systems are more productive than natural forests or grasslands. Of course, this rise in productivity is often gained through high levels of fertilization and pest and disease control. The landscape itself is not necessarily damaged by agricultural development, but some plants and animals can be harmed by fertilizer and pesticides.

Industrialization and urbanization have modified the environment more obviously, if less extensively, through dredging, leveling, draining, and filling; through the channeling and banking of streams, the bulkheading of shores, and the construction of reservoirs; and through the paving of large areas with materials that do not absorb water (figure 3.9). Some of these measures increase the probability of erosion; others reduce it. On balance,

3.9. Landscape modification. To construct a water pumping station in Geesthact, Germany, a pathway is being carved to the Elbe River and an artificial lake created at the summit of the plateau. (Photo from International Labour Office.)

society's immense investment in modifying the physical environment is designed to control and stabilize natural forces. In the process, however, environmental systems may be fundamentally altered. Flood control systems and power dams, for example, reduce erosion by harnessing the flow of rivers. But they may also unintentionally destroy fisheries or diminish bottomland habitats.

The probability of environmental degradation rises not so much from the extension of the human landscape as from the concentration of people and activities in fewer, larger places, especially on or near bodies of water (see figure 1.3). The concentration of urban and industrial development along rivers and shorelines severely strains those limited but productive environments. Industrial and human wastes, unless treated and purified, create pockets of pollution that endanger plant and animal life. On a larger scale, the burning of fossil and nuclear fuels may be capable of diminishing the amount of solar energy reaching the earth. Air pollution has already been shown to influence the local climatic conditions—known as the **microclimate**—of cities and may ultimately affect the productivity of agriculture, for example, by increasing soil acidity.

It would be an exaggeration to claim that human activities are causing irreparable harm to the physical environment's capacity to sustain life and recycle wastes. After all, human efforts are negligible compared to the cataclysmic power of nature. The energy released by a hurricane in a single hour through condensation is equivalent to the force of sixteen twenty-megaton hydrogen bombs. A small portion of this force, harnessed in hurricane winds, can devastate entire coastal communities. Yet in some places, particularly in urban areas, the impact of human development on the environment has been damaging enough to arouse the concern of citizens and government officials. Increasingly laws are being passed to protect natural systems—and human life.

THE EVOLUTION OF THE CHICAGO LANDSCAPE

To summarize our discussion, it would be useful to review the evolution of the Chicago landscape—first that of the region, then of the city. For convenience, let us say that the region is the area encompassing all of northern Illinois. The prime agent in the early evolution of this region was, of course, nature.

Far back in the geologic past, the Chicago region was submerged under a tropical sea covering the midcontinent. Over millions of years, the shells and skeletons of creatures that swarmed in its waters settled to the bottom. They eventually formed the limestone now underlying most of Illinois. As eons passed, the sea, bordered by swamps and marshes, receded. The dense jungle vegetation around its shores slowly decayed, compressed under pressure, and developed into the largest beds of bituminous coal existing in the United States today.

The steamy tropical climate cooled and was followed by at least four ice ages. These gave the Chicago region much of its present character. A succession of great ice sheets, some more than a mile high, scraped and leveled the terrain. They flattened most of northern Illinois and deposited material that evolved into fertile black soil. Natural erosion from wind and rain added finishing touches to the terrain. Filling a large depression made by former glaciers, the meltwaters of the last glacier became the ancestor of Lake Michigan. Today the city of Chicago rests on the bed of that larger postglacial lake.

Toward the end of the last ice age, human beings began to occupy northern Illinois. The earliest known tribes were the Mound Builders, who created a human landscape characterized by thousands of burial and temple mounds, many of which still exist. For several hundred years, the impact of human development on the landscape was relatively minor. Periodic burning of vegetation may have helped to maintain grasslands where forests might otherwise have prevailed; but tribes generally lived within the constraints imposed by their surroundings.

The arrival of European explorers and fur traders in the late 1600s heralded a new stage in the evolution of the Chicago landscape. Initially the Europeans' impact was limited to the building of trading posts and the exploitation of wildlife, especially beaver. But the landscape changed rapidly at the end of the eighteenth century when Illinois opened to farm settlement.

The early settlers built homes along the Illinois and Chicago rivers. With the construction of the Illinois and Michigan Canal, linking the Great Lakes to the Mississippi River, and the extension of the railroad in the 1850s, settlement soon spread into the fertile interior. The tall-grass prairies were plowed and planted to corn, a vegetable that European settlers had adopted from the Indians. Illinois, with its comparative advantage for corn production, became a major granary for the Northeast. After the Civil War and the settlement of the plains and mountain states to the west, Illinois agriculture evolved into the more intensive corn-hog combination that predominates there today (see chapter 5).

When the industrial revolution spread to North America in the late 1800s, Chicago and other cities rapidly industrialized, serving the growing farm population of their local regions. The thick beds of coal formed millions of years earlier fueled the economic expansion of the Chicago region. Settlement diffused across northern Illinois as easterners and European immigrants learned of the region's many advantages. Railroads and factories were built in and around the smaller cities and towns. By 1900, the Chicago region had evolved from a sparsely populated prairie wilderness to a highly developed urban and industrial center, surrounded by a rich agricultural hinterland.

Further changes took place as the twentieth century rolled on. Farms grew larger and more mechanized. Population shifted in large numbers from the farm to the city or to California and other western states. As the automobile and truck supplanted the train, a dense network of highways began to compete with the railroad.

Central to the development of the Chicago region during the past 150 years has been the evolution of the city of Chicago from a frontier outpost—Fort Dearborn—to a modern industrial metropolis. Founded as a fortified trading post in 1803, Chicago grew slowly at first. The threat of Indian attack discouraged potential settlers. But in the 1830s, after the Indians were driven west of the Mississippi River, Chicago emerged as the controlling center of commerce and transportation for the surrounding farmland. By the end of the Civil War, Chicago was the major transport and industrial center for a much greater region. As the hub of a vast network of railroads and water routes, it served the entire nation.

The city grew despite the typical American frontier pattern of boom and bust. It survived a disastrous fire in 1871, rising from the ashes "bigger and better" than before. Factories and residences were built continuously out along access railways, forming a classic star-shaped pattern. The expanding metropolis buried local dairy and truck farms under acres of concrete and asphalt. After World War II, automobile-oriented suburbanization and exurbanization extended the commuter zone to fifty miles or more.

The Chicago landscape is clearly a *modified landscape*; the local environment has been changed to serve human needs and interests. At the same time, the balance of nature has been upset by the concentration of nearly seven million people and countless activities in a single metropolitan area. Air, water, and land pollution is a serious problem that can be solved only through careful management and expensive pollution-abatement measures. Such is the price society must pay to enjoy the benefits of city life—of agglomeration—in a setting favored by nature and made even more favorable by human investment.

CONCLUSION

In many respects, the evolution of Chicago and its surrounding region mirrors the changes that have occurred in other parts of the world: the spread of specialized, commercial agriculture; the development of trade and transportation; the growth of towns, cities, and industries; and the increasing concentration of people in large urban centers. These processes and the spatial patterns they produce will be examined in greater detail in the next seven chapters. Chapter 4 looks at the changes that are now taking place in low technology societies.

SUGGESTED READINGS

Brown, R. *Historical Geography of the United States.* New York: Harcourt, Brace, 1948.

Glacken, Clarence J. *Traces on the Rhodian Shore.* Berkeley, Calif.: University of California Press, 1967.

Isaac, E. *The Geography of Domestication.* Englewood Cliffs, N.J.: Prentice-Hall, 1970.

Lampard, Eric. *The Industrial Revolution.* Washington, D.C.: American Historical Association, 1957.

McManis, D. R. *Historical Geography of the United States: A Bibliography.* Ypsilanti, Mich.: Eastern Michigan University Press, 1965.

Mumford, L. *The City in History.* New York: Harcourt Brace Jovanovich, 1961.

Pred, Allan. *The Spatial Dynamics of U.S. Urban-Industrial Growth, 1800–1914.* Cambridge, Mass.: MIT Press, 1966.

Sauer, Carl O. *Agricultural Origins and Dispersals.* New York: American Geographical Society, 1952.

Sjoberg, G. *The Pre-Industrial City.* Glencoe, Ill.: Free Press, 1960.

Smith, W. *An Historical Introduction to the Economic Geography of Great Britain.* London: Bell, 1968.

Thomas, William, ed. *Man's Role in Changing the Face of the Earth.* Chicago: University of Chicago Press, 1956.

Ward, David. *Cities and Immigrants.* New York: Oxford University Press, 1971.

Webb, W. P. *The Great Frontier.* Austin, Tex.: University of Texas Press, 1964.

4

Low Technology Rural Landscapes

Characteristics

Land, Markets, and Exchange

Ways of Life

Development and Change

4 Low Technology Rural Landscapes

Location theory was developed primarily to explain spatial patterns formed in high technology, interdependent, commercial and industrial societies. These patterns have resulted in part from society's desire to maximize the net utility of places at least cost: generally individuals and groups in high technology societies act to maximize productivity and efficiency through the best possible spatial arrangement of activities. Economic considerations are paramount. In contrast, spatial patterns in low technology, agrarian societies seem to be determined more by cultural preferences and environmental conditions than by economic considerations. Yet the principles underlying all human spatial organization may well be universal.

Within the constraints of their cultures, virtually all societies attempt to use space as efficiently and productively as possible—first, as a matter of survival, and second, to improve their quality of life. Pursuing these goals may result in landscapes that are very different from those of technically advanced societies, especially since local areas in low technology societies tend to be self-sufficient rather than interdependent. This will become evident as we discuss the spatial patterns developed by hunting-gathering societies, nomadic herdsmen, shifting cultivators, and peasant farmers. We will also consider the impact of modern technology on these rural landscapes, particularly with respect to the commercialization of agriculture.

In less developed societies, efficient spatial organization of the kind that results from specialization, dependable trade flows, high levels of production, and the like is hindered by many conditions, including small-scale operations, lack of modern technology, limited capital or credit, poor transportation, illiteracy, and traditions that impede efficiency (religious restrictions, limited roles for women, and complex patterns of land ownership are a few). Yet the desire to maximize productivity and efficiency motivates the landowner or social group in low technology landscapes as well as in high technology ones. We study low technology landscapes not only because they are interesting in themselves but because most of humanity lives in such settings.

Spatial behavior in low technology societies is reminiscent of now-vanishing ways of life that prevailed not so long ago in North America. It brings to mind the farming and hunting-gathering activities of native Americans before the arrival of the Europeans. Some Indian tribes dwelling in fertile river valleys and woodlands planted maize, beans, and squash (which had first been domesticated in what is now Mexico), using simple farming techniques. But in the short-grass plains west of the Missouri River, conditions were not suitable for farming. The drier, more compacted soil and semiarid climate did not lend themselves to primitive farming. Indians in that region survived by hunting buffalo during the long, harsh winter when other wild foods were scarce.

Occasionally Indian farmers from the woodlands and Indian hunters from the plains met at trading centers—villages located on high bluffs overlooking the Missouri River. Here they exchanged their small surpluses of grain and vegetables, meat and hides. The Missouri River and its tributaries provided an important means of access at a time when footpaths and navigable rivers were the only transport routes. (The horse, which revolutionized the plains Indians' way of life, was not introduced to North America until the sixteenth century.)

The Indians used space to increase their chances for survival: they adapted to the physical environment, farming in the river valleys and hunting buffalo on the plains. They produced enough surplus to engage in some trade, using navigable rivers and footpaths for this purpose. Many Indian trails were later developed into roads by European settlers. The city of Chicago is located at the convergence of several former trails, evidence that the centrality of the site was long recognized.

CHARACTERISTICS

Early native Americans formed societies that were largely self-sufficient. They lived off the land when they could by hunting and gathering wild foods. Some produced a limited amount of agricultural surplus for exchange. Until the Indians acquired horses, people and goods usually did not travel far.

Low technology societies in Asia, Africa, and Latin America are rather similar today. Much of the population engages in **near-subsistence farming,** producing only a small surplus, if any, for exchange. The poor roads and slow transportation discourage long-distance travel (hence the historic importance of Asian and African caravan routes, which provided overland passages for trade). The local and neighboring villages constitute the **social space,** the area in which most activities take place.

Spatial patterns differ in scale from those of advanced societies. Farming, commerce, and manufacturing are organized on a small scale. The landscape seems miniaturized: fields are small, roads are narrow, and most settlements are modest in size (figure 4.1). Patterns of land ownership vary. In some areas, notably in Africa, farmland is owned communally but is periodically allocated to individuals. More commonly, most of the land is held by a few landowners, and farmers must rent most or all of their fields. The community usually achieves a moderate degree of self-sufficiency in food production and handicrafts or reaches it through exchange with neighbors. The production and trade of specialty goods, requiring wider markets, is limited in scope, volume, and variety.

Patterns of spatial organization in low technology societies tend to be highly repetitive over time; the same ways of using space persist from one generation to the next. (In contrast, patterns in high technology societies

4.1. Village of Paniquita, Colombia. The rural landscape is generally organized on a small scale in low technology societies, with small settlements, small, fragmented fields, and narrow, unpaved roads (the road in this village is wider than the typical village road). (Photo from United Nations.)

tend to be highly repetitive over space, as from city to city.) But as agriculture becomes increasingly commercialized, governments encourage farmers to produce crops for export. Such production competes with, and may gradually replace, that of outside-owned estates. While the long-term benefits of this shift may be significant, the short-term effects may include widespread displacement of peasant farmers and, in some areas, food shortages, as occurred in Europe in the past.

Generally, however, land is used in much the same way from generation to generation and from village to village. Most farmers plant fields as their ancestors did, raising the same kinds of crops and using many of the same techniques. Life goes on much as it has for centuries. Local traditions and the need for certainty in a marginal economy, with a low net productivity per person, where unproven innovations may prove disastrous to the year's food supply, favor the repetition of safe, familiar methods. Little emphasis is placed on innovation, which implies risk. Illiteracy, particularly among women, high rates of infant mortality, and the prevalence of malnutrition and disease also inhibit the acceptance of new ideas (such as birth control) and the achievement of higher levels of well-being. A family's emotional ties to its lands are deeply rooted and act as a magnet. Migration tends to be circular, ending back in the home village once the desired objective—a bride from another village, a temporary job, or a visit with friends and relatives—is attained.

Dialects, customs, and character peculiar to the region often develop in areas isolated by poor roads and inefficient means of communication. Note, for example, the many language areas contained in a small section of the New Guinea highlands (figure 4.2). Language and other cultural barriers discourage interaction and delay the transmission of new methods and new ideas, as they did in Europe in centuries past.

LAND, MARKETS, AND EXCHANGE

Societies at different stages of economic development vary in level of *commercialization,* the degree to which goods and services are produced for exchange. The typical farm in a high technology Western society is a business; before planting, farmers carefully consider the costs of producing the crop and the price they can expect to receive for it. They are usually willing to innovate and take risks that might increase their income, such as shifting to a more productive crop. Agriculture is organized on a large scale, and the average farmer produces enough food to feed many families.

In contrast, the average farmer in a low technology society raises barely enough food to feed his own household and is fortunate if any surplus remains to be sold. Agricultural activities are largely governed by the needs of the family, tribe, or clan, not by conditions of national or world supply and

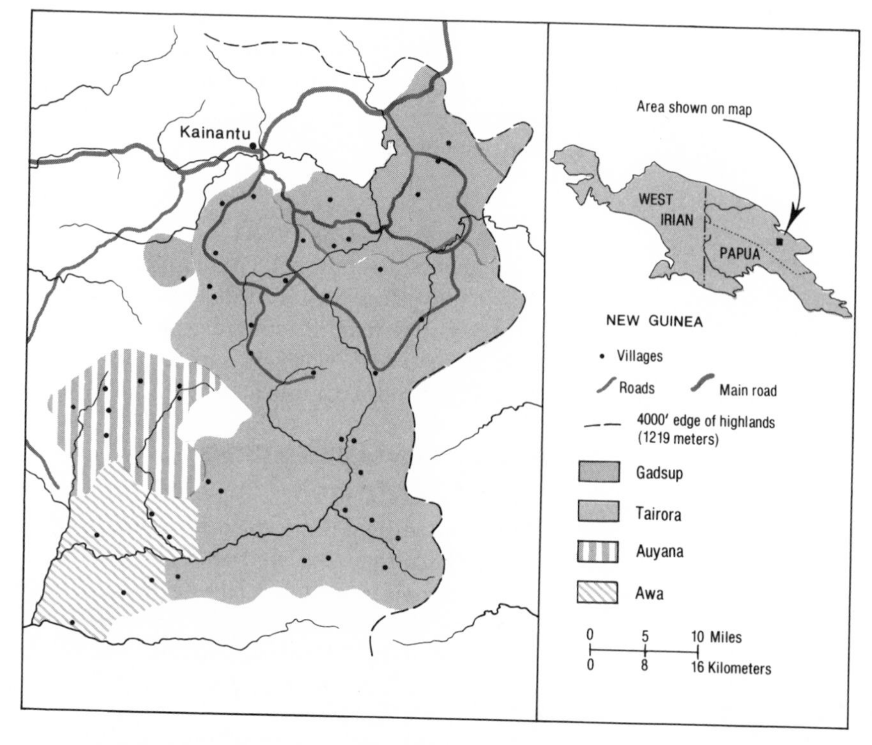

4.2. Language regions in the eastern New Guinea highlands. Poor transportation and physical isolation have contributed to the development of diverse languages—Gadsup, Tairora, Auyana, Awa—in this small area.
Source: Data from a map by K. Pataki.

demand. Increased productivity results more from the farmer's desire to improve the well-being of the household than from outside demand. The change from relatively primitive to modern agriculture entails a radical shift in attitude as well as in method. Farmers begin to think not merely of satisfying the needs of their family but of meeting larger scale external demand. And when this occurs, the local food supply may suffer.

As technology improves and external demand rises, almost pure subsistence farming gradually gives way to a partly subsistence, partly commercial *peasant agriculture.* Exchange takes place in local markets or in the form of compulsory rent or tax, payable to landlords as a share of the crop—one-quarter to one-half the crop is a common rent for tenant farmers. Any surplus from this exchange is small but sufficient to introduce markets into the culture.

Many peasant agricultural societies have a semifeudal social and economic system that supports a small upper class, including the landlords. Members of this group are better educated, comparatively wealthy, and enjoy more frequent contacts with the outside world. In spite of the inequities, such a system can help bring about the transition from a near-subsistence economy to a commercial one. The large landowner is in a position to concentrate capital and introduce modern technology. Yet many landowners are reluctant to institute change, for they can maintain their high levels of living merely by owning extensive land and by exploiting peasant labor.

The landlord-peasant tenant system may be broken up when land is distributed to peasants through social reform measures. This policy raised productivity in Taiwan and Japan in the 1950s where there were nonagricultural jobs for the displaced peasants, and ready urban markets. In less developed areas, land reform may lead to higher production but also to a decline in export specialties. (Tenancy and land reform are discussed further later in this chapter.)

WAYS OF LIFE

Rural economies in low technology societies throughout the world fall into four major categories. In order of developmental level, these are: hunting-gathering and fishing, nomadic herding, shifting cultivation, and peasant agriculture. Their relatively simple patterns provide a useful starting point to study the more complex patterns that will be analyzed in other chapters.

Hunting-Gathering and Fishing

Early human beings survived by hunting wild animals and gathering wild plants. Once they depleted an area of its food resources or the population exceeded the carrying capacity of the local environment, some had to move on to open land or conquer the territory of their neighbors. Driven by necessity, primitive hunter-gatherers slowly dispersed over much of the world. There were probably no more than ten million at any one time, because few human beings can live simply by exploiting wild foods. Hunting and gathering prevailed in environments rich in game or fish a few thousand years ago in much of Europe and until the late nineteenth century in the midwestern prairies and along the Pacific Northwest coast of North America. More intensive activities, such as farming, commercial fishing, and forestry, have now preempted such areas.

Today remnants of hunting-gathering societies survive in remote places. These fast-disappearing societies include the Inuit (Eskimos) of North America, the aborigines of Australia, the Bushmen of the Kalahari Desert in south Africa, some Indians of the Amazon Valley in Brazil (figure 4.3A), and the Hillmen of India. Some dwell in isolated areas untouched by the technology of agriculture or the domestication of animals. Yet even though these regions may be physically marginal (too cold, hot, wet, or dry for adaptation to the agricultural technology thus far developed), they can and may be developed under certain conditions—for example, if they are found to contain gold or oil, or if they have forests that can be exploited commercially.

Hunting and gathering is a spatially **extensive** activity. An average unit of land provides only a small amount of food, so a very large area is required to support each human life: as much as 2.6 square kilometers (1 square mile) or more per person. Small hunting groups of fifty members need an area with a radius of 10 kilometers or more and a food margin (ratio of potential to needed food) of as much as forty to one.

The general location of hunting-gathering societies is governed by the lack of competing, more productive systems, such as ranching, forestry, or sedentary (permanent rather than shifting) agriculture. The specific location of a hunting-gathering unit is determined mainly by the availability of food or the strength of competing units. Tribes, which constitute the basic social unit, pitch semipermanent camps wherever they can find unoccupied space and sufficient food. Often the group will circulate within a large area, taking advantage of different food or game resources at different seasons or in different environments. The Inuit hunters of northern Canada, for example, may cover several thousand square miles in pursuit of caribou or polar bear. Some tribes survive primarily by fishing and live in permanent but often isolated villages, as along parts of the African coast and in Southeast Asia and Oceania.

Certainly all these peoples, for whom survival is an endless struggle, try to make the best use of space within the limits of their technology. Their generally efficient spatial behavior is apparent in the carefully selected campsites, in the hunting circuits, and in the small number of people in each tribe. The plight of most hunting-gathering societies, however, is worsening. Many of the more productive areas are being turned into commercial grazing lands and, in India, into farmland to feed the expanding national population. The Amazon hunter-gatherers of Brazil are being displaced by the Trans-Amazon Highway and related development. As we mentioned earlier, hunting-gathering societies used to prevail in far more

A
B

C

4.3. Low technology societies. Photograph A shows an Indian woman in Brazil's Amazon Valley fashioning a bowl from a mixture of clay and ashes. In photograph B, camels and donkeys provide transportation for Bedouin nomads in northern Amman, Jordan. In photograph C, Peruvian slash-and-burn cultivators have recently cleared this field. Photograph D shows peasant farmers in Thailand planting rice in the traditional way. The woman in the foreground is transplanting clumps of rice into a paddy, while the man in the background is plowing with a water buffalo. (Photo A by Emil Schulthess from Black Star; photo B from United Nations; photo C by Kensinger from Anthro-Photo; photo D by Picou–A.P. from De Wys.)

D

generous environments before competing activities took over. The Indians of the Pacific Northwest coast, for example, subsisted easily on the rich marine resources of the region. Now most primitive societies are relegated to the most desolate, least productive corners of the earth.

Nomadic Herding

Human beings lived in hunting-gathering societies until about 8000 B.C. Eventually some groups learned to tame and raise wild sheep, goats, and, later, pigs and cattle. This domestication appears to have accompanied or followed the beginning of crop selection and cultivation. The innovations of animal husbandry and crop cultivation diffused from one society to another. Groups dwelling in fertile river valleys and plains were able to cultivate crops more successfully than those occupying hilly, arid, or semiarid regions. The latter, therefore, concentrated on maintaining herds of domesticated animals that lived off grasses, shrubs, and occasional watering holes.

Like hunting-gathering, herding is a spatially extensive activity; productivity per unit area is low (only one or two persons can be supported per square kilometer). This is because herding is now typically practiced in regions most subject to drought and diminishing grass cover and also because animals are inefficient producers of food. It takes three pounds of plant, for example, to produce one pound of beef. The size of the herd that can be maintained and the number of people supported varies from region to region and from one time to another. The carrying capacity of the land depends on temperature, moisture amount and variability, amount of territory (which controls how intensely a given area must be grazed), and availability of supplemental crops.

Considering their limited technology and the uneven quality of grazing lands, nomadic herdsmen have developed remarkably efficient spatial patterns. Tribes and clans often migrate in a regular, cyclical fashion (figure 4.4). This pattern may be altered by the vagaries of weather and by competition with other groups and activities. Tribes usually occupy their traditional territories and rarely cross paths accidentally.

Migratory movements may be both horizontal and vertical. The tribal herdsmen of Iran and Afghanistan migrate hundreds of kilometers horizontally each year and two or three kilometers vertically. In vertical migration (transhumance), herds are driven up to cool mountain pastures in summer and down to protected valleys in winter, an efficient spatial adjustment to the changing seasons and local terrain. Normally when an area is fully grazed, the herds are driven to new pastures, giving the range time to recover (though in times of drought, as is often observed in Saharan Africa, the vegetation may be exhausted in the fight for survival).

Although in steady decline, nomadic herding remains a practical way of life (see figure 4.3B). The livestock provide basic necessities: meat and milk products for food, hides for clothing and tent coverings, dung for fuel, and muscle power for transportation.

Most midlatitude arid regions are dotted with at least a few green and

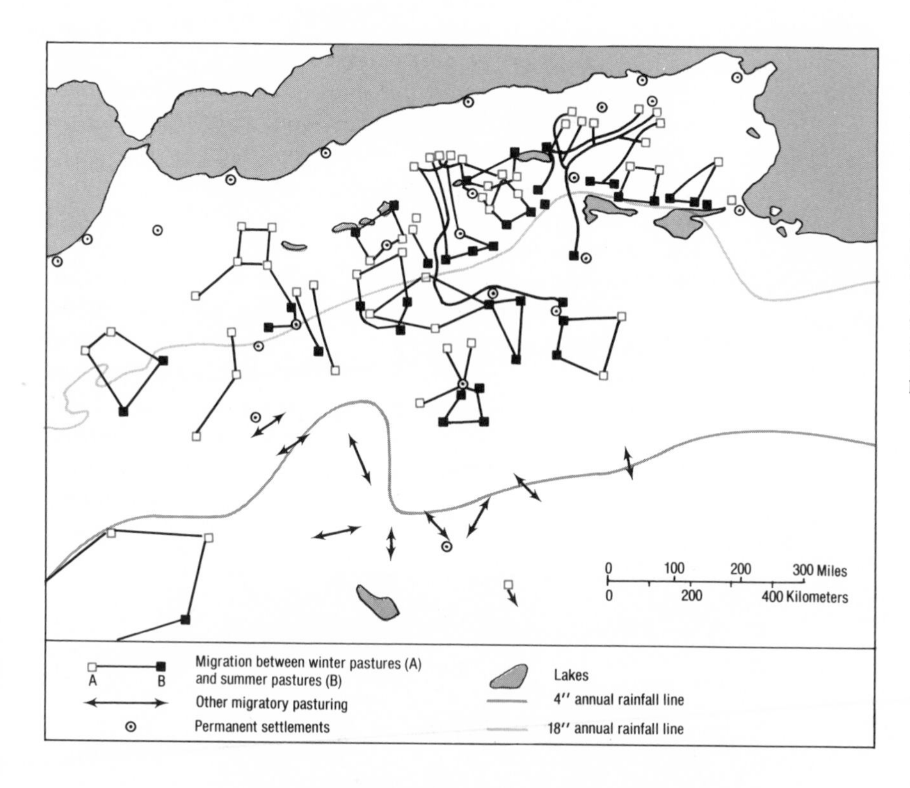

4.4. Patterns of nomadic migration in far northwest Africa. The range of most nomadic tribes is small, but some travel fairly long distances. Many tribes migrate vertically from lower arid zones in winter to more humid mountain pastures in summer.

Source: Redrawn with permission from Hans Boesch, *A Geography of World Economy* (Princeton, N.J.: Van Nostrand, 1964).

fertile oases, which support crop agriculture, towns, and larger concentrations of people. They provide a convenient market for the nomads' small surpluses of animal products and are a welcome source of fresh fruits and vegetables, grain, water, and tools.

Nomadic migratory movements are fading from the contemporary landscape. Herdsmen searching for fresh pastureland have traditionally wandered freely across national boundaries. But in recent years, increasing nationalism and a concern for boundary defense have curtailed many of these movements. Nomadism is further weakened as modernizing nation-states attempt to stabilize agriculture and to concentrate livestock on ranches to ensure reliable food supplies for urban markets. Thus sedentary livestock ranching is gradually supplanting nomadic herding in many parts of the world.

Shifting Cultivation

Shifting cultivation (successive clearing, planting, and abandoning of fields) is a technologically primitive form of agriculture practiced by some of the earliest farmers and by perhaps 50 million people today (see figure 4.3C). Once widely employed over much of the world, shifting cultivation, also known as slash-and-burn cultivation, is now limited to remote or sparsely populated regions in Asia, Africa, and South America.

Shifting cultivation is another spatially extensive activity. Total productivity is extremely low, although output per man-hour can be fairly high since farmers (often the women of the tribe) may work only a few hours a day. No more than three to fifteen persons can be supported per square kilometer. Villages in richer areas will support three hundred persons or more, but in poorer areas the number may drop to thirty. Long periods of drought and disease may ravage the larger villages, reducing their population. Living at a subsistence level, shifting cultivators have little contact with the commercial world. Their tools are generally limited to digging sticks, hoes, shovels, and axes or machetes.

Given such primitive technology, shifting cultivators may well use space efficiently and in a manner ecologically appropriate for tropical environments, provided that the population remains small and land plentiful. Figure 4.5 depicts a typical pattern of shifting cultivation. First the tribe sets

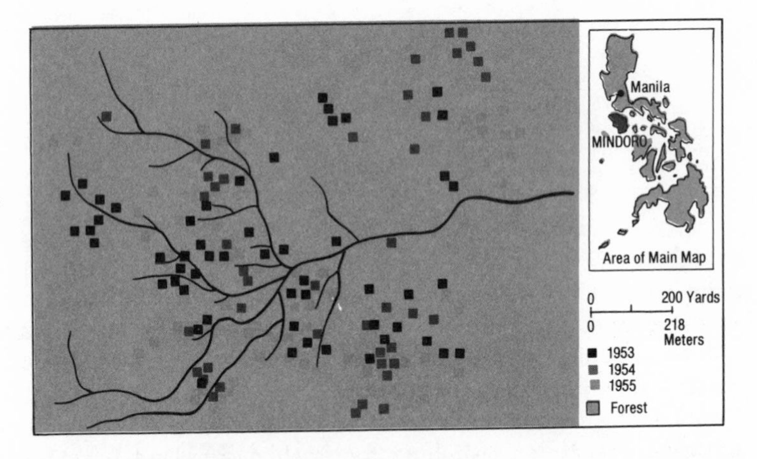

4.5. Patterns of shifting cultivation. The Hanunoo of Mindoro, an island in the Philippines, clear new fields every year. The fields are reused about every twelve years in this area. Village locations are not given.

Source: Redrawn with permission from Donald W. Fryer, *World Economic Development* (New York: McGraw-Hill, 1965). Inset map added.

up a new village in virgin forest. Then the members cut patches of communally held woodland around the village, burn much of the vegetation for nutrients, and plant a variety of crops in the clearing. Since they use no fertilizer and keep few animals, whose manure would enrich the soil, the fertility of the soil rapidly declines. Shifting cultivators can raise satisfactory crops in a single area for only three or four years in succession (yields may drop 30 percent the second year and 50 percent the third). They leave worn fields to fallow and cultivate other ones, often farther from the village. If there is enough land around the village to allow long-term rotation among fields, the tribe may remain at a given location. Otherwise the entire village may migrate to a new area, usually no more than a few kilometers away, and allow the old land to recover its fertility over a long time—sometimes as much as forty years. Only by frequent shifting to more fertile land can the village population support itself at a consistent level. Part of the population may form a new village if the original one becomes overcrowded or is disrupted by internal feuds.

More advanced kinds of agriculture or forestry are displacing shifting cultivation. (Interestingly enough, however, shifting cultivation is carried on in modern form in the practice of clear-cutting stands in commercial forestry.) As village populations expand, some members migrate from their traditional lands to settle in more intensively farmed regions. Eventually a high rate of population growth and a lack of unoccupied land will force primitive cultivators to find permanent locations for their villages and fields (this has occurred in parts of the Philippines, for example). Such changes often occur after the tribe comes into contact with more advanced cultures and learns new farming methods, such as the use of the plow and fertilization. Whether population growth forces shifting cultivators to become sedentary farmers or commercial farming (coffee plantations or cattle ranching, for example) is introduced into the area, the ecological balance may be upset and the land damaged. A post–World War I attempt to develop rubber plantations in the Amazon, for example, resulted in the destruction of natural vegetation and in massive soil erosion.

Shifting cultivation illustrates the relative nature of what is considered optimal spatial behavior. For societies whose numbers are small, whose spatial experience is limited, and whose technical skills are primitive, the pattern of shifting field and village locations may be optimal—it is the best use of the environment for them. The burning of vegetation clears the land and deposits a ready-made layer of nutrients over the ground. This practice minimizes the need to weed and fertilize and is thus an important labor-saving technique for a society whose members are few and often physically weak because of poor diet. Yet in the Western view it seems inefficient and wasteful. If population pressures are too great, forcing the cultivation of slopes and hillsides, erosion can be severe. Once the topsoil is stripped of its protective cover of vegetation, heavy rainfalls may quickly leach nutrients from the soil. Moreover the greater amount of burning necessary to clear larger fields may lead to widespread air pollution and the destruction of forests needed for flood control and timber resources.

Environmental problems aside, levels of living are steadily declining for shifting cultivators as regional or total population grows and available lands diminish. It is difficult for such societies to survive even at a subsistence level. Shifting cultivators may be willing to change their mode of existence, but, like the marginal producers of Appalachia, they do not always know how. In some areas, as in sub-Saharan Africa, physical conditions may be favorable enough to permit tribes to change to sedentary agriculture, provided that they adopt more advanced farming methods. In other areas, as in parts of Brazil or Southeast Asia, governments may seek to curtail shifting cultivation in favor of commercial forestry, thereby destroying traditional ways of life.

Peasant Agriculture

The three spatially extensive activities previously discussed are practiced by relatively few societies today. Far greater numbers—about one-quarter to one-third of the world's total population—are engaged in a more productive form of agriculture: intensive peasant farming. (**Intensive** refers to the application of high inputs, such as fertilization, irrigation, and human labor, and the realization of high yields per unit of land.) In level of technical development, peasant farming stands midway between the shifting cultivation of primitive societies and the commercial agriculture of advanced societies. It is often as productive per unit area as commercial agriculture but not as productive per unit of labor. Peasant farming is distinguished from shifting cultivation by the greater modification of nature, the permanent location, the heavier dependence on human labor, the higher productivity per unit area, and the incipient (developing) markets. It differs from commercial agriculture (see chapter 5) by the less efficient use of labor, the lack of mechanization, the smaller scale of operations, and the limited surplus per farmer.

As we noted in chapter 3, the rise of civilization was made possible by the development of successful intensive farming (the agricultural revolution), which was capable of supporting larger concentrations of people. Land that could support only one person in hunting-gathering economies could, with improved agricultural technology, support many families. Agriculture became productive enough so that one family could help feed others. This abundance freed some people from tilling the soil and enabled them to engage in other pursuits, such as weaving, metalwork, and religious or military activities.

Intensive peasant agriculture was both a cause and a result of population growth: higher productivity made possible the survival of greater numbers of people, and booming population growth created a demand for higher agricultural productivity. Peasant agriculture was the backbone of European economies until the late seventeenth century, when the industrial revolution began. Today it still supports the traditional economies of much of East and South Asia as well as parts of Africa and Latin America.

Peasant farming is important in terms of both size of population supported and level of development. After briefly describing peasant farming life and settlement and reviewing the role of peasant agriculture in European history, we will examine the system as it exists today. We will also discuss the transition from peasant to commercial agriculture and the influence of Western forms of development.

Farming Life and Settlement

In most peasant farming societies, the village is the principal social and economic unit. As a political unit it may include several small settlements and their surrounding fields. Villages tend to be distributed in a fairly efficient and regular pattern. Their spacing was originally determined by the distance easily covered in a day: farmers had to be able to walk from the village to their fields, work most of the day, and return home by dusk. Later new villages sprang up to relieve population pressures in the older ones. Larger consolidated and fortified villages were built in some areas for defense or for ease of administration. Some villages in Yugoslavia and Turkey, for example, contain several thousand people, almost all farmers.

Life in a peasant agricultural society was and still is repetitive. The seasons enforce a rigid regimen of work. Agricultural tools and methods remain much the same from one generation to the next. Changes may occur gradually through trial and error or more dramatically when highly rewarding methods are introduced. An example is the Green Revolution, the rapid rise in agricultural productivity that began in the late 1960s when techniques of growing "miracle rice" (a variety developed by the International Rice Research Institute in the Philippines) and other fast-growing, high-yield grains were taught to farmers in parts of Asia and Latin America. Such innovations enable a society to intensify production further and support larger populations. The adoption of radically new production methods, however, is rare. More typically, the existing technology is slowly improved. The wealthier classes may benefit from maintaining the status quo, and local traditions tend to discourage change. Hence while the commercialization of agriculture will spread as urbanization and industrialization occur, peasant agriculture continues in less developed societies, just as it persists, to some degree, even in the United States (Appalachia) and in parts of Europe.

Agriculture in European History

Peasant agriculture was the precursor of the more productive forms of agriculture practiced in Europe today. The typical farming unit in medieval northwestern Europe consisted of family-held strips of arable land and meadow surrounding a small village and communally held woodland and pasture. Each family was given the use of several strips of diverse quality at various distances from the village. The total land per family amounted to about five to fifteen acres. The plots were sometimes held as private property but could be reallocated by the village leaders. The inner strips, conveniently located near the village, were more intensively and frequently cultivated. A traditional fallow rotation system was used: a strip was planted for two years and then allowed to lie fallow for one year. In many areas the village lands belonged to a nobleman's estate. Rents due to the landowners, plus taxes and interest on loans, usually claimed from one-quarter to one-half the crop, leaving very little for the peasants to sell after providing for their own needs.

The population explosion and industrial revolution of the seventeenth to nineteenth centuries altered traditional patterns of land use in Europe. (Similar changes are occurring in developing countries today.) Improvements in sanitation and health services reduced the death rate, and families multiplied from a continuing high birthrate. Parcels of land divided among family heirs became smaller and less productive. No longer fully employed, farmers found it increasingly difficult to support their households. Moreover the demand for food in the expanding cities could not be met by the existing fragmented farm organization. Governments and large landowners finally forced the enclosure and consolidation of strips and communal lands into larger, separate farms. But many of these farms were still too small, and the more successful and ambitious peasants acquired the holdings of the poorer ones. The new landowners made the transition to commercial farming, while many of the dispossessed either migrated to the New World or joined the slums in the growing industrial cities. (The parcelization, or fragmentation, of land is still an important factor in European farming.)

Agriculture Today

Intensive peasant farming continues to be practiced over large parts of East and South Asia and in portions of Latin America and Africa. As an example of such agriculture, we will look at rice cultivation in Asia, which differs in several ways from the practice of commercial agriculture in high technology societies.

Though climate and terrain vary greatly in East and South Asia, physical conditions over wide areas favor the cultivation of rice, the preferred staple grain of hundreds of millions of Asians. The warm, wet summer monsoon (seasonal wind) drenches the coastal zones, providing the conditions needed for growing this crop. Farther north and inland, where there is less moisture and a shorter growing season, hardier grains such as wheat, grain sorghum, and millet are grown.

The distribution of high-density population in East and South Asia closely follows the concentration of rice production in the lowlands and floodplains. Land is at a premium in this heavily populated region. Steeply sloping land can be cultivated intensively, but only at the great cost of human labor in systems of terraces (see figure 14.1B). Most farmers prefer floodplain locations, where the ground is level and water is close at hand. As figure 4.6 shows, rice is planted near water. Vegetables are cultivated in fields close to the village, where they may be fertilized by human and animal wastes. The pattern of fields reflects variations in the quality of the terrain and in the availability of water. The fields are small because of the practice of breaking up larger units into smaller ones as population expands, or carving level fields in rolling terrain.

Intensive rice cultivation is highly productive. Since yields per unit area are much greater than yields from **extensive** (low input, low yield) forms of agriculture, denser populations can be supported for a given area size. (Note, however, that yields per man-hour are normally quite low under crowded conditions.)

The innovation responsible for the development of intensive rice cultivation was paddy irrigation. A paddy is a carefully constructed level field bordered by dirt walls, or dikes. In paddy irrigation, farmers utilize the annual floodwaters of rivers or divert small streams onto terraces. (This practice existed in Egypt over four thousand years ago; irrigation permitted the development of sedentary agriculture, a major advance from the shifting

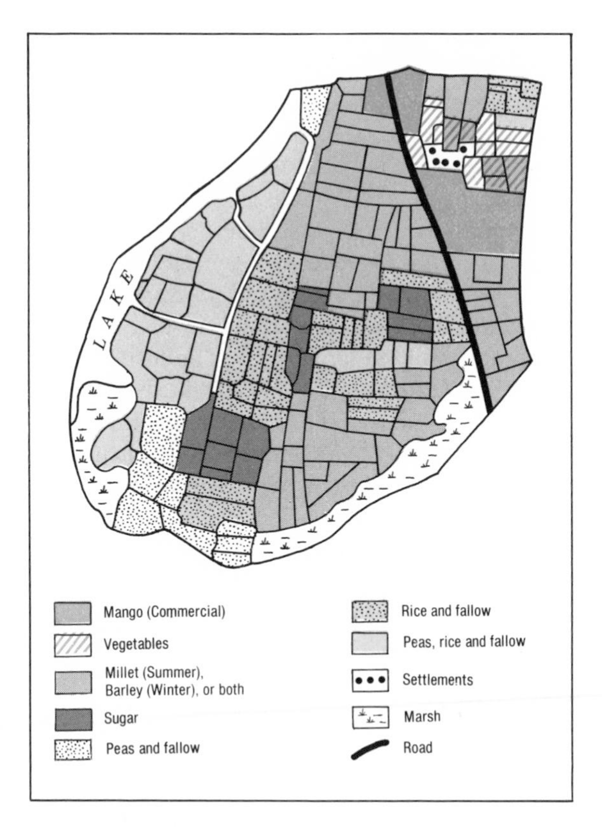

4.6. A village and its fields: Bauria, India. The more intensive vegetables, requiring more labor, and mangos are concentrated around the village. Rice is grown near the lake and canals for ease of irrigation. The entire area is only about two miles long.

cultivation of more primitive societies.) Fields are flooded at critical times in the growing season to obtain maximum yields. Productivity can be increased even more if seedlings are started in separate fields while the more mature crops are cultivated in the main fields. *Double-cropping* (two crops of rice per season) is possible in areas where moist tropical conditions prevail most of the year. With the shorter growing time required for the new miracle rice, even three crops can be produced per season. Where the climate is alternately wet and dry, both rice and a hardier small grain crop are raised: rice in the summer, and wheat, millet, or other grains in the cooler, drier winter.

The soil must be heavily fertilized as well as irrigated to keep it in continuous production and to obtain consistently high yields of rice. In China and Japan, human waste (night soil) has traditionally been the main source of fertilizer. In addition, the Chinese often use mud from ponds and canals. Very early, the Japanese have in part adopted Western practices of applying synthetic fertilizers. Now such fertilizers are used in much of Asia, particularly in areas where the miracle strains of rice and wheat are grown.

Rice cultivation is labor intensive (see figure 4.3D). Many man-hours are spent irrigating, transplanting seedlings, fertilizing, and weeding. Mechanization would help, but the small size of the fields makes the use of heavy equipment technically difficult and economically impractical. Where fields are very small, even animals cannot compete with human labor. But in larger fields, water buffalo and other draft animals can be used profitably. The marginal productivity, or extra output, more than pays for the animal and its maintenance. Japanese farmers use hand tractors even in small fields, but Japan is a special case: the machinery is reasonably priced and appropriate; roads are in good condition, so farmers can get their surpluses to market; rice commands a high, subsidized price; and typically more than half the crop is sold (the other part is consumed by the producers).

Although labor is used intensively during the spring transplanting and the fall harvest, it is often underutilized at other seasons. The result is low productivity per farmer. Both labor productivity and farm income would rise, at least in the active season as economies of scale take effect, if farmers were able to acquire more land. Motivated by the desire to raise their family's level of living, most farmers work strenuously to increase productivity from their small, heavily pressed farms. Those with sufficient land are usually eager to take advantage of commercial demand.

Some Problems

The low net productivity per person in peasant agricultural societies results from several factors, including the large population directly dependent on farming, the practice of dividing land among family members as population grows, inequities in the distribution of land, unequal access to capital and information, and poor transportation.

Land Division. In all areas of intensive peasant agriculture, the large population and the small size of farms prevent high levels of per capita productivity. When land is divided among heirs or transferred as payment for debt or as a dowry, smaller farms and complex ownership patterns result. Reduced farm size tends to foster *tenancy* (dependence on rented land), as farmers seek to add to their output by renting additional fields from wealthier landowners. Since payment demanded for rent may consist of from one-quarter to one-half (normally about one-third) of the crop, the marginal return (extra income from the renting) to the tenant farmer can be exceedingly small.

For various reasons, fields belonging to one farmer are often widely dispersed. In Asia, for example, this *fragmentation* may arise from land division, the accidental location of land available for renting, the adaptation of paddy fields to local terrain, and the fact that some families own a combination of some irrigable paddy and some upland fields. The unfortunate consequence of fragmentation is increased labor for reduced yields on a given quantity of land. The farmer simply cannot give as much attention to the more distant fields, especially for the fertilizing so essential to productive farming (figure 4.7).

Land Tenure. Most peasant societies existing today slowly developed

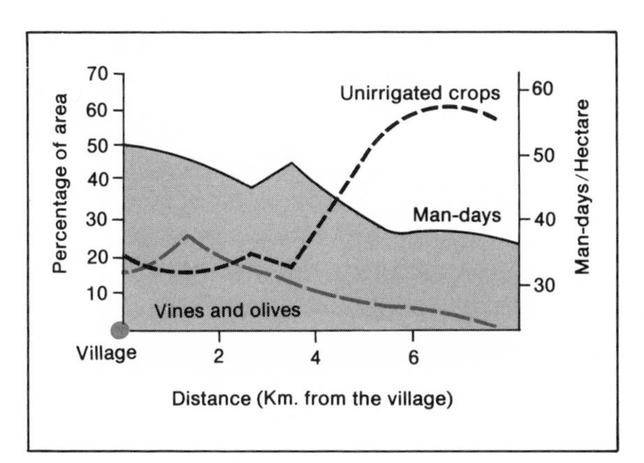

4.7. The impact of fragmentation on farm labor in Canicatti, Sicily. The left-hand scale shows the percentage of farm area devoted to vineyards and olive groves and to unirrigated crops. The right-hand scale shows man-days per hectare (labor per unit area). Crops requiring more attention—fertilization and irrigation, for example—are located close to the village. Labor per unit area decreases with distance from the village.

Source: Data from M. Chisholm, *Rural Settlement and Land Use*, Hutchinson University Library, 1962.

under a feudal social and economic structure. Land came to be concentrated in the hands of a small landed aristocracy. Most of the peasants owned little or no land, subsisting instead as tenants or sharecroppers on the landlord's estate. Typically paying from one-third to one-half their crop as rent, the peasants had little incentive to improve their methods and little opportunity to escape subsistence.

In some societies that are rapidly shifting from peasant to commercial organization, such as Taiwan and Japan, land has been redistributed to the peasants. Productivity in most cases has risen markedly. The transition to commercial agriculture under such conditions can be surprisingly swift, especially if there are nonagricultural jobs to drain off the excess rural labor. On the other hand, governments have found that land reform is extremely difficult to accomplish, given the landowners' power, the expense of compensating landowners for their land, and the costs of surveying the land. Where population pressure is too great and infrastructure too poor (lack of roads, capital, markets, and alternative opportunities), productivity may fall, at least temporarily.

Inequality in land ownership is accompanied by unequal access to capital and information. Often it is not conservatism or unwillingness that prevents change but the fact that poor peasants have no access to credit for possible innovations.

Transportation. Transportation in most low technology societies is local, slow, and irregular. Footpaths and small canals are used for short-distance movements and rivers and canals for longer trips. Local access to transportation is generally so poor that most people rarely plan trips beyond the district capital (figure 4.8).

The poor quality of the transport system is one reason why commercialization is slow to develop in low technology societies. Surplus crops cannot easily be brought to market. In contrast, farmers in advanced societies can quickly ship and market fresh produce. Indeed efficient transportation

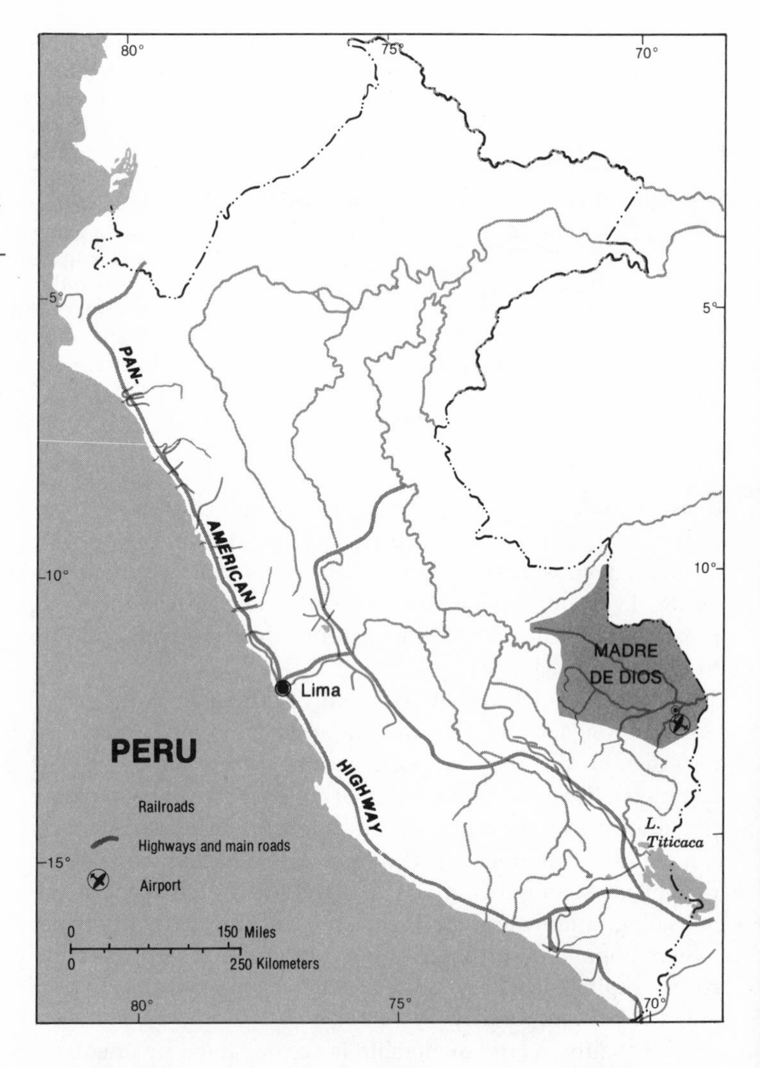

4.8. The transport network of Peru. Peasants living in a peripheral region like Madre de Dios, about 640 kilometers (400 miles) from Lima, the capital of Peru, would have trouble getting to Lima if they had to visit a government agency (the government system in Peru is highly centralized). Although an airport is nearby, most peasants cannot afford to fly. No railroads run between Madre de Dios and Lima, no river system connects the two places, and roads are few, poor, and indirect. The Andes Mountains (not shown in this map) can be crossed only in some seasons, and even then with difficulty.

is essential to commercial economies, as demonstrated in Brazil by the almost instantaneous expansion of farm settlement and marketing with the construction of the Trans-Amazon Highway during the past decade.

DEVELOPMENT AND CHANGE

The change from peasant farming to commercial agriculture is a good example of development from both internal and external forces. A gradual evolution takes place internally as transportation improves and markets develop. But in addition, enclaves of commercial agriculture have been imposed on less developed societies since the time of European colonial expansion in the sixteenth century, when Western estates were formed. Western-run mines have also helped to commercialize near-subsistence economies.

The Growth of Markets and Commercial Agriculture

Peasant farming societies have weak local or regional markets rather than national ones. The exchange of goods often arises chiefly from legal and social obligations, such as rents owed to landlords, dowries, or tithes paid to the church. Yet strictly commercial exchange also occurs in these societies. There is a collective demand for the exchange of grain, meat, vegetables, and handicrafts, as well as for locally unavailable food products and manufactures. Traveling merchants, seasonal fairs, and occasional village fairs provide the market setting. Market towns also exist, perhaps one for every 200,000 persons (based on studies made in India and China). Because of the high density of most rural populations, market towns may be within 15 to 25 kilometers (9 to 15 miles) of most families.

Furthermore most peasant societies support some fairly large cities and towns. These are vital to the commercialization of agriculture. They provide important markets and promote development of the transport systems essential for commercial exchange. In addition, cities foster the growth of industry and thus provide jobs for surplus rural labor.

Many peasant farming regions are making the transition to commercial farming. In Asia, the proportion of rice and other crops sold commercially is rising. Special crops are being produced for export, such as cotton in China, jute in Bangladesh, and sugar cane and cotton in India. In Japan, the commercialization of agriculture has accompanied urban-industrial development. More Japanese farmers are specializing in meat, poultry, and even dairy products, besides the traditional rice. Cash sales have enabled farmers to buy more fertilizer and to invest in hand tractors and other modern equipment, raising productivity to the highest levels in Japan's history. But yields per family are still relatively low because the average farm is so small. Hence an *economic dualism* has persisted in Japan: a fully modern urban-industrial society coexists with a half-peasant, half-commercial agricultural society.

Western Estates

Enclaves of commercial agriculture have long been superimposed on low technology landscapes through colonization and the establishment of Western (European- or U.S.-managed) estates or plantations. After the voyages of discovery in the fifteenth and sixteenth centuries, European demand for tropical products like sugar cane and spices rose sharply. When supplies purchased from local rulers in Africa, the Caribbean, and South America proved inadequate, European companies founded estates specializing in the efficient, large-scale production of sugar cane, rubber, coffee, tea, cacao, copra (dried coconut meat), oil palm, bananas, fibers (hemp, abaca, sisal, cotton, henequen), and spices. Though these estates helped to commercialize the local economy, they were primarily part of the agricultural economy of the mother country.

Four conditions governed the location of estates: suitable climate and soil; availability of a large labor force, either native or imported; protection by the colonial power; and access to port facilities. Coastal regions generally had a comparative advantage for estate agriculture, so these were the first to be developed (figure 4.9). As railways and roads were constructed, fertile areas further inland were also exploited.

Modern estate production tends to be concentrated in certain specialized districts. Although many estates are now locally owned or divided into small farms, many are still operated by American and European multinational corporations or, as in the case of new developments in the Sudan, by

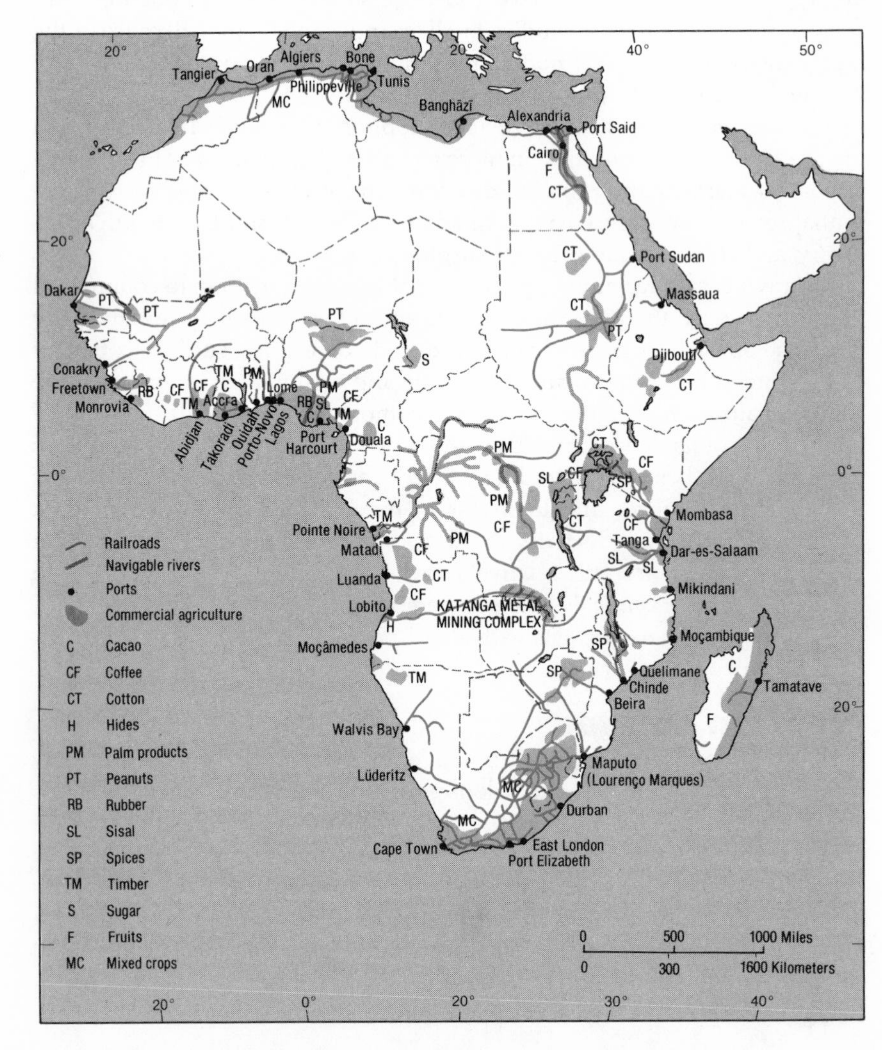

4.9. External orientation of commercial agriculture in Africa, c. 1955. Most commercial farming is concentrated in estates managed by Europeans or Americans. The estates depend on foreign markets and are located on the coast or in interior areas linked to the coast by railways. The transport routes provide external access rather than internal connections.

Arabian investors. Sugar and cotton are produced on the coast of Peru, coffee in the highlands of Central America, tea in Assam and Sri Lanka, and sugar cane in the Philippines. If physical conditions were the only constraints, a given crop could be produced throughout a much greater area. But lack of markets, or oversupply, limits production to particularly favorable areas. While a sufficiently large labor force is essential, estate developers have found it more practical to acquire cheap, underutilized land in low-density areas and to bring in the necessary labor.

In the past, workers were imported from distant lands either on contract (from India to Malaysia, and from Japan or the Philippines to Hawaii, for instance) or as slaves (blacks from Africa to the Caribbean sugar plantations and the American colonies). The need for large labor forces, local or imported, helps explain the particular and limited distribution of modern estates.

Mining

Western commercialism and technology have also been introduced into less advanced societies through mining activities. During the period of European colonization, commercial mining enclaves in Africa and Latin America developed wherever high-value ores were discovered. Today as in earlier times, easily accessible locations are preferred, particularly if cheap local labor is available (figure 4.10). But the discovery of high-quality resources

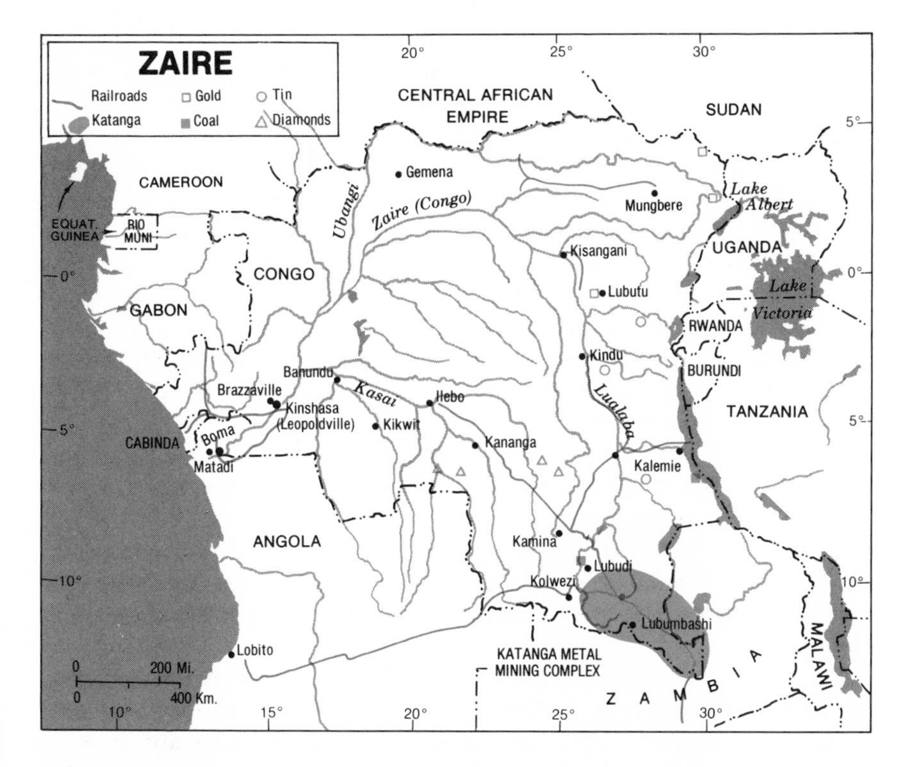

4.10. External orientation of commercial mining in Zaire. Zaire (the former Belgian Congo) contains concentrated deposits of scarce resources, including the world's largest reserves of cobalt and industrial diamonds, and rich deposits of copper, tin, and zinc. The rich Katanga mining complex was connected by rail and water to ports during colonial times.

usually justifies the creation of access through the construction of railroads and other transport lines.

The Impact of Estates and Mines on Development

Estates and mines have helped to commercialize low technology societies by paying wages, no matter how low. Wages provide workers with cash to buy goods and services, fostering a commercial economy. In some countries, local farmers have been able to compete with the foreign-owned estates—for example, cacao farmers in Ghana and rubber growers in Malaysia. Estates and mines have also induced development through the construction of railways, bridges, and roads. The process continues today.

Outside investors, generally Europeans, have traditionally emphasized the development of a colony's primary activities (agriculture and mining). Secondary development (manufacturing) has generally been minimal, since investors prefer to locate in their own countries the **processing activities** that produce the greatest wealth. (Processing activities convert raw materials into finished goods.) In the past, the aim was to enrich the colonizing nation, not to develop the economic base of the colony itself. Railways in Nigeria and other African countries were built to penetrate the fertile or mineral-rich interior of the colony and to facilitate export. Improving internal transportation for the native population was incidental. The external orientation of the rail system has tended to create a bias toward relations with—if not dependence on—the mother country.

The **colonial** trading pattern, in which most of the cash income of the colony derived from export activities, did little to promote balanced indigenous development. The colonial administration controlled power, banking, investment, and trade. Typically the income earned from exports was spent on expensive capital and consumer goods from the home country, not on developing commerce and industry within the colony. In fact, official policy usually discouraged industrialization. The colonies were more useful as captive markets for manufactured goods from the mother country. The gap widened between the small numbers of wealthy, landed upper-class families, who often allied themselves with the colonial administration, and the large mass of peasants who struggled for subsistence (exemplified by conditions in prewar Vietnam).

Colonial administrations in Asia and Africa introduced such benefits as formal education, improved sanitation facilities, and public health measures. The last, ironically, may have worked to the long-range detriment of the colonial economy. Declining infant mortality gave rise to a spectacular increase in population. The traditional agricultural system remained unchanged, however, and was incapable of adequately supporting these larger numbers.

Colonization has also been a leading factor in the redistribution of population in low technology societies. Many people became concentrated in estate regions and associated towns and ports, such as the banana-growing districts in coastal Honduras and Guatemala. In these areas, the local population came to depend on the estate system for livelihood. When estate managers moved to escape the impact of deteriorating soil, plant diseases, and other problems, severe social and economic dislocation swept through the community. Finding themselves suddenly jobless, the laborers had to decide whether to follow the estate (managers encouraged the skilled labor to do so), migrate to the city in pursuit of possibly nonexistent jobs, or fall back on traditional peasant farming.

Most developing countries are now politically, if not economically, independent. They are devoting much effort and investment to building an internal network of transport and services and an industrial base that will free them of excessive dependence on one or two export products.

Urbanization and Rural-Urban Migration

Widespread urbanization has accompanied or even preceded the commercialization of agriculture in many low technology societies. Contrasts between urban and rural life are often striking. Consider the cultural landscape of Swaziland, a small, landlocked nation in southern Africa. Once a British protectorate, it is now an independent state. Swaziland's landscape is dominated by large European-run plantations and asbestos mines. About 14,000 people out of a total population of 412,000 live in the city of Mbabane, the administrative capital. Here modern stores sell the latest appliances; expensive sportscars speed along smoothly paved streets; and comfortable split-level homes cluster around the city's outskirts. But farther into the countryside, an entirely different world exists, composed of tribal villages and fields, where primitive tribesmen guard their cattle (a symbol of wealth) and practice subsistence farming. Very few Swazi farmers raise cash crops, though more may be expected to do so in response to growing urban markets.

Cities like Mbabane usually support some modern industry as well as extensive small-scale commercial, service, and handicraft activities. But generally the population exceeds available employment opportunities. Although some permanent rural-urban migration occurs, most of the workers simply circulate between urban jobs and rural homes. Urban workers in Swaziland and other low technology societies tend to maintain strong tribal, kinship, and village loyalties and social ties. Such relationships help them adjust to the radical change from a subsistence or peasant village culture to a commercial urban one.

In most developing countries such as Swaziland, the capital cities are primarily cultural and administrative centers, similar to the Western preindustrial city. More representative of the type of urban development that has taken place in advanced countries is that of cities in Kenya, based on commercial agriculture and some industry, or in the Katanga (Shaba) region of Zaire, based on mining. As we will see in chapter 6, most cities in advanced nations developed as a result of commercial agriculture and industry.

CONCLUSION

The spatial organization of low technology societies—compared with that of the United States, for example—is characterized by repetitive and miniaturized landscape patterns. Though languages, customs, and other cultural traits may differ widely, patterns of village organization are similar over much of the less developed world. Activities are organized on a small scale because landholdings are small, investment capital for modern equipment is often lacking, and cultural traditions and attitudes tend to discourage modernization. In many areas, tools and methods have changed very little over the centuries.

The low level of technology affects spatial organization in several important ways. Because transportation is generally poor, the friction of distance impedes long-distance trade and movement far more than it does in advanced societies. Hence most basic necessities must be produced in the local area. Relatively self-sufficient and independent cultural groups may coexist but often remain isolated, even within fairly small regions. Advanced societies, in contrast, are culturally more homogeneous: the same transportation and communication system that sustains the economy helps to unify the culture. Another consequence of limited technology is that land use is strongly influenced by cultural preferences and physical conditions.

The goal of maximizing the utility (productivity) of places at least cost is pursued in low technology societies, just as in high technology ones. But in many areas, the lack of modern tools and methods, the high degree of tenancy and land division, and the backward social and political institutions prevent the achievement of this goal.

While in a commercial society productivity is increased by specializing and differentiating land use and labor, in a peasant society it is increased by intensifying local production. The effect of intensification has been, in some cases, to support a larger peasant farming population rather than to free some of the people for employment in nonfarming activities or to raise general levels of living. The lack of efficient technology and the high rate of population growth typical of peasant agricultural societies result in a relatively low level of man-hour productivity compared with that of commercial societies.

Low technology societies strive to organize their space in the best way possible under sometimes severe technological, cultural, and physical constraints. It is clear that in order to feed growing numbers of people and to raise general levels of living, farm productivity must be improved. If a higher quality of life is to be achieved, lower fertility rates and a shift from peasant to commercial agriculture (but not necessarily the capital-intensive form practiced in the United States) are essential. The agricultural systems examined in the next chapter are likely to become more prevalent as low technology societies continue to evolve.

SUGGESTED READINGS

Buchanan, Keith. *The Transformation of the Chinese Earth.* London: Bell, 1970.

Clark, C., and Haswell, M. *Economics of Subsistence Agriculture.* New York: Saint Martins Press, 1954.

De Souza, A. R., and Porter, P. W. *Underdevelopment and Modernization of the Third World.* Resource Paper 28. Association of American Geographers, 1974.

English, Paul. *City and Village in Iran.* Milwaukee: University of Wisconsin Press, 1966.

Hance, W. A. *Population, Migration and Urbanization in Africa.* New York: Columbia University Press, 1970.

Johnson, D. *Nature of Nomadism.* Research Paper 118. Chicago: Department of Geography, University of Chicago, 1969.

Johnson, Edgar. *The Organization of Space in Developing Countries.* Cambridge, Mass.: Harvard University Press, 1970.

Kimble, G.H.T. *Tropical Africa.* New York: Twentieth Century Fund, 1960.

Myrdal, Gunnar. *The Asian Drama.* New York: Twentieth Century Fund, 1968.

Redfield, R. *The Primitive World and Its Transformation.* Ithaca, N.Y.: Cornell University Press, 1953.

Riddell, J. B. *The Spatial Dynamics of Modernization in Sierra Leone.* Evanston, Ill.: Northwestern University Press, 1970.

Rudolph, L., and Rudolph, S. *The Modernity of Tradition.* Chicago: University of Chicago Press, 1967.

Spencer, J. E., and Stewart, N. "The Nature of Agricultural Systems." *Annals of the Association of American Geographers* 63 (1973): 529–544.

Trewartha, Glenn. *The Less Developed Realm.* New York: Wiley, 1972.

White, Gilbert et al. *Drawers of Water: Domestic Water Use of East Africa.* Chicago: University of Chicago Press, 1972.

5

High Technology Rural Landscapes

5 High Technology Rural Landscapes

The rural landscapes that we will now explore differ in scale, use of technology, level of well-being, and organization from those in low technology societies. In the Philippines, the average farmer owns (or rents) only three or four acres of land. The family works strenuously to produce crops using hand tools and draft animals. In contrast, the average farmer in Iowa owns three hundred or four hundred acres of land and uses modern, mechanized equipment. Productive farming methods and the relatively large size of the farm permit this family to enjoy a far higher level of living than is possible for the Philippine family.

The average family farm in the Western world is perhaps ten to twenty times larger—and more prosperous—than its counterpart in Asia, Africa, and Latin America. Transportation is modern and efficient, and most agricultural produce is marketed. The high technology rural landscape is more specialized and differentiated. In contrast to the village-and-field pattern of traditional peasant agriculture, with its self-sufficient variety of crops and animals, commercial agriculture displays large-scale regularity or zones of different kinds of specialized production, ranging from truck farming (vegetable and fruit production) and dairying near market centers to livestock ranching and commercial grain farming at the periphery. Classical agricultural location theory partially explains this fascinating geographic distribution.

This chapter focuses on the spatial organization of large-scale, commercial agriculture. We have divided the discussion into two parts. The first deals mainly with theoretical models of agriculture but also looks at rural settlement patterns and some trends and problems of the contemporary rural landscape in high technology societies. The second examines specific agricultural systems and briefly explores three activities of the rural or small-town landscape—commercial forestry, fishing, and mining.

Part I: Agricultural Location in Theory and Practice

THE ROLE AND IMPORTANCE OF AGRICULTURE

The agricultural landscape provides the most obvious evidence of a society's interaction with its physical environment (figure 5.1). As it has been for millennia, agriculture is still society's most fundamental economic activity. Its impact on the natural landscape has been spectacular: for this space-consuming activity, whole portions of continents have been deforested, cultivated, and reordered.

In the Western world, powerful and efficient machines have eliminated back-breaking toil for many farmers, and productivity has risen to levels undreamed of fifty years ago. Machines have replaced most human and animal labor, permitting as many as nine out of every ten people to engage in nonfarming, urban activities.

Yet compared to industrial activities, agriculture has not responded to technological advances as well as one might expect. Some farmers will not adopt mass-production techniques or cannot because they lack information or because their farms are too small. Lack of effective political organization often weakens their ability to compete. The uncertainty of unstable prices and short-term weather conditions discourages others from taking potentially profitable risks. Like peasant farmers, many cannot acquire sufficient land. In the United States, Europe, and Japan, government has been willing to subsidize small, less efficient farms. For these reasons, among others, overall income levels are lower for most farmers than for people who work in the city, except perhaps during periods of unusually short supply and subsequent high farm prices.

The role of agriculture as an employer in advanced societies has been eclipsed by the urban activities that provide a livelihood for most of the population. But agriculture is no less essential for the supply of food and some industrial raw materials. Directly or indirectly, agriculture supports as much as half the economic activities in most advanced societies. Farm products are used in the manufacture of such commodities as textiles (cotton, wool, flax), paper (plant cellulose), paints (linseed and soybean oil, milk casein), and drugs (glands of cattle, hogs, and sheep). The railroad and trucking industries serve farmers. Manufacturers and distributors of farm machinery, tools, fertilizer, and fuel sell to farmers. Food processors, packers, and marketing firms depend directly on farm produce. Indeed thousands of businesses and service centers owe their existence to farmers.

A

B

5.1. Society's use and abuse of the land. Contoured strip cropping in photograph A has helped prevent erosion and increase the productivity of the land in a portion of Pennsylvania. In contrast, erosion and lowered productivity—probably the result of careless land management—are evident in photograph B showing gullied, abandoned farmland in Wyoming. (Photos by Grant Heilman.)

THE SPATIAL ORGANIZATION OF AGRICULTURE

Before commercial agriculture evolves from or replaces traditional peasant farming, physical and cultural forces influence farm organization and land use more than does the prospect of market sales. As transportation improves and regional and national markets become concentrated in urban areas, specialization becomes possible, and access to major markets becomes very important. Spatial and economic forces—particularly transport costs and profitability—gain significance as commercial agriculture develops. But physical, cultural, and other noneconomic factors still influence farm organization.

Because of these modifying, noneconomic factors, the actual agricultural landscape is often confusing. The simplified models of location theory clarify reality. In essence, the theory of agricultural location states that an orderly pattern of crop production would result if location decisions were based on economic (spatial) factors alone and if decision makers behaved in an economically rational manner. Thus the distinctive theoretical distribution of commercial agriculture is based on the productivity and profitability of various crops and the costs of transporting them to market. These factors are discussed below. We will see later on that agricultural location in the real world is substantially affected by physical environmental variation, technology, government policies, and noneconomic human behavior. Yet a discernible order still exists.

Factors Influencing Agricultural Land Use and Productivity

Physical Conditions

Terrain, soil quality, precipitation (and its intensity, seasonality, and variability), temperature, and length of growing season directly affect crop distributions, production costs, and crop yields, which in turn influence the spatial organization of agriculture. Most crops survive within a broad range of physical conditions but attain peak productivity within a relatively narrow range. Citrus fruits grow best in subtropical regions and white potatoes in cooler regions. For each crop, there are physically optimal locations where the value of the harvested crop exceeds its production costs by the widest figure. The comparative advantages of different regions vary enough to allow a diversity and complementarity (potential for interregional trade) of crop production. In some areas only one activity may be carried on, while in others several can compete. For example, some parts of Montana and Nevada are at present suitable only for livestock ranching, whereas California favors the production of grain, fruits, and vegetables as well as dairy farming.

Demand is naturally highest for land in areas best suited to agriculture—areas with relatively level terrain, adequate and reliable precipitation, and a long growing season. Production under less than optimal conditions means extra work and higher costs for the farmer, or lower output and profits, from a given acreage. Such constraints, however, will not necessarily discourage farmers. Some tomato growers in Ohio create their own productive environment by means of greenhouses, some of which span ten acres. Irrigation systems and land terracing are larger-scale examples of human adjustments to physical environmental limitations.

Crop Characteristics

Some crops are more valuable or more costly to produce than others. The qualities that affect value and costs include inherent productivity (volume of output for a given effort), labor requirements, perishability, transportability, and adaptability to mechanization. Under equally favorable physical conditions, vegetables produce a much greater usable volume of food per unit area than grain does. Given sufficient demand, therefore, vegetables are inherently more valuable, economically, than grain. Higher value but less transportable crops, such as garden vegetables, eggs, and fluid milk, can usually compete for quality land closer to markets. Grain is less perishable, less fragile, and thus cheaper to transport than fresh vegetables, eggs, and milk. Less productive, more transportable crops such as wheat tend to be raised in regions farther from market, where land is cheaper. Specialized wheat regions are found in the western United States, southwestern and southeastern Australia, and the Ukraine, southwestern Siberia, and northern Kazakhstan in the Soviet Union.

Response to Inputs

Investment in such inputs as fertilizer, pesticides, herbicides, and fast-growing, disease-resistant, or higher yielding seed varieties usually pays off in higher productivity. But after a certain point, it is no longer practical to add more inputs, especially fertilizer. When farmers overfertilize their fields, crops can be damaged and yields may diminish.

Crops vary in their tolerance of extra inputs. Their response governs how intensively certain crops can be cultivated and how well they can compete for good land and accessible locations. As long as the inputs raise revenues more rapidly than they raise costs, intensification is worthwhile. Price is also an important consideration because producers of low-value crops cannot afford to invest as much in fertilizer and other inputs. Thus garden vegetables, which command high prices, are more intensively cultivated than relatively cheaper crops such as grain sorghum, barley, and wheat.

Mechanization and Farm Size

Mechanization has dramatically raised productivity. Farmers can work as much as a hundred times the land they cultivated in the horse-and-plow era. Specialized machines have been developed for plowing, cultivating, weeding, spraying, drilling, harvesting, and picking. These machines can be used to full advantage only on large, specialized farms, however. Tomato harvesters, for example, can pick and bin fifteen tons of tomatoes an hour. But this machine costs $23,000, so only farmers who raise a great many tomatoes can use it profitably.

Not all farmers have the capital, the land, or the information necessary to bring production to its optimal level. Moreover economies of scale do not always increase at a constant rate with size. At the point where a hired hand must be employed on a family-run farm, for example, net income may fall unless more acres are utilized and crop production rises markedly. But the family farm tends to be rather more productive per acre than the large corporate farm in the United States or the collective or state farm in the Soviet Union, probably because of the incentive that ownership provides.

Consumer Demand

The demand for agricultural products reflects consumer preferences, prices consumers are willing to pay, and available purchasing power. If the prices of certain foods exceed production costs by a wide enough margin, farmers in that economy will produce those foods. Thus the demand for meat in the richer countries is great enough to induce farmers to use the bulk of their grain production as livestock feed. (In contrast, most of the grain produced in poorer countries is consumed directly by people.) Changes in price, caused by fluctuations in level of demand or by surplus or deficit production, lead to an increase or decrease in the acreage planted to given crops and even to shifts in the location of production. In nineteenth-century Europe, the rise in consumer demand for higher valued foods, such as milk products, meat, fruits, and vegetables, prompted Danish farmers to change from wheat to dairy farming and swine raising.

Try to relate the variables discussed above to crops produced in your home state. How do local physical conditions influence the kinds of crops produced? Is agriculture an important activity in your state? Why or why not? Think in terms of competing activities, accessibility to markets, physical conditions, average farm size, and consumer demand for local specialties.

Agricultural Location Theory

Given the many variables described above, one might expect to find a bewildering array of agricultural practices in high technology societies. Yet the landscape exhibits a basic spatial pattern that is quite orderly. Consider the farmlands extending from the Chicago area to the Rockies (figure 5.2). Around Chicago and throughout southern Wisconsin, farmers specialize in intensive dairy and truck farming. To the west, in Minnesota and Iowa, they specialize in intensive livestock-grain production—hogs, cattle, corn, soybeans, and oats. Beyond the Missouri River, in Nebraska, Kansas, and the Dakotas, a somewhat less intensive livestock or grain production, or both, predominates. And still farther west, in Montana and Wyoming, large wheat farms or extensive livestock ranching prevail. Other crops are also produced in these zones, but there is a clear progressive decline, or **gradient,** in intensity of land use, ranging from the most to the least intensive systems: dairy and truck farming near Chicago, livestock-grain production in the U.S. corn belt heartland (Iowa, Illinois, Indiana, Minnesota, and Nebraska); wheat farming on the high plains (Kansas, North Dakota, Montana, Colorado, and Oklahoma); and livestock ranching farther west (Montana, Wyoming). A similar pattern has evolved in the Argentine Pampas, around Buenos Aires, whose physical environmental conditions and settlement history have many parallels to the landscape around Chicago.

Figure 5.2 suggests that location relative to major markets influences production in the various zones. Indeed relative location, or distance from market, is the key to one of the basic elements of classical location theory: the *intensity gradient,* the progressive decline in intensity of land use and density of population away from the market.

The underlying spatial structure of the agricultural landscape was first recognized in the nineteenth century by a German economist and land-

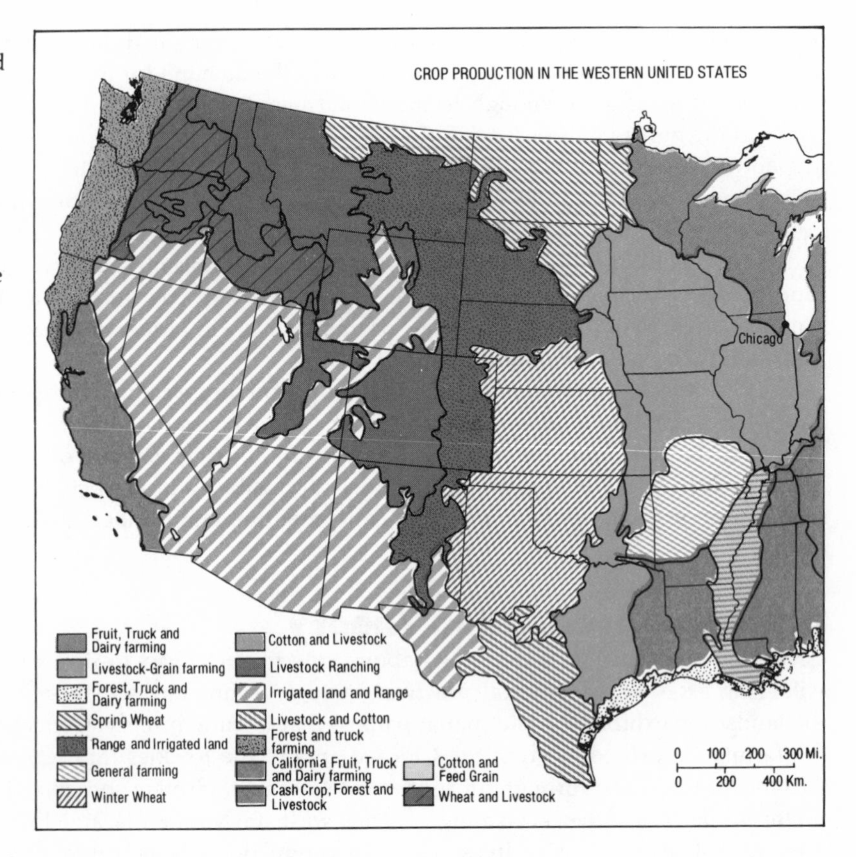

5.2. Crop production zones in the western United States. The most intensive kinds of production are concentrated near Chicago, the major market, and the least intensive are located at the periphery, forming a gradient of intensity. This schematic map suggests the orderly pattern that underlies crop production. (Note that California is also both a major market and a highly specialized farming region.)

owner, Johann Heinrich von Thünen. He formulated classical agricultural location theory, which defines the gradient aspect of spatial organization. At its simplest, the theory describes how space would be organized in the absence of environmental and cultural variation and assuming that farmers want to maximize their profits. It presupposes a uniform, homogeneous plain, where land use is determined only by spatial and economic factors. Entrepreneurial ability is assumed to be equal, technology is comparable at all points, and access to transportation facilities is equal (transport costs vary, however, according to distance from market, perishability of the crop, and so on).

Given these simplifying assumptions, including the existence of a single, large market, von Thünen's theory states that differences in the ability of agricultural activities to compete for land close to market result in a spatial ordering of crops. In real environments, of course, this differential competitive ability is partly determined by variations in physical conditions. But the most important factor, according to the theory, is the differential cost of transporting different crops to market, a cost that increases steadily with distance from market. The agricultural gradient typically ranges from the high value, more perishable fresh milk and vegetables nearest the market to the lower value, less perishable wheat and livestock farthest from market (see figure 5.2).

To illustrate the effect of transport costs on production, consider two dairy farms with equally good land located 10 and 100 kilometers from a large market. From a given unit of land, each farmer can market either one gallon of milk or one pound of cheese. Suppose that the price of milk is $1 a gallon in the city (wholesale) and the price of cheese is $1 a pound. The farmer 10 kilometers away pays $.20 a gallon to ship milk and $.16 a pound to ship cheese (cheese is somewhat easier to ship longer distances than milk). Assume that for both farmers the cost of producing a gallon of milk is $.40, and a pound of cheese $.50. But they can produce the same total value of milk or cheese from the same unit of land. The farmer at 10 kilometers will certainly choose to market milk, since his net revenue is $.40 per unit—$1 less production cost of $.40 and transport cost of $.20, whereas for cheese it is $.34 ($1 – $.50 – $.16). The farmer at 100 kilometers will choose cheese because the net revenue is $.20 ($1 – $.50 – $.30), while for milk it is only $.10 ($1 – $.40 – $.50). From a production-cost point of view, both would prefer milk. But it is more profitable for the more distant farmer to market the cheese simply because the transport cost for milk rises so rapidly with distance from market. (This is why Wisconsin dairy farmers near Milwaukee tend to produce milk, and those farther away produce cheese.) The more distant farmer will not be poorer because of a net return of $.20 rather than $.40 because his land is less costly in proportion to its lower profit per unit area. The more distant farmer will, on average, be able to afford twice as much land and therefore will achieve the same net income.

The Agricultural Gradient

Generally as one goes out from the market center, production becomes less intensive, fewer inputs are added, and returns per acre fall. These conditions define the **agricultural gradient** (figure 5.3). Farmers producing crops that have a high value per unit of weight, that cannot be easily transported, or that respond unusually well to inputs compete for the limited space around market centers. Producers of fresh milk, eggs, and garden vegetables, for which the transport rate is highest relative to net revenue, tend to bid a very high price to buy or rent land close to market. They cannot afford much land, so they usually employ intensive methods to maintain high productivity; fertilizer and fast-growing seed varieties in a sense replace land for these farmers.

Producers of crops that are relatively inexpensive to transport, such as wheat, corn, soybeans, and cotton, can buy more acres of cheaper land farther from market. The precise distance at which a particular crop is competitive depends on prices, yields, and transport costs. As distance from market increases, transport costs become greater and greater and net revenue lower and lower until a point is reached—the **economic margin**—where revenues equal costs. In the example given above, the economic margin for

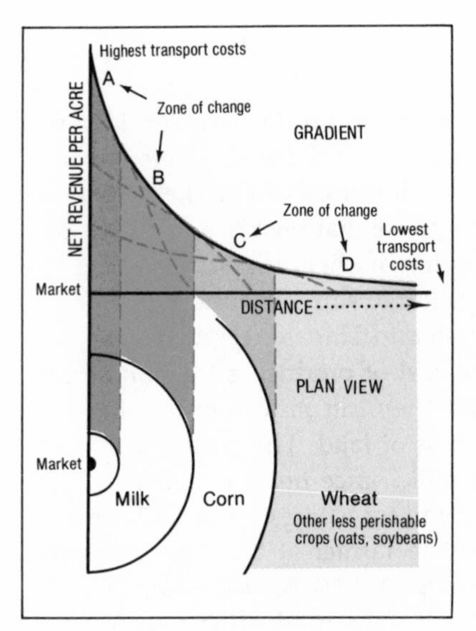

5.3. The agricultural gradient. The horizontal axis indicates distance from market, the vertical axis net return per acre. Products that are most costly to transport, such as vegetables, bid the highest price (rent) for land and obtain land closest to market, in zone *A*. Within zone *A*, increasing transport costs rapidly lessen net return per acre. A point is reached (*A*/*B* boundary) where the return per acre for vegetables is less than that for products less costly to transport, such as milk. Within zone *B*, milk, for example, becomes most profitable. But transport costs, which increase with distance, then reduce net return per acre below that of a third activity with yet lower transport costs, such as corn, which is most profitable in zone *C*. Activities with the lowest transport costs do not need to compete for land close to market, so they are displaced to the periphery.

milk would occur at 130 kilometers if transport costs were $.60 a gallon ($1 – $.40 production costs – $.60 transport costs). Thus milk producers (except those producing for local markets) would tend to fade from the landscape beyond 130 kilometers from the single major market assumed in the model.

Figure 5.3 depicts the net return per acre for different crops relative to distance from market. Note that products in zone *A* closest to market have a very high return initially but that high transport costs rapidly reduce the net return. Rather abruptly (in theory) the profitable crop changes. Although milk is less profitable near the market, it becomes relatively more profitable in intermediate zone *B* because its transport costs reduce net return at a slower rate. The less perishable and more transportable corn does not require such convenient access to market. Thus it is located farther away, where it still remains profitable. The agricultural gradient extends outward until no activity exists for which transport and production costs do not exceed revenues. At its simplest (assuming physical and cultural homogeneity), the ideal pattern of agricultural production consists of concentric rings surrounding a market, with decreasing returns per acre in each ring as distance from market increases (see figure 5.3).

Theoretically the pattern of farm production reflects the agricultural gradient. Size of farm and amount of inputs will vary with the return per unit area and the distance from market. Strawberry growers who own comparatively few acres in the intensively farmed fruit and vegetable region around Los Angeles rely on the inherent productivity of their crop and its high response to fertilizer and irrigation to compensate for the high cost of a small amount of land. Distant producers, such as wheat farmers in Nebraska, prefer to maximize their profits by cultivating more land rather than by intensifying production. In short, where land is cheap (far from market), farmers can buy more acres and use fewer inputs; where land is expensive (close to market), more inputs must compensate for fewer acres.

An intensive and perishable crop can sometimes be produced profitably farther from market if its transportability is improved by an intermediate processing step. Dairy farmers can process less transferable fluid milk into cheese or butter. Canning, freezing, drying, and refining may in effect place distant producers of fruit, vegetables, and sugar beets a thousand kilometers closer to market. Processing techniques thus enable farmers to take advantage of distant but high-quality land, thereby modifying the theoretical agricultural gradient.

The pattern of agricultural production is relatively slow to change. Tomato country, such as New Jersey farmland, remains tomato country for a long time. Short-run stability in the location of crops is maintained by productivity (value of output per unit of land) and transport costs working together. Suppose that a New Jersey farmer decides to raise wheat rather than more productive tomatoes. By making this change, the farmer sacrifices the higher price that could have been received for tomatoes. This **opportunity cost** (the opportunity that a person passes up when making an alternative choice) jeopardizes his chances for survival. He will be forced to turn to a more profitable kind of production or to sell his land. If, on the other hand, a Kansas wheat farmer attempts to raise a more productive, higher priced, but less transportable crop, such as garden vegetables, he may find that the transport cost to the northeastern urban market is greater than the market price he could receive.

In reality, long-term stable **equilibrium** between (a) productivity and transportability and (b) the location of various crops is never reached, since ownership, technology, and demand constantly change. The relative price and profitability of two or three crops, such as wheat and barley, may fluctuate faster than farmers can alter their planting and marketing schedules. Change in the most profitable crop and the inevitable slowness of farmer response help explain, along with physical and cultural variation, the lack of precise and stable crop-zone boundaries. Yet the model works fairly well for the intermediate term, since most farmers base their decisions on average, long-term prices, which may also be stabilized or even guaranteed by government actions. These actions, in some ways noneconomic, tend to perpetuate the geography, or distribution, of specialized agricultural regions.

Historically in both North America and Europe, there has been a shift from less to more intensive land use. In the nineteenth century, the Los Angeles basin evolved from a region of extensive livestock-hide production to one of the world's most intensive fruit-vegetable-dairy regions. The ability of this region to compete for the large northeastern market was dramatically increased by irrigation, the construction of the transcontinental railroad, and the development of refrigerated freight cars.

The entire agricultural gradient has moved upward and outward because of the growth of the national and international market. As the northeastern market expanded, it became unprofitable for relatively nearby farmers to confine their production to wheat. The U.S. wheat belt thus shifted from New York to Ohio, then to Illinois and Wisconsin, then to Iowa and Minnesota, and finally to Kansas, the Dakotas, Montana, and Washington (figure 5.4). The displacement of the wheat belt westward was paralleled by a similar westward migration of the Canadian wheat frontier. In Europe, particularly in England and the Low Countries, farmland came to be used primarily for dairy production and horticulture. The wheat and livestock belt shifted eastward to Russia and overseas to the Americas, Australia, and New Zealand.

5.4. Shift in U.S. wheat production. This map traces the westward movement of wheat production from New York through the Great Lakes region to the plains and the West over a century, as new lands were opened up and older ones were used more intensively.

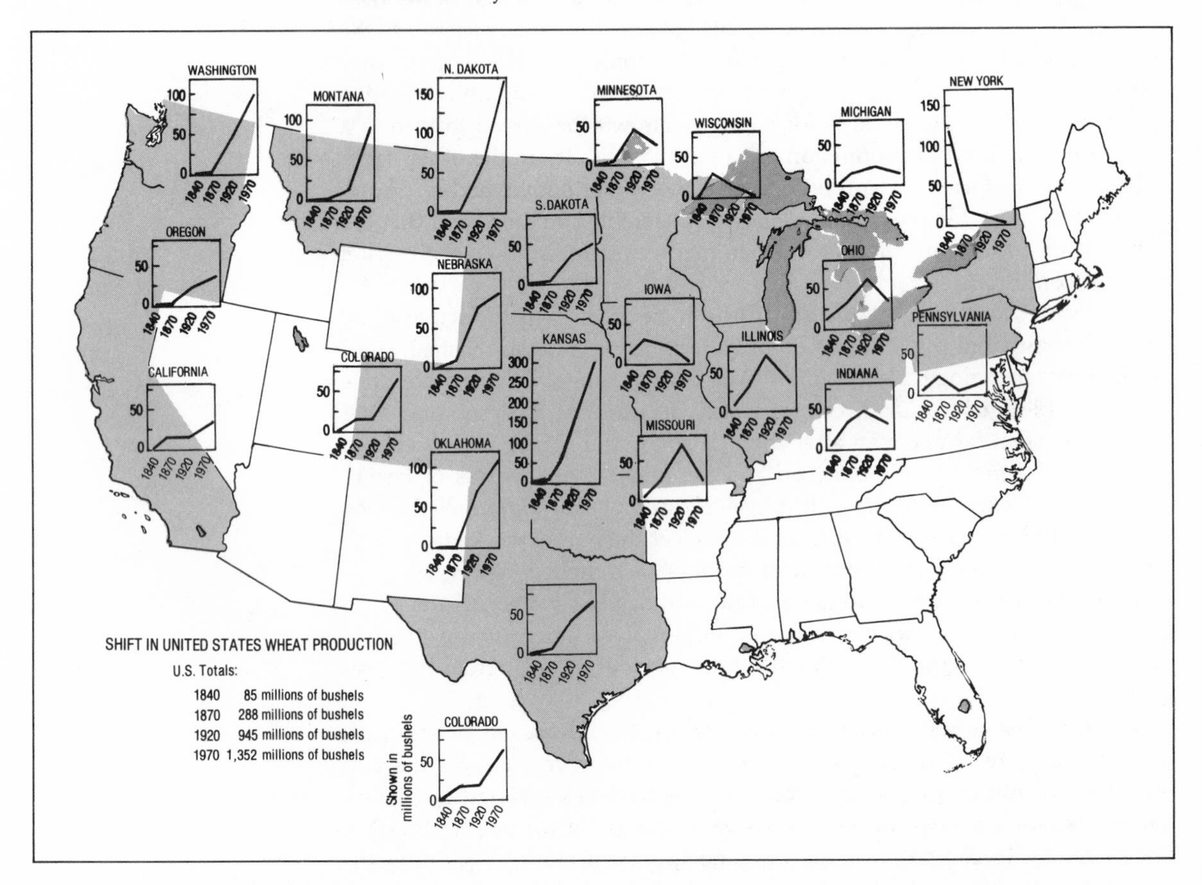

Rent

The agricultural gradient affects the value of agricultural land. This value, in the absence of government intervention, psychological influences, and other factors, should closely reflect the average net return per acre. Net return, as you recall, is the value of the crop minus production and transport costs. It includes both compensation for the farmer's labor and the rent paid for using land he does not own. Normally rent amounts to about one-quarter of net return. Rent is very high per unit area for property near markets because there is only a limited amount of such land.

Factors Affecting the Agricultural Gradient

We noted earlier that agricultural location theory is based on a simplified landscape in which access to market (measured by transport costs) is the key factor. Such factors as physical, technological, political, and behavioral conditions are assumed to be constant. In the real world, of course, these conditions vary considerably and act as modifying influences on theoretical spatial patterns.

Additional Markets. The theoretical patterns of agricultural location described in the previous section assume the existence of a single large market. Typically, however, many smaller additional markets develop within an economy. These smaller markets usually handle a more limited range of products and create landscapes like the theoretical one shown in figure 5.5. The markets may also be more limited or specialized in their demand. Such landscapes reflect the repetitive and dispersed spatial pattern of urban centers (see chapter 6). **Rent gradients** (progressive decline in the value of land) and a zonation of crops by decreasing intensity will also surround these secondary markets. But the lower demand will probably result in a steeper gradient and a smaller market area. The intersections of the gradients are the points at which rents are equal from the major market and any smaller markets. Theoretically circular market areas will develop around the secondary centers. In reality, this circular pattern is often modified by such physical variations as differences in soil quality or terrain and inequities in transport rates.

Physical Variation. The physical environment varies markedly even in small areas. Differences in topography, soil quality, and climate affect the costs of production and thus alter the theoretical gradient (figure 5.6). An enclave of above-average land relatively far from market permits farmers either to realize greater profits on a crop typically grown in their region or to produce a more intensive crop. Instead of cotton, for example, they could raise garden vegetables or fruit, provided that the greater yield or value of the more intensive crop can offset the increased transport costs for special handling, refrigeration, and the like. Irrigated land in a distant but superior environment may yield up to five times as much cotton as could land of only average quality. It can thus compete successfully with cotton-produc-

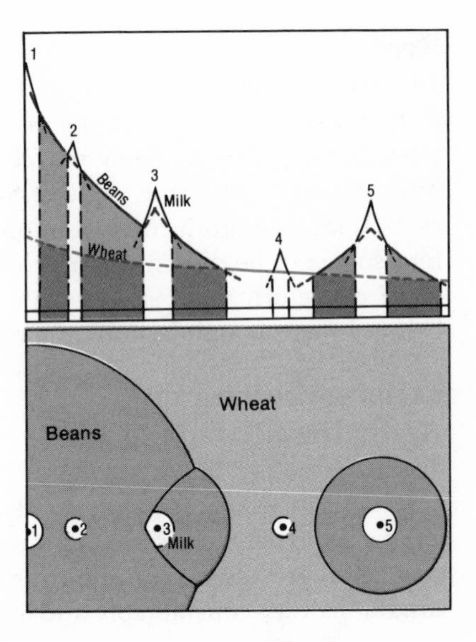

5.5. The effect of additional markets on the agricultural gradient. The upper portion of the figure is a graph of the gradient in return per acre out from a large market. The lower portion is a map, as from the air, of the resultant zonation of crops. In this model smaller markets 2 through 5 disrupt the land use pattern around the main market, 1. The smallest markets, 2 and 4, have a demand only for milk. Markets 3 and 5 have a demand for beans as well as milk. The dashed cones under 3 and 5 indicate what the return would be if the farmer produced beans instead of the more profitable milk close to market. The demand in the smaller markets is sufficient to induce nearby farmers to supply them rather than the main market. Still the markets are incomplete: they are surrounded by farmers shipping directly to market 1. The lower demand at the smaller markets also means that there is less competition for access to them. Thus land prices near these markets are lower, intensity is reduced, and return per acre is less. (Dollar return per acre is indicated on the vertical axis.)

Source: Redrawn with permission from Edgar M. Hoover, *Location of Economic Activity* (New York: McGraw-Hill, 1948).

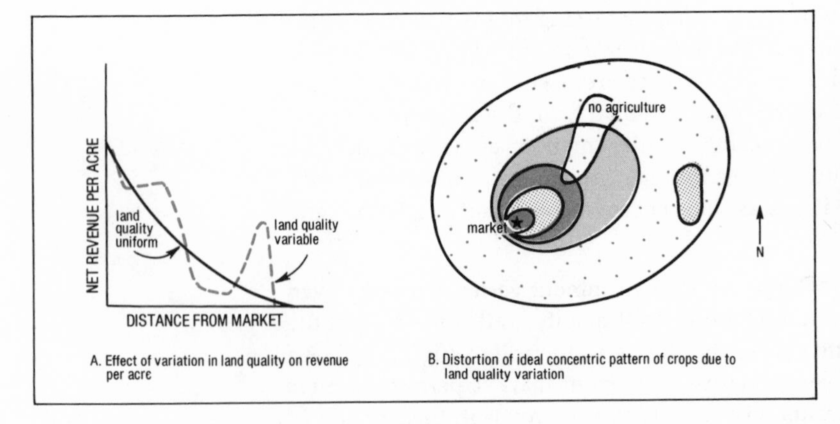

5.6. The effect of physical variation on the agricultural gradient. In A, the dashed line shows that variation in land quality can cause net returns per acre to rise above or fall below the theoretical gradient based on uniform land quality. B shows how variation in land quality can modify the theoretical concentric pattern of crop production. In this figure, land quality declines sharply in the southwest. To the northeast, an enclave of inferior land reduces crop intensity—in fact, one area is not farmed at all. Farther to the east, an enclave of high-quality land creates a zone of very intensive production.

ing districts much closer to market. Conversely an enclave of inferior land will entail lower returns or extra production costs. The farmer, therefore, may choose to produce a less intensive crop, replacing wheat with livestock grazing, for instance. If cost variations caused by physical conditions are combined with the theoretical gradients around markets, an idealized crop pattern like the one shown in figure 5.7 might be formed.

Physical conditions sometimes account for the concentration of specialized production in local areas. Only certain parts of Massachusetts, Wisconsin, New Jersey, and Washington have the right combination of cool climate and marshy soil needed to produce cranberries. Cranberry growers, therefore, are naturally concentrated in these areas.

More often, economic rather than physical advantages bring like producers together. With the possible exception of local dairies, which often have the equivalent of monopoly control over their markets, isolated farms producing any kind of crop are seldom profitable. Competitive strength seems to require the agglomeration of like producers, directly or indirectly sharing techniques and marketing procedures. Advantages gained from ag-

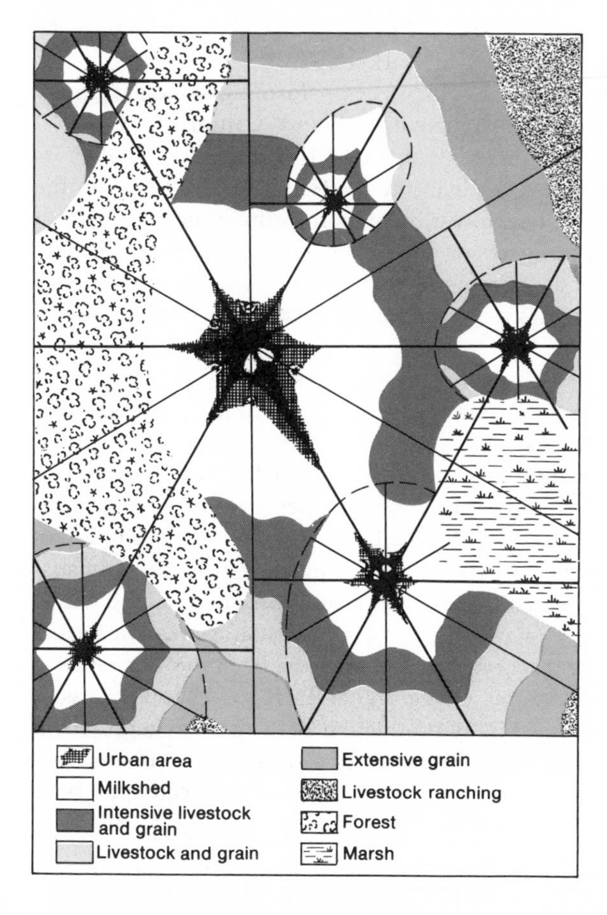

5.7. A composite agricultural landscape. This idealized schematic diagram of agricultural land use in a region shows radial transport routes and large concentric zones of activities extending out from the major urban center, and smaller patterns around the smaller centers. Activities decrease in intensity with distance from the center.

Source: Adapted from *Location and Space Economy* by Walter Isard by permission of The MIT Press, Cambridge, Mass. © 1956 The MIT Press.

glomeration help to explain why the production of specialty crops is often concentrated in local areas. Such specialties as peppermint and hops are highly concentrated in small regions, and even common crops like rye and potatoes are grown in relatively limited areas within a larger agricultural region.

Modern Technology. In recent decades technological advances have made distance from market less decisive a factor in agricultural location. (Not until the 1950s, for example, did most U.S. farmers have access to markets via all-weather roads, let alone electricity or telephones.) Improvements in the quality and speed of transportation have reduced marketing costs. The development of cheap but fairly rapid ocean transport, for instance, enabled New Zealand farmers to sell butter competitively in the British market. New developments in agricultural technology and plant and animal breeding have raised productivity. All these technological advances have allowed farmers to locate farther from market. Many now consider environmental quality to be more important than accessibility in their choice of location. Indeed as distant producers in superior environments become competitive with nearer producers, farmers tend to abandon low-quality land that may exist near markets. (Such land may be far more valuable subdivided into exurban homes for urban workers.)

At the same time, modern technology is transforming some seemingly lean environments into highly productive farmland. California's Imperial Valley is an outstanding example of latent productivity released through irrigation. Though most irrigated lands in the United States are found in the arid and semiarid West and Southwest, irrigation in the humid East is not uncommon (figure 5.8). It mitigates or offsets the year-to-year variability of rainfall. While usually not as extreme as that farther west, such variability can nonetheless plague farmers, who cannot grow crops on statistical average rainfall.

Irrigation systems are costly, major investments; usually they are subsidized by the entire society. Successful projects pay off in increased productivity and in indirect ways as well. Agglomeration economies, such as shared markets and distributors, may develop from the concentration of crop production in irrigated areas. Marketing cooperatives for fruits, vegetables, or other crops often evolve from the teamwork of farmers sharing in the project. Furthermore crops raised in dry, irrigated regions are less susceptible to attack from diseases that spread rapidly in more humid areas.

Many irrigation projects, however, require heavy subsidization by the general public in their construction, which may have been motivated more by the desire to induce economic development than by response to economic demand. Even the large, highly productive Columbia basin project in Washington State had to be financed by sales of electric power. Some areas that have been selected prove unsuitable for irrigation because of poor drainage and subsequent salinization or too short a growing season.

Variations in Human Behavior. Individual farmers who operate their farms profitably are faced with many hard decisions. They must be skilled business managers as well as farmers. By evaluating the quality of their land and its distance from market and by applying their knowledge of crops and the effect of inputs, farmers should be able to determine an ex-

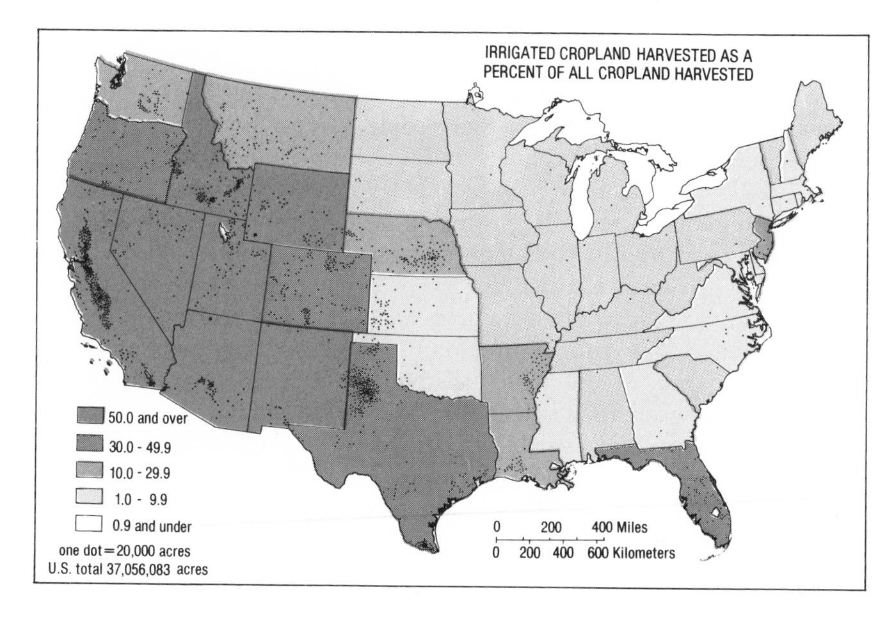

5.8. Irrigated cropland in the United States. Irrigation is relatively more important in the Southwest and West, but it is not uncommon in regions with humid climates, as in Louisiana, Arkansas, Florida, and New Jersey.

pected level of return from specific crops. Modern farm management has become so complex that farmers with large landholdings or large capital investments often use computers or consultants to help them make the most profitable decisions.

Despite such efforts, the goals of efficient spatial behavior are seldom achieved. Many farms are operated somewhat inefficiently owing to farmers' lack of knowledge or their satisfaction with moderate rather than optimal levels of profitability. Moreover the element of uncertainty discourages risk taking. An unexpected frost or severe drought may ruin a crop; a sudden drop in consumer demand may cause price declines and heavy losses; fuel shortages may raise operating costs to exorbitant levels. Thus farmers tend to behave conservatively, especially those who cannot afford to gamble with their income. A cautious response to uncertainty may minimize year-to-year fluctuation, but at the same time, it tends to reduce long-term profits. Farmers are likely to reject risky but potentially profitable alternatives. Instead of specializing in one product, such as beef cattle, they may decide to raise both cattle and hogs to prevent possible failure should demand for one or the other weaken. This kind of decision is often made in spite of advice from agricultural experts that specialization in one product is most profitable in the long run. Then, too, many families farm because they prefer it as a way of life. They are not particularly interested in maximizing the utility of their land. They may produce a wide variety of crops and animals rather than specialize in one or two cash crops, perceiving much of their real income to be outside the market.

RURAL SETTLEMENT PATTERNS

The rural landscape encompasses not only farmlands but farm residences. About one out of every three or four people in most advanced, urbanized societies still resides in the country. In Europe, the birthplace of the industrial revolution and hence of high technology rural landscapes, rural settlement has typically been village oriented. Village residence, initially preferred for reasons of protection, now suggests human contact, conservation of space, convenience, and tradition.

The size and spacing of most European villages was evidently governed by population density, cultural preferences, transport conditions, and environmental variation. Within these constraints, the patterns were fairly even or regular. Cultivated fields, which were part of the general farming system of traditional agriculture, usually surrounded each village. Grazing land and woods often separated one village from the next and were often communally owned. The spacing between settlements was governed by such factors as the need for reasonable access to village fields and for enough land and a sufficiently large village to support a church and priest. When oxcarts were the fastest mode of travel, any field more than a few kilometers away received little attention. In the twentieth century, however, in England and Scandinavia, as in much of the United States, the individual farmstead rather than the village became the norm (figure 5.9).

For Americans, the great emphasis on individualism, the wide-open spaces, and the larger average farm size made residing on separate farmsteads more desirable than village settlement. The pattern of farmsteads around early centers of population reflected the irregular distribution of arable land. But in the interior of the country, most of the land was arable and open. Here settlement patterns were determined by the Land Ordinance of 1785, which established the rectilinear Range and Township Survey System (see figure 7.10), and the Homestead Act of 1862, which gave each family head 160 acres of public land (often a quarter section in a square township) after five years' residence and use. As a result, farmstead spacing became remarkably uniform throughout the Middle and Far West; indeed regularity was often imposed even where the terrain was obviously unsuited. Thus the settlement pattern from the Atlantic seaboard to the Rocky Mountains demonstrates a change from early responsiveness and accommodation to terrain to the later disregard of terrain in favor of an orderly geometric pattern.

Rural hamlets in the United States evolved primarily as trade centers for the agricultural population, with a school, post office, church, and general store. In these centers farm produce was collected for shipment to industrial cities in the Northeast, and manufactured goods were made available to farmers in the interior. Over the past few decades, many hamlets have declined as economic and social activity shifted to larger villages and towns. Since World War II, millions of farm workers have left rural areas because increasing farm mechanization and efficiency have made their labor unnecessary, and many have been drawn to the city by a desire to improve their standard of living. Between 1945 and 1970, an average of 590,000 people a year abandoned their farms for the city. The farm population has plummeted from one-third of the total U.S. population in 1910 to less than one-twentieth today. Yet the total rural population in the United States, Europe, and Japan has often increased in the general vicinity of metropolitan areas. Most of these rural people are not primarily farmers. Even many farm families may earn more from the urban jobs of commuting family members than from farming itself. And much of the hamlet and village population may identify with and depend upon a nearby city more than upon the surrounding farmland.

SOME TRENDS AND PROBLEMS IN AGRICULTURE

Farm Ownership and Tenancy

The declining farm population reveals a long-term trend toward fewer and larger farms in high technology societies (figure 5.10). In recent years urban investors in North America and Europe have bought many farms and transformed them into highly mechanized and specialized *corporate farms.* This development may enable agriculture to compete more successfully with industry as a generator of income, although it is sometimes deplored for its possible social disruption and rural depopulation, as well as for the very high use of energy and agricultural chemicals.

The increasing specialization and scale of agricultural production is creating difficulties for those who are unable to cope with the complexities of modern agriculture. To survive, many are forced to find part-time nonfarm employment or to abandon farming altogether. About 20 percent of all U.S. farms are operated on a part-time basis. In some European countries and in Japan, the proportion is even higher. Though they participate in an exchange economy, most part-time farmers, like the peasant farmers discussed in chapter 4, are tradition bound. They can do little more than produce enough food for their families and must supplement their income with nonfarm jobs, as in mining, forestry, or industry. Part-time farming is fairly common in the remote uplands of the Appalachian and Ozark mountains.

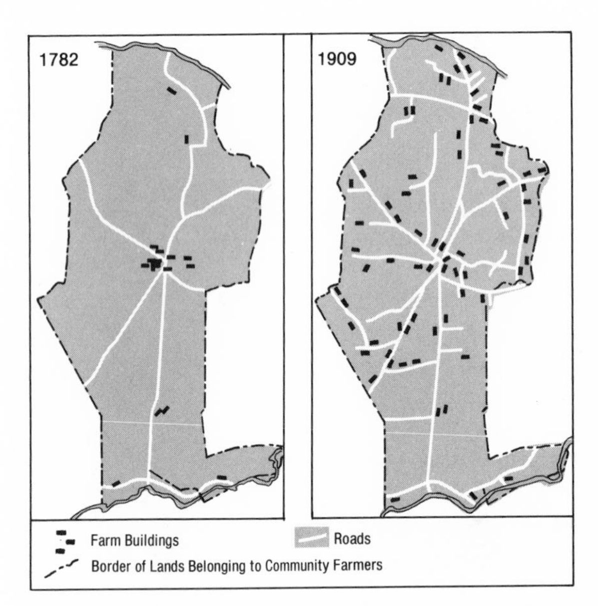

5.9. Dispersion of settlement. Before the twentieth century, villages in most European countries were clustered; but when the Danish government, for example, allocated community land to individual families, the settlements often dispersed as shown here.

Source: After East and Vahl from W. Gordon East, *An Historical Geography of Europe,* 5th ed. (London: Metheun, 1967).

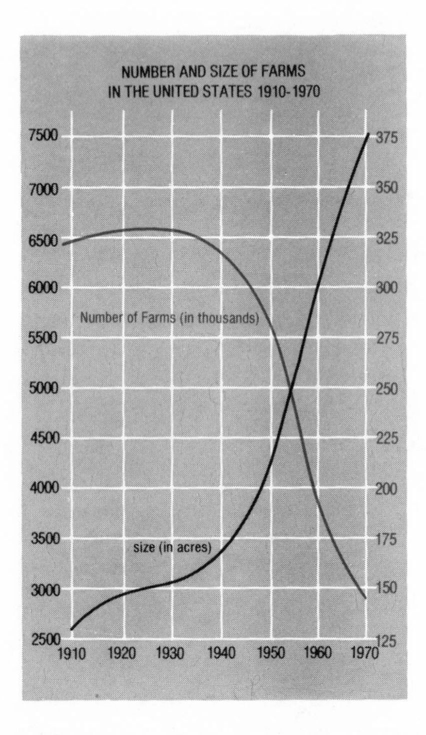

5.10. Changes in the number and size of farms in the United States, 1910–1970. During the first third of the century, both number of farms and average farm size rose as settlement continued and mechanization began. Since the 1930s, the amount of farmland has remained fairly stable, but many farms have been consolidated into fewer, larger, more mechanized units.

Another, very different kind of farm is the *suitcase farmer,* who lives and works in the city but also spends the minimum time necessary to produce a cash crop, typically wheat, on rural acreage (whether inherited or bought as an investment).

Fifteen percent of U.S. farms are operated by **residential farmers,** who commute to regular jobs in the city and farm mostly on weekends. In some cases, part of the family may continue to farm while other members work in the city. This arrangement is rather common in Western and Eastern Europe and in Japan. Like part-time farmers, these are marginal producers. For some, farming is only a hobby. Often inefficient and unproductive, they may withhold land from more intensive agricultural use.

Residential and part-time farms together comprise one-third of all U.S. farms but contribute only 2 percent to total farm production. Working farms on the outskirts of cities are quickly bought up by developers for possible subdivision or by families who turn them into residential farms. Between 1960 and 1970, about 3 million acres of farmland close to cities were sold to suburban developers. (About 9 million acres of new land farther from market, however, were put into agricultural production during the same period through irrigation, drainage, and clearing.) In more peripheral areas, many smaller and part-time farms are either lost to farming or absorbed into larger neighboring operations.

Many farmers double their acreage by renting land held as an investment by banks, businesses, and individuals. The average individual landholding of fifty years ago would be much too small to compete in the modern economy. **Tenant farmers** rent all their land, paying as much as one-third of the net return on their crops for rent. Tenancy as such is not inefficient, but when farm prices drop, the shift of farm income through rents to city landlords can only hurt low-income farmers.

Today there is a growing interest in the urbanized U.S. Northeast to preserve small, part-time, and residential farms for aesthetic reasons and to preserve open space and keep residential densities down. There is popular interest in releasing farmers in these areas from the pressure of urban-level taxes by various means, such as taxation by actual agricultural use, not "highest and best" (potential residential) use.

Government Policies in Free Market Economies

In spite of rapid decline, farm population in most developed countries remains larger than is warranted by typical farm income. Family tradition, lack of marketable skills, preference for rural life-styles, and the desire to own land induce large numbers of people to remain farmers, even when they might be economically better off in urban activities. As a result, family income is often low, property values are inflated, and production costs are high. The dispersed spatial distribution of farmers has hindered the formation of organizations that could effectively reduce competition, restrict entry into farming, and improve the farmer's bargaining position with the larger industrial buyers. Government agencies have attempted to alleviate these problems through policies designed to stabilize farm production and to guarantee a higher income to farmers. These policies have greatly influenced farm organization—for better or worse.

Probably no country exists in which government does not significantly affect commercial agriculture. Governments authorize road and reservoir construction and drainage and flood control projects. They also provide generous research grants and other stimulants to farm productivity. In the United States and other countries, government-supported agricultural experiment stations and extension agencies introduce farmers to technological innovations, superior plant and animal breeds, and new methods of disease and erosion control. Less directly but very effectively, the national government influences commercial agriculture by regulating transport rates on railroad and truck carriers. Because theoretical patterns of agricultural location are based on accessibility to market, which depends on transport rates, any arbitrary rate structures interfere with the ideal patterns described earlier.

Policies enacted to protect farm income include tariffs or quotas on foreign imports, production restrictions, and price supports (subsidies). High support prices unfortunately encourage farmers to raise production even when demand is low. In Japan the subsidized price of rice is so high that, despite limited acreage and a large population, there is a large rice surplus. (The real cost of production is higher than the world price of rice.)

To control surpluses resulting from price supports, acreage restrictions may be imposed. These restrictions tend to reduce proportionately the cultivated land of all farmers—the less efficient **marginal farmers,** whose net return per man-hour is very low, as well as the most efficient producers. Restrictions on the basis of acreage rather than production have played an important part in increasing yields per acre in the United States. But often the result is inefficient land use and unnecessarily high prices. Price supports and acreage restrictions are usually rescinded or inoperative in periods of rising demand and prices for grain or meat. The United States restricted acreage in the surplus, low-price years of the 1960s, removed restrictions during the high-price early 1970s (when Soviet crops were poor), and reimposed constraints in the late 1970s as prices fell.

The model of agricultural location described earlier assumes the free operation of a competitive economy. But actually the pattern is modified by such government interventions as irrigation subsidies, differential production and export subsidies and controls, discriminatory transport rates, and large government purchases. Irrigation subsidies and transport rate structures, for example, tend to discriminate in favor of the more distant producers, thereby weakening the effect of distance from market in determining location.

Government policies in free market economies are often ambiguous. Many appear to favor large corporate farms over family farms. Others are designed to improve the income of small farmers. Although a massive move out of farming has occurred over the past few decades, many remaining farms are still unprofitable. Yet whether from age, lack of other skills, or preference, many farmers would rather stay on their own land, despite low income, than move to the city and face an uncertain future or live in what they perceive to be unpleasant surroundings.

Government Policies in Centrally Planned Economies

Government in centrally planned systems exerts far greater control over agricultural location than in free market systems. Although power concentrated at the top of the political hierarchy theoretically should generate high efficiency and productivity, centralization does not automatically guarantee success. Planning errors applied to an entire agricultural system may lead to results that seem disastrous when compared to an individual farmer's mistakes. Examples are the failure of Premier Khrushchev's costly scheme to bring the Soviet Union's virgin lands into continuous agricultural production in the 1950s and the overambitious extension of corn (maize) production in the late 1960s.

One might also expect the spatial imprint of efficiency—the agricultural gradient—to be more clearly evident in the Soviet Union and other centrally planned societies. Yet the Soviet policy of regional self-sufficiency, which makes sense for a country that is so large and lacks adequate transportation, has impeded the development of specialized crop production based on comparative advantage. Thus the strict theoretical zones of production defining the agricultural gradient cannot be observed in the Soviet Union.

In the Soviet Union and much of Eastern Europe, private family-owned farms have been supplanted by collective and state farms. All land is owned by the government. *Collective farms* are operated somewhat like a business: farmers in each collective work as a unit and share the profits. In *state farms* they are paid wages by the government. For a better understanding of the present system, it may be helpful to review briefly the history of agricultural collectivization in the Soviet Union.

Immediately after the Revolution of 1917, the Soviet Union broke up the large estates of the aristocracy and distributed the land to the peasants. During the late 1920s and 1930s, about 200,000 collectives were created by amalgamating peasant holdings. Collective lands were allocated to different varieties of crops. Individuals received a share of the earnings in proportion to the work they did. To modernize agriculture as rapidly as possible, the limited number of available machines were distributed to specialized machine tractor stations and later to collective farms rather than to individual farm families. The present trend toward larger collective and state farms (and toward greater specialization) technically matches the U.S. trend toward large corporate farms.

Soviet agriculture, like its American counterpart, relies partly on subsidies. The most significant represent compromises with the past, that is, with traditional peasant farm organization. Most collective farmers are also given small parcels of land for private use. On these intensively worked gardens, they produce and sell on the free market a high proportion of the nation's meat, poultry, and vegetables. In one sense, the man-hours, fertilizer, and equipment used to intensify production on these inefficiently small plots constitute a generous indirect subsidy from the collective sector of the Soviet economy to the private. But since the nation would probably have to import much more food in the absence of private plots, the private sector may actually be subsidizing the collective sector.

Agricultural productivity in the Soviet Union lags far behind that of other industrialized nations. About a third of the Soviet population is employed in agriculture, a much higher proportion than should be necessary in a technologically advanced society, and the return per man-hour is very low. The low productivity of Soviet agriculture results in part from physical environmental problems such as aridity, short growing seasons, cool temperatures, and inferior soils; an imbalanced labor force (relatively elderly and untrained workers); and, especially, the centralization of decision making, not only from the family to the collective, but from the collective to the state, removing the incentive to excel.

THEORY AND REALITY

In the theoretically ideal world, all activities would be concentrated around a single point, minimizing the distance between related activities. In the real world, such activities as agriculture occupy vast areas, physically separating sources of production from centers of consumption. The competition of activities for access to these centers leads to a spatial ordering of activities and a consequent decay (decline) in intensity of land use, value of land, and density of population with distance from the center. This gradient, which satisfies the basic goals of human spatial behavior—to maximize both the utility of places and spatial interaction at least cost—is modified by environmental, cultural, political, and behavioral factors.

How closely does the real world follow theoretical spatial organization? The answer can best be found by observing the patterns of agricultural production in high technology societies. The largest concentrated markets are in northwestern Europe, the western Soviet Union, and the northeastern United States. The areas surrounding these market centers produce the most intensive yields and command the highest rents, even if the land itself is not superior. Dairy farms (fluid milk), truck farms, and feedlot operations dominate the regions nearest the market. A generalized zone of intermediate productivity, emphasizing animal-crop combinations, occurs next. Beef and pork are the major products of this zone, but butter and cheese replace them in more suitable environments. A predominantly specialized cash-crop zone appears next, such as wheat in the plains states or in the Ukraine. A fourth zone concentrates on livestock ranching, completing in a very general way national and continental patterns of von Thünen rings. On a regional scale, separate supply areas, especially for milk, can be distinguished for major regional urban markets.

Location theory provides a useful framework for a partial understanding of the agricultural landscape. The key point is not the precise pattern of geometrically concentric rings but rather the importance of accessibility. Particularly in free market economies, relative access to markets (as measured by transport costs) strongly conditions what farmers can profitably produce—regardless of physical conditions, government policies, or personal preference. As we will see later on, urban activities also compete for access to centers of consumption, creating a similar intensity gradient.

Part II: Agricultural Systems and Other Primary Activities

AGRICULTURAL SYSTEMS

Agricultural location theory permits an understanding of the location of actual agricultural systems in high technology societies. To demonstrate its relevance—and the influence of nonspatial factors as well—let us begin far from market at the most extensive end of the gradient and look at livestock ranching, then proceed toward market through progressively more intensive systems such as truck farming (figure 5.11).

Extensive Agriculture

Extensive activities use large land areas and relatively little fertilizers, irrigation, and other inputs to achieve the desired yield. The most extensive forms of agriculture in high technology societies include livestock ranching and commercial grain farming. Cattle and sheep ranches are generally located at the outer reaches of the gradient. Commercial grain farming normally occupies the next farthest zone.

Livestock Ranching

Livestock ranching is the only agricultural activity possible in areas that are too arid for crop production or too rugged for mechanized farming. Critical factors in livestock production are distance from market, market prices, and quality of the rangeland. The farther the ranch is from market, the higher the transport costs. When transport and operating costs exceed the market price of the livestock, ranches can no longer survive. Beyond the economic margin, land that might be suitable for ranching lies idle (figure 5.12).

Consumer demand can modify the characteristic location of livestock ranching. Because of both local cultural preferences and high overseas demand, cattle in Argentina and Uruguay are grazed on land that could well be used for crops. In the United States, most of the better rangeland is used for beef, and sheep ranching is rarely profitable. As a result, world wool-export production is now concentrated in Australia and New Zealand,

5.11. Extensive and intensive agricultural systems. Livestock ranching in Argentina (photograph A) represents the more extensive end of the agricultural gradient, where much land is used and few inputs are required. A truck farm in Alabama (photograph B) belongs to the more intensive portion of the gradient: fewer acres are planted but considerable human labor, fertilizer, irrigation, and other inputs are needed to maintain high productivity. (Photo A from United Nations; photo B from U.S.D.A. Soil Conservation Service.)

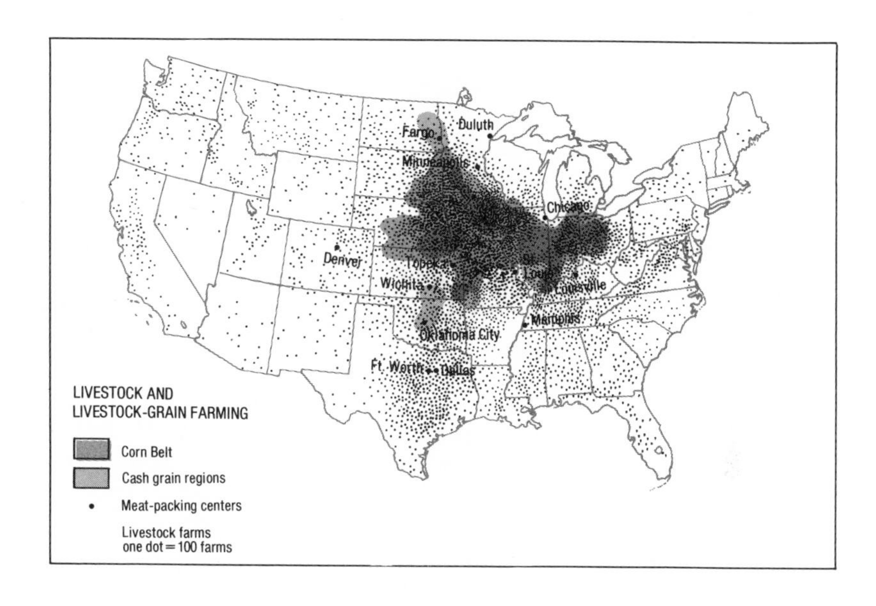

5.12. Livestock and livestock-grain farming in the United States. Livestock can be raised successfully in the Far West in areas unsuitable for crops or mechanized farming; but increasing distance from market eventually raises transport costs to an unprofitably high level, and livestock operations decline. Livestock-grain farming occupies a broad zone in the central Midwest. Meat packing reflects this distribution, but, like cash corn, it tends to be concentrated in particular areas or points of high transport accessibility.

where demand for beef is lower and cattle raising does not require so much land.

During the past century, the pattern of livestock production in the United States has changed. In the early days of American ranching, roads and transportation were so poor that only highly concentrated and lightweight products, such as hides and dried meats, could be shipped profitably from frontier ranches to population centers in the East. With the expansion of railroads and refrigeration around 1880, it became possible to transport livestock and fresh meat over long distances. A complementary system of livestock production eventually developed: ranchers found it more profitable to specialize in breeding and yearling production and to ship the young cattle to feed-grain centers near large markets for fattening. Through this system of two-stage ownership, nearly one-third of all cattle in the United States is now finish-fattened on feedlots. The range, freed from the burden of maintaining mature stock, can be used more flexibly and at fuller capacity. Depending on market conditions, ranchers can raise either a larger number of smaller animals or animals of varying ages and numbers.

Commercial Grain Farming

The production of grain for human consumption (as opposed to livestock-feed grains) has become a separate specialty in high technology societies. Although a variety of grains are produced in the Western world, we will focus on wheat, the world's primary cash grain.

Wheat has a dual location. Somewhat over half of the wheat produced in the world is raised on a specialized basis in low-density, subhumid plains, particularly in the United States, Canada, Argentina, the Soviet

Union, and Australia. Somewhat less than half is grown as part of a crop-livestock agriculture in higher-density, more humid areas, as in the eastern part of South America and particularly in Europe (the farmers sell both wheat and meat).

Different varieties of wheat are produced in different regions. Spring wheat, planted in spring and harvested in summer, is grown in areas where winters are extremely cold. Most of the land planted to spring wheat is located far from market, where land values are lower (North Dakota, the Canadian prairies, western Siberia). Winter wheat, planted in fall and harvested in spring or summer, dominates in subhumid plains (30–50 centimeters [12–20 inches] precipitation), where winters are not so severe. (Texas, Kansas, Argentina, the Ukraine). Americans prefer the hard winter and spring wheats grown in the Great Plains for bread flour and the soft wheats produced farther east—in Ohio, Indiana, and Pennsylvania—and in Oregon and Washington for pastry flour.

Because of its hardiness, wheat thrives in marginal locations. It tolerates a wide range of physical conditions, stores well over long periods, and is fairly cheap to transport. It is not especially responsive to fertilizer or extra moisture, so rather than investing in fertilizer or irrigation, wheat farmers may increase production by cultivating more acres.

Most commercial grain farms consist of wide, level fields suitable for large-scale, specialized equipment. Many wheat farmers plant thousands of acres in marginal locations far from market. Successful farmers use their land productively, employing special moisture- and soil-conservation techniques—for example, *strip cropping* (planting different crops in alternate strips or leaving alternate strips fallow), which reduces wind erosion; *fallowing,* or *dry farming* (cultivating only half the land at one time, then planting only the other half the next year, a technique developed in the Mediterranean during the Middle Ages), which helps to retain moisture; and *bench terracing* (plowing rows horizontally across slopes in a stairstep fashion), which prevents soil from eroding into streams and other waterways by letting natural erosion fill gullies and level fields. Bench terracing is sometimes used as an alternative to contour plowing. Many acres must be put into production to compensate for the low per-acre yield. But excessive planting on marginal lands in dry years can damage the soil, as demonstrated by the Dust Bowl of the 1930s, when strong winds blew away the topsoil, and similar disasters in Kazakhstan in the late 1950s.

Commercial wheat farms tend to be more profitable per man-hour than the average farm. Wheat farmers located far from market (in Montana, North Dakota, or Kansas, for example) can compete successfully with more intensive producers, such as truck or dairy farmers, closer to market. Their success despite marginal location and restrictive physical conditions results from their informed organization and agricultural methods. The per-acre yield may be low, but thousands of acres of land far from market are planted, few inputs such as fertilizer are needed, labor requirements are low because of the use of mechanized equipment, and consumer demand for wheat is high.

Wheat zones are also used for other small grains, particularly rye, barley, and sorghums. Barley and rye tend to be grown in cooler spring wheat regions and, partly because of dietary preference, across the north European plain from Germany through the Soviet Union. Sorghums do well in the warmer, drier margins of the winter wheat zone, as in Texas, Oklahoma, and Kansas. Less tolerant crops, such as corn, oats, and soybeans, tend to preempt land closer to market for the priority purpose of providing feed grain for livestock.

Intensive Agriculture

Compared with extensive systems, intensive agriculture in high technology societies entails more labor and more fertilizer, pesticides, herbicides, and other inputs to ensure high yields per unit area. Except where unusual opportunities exist, intensive farms are located closer to market centers, where land rents are higher.

There is, of course, no sharp dividing line between extensive and intensive agriculture. The agricultural gradient ranges from less to more intensive forms of agriculture as one moves closer to market. The least intensive systems—livestock ranching and commercial grain farming—have already been discussed.

General (Mixed) Farming

American and European agriculture evolved from *traditional general,* or mixed, *farming.* The variety of crops and animals raised on each farm provided families with most of their basic necessities—vegetables, fruits, eggs, dairy products, meat, leather, and fiber. The production of fodder for livestock was included in the regular crop rotation. As we have seen, the productivity of medieval European agriculture rose sharply with the development of intensive general farming. However, rapid population growth and excessive land division, largely due to inheritance laws, weakened the system. Small farms and lack of specialization limited the surplus available for sale. General farming began to decline as large urban markets developed and transportation and technology improved. By the end of the nineteenth century, the age of specialization had arrived.

Yet general farms are not uncommon today. Self-sufficient and only partly commercial, they are similar to peasant farms. In the United States and Canada, about one-third of all farmers are general farmers, although they account for less than 5 percent of farm output. In Europe the proportion is still higher. The persistence of general farming, despite the trend toward large-scale regional specialization in advanced societies, reflects in part the conservatism of many farmers, particularly the older ones. In the United States general farming is practiced chiefly in Appalachia and the Ozarks and in other small pockets of land where physical and cultural isolation has helped preserve the ways of the past. It also persists on a part-time basis in the vicinity of large cities. In Europe the general farm is still common in several areas in some Mediterranean countries, including France, and in Germany and Poland.

Many general farmers find it difficult to survive in an age of specialization, efficiency, and bigness. They often lack the capital, land, information, and business skills to manage their farms profitably. Many are forced to take part-time, nonfarm jobs to make ends meet. The contrast between small-scale, traditional general farming and large-scale, specialized agriculture underscores the striking changes that have taken place within just the past twenty or thirty years. (However, although productivity is higher on large farms, production per acre may well be higher on smaller ones.) It also reminds us that disparities exist in patterns of organization and levels of development within advanced as well as less advanced societies.

Commercial Dairying

During the past century, general farming evolved into dairy farming in many parts of North America and Europe. Dairying is an efficient response to cool, temperate climate or to location near market, or both, given a sufficient demand for milk and other dairy products.

The densely populated, urbanized areas of northwestern Europe and the northeastern United States provide both a market and the proper physical conditions for dairy farming. The year-round humidity and moderate-to-cool summers characteristic of these regions are ideal for producing the hay, pasturage, and small grains fed to cattle. Land that is too rugged or poorly drained for crop agriculture can be used as permanent pasture. But in extremely urbanized areas, open land is so limited and the cattle population so dense that feed must be imported rather than produced locally. The high demand for dairy products in urbanized regions, where population density often exceeds five hundred people per square kilometer, gives dairy farming a comparative advantage over other crop-livestock combinations close to market. Moreover since the distance fresh milk can be transported profitably is more limited than for many other products, dairy farmers must bid for land near market.

The spatial variation within dairying is striking. Around cities, particularly the larger ones, there are rather clearly delineated **milksheds,** or milk-supply areas. Their extent is governed by such factors as the total urban demand for milk, the price commanded by milk, and any legal restrictions imposed on production, such as quality control, negotiated prices, and marketing channels. Beyond the zones from which it is most profitable to ship fresh milk, the principal dairy products are cheese, butter, and yogurt. These are processed at numerous local creameries rather than at the farm. Minnesota and Wisconsin are major suppliers to much of the United States, as are Denmark and the Netherlands to northwest European markets.

Although butter and cheese are commonly produced in dairy regions, cooperative production arrangements and local traditions often result in separate butter and cheese districts (figure 5.13). In the United States and Europe, the production of milk for butter and cheese occupies humid zones of intermediate to high settlement density fairly near markets. But if physical conditions are especially suitable and transport costs are low, such production will also be common in regions farther from markets. The low cost of ocean shipping permits New Zealand dairies to compete in North American and European markets (subject to tariff barriers).

Dairy farms in the United States have on average been less profitable than wheat and beef or hog farms, probably because the average dairy farm is smaller in both acreage and stock, though state laws governing quality controls, usually more rigid and expensive for dairy farmers, may be a factor. (On the other hand, state laws and federal subsidies tend to keep dairy prices high; in Minnesota and Wisconsin, laws forbade the retail sale of colored margarine.) Limited size makes it difficult to realize scale economies on the costly equipment needed for refrigeration, milking, and other operations. Labor requirements are high, and man-hour productivity is fairly low. Productivity may be further restricted by rolling terrain, inadequate cropland, and small fields characteristic of many dairy regions. However, if the farm is large enough to permit economic use of labor-saving equipment, dairying can be quite productive and profitable.

Intensive Livestock-Grain Farming

The classic form of agriculture in Europe and North America is livestock-grain farming. It evolved as the more commercialized and specialized offshoot of the general family farm, after specialized commercial dairying and grain farming were developed. Livestock-grain farming is the most self-contained, self-sufficient commercial farm system today. The farmers pro-

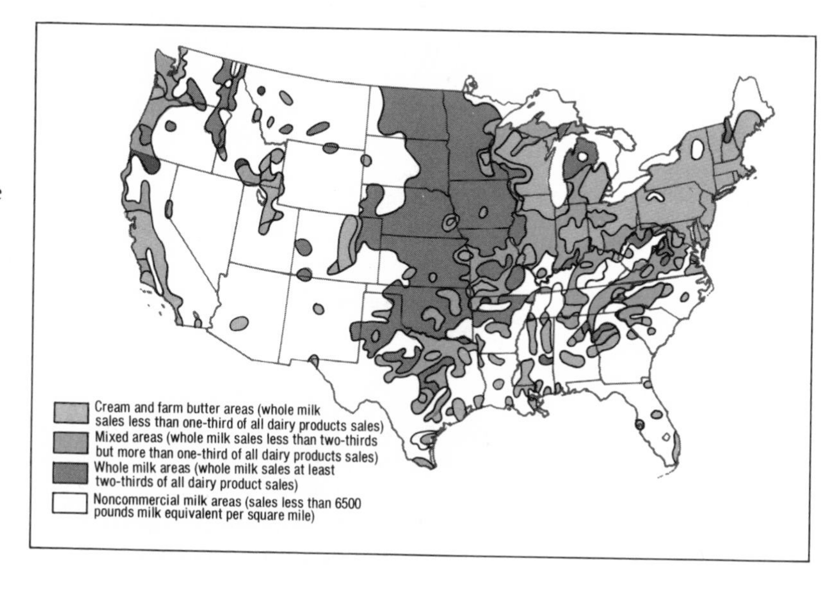

5.13. Dairy production in the United States. Areas farther from large markets sell more cream and butter than whole milk because the concentrated products are relatively less expensive to transport.

Source: Data from U.S.D.A., Bureau of Agricultural Economics.

duce most of their own feed grain and breed most of their own hogs and cattle. In addition, many farmers raise a few chickens, one or two milk cows, and enough vegetables for the family table. Competitively livestock-grain farming occupies the middle zone between the extremely market-oriented city milk suppliers and the more export-oriented commercial grain farmers.

The epitome of livestock-grain regions is the U.S. corn belt (see figure 5.12). This region is endowed with a superior combination of level or slightly rolling terrain, moderate and even precipitation, hot summer days, warm summer nights, and rich soil. Because of these highly favorable conditions and an ideal location between western cattle ranches and eastern urban markets, the region has a comparative advantage for both corn, the dominant feed crop, and livestock (cattle and hogs).

The **monoculture** (long-term cultivation of a single kind of crop) of corn or other crops gradually depletes the fertility of the soil. Hence crop rotations are used, providing feed variety or extra cash and conserving the soil at the same time. Common rotations in a single field over a four-year period are corn-corn-wheat-hay and corn-corn-soybeans-hay. Soybeans have become such a staple that the corn belt is sometimes referred to as the corn-soy belt. Soybeans enrich the soil with nitrogen and give farmers extra cash income. Soybeans are not only the chief source of edible oil (used in margarine and other products) in the United States, but they are also a major export crop, particularly to Europe and Japan. Clover, alfalfa, and other hays are alternative sources of feed. When their residue is plowed under, they, too, improve the soil.

The typical livestock-grain farmer raises beef cattle or hogs, or both. This combination is often preferred because it brings in a more stable income and maximizes the efficient use of feed. As much as one-third of the cattle produced in the United States are purchased from western ranches as yearlings and are finish-fattened on corn belt farms. A feeder-in-transit privilege provides a through-rate for cattle that are stopped and fattened on their way to livestock markets in Omaha, Sioux City (Iowa), Kansas City, and other market centers. (A *through-rate* is a lower rate, as if from the farm to the meatpacker, rather than the higher rate for the two shorter trips actually made.) The transit privilege in effect reduces distances between places and has promoted the practice of finish-feeding cattle on midwestern farms.

Feed grains sell for a lower price than commercial grains; yet the transport rate is about the same for both. Thus farmers primarily raise corn as feed for their own livestock. If they market the corn, it is usually shipped only short distances. Corn does become a cash crop in regions such as north-central Iowa, where the level terrain permits highly mechanized farming, and near some major feedlot areas, such as Omaha.

Compared with American livestock-grain farms, European farms are generally more self-sufficient and varied. Particularly in areas where summer temperatures are lower, hay and small-grain production (wheat, rye, oats) may be supplemented by the intensive cultivation of root crops (turnips, beets, potatoes), which are used mainly as animal feeds. Cattle and hogs are the principal animals, just as they are in the U.S. corn belt. Poultry are often raised on the side (specialization in poultry and egg production is less common in Europe than in the United States).

European agriculture is highly productive and is becoming increasingly mechanized. Though average farm size is growing, many farms are still inefficiently small and landholdings are fragmented. The average European livestock-grain farm is about one-fourth the size of the average American farm (100 acres versus 380 acres). The small size and fragmentation are due to population pressure and a tradition of family farming, which tends to keep excessive numbers of people in farming. Many European farmers earn a reasonable income only through subsidized prices, the nonfarming jobs of family members, the insulation of protective markets such as the European Economic Community, and more intensive production.

Horticulture (Truck Farming)

Other agricultural specialties tend to reflect unusual local or regional comparative advantages or exceptional entrepreneurial skill. *Truck farms* near large urban markets, as in the urbanized belt between Hartford, Connecticut, and Washington, D.C., specialize in flowers, vegetables, fruits, and seeds. These farms are cultivated intensively. To ensure profitable yields, farmers apply huge amounts of fertilizer—over three times the national average per acre.

American and European truck farming has shifted in part from traditional market areas to more remote but climatically favorable locations. A year-long growing season has given California, the Rio Grande Valley (Texas), Florida, and the Mediterranean coast a winter monopoly on fruits and vegetables. In these areas, farmers—particularly on large corporate farms—have been able to specialize, increase their yields, and reduce their costs. Cheaper labor has been an important factor. Mexican bracero labor (workers temporarily imported on contract) helped to make California and Texas competitive with eastern states. Although far from final markets, Florida, California, and Texas have come to dominate the production of many fruits and vegetables in the United States. They produce distinct specialties, such as avocados, oranges, and grapefruit, which cannot be grown in other regions. Low production costs have enabled farmers in those states to compete nationwide against local in-season suburban producers of other fruits and vegetables, despite higher transport costs (figure 5.14). The growth of the canned and frozen foods industry has also contributed to their success.

Because of cooperative marketing, research, disease control, and other agglomeration advantages, the production of specific kinds of vegetables, fruits, or nuts is often highly concentrated in very small areas—for example, hop production in the Yakima Valley of Washington and artichoke production around Watsonville, California.

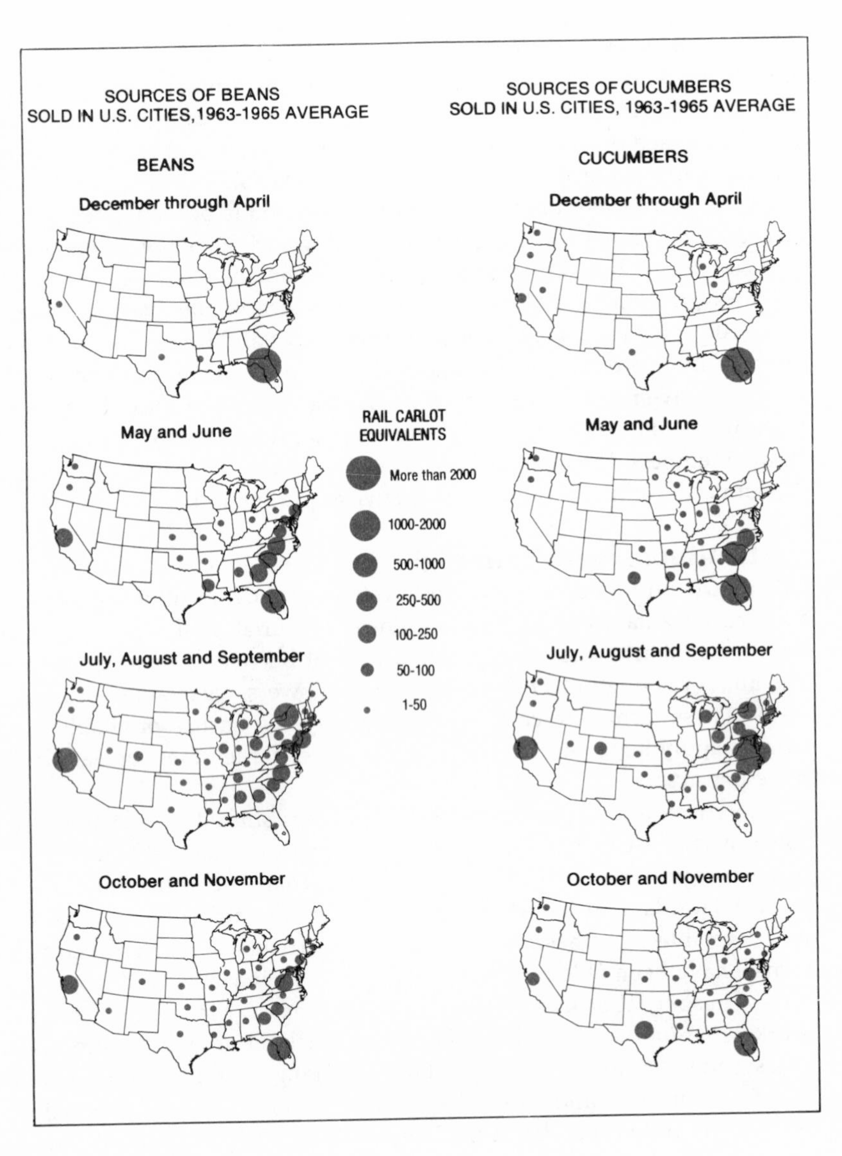

5.14. Horticultural production in the United States: bean and cucumber sources, 1963–1965. These maps show the importance of seasonal differences and the advantages of large-scale, efficient production in subtropical areas. Florida, Texas, and California, which have long growing seasons, are particularly large producers. Volume discounts in transportation help farmers in these areas compete successfully with farmers close to the northeastern market.

Viticulture, the production of wine grapes, a specialty dating back many centuries in Europe, is practiced in areas of favorable climate and soil. These include the German Rhineland, parts of France, Italy, Spain, and Portugal, and comparable zones in California and Chile.

Whether in truck gardens near cities or on specialized farms in more remote locations, intensive horticulture contrasts sharply with traditional, small-scale fruit and vegetable production on countless general farms in Europe and America.

Cotton Production

Cotton is highly productive and very transportable. The distribution of cotton production in the United States reflects both the persistence of past patterns of location and the development of new patterns resulting from environmental and technological change (figure 5.15). Cotton production began as a labor-intensive activity on southern plantations, with principal markets in Europe. Cotton could readily have been cultivated on larger, more mechanized farms farther from urban markets, given adequate moisture and warmth. But through World War I, production remained concentrated in specialized districts of the South. It was only after the boll weevil ravaged crops in the Carolinas and other districts that cotton production shifted westward. Lack of mechanization and innovation in some areas, such as the Mississippi alluvial valley, also contributed to the change. Now farmers in the western United States produce cotton on a large scale, using irrigation. Cotton is also raised under irrigation in central Asia by the Soviet Union, another major producer.

Because the acreage available for cotton far exceeds the amount required to meet demand, the risk of overproduction is great. Government price supports have unintentionally encouraged excessive planting and have increased competition from cheaper foreign cotton for traditional American markets. Acreage restrictions have been imposed to reduce production (see figure 5.15). (The land released from cotton is used for hay,

5.15. Shift in U.S. cotton production, 1840–1970. This map traces the shift in cotton production from the Atlantic coastal states to the Southwest (Mississippi, Louisiana, Arkansas, Texas) and the far Southwest (Arizona and California), reflecting the opening up of new land, the impact of disease, irrigation, and other factors.

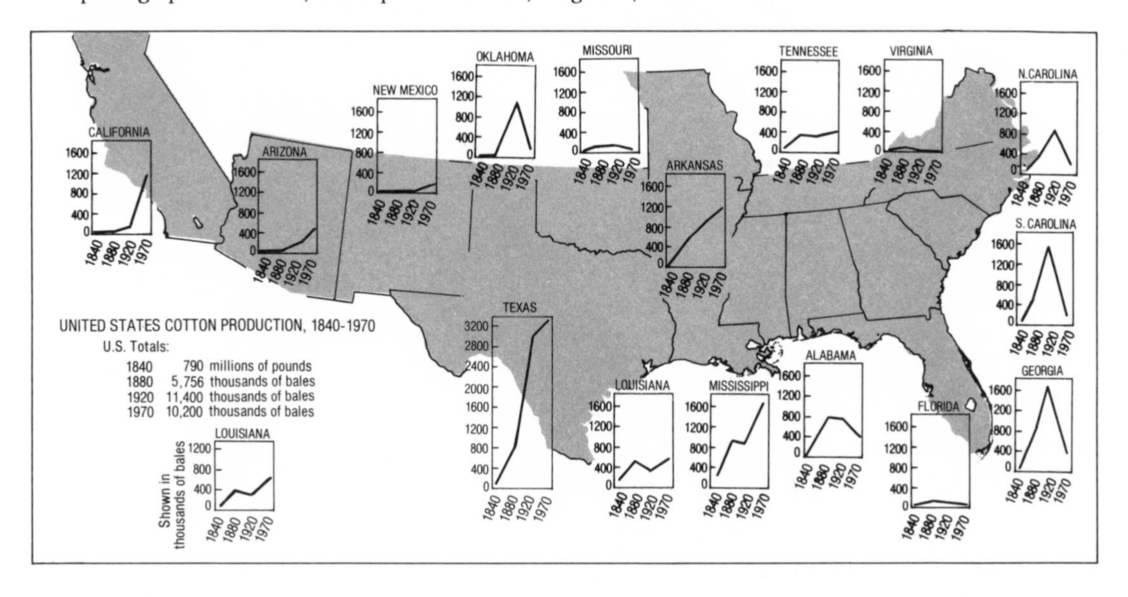

corn, and other grains.) Farmers may be limited to the use of as little as one-fourth of their available acreage for cotton. To get the highest possible yields from the restricted allotment, farmers tend to plant their best fields to cotton. Such fields are often small, dispersed, and unsuitable for the efficient use of large-scale equipment. Thus efficient producers, who would normally apply mass-production techniques to extensive areas, are penalized by this system. Large irrigated cotton farms in Texas, Arizona, and California have been able to avoid dispersion and, because of their greater productivity, have gradually gained a larger share of the market and of crop allotment (see figure 5.15).

Poultry Production

Poultry production in the United States has become a specialized industry, displacing poultry raising as a sideline on the general, dairy, and livestock-grain farm (figure 5.16). A trend toward specialization in poultry has also developed in Europe. Once a Sunday luxury, chicken has become one of the cheaper foods—except when grain shortages force up the price of feed.

Poultry farmers near large urban markets often specialize in egg production. Because space in such areas is limited, the farmers usually import feed from more extensive zones. Egg production is a good example of agricultural response to economies of scale; some egg farms produce as many as a million eggs a day. Figure 5.17 shows the significant variation in egg prices received by farmers in the United States. Prices are lower in regions of surplus production, where supply is greater than demand, and they are higher in regions of deficit (but not necessarily low) production, where supply is lower than demand.

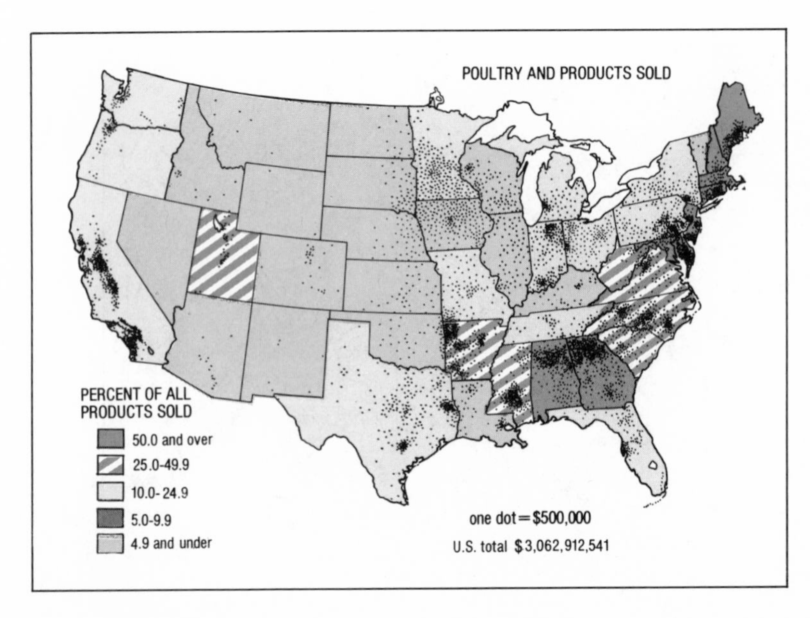

5.16. Poultry production in the United States. Poultry production is both oriented toward very large urban markets, as in the megalopolitan fringes in the Northeast and in California, and concentrated in specialized districts, as in the South.

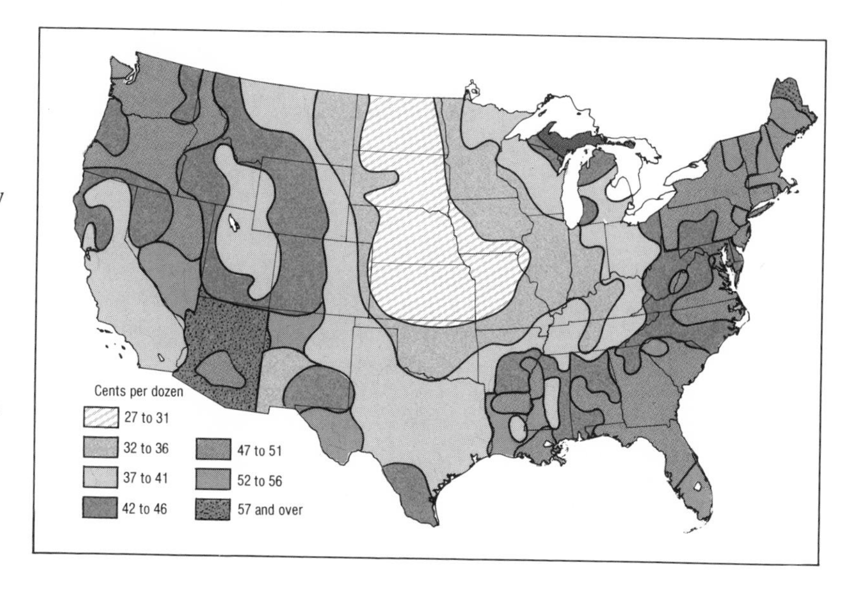

5.17. Variations in average egg prices received by farmers in the United States. This map from a study conducted in the mid-1950s shows that farm prices range from fairly low levels in regions of greatest relative surplus, such as Nebraska and Kansas, to fairly high levels in regions of greatest relative deficit, such as Arizona. Price differentials reflect the cost of transportation from surplus to deficit regions.

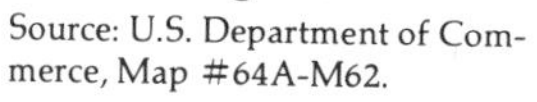
Source: U.S. Department of Commerce, Map #64A-M62.

In some areas, particularly in the South, poultry production has developed on a contract basis on some formerly poor, marginal general farms. Feed companies supply the chickens and most of the feed, and they guarantee to buy the full-grown birds. There is more security for both farmer and distributor under *contract farming,* but prices remain low because of excessive competition. The farmers typically remain poor, while farm production becomes increasingly concentrated in the hands of a small number of skillful managers.

Agriculture in Non-Western High Technology Societies

The agricultural gradient that we have analyzed in both theory and practice describes the organization of specialized farm production in high technology societies in the Western world. It reflects Western cultural preferences for foods such as wheat, dairy products, and beef. The same generalizations cannot always be made about non-Western high technology societies—Japan, for example. The Japanese do not share our food preferences, and agriculture in Japan, as in Europe, is less specialized than in North America. Nine out of ten farms in Japan are general farms. (Eighty percent of Japanese farm families have members in nonfarming jobs; almost two-thirds of these families earn more from outside activities than from farming.) The typical general farm in Japan produces rice, the preferred grain, and poultry, vegetables, or fruits. Specialized fruit and vegetable production is becoming more prevalent, but there is still little demand for dairy products. The rising demand for beef and pork must be met largely by animal or feed-grain imports, owing to the lack of farmland in Japan itself.

The differences between Western and Japanese agriculture are great enough to prevent the strict application of classical agricultural location theory to Japan. But as Japanese agriculture continues to evolve from traditional peasant to commercial farming, we can expect greater specialization and differentiation of crop production. The kind of crop produced may differ from Western agriculture, but the gradient of intensity will be similar. For the reasons given earlier, more intensive forms of agriculture are likely to be practiced closer to market centers such as Tokyo, Osaka, and Kobe, and less intensive crops farther away, as in Hokkaido in the north, or imported from overseas (such as wheat and barley from the United States and Australia).

FORESTRY, FISHING, AND MINING

A brief look at the organization of commercial forestry, fishing, and mining will complete our study of high technology rural or small-town landscapes. These three activities do not figure importantly in the employment statistics of most advanced nations, but they are regionally very significant, especially in a human geographic sense. Forestry and fishing occur over broad areas, and mining is widely dispersed. As resource-exploitation activities, they have a far greater and more visible impact on the physical environment than the number of people involved might suggest.

Commercial Forestry

About five thousand years ago forests covered perhaps one-quarter of the earth's land surface. Nearly one-third of this area has been cleared for agriculture; the rest is generally too remote or rugged for farming. Although forestry alone cannot support dense populations, it can generate enough capital to nurture secondary industries, such as shipbuilding in colonial New England and pulp and paper manufacturing in British Columbia today.

Location and Production

In von Thünen's day (and in many less advanced countries even now), small forests used as a source of firewood occupied a fairly intensive zone near cities and towns. But today in Europe and the United States, commercial forestry occurs at the low-intensity end of the agricultural land gradient. It is comparable to wheat or livestock ranching in the return per acre but occupies areas that are more humid and rugged.

In Europe and the United States, high-quality forests near settlements were the first to be exploited. Lumbering gradually spread out toward the periphery, where it predominates today (figure 5.18). To be profitable, forestry must be able to compete with alternative uses of land, labor, capital, and raw materials. Forestry is spatially extensive (large individual or corporate landholdings and low to moderate net returns per acre). The return is

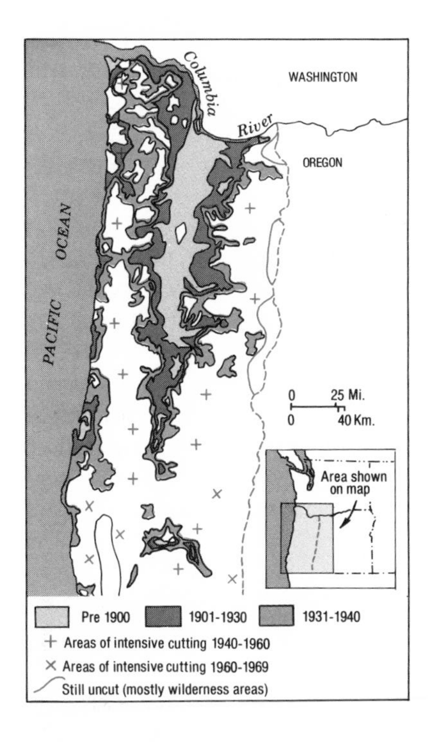

5.18. Diffusion of lumbering in Oregon. The more accessible forests are exploited first. In recent decades, the area of active logging has shifted to southwestern Oregon, continuing the process of diffusion.

generally high enough to be competitive if the forest is fairly close to market and/or the lumber is of high quality and quantity per unit area.

Even though excess material is removed at local sawmills, lumber is so heavy and bulky that long-distance transport is often prohibitively expensive. But logs, like fruits and vegetables, can be processed locally to remove extra bulk making the products—paper, for instance—more transportable. Because of high transport costs, some large but remote forests of high quality are left untapped as long as forests nearer market continue to satisfy demand.

On the other hand, some lower-quality forests close to major markets are bypassed in favor of distant but superior ones, such as those in the Pacific Northwest. The lower production cost of the better lumber overrides the higher transport cost. Production costs for lumber vary with labor costs, accessibility, and quality of stands. Most logs from Pacific Northwest forests are large, and production costs per board foot are low. Thus it is profitable to ship the lumber to eastern U.S. markets.

The old-growth forests of the Pacific Northwest have been almost completely harvested. Twice as much Douglas fir is being cut as is now growing. The harvest has included trees that have taken fifty to one hundred years to mature. Since these forests take long to recover, the center of U.S. commercial forestry may shift from the Northwest to the South, where the warm, humid climate permits new forests of fast-growing species to reach marketable size in twenty years or less. The technology of forestry, however, is continually improving. Scientists have developed new breeds of high-yield, fast-growing, standardized trees, which can be planted like crops. Thus forests in the Pacific Northwest, like those in Scandinavia and Finland, may be able to maintain high yields indefinitely.

Logging Patterns

Logging patterns reflect the cutting methods employed. *Selective cutting*—a method in which only mature trees are cut—is perhaps the most efficient way to maintain a sustained yield of pines and hardwoods because the stands tend to be of trees of mixed age and size. *Clear-cutting,* where entire portions of forest are stripped, is the most efficient way to harvest Douglas fir and hemlock because the stands tend to be of trees of uniform age and size (figure 5.19). But clear-cutting is unsightly and increases erosion. An ecologically sounder method, which allows quick reseeding and causes minimum erosion and watershed damage, is to clear-cut small alternate sections or strips of forest, creating a checkerboard pattern. Most lumbermen,

5.19. Clear-cutting in Washington State. This used to be a stand of Douglas fir. Removal of the soil's protective covering through excessive clear-cutting not only spoils the beauty of forested areas but also hastens the erosion of slopes. (Photo by George H. Harrison from Grant Heilman.)

however, prefer the cheaper technique of clear-cutting large stands. Clear-cutting is technically and ecologically comparable to shifting cultivation. In both cases, the land must rest twenty to forty years before another crop can be harvested.

Competition for Land Use

Commercial forestry must compete with both agriculture and with recreation and wilderness. In North America and Europe, commercial forest acreage is growing as marginal farmlands are abandoned. In Vermont, where this has occurred, moose and mountain lions have returned to the forest since the turn of the century. The demand for lands for recreation, wildlife protection, and scenic regions is removing some timberland from commercial exploitation and creating major land-use conflicts. Recreation may be compatible with forestry, but scenic and wilderness protection is not. Often the conflict is between the local landowner and the need for local jobs versus the open-space demands of environmentalists.

Commercial Fishing

Commercial fishing, like agriculture, forestry, and mining, is a primary, resource-exploiting activity. Fish and shellfish are the chief economic resource now extracted from the three-quarters of the earth's surface covered by oceans, rivers, and lakes.

Classification, Location, and Environment

There are three major kinds of fisheries: *open-ocean* (pelagic), *coastal-bottom* (demersal), and *river or lake freshwater.* Freshwater fisheries are widespread but not highly productive or commercially significant. Exceptions include those in the Volga-Caspian area, where that expensive delicacy—sturgeon caviar—is produced, and the fish farms of China and other parts of South and East Asia, where fish raised in ponds and flooded paddies are an important source of protein for the population's diet.

The world's largest ocean fisheries are located in the North Atlantic and Pacific, off the western coast of South America, and in the seas off Japan, China, and Siberia. These relatively limited fish-producing areas are quite close to the shoreline. They contain cool upwelling or mixing waters rich in the nutrients that form the base of the ocean's food chain. As demonstrated by the temporary decline of Peruvian anchovy fisheries in the 1970s, variability in the behavior of upwelling currents can displace ocean fisheries.

The most productive fisheries are found in the shallow waters of the *continental shelf,* the shallow, submarine plain bordering the continents, notably in areas where freshwater nutrients flowing in from river mouths, combined with upwelling currents, provide an exceptionally rich plankton and marine-flora food base. Commercial coastal-bottom fisheries are moderately important along the coastlines of Europe, Japan, and China, and off the Pacific, Gulf, and Atlantic coasts of North America.

Traditionally fishing was the chief economic activity of hundreds of small coastal and island villages in Europe, Japan, and North America. There were dozens along the indented coast of Norway alone. But the development of larger, mechanized fishing vessels and modern fish-processing plants changed the industry. Now many fishing fleets are concentrated in larger port cities, such as Bergen and Stavanger in Norway.

Depletion and Conservation

Fish are a renewable resource, but they can be only partially replenished by human effort. Often human activities deplete the resource. For example, chemical pollution, especially in coastal waters, has imperiled many species that tend to absorb and concentrate poisonous substances. A more common problem is the threat of depletion through overfishing. Declining runs of Pacific salmon have prompted government agencies to regulate the North Pacific fishery. Atlantic salmon were overfished long ago. Because the seas are erroneously considered a common free resource by maritime nations, competing fleets try to maximize their own share of the total catch beyond coastal boundaries. Coastal-zone boundaries have expanded from a traditional three-mile limit to a more common twelve-mile limit. The two-hundred-mile limit now claimed by the United States and other nations may become the norm in the future. International regulation is essential if the world's fisheries are not to be destroyed (some species of whales are already nearly extinct.)

Mining

Mining differs from the other primary activities in that it entails the extraction of finite, exhaustible mineral resources. Mining operations supply not only most of the energy needed to power industry and transportation but also raw materials for industry and construction, as well as for synthetic fertilizer for agriculture. The minerals important to modern society include coal, petroleum, natural gas, iron, bauxite, copper, and, in smaller quantities, uranium, tungsten, diamonds, and manganese, to name only a few.

Distribution and Location

Minerals valuable enough to be exploited are widely but unevenly distributed. Hence the mining industry displays a pattern of *sporadic concentration.* Any country, particularly a small one, may have a large deposit of one kind of mineral, but it is not likely to be self-sufficient in a wide range of minerals. This uneven distribution is the basis for a major portion of international trade on the one hand, and international and private-public conflict over control of resources on the other.

The location of most mineral deposits is not obvious. As the more easily discovered resources become depleted, an ever larger investment must be channeled into exploration and recovery. Increasing scientific knowledge of the physical processes involved in mineral formation helps to predict where new deposits may be found, but only a small fraction proves to be commercially exploitable. Fortunately new sources of nearly depleted minerals crop up from time to time, such as the large petroleum field in the

North Sea between Scotland and Norway. There is no way of knowing whether such luck will continue.

Once discovered, the resources that can be delivered to market at the lowest price are utilized first—either leaner deposits close to markets or very rich ones farther away. The industrialized nations, the major consumers of the world's mineral resources, at first depended on known local reserves, many of which are now exhausted. Today the most actively exploited reserves are found in the less industrialized nations. But the recent rise in the price of many minerals has made it profitable for mining companies to rework old deposits with new mining technology and has encouraged recycling efforts as well.

Mining and the Landscape

The impact of mining on the landscape has far exceeded the small proportion of people involved. Just since World War II, *strip mining* (a form of surface mining in which minerals are uncovered by removing top layers of soil and rock) has destroyed some agricultural lands in the Midwest and increased the rate of erosion in large portions of the Appalachian and Rocky mountains. In some states, such as West Virginia and Montana, mining firms are compelled by law to restore damaged terrain. A visible landscape feature in many mining areas is the pollution plume emitted from power plant smokestacks.

Mining and Human Settlement

The pattern of human settlement chiefly reflects the distribution of farmland rather than mineral resources. But when coal dominated the industrial revolution in the nineteenth century, population shifted in large numbers from farming areas to coalfields. Mining also induced settlement and urbanization in the mountain and desert regions of the United States, as well as the tundra and muskeg (bog) of Canada, Alaska, and Siberia. More recently, the development of petroleum reserves has contributed to settlement and growth in the southwestern United States.

Because mineral resources are widely dispersed and often remote, their exploitation may lead to the development of small rural settlements. These typically form a linear or clustered pattern, depending on the geology of the mineral (linear along a coal seam, clustered around a gold mine). Mining seldom provides a sufficient economic base for the growth of a settlement into a city. Frequently the depletion of the mineral causes the settlement to decline and disappear. Now that mining and transport technology has reduced the need for human labor, large-scale urbanization is unlikely to develop in harsh, remote regions such as Alaska's petroleum-rich North Slope.

CONCLUSION

Advanced societies in the Western world generally organize agriculture and other primary activities in an economically rational manner. Farmers who want to make the most profitable use of their land will consider such factors as distance from market, transport costs, the value of potential crops, environmental conditions, and consumer demand when deciding what crop to raise. Of course, many farmers are satisfied with less than maximum profits. Others may lack the expertise to make the most rational decision, or their farms may be too small to be operated efficiently. If agricultural location decisions fall too wide of the mark, the farm will not survive. The validity of agricultural location theory is supported by the existence of an agricultural gradient in the real world—modified though it may be by environmental variation, government policies, technological developments, and noneconomic human behavior.

The location of forestry, fishing, and mining activities is linked more closely to physical environmental factors than to spatial ones. But distance and transport costs still influence the pattern of forest and mineral exploitation (resources closest to population centers are usually tapped first). Even fishing is affected by spatial factors, evidenced by the concentration of fleets and fish-processing plants in fewer, larger ports.

In the next chapter we begin our study of the urban landscape. We will return once again to the gradient theory of land use when we discuss the organization of activities in the city.

SUGGESTED READINGS

Chisholm, Michael. *Rural Settlement and Land Use.* London: Hutchinson University Library, 1962.

Dunn, E. S. *Location of Agricultural Production.* Coral Gables, Fla.: University of Florida Press, 1954.

Found, W. C. *A Theoretical Approach to Rural Land-use Patterns.* Toronto: Arnold, 1971.

Gregor, H. F. *Geography of Agriculture.* Englewood Cliffs, N.J.: Prentice-Hall, 1970.

Grigg, D. *The Harsh Lands: A Study in Agricultural Development.* New York: Macmillan, 1970.

Hall, P., ed. *Von Thünen's Isolated State.* New York: Oxford University Press, 1966.

Tarrant, John. *Agricultural Geography.* New York: Wiley, 1974.

Thrower, Norman. *Original Survey and Land Subdivision.* Chicago: Rand McNally, 1966.

6

Urban Settlement

6 Urban Settlement

During the past 150 years, the pattern of human settlement in many areas has changed dramatically. Many people used to live in countless villages and farmsteads. They now live in cities. Formerly dispersed settlement patterns have become concentrated. Consider the changes that have taken place in the United States. In 1800, we were a rural nation. Only 5 percent of the population lived in the small cities that existed then. Now 70 percent of the U.S. population is urban, and more than 90 percent is employed in nonfarming jobs. Many people are crowded into a few, enormous metropolitan centers. Greater New York City contains nearly 17 million persons—an average of more than 21,000 persons per square mile (figure 6.1). Some European countries are still more urbanized: Great Britain's population is 78 percent urban, and Belgium's is 87 percent urban. Even in less developed countries, as much as half the population may live in cities.

The concentration of population in cities is made possible by the production of agricultural surpluses by those who do farm. A subsistence farming economy can support few cities and towns. As agriculture becomes more productive, larger places evolve. Of course, agriculture has also depended on the city. The demand of city dwellers for food, the technology developed by inventors who lived in the city, and the protection afforded by the city first enabled farmers to produce and market a surplus.

Urban civilization first arose in river valleys, where adequate water supply and the technology to use it permitted surplus crop production. Thus part of the population was freed for other, more specialized activities, such as trade, handicrafts, government, and art. Nonfarming people bartered their products and services for a portion of the surplus food. Specialized nonfarm workers found it more efficient to live in larger settlements near other people and activities. Thus began the growth of cities and the development of the urban system. The term *urban place*, or *city*, will refer here to an entire built-up agglomeration, not just the largest incorporated place. It includes the area of urban development sprawling beyond the legally defined boundaries of the central city. For example, New York City as a legal entity—the central city—has about 7.5 million persons, while the physical agglomeration contains over 17 million. In the United States, the term *urbanized area* best measures this idea. The term *metropolitan area* refers not only to the agglomeration but to the rural–small town areas around it, from which many people commute to jobs in the city.

URBANIZATION

The urban system defines the spatial organization of high technology societies such as the United States. Urban activities, particularly manufacturing and services, support and dominate advanced economies. Their eco-

6.1. New York City. This view of New York's financial district, dominated by the World Trade Center's twin towers, suggests the great concentration of population and activities in metropolitan centers today. The population of New York's metropolitan area exceeds the population of 75 percent of the world's nations. (Photo by Bob Takis from American Airlines.)

nomic value far surpasses that of the rural activities described in chapter 5. And even in primarily rural, low technology societies, the few urban places have an important role.

From a global perspective, cities are but tiny points scattered across vast continents. The amount of space they consume is relatively small. Cities cover only 1.5 to 2 percent of the total land surface in the United States. Nonetheless cities, linked by a web of roads and railroads, form intense concentrations of activity that can be measured by heat, light, and pollution and can easily be seen in satellite photographs (figure 6.2).

At close range, many cities are enormous. The population of some cities, like New York and Tokyo, exceeds that of most countries. These concentrated centers of people and activities are spreading into the rural fringe, transforming farmland into areas of high-intensity use. To accommodate growing urban populations, wetlands are drained, shorelines filled,

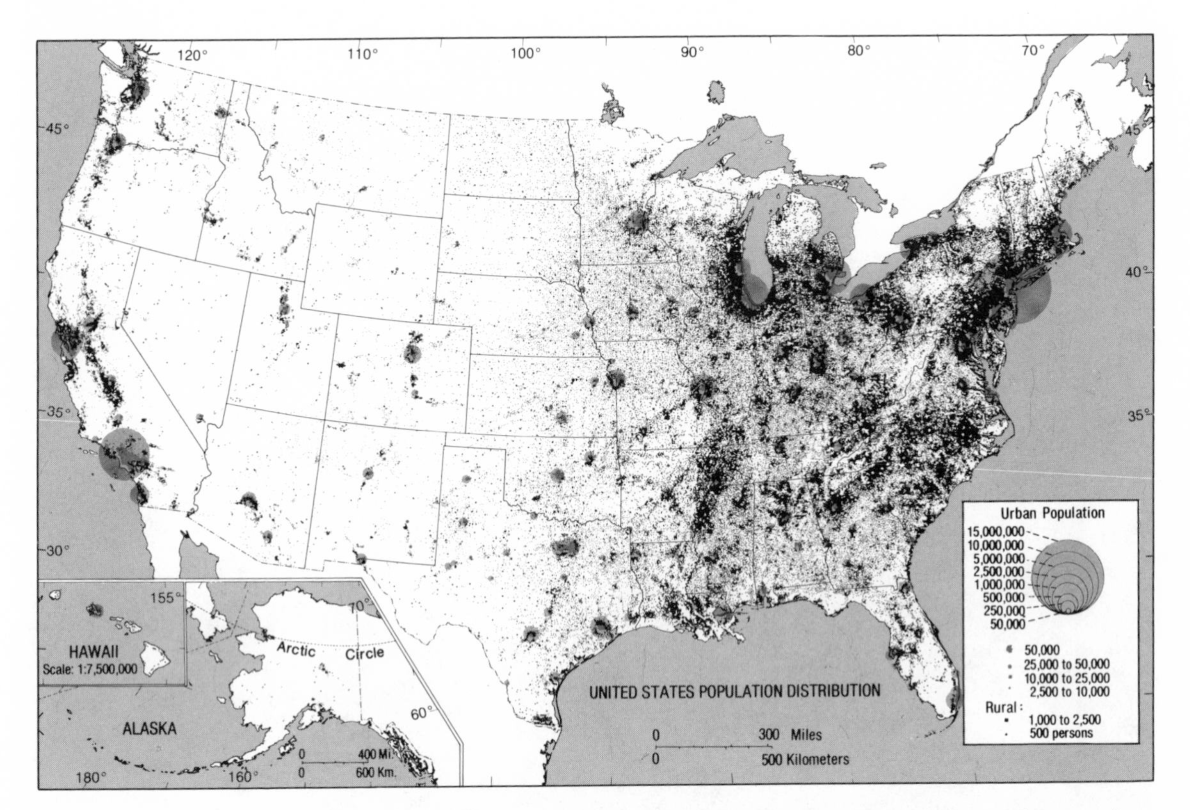

6.2. Distribution of population, United States (above). City lights in this remarkable nighttime satellite photo of the United States (opposite) reflect the distribution of population shown in the map. (Photo from *Aviation Week and Space Technology*, May 1974.)

hills leveled, and mountain slopes terraced. While the largest cities are exciting, innovative, and productive places in which to live, they are also beset with serious problems of pollution and human organization.

History

The first well-organized cities probably appeared about fifty-five hundred years ago in the Tigris-Euphrates Valley of Mesopotamia, the Cradle of Civilization. Other early cities emerged in the river valleys of Palestine, Egypt, India, and China, where agriculture was most productive (figure 6.3). Favored sites also included hilltops and other protective landforms for defense, and river junctions and fordable or bridgeable river sites for ease of transportation.

Cities have always been centers of political and religious control, economic exchange (markets), and production. In most ancient and medieval cities, the political and religious roles were especially important; palaces and temples dominated the city. Markets were local, and production (handicrafts) was limited. A granary within the walls indicated the importance of

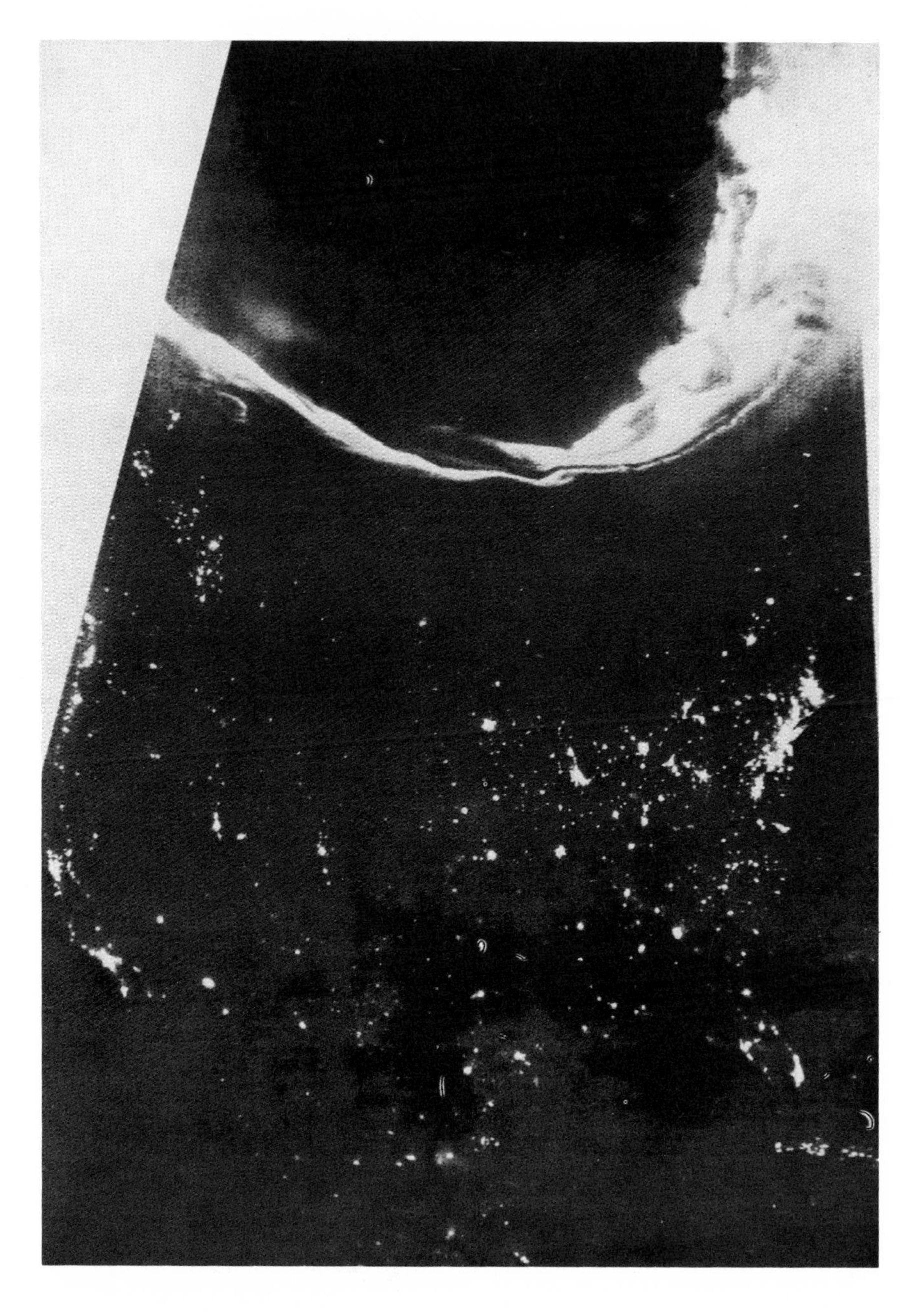

maintaining a farm surplus. Today, in contrast, while political and cultural functions remain significant, economic exchange and production are paramount.

The landscape in areas of early intensive agriculture was typically divided into many small states, each dominated by a small city. Local rulers protected the peasants, maintained the irrigation systems, and extracted a

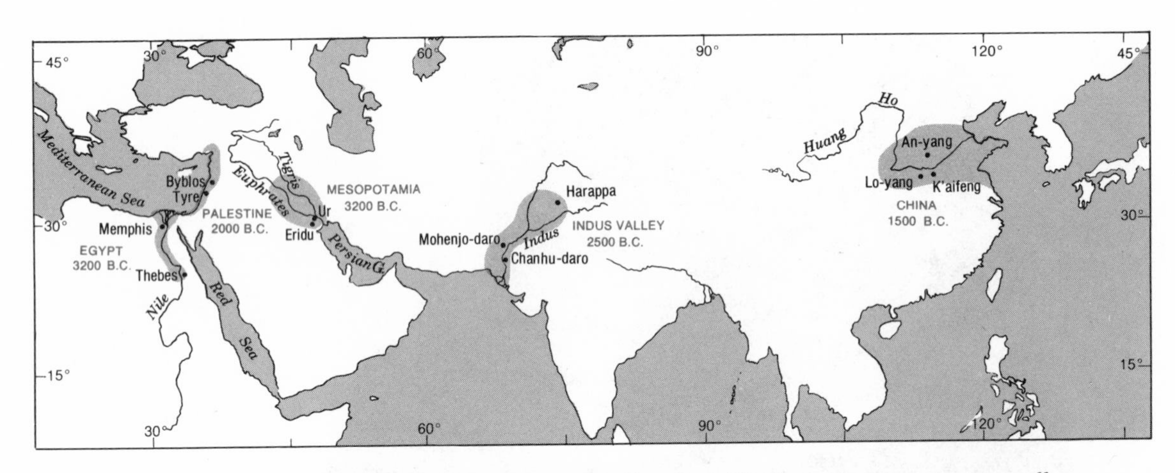

6.3. The earliest cities and civilizations. The first cities arose in river valleys, where enough surplus food could be produced to support nonfarming people. It is still not known whether the innovation of cities diffused from a single point of origin, perhaps Mesopotamia, or arose separately in most of these regions.

surplus from the harvest that supported priests, artisans, merchants, and soldiers—a division of labor that fostered the rise of civilization. At times, the more aggressive and successful cities conquered their neighbors and created empires. As long as the cities could control the surplus of a wide area, a more elaborate civilization was possible.

Most ancient and medieval cities were small in area and population. Ur, Babylon, and Jerusalem each contained fewer than 10,000 persons. Only a few cities were able to control food and material supply lines of sufficient size to support large populations. Imperial Rome may have had a million inhabitants. The supply lines were vitally important; when they were cut, the cities could no longer survive. Even Rome dwindled to a small town after barbarians destroyed its supporting network of roads and trade routes.

After the fall of the Roman Empire in A.D. 476, exchange and commerce declined. People abandoned the stagnating cities and returned to farming. For several hundred years, few European cities supported more than three thousand or four thousand persons. Similarly in India, China, Africa, and elsewhere, cities periodically flourished, declined, and grew again, depending on the size, strength, and stability of the empires. European cities were at a low ebb from about the eighth to the tenth centuries, but Peking and Hangchow (China) are estimated to have had about a million inhabitants at this time.

While most cities of the former Roman Empire were declining, Constantinople (modern Istanbul) maintained its importance as the capital of the Byzantine (Eastern) Empire. Increased trade and urbanization accompanied the spread of Islamic civilization across the Middle East and southern Mediterranean. Between 650 and 1050, many cities thrived as centers of commerce, handicraft production, and science as well as administration and religion. These included not only the holy cities of Mecca, Medina, and Jerusalem but also Damascus, Antioch, Baghdad (the world's largest city in the tenth century, with perhaps a million inhabitants), Cairo, Toledo and Cordoba in Spain, and many others.

The rise of commercial and industrial cities, as contrasted to primarily political and religious ones, began along the coast of Europe during the eleventh and twelfth centuries, when international commerce and long-distance transportation were reestablished. Bruges, Lübeck, Genoa, and Venice were some of the more important cities. Capital was also invested in local communication networks and internal manufacturing. This activity stimulated the rise of both market towns that provided services and handled exchange, collection, and distribution for farmers, and small industrial settlements that processed wheat, timber, iron, and other raw materials into simple demanded products such as flour, lumber, and tools.

The gradual emergence of nation-states (twelfth to sixteenth centuries), the creativity unleashed by the Renaissance and Reformation (fourteenth to sixteenth centuries), and the opportunities opened by exploration and colonization (fourteenth to seventeenth centuries) stimulated trade, manufacturing, and the revival of cities. It is either a tribute to the wise location decisions of the Romans (and of local people before or beyond Roman occupation) or to the forces of inertia, or both, that many of the emerging cities developed from medieval towns that had succeeded Roman or pre-Roman settlements. The cities, as communities of free citizens, were vital to the evolution of democratic principles, the rise and success of capitalism, and the emancipation of the serfs.

Modern Urbanization and Urban Character

The transition to a society dominated by cities began with the industrial revolution in Great Britain, Belgium, and Holland in the mid-eighteenth century. Machines powered by inanimate sources of energy raised the productivity of both agriculture and manufacturing. As farming became increasingly mechanized, millions of unemployed farmers and peasants poured into city factories and mills, despite the long hours, crowded conditions, filth, noise, and low wages. Manufacturers and traders took advantage of this growing supply of cheap labor. In addition, they were usually able to keep the price of agricultural products low and of manufactured products high. The cheap labor and favorable price structure contributed to the rapid growth of cities and of urban wealth.

Service and manufacturing activities began to reinforce each other and worked together to generate further growth. Before the industrial revolution, for example, most towns in Great Britain were small. But after 1750, formerly small towns like Birmingham and Glasgow, which were located near coal and iron deposits, developed into large cities (see figure 3.4). Many medieval cities, like London, Paris, and Milan, also grew rapidly and became centers of commerce, government, and finance as well as manufacturing. Other places important in medieval times, such as the French towns of Troyes and Lagny (famous for their annual fairs), did not adapt, were not served by the new railways, and remained stagnant. As cities evolved, their character changed. The community spirit, which stressed cooperation, was gradually supplanted by a greater emphasis on private property and individual responsibility.

For the next two centuries in Europe and North America, the growth of cities, the strengthening of the middle classes, and the unprecedented expansion of wealth were fueled by the transfer of labor from the countryside or continuing improvement in productivity and the unending innovation of new products and services.

The urbanization process was harsh. Millions of people were uprooted from the often wretched rural life to which they had become accustomed and were subjected to an urban poverty that was not only wretched but unfamiliar. The urban-industrial landscape of mid-nineteenth-century England was vividly described by Charles Dickens:

> The water had become thicker and dirtier . . . the paths of coal-ash and huts of staring brick, marked the vicinity of some great manufacturing town. . . . Now, the clustered roofs, and piles of buildings, trembling with the working of engines, and dimly resounding with their shrieks and throbbings; the tall chimneys vomiting forth a black vapour, which hung in a dense ill-favoured cloud above the housetops. . . . A long suburb of red-brick houses,—some with patches of garden-ground, where coal-dust and factory smoke darkened the shrinking leaves . . . and where the struggling vegetation sickened and sank under the hot breath of kiln and furnace . . . they came . . . upon a cheerless region, where not a blade of grass was seen to grow, . . . where nothing green could live. . . . On every side . . . tall chimneys crowding on each other, and presenting that endless repetition of the same, dull, ugly, form . . . poured out their plague of smoke. . . . Dismantled houses here and there appeared, tottering to the earth, propped up by fragments of others that had fallen down, unroofed, windowless, blackened, desolate, but yet inhabited.*

Many cities in Dickens's time were dirty, dreary, and unhealthy. But over the generations, urbanization has raised the economic level of city dweller and farmer alike. It has also led to the mixing of different cultural groups and to dramatic social change. During the twentieth century, urbanization spread to nearly all countries, often outpacing the growth of an industrial employment base. By 1970, many of the world's largest cities were in developing countries. Figure 6.4 shows the different levels of urbanization around the world. The highest levels of urbanization tend to be found in regions that had an early start and have an industrial economy—Europe, North America, Japan, and Australia. In many countries in the socialist bloc, levels of urbanization (and services) lag behind industrial growth,

* Charles Dickens, *The Old Curiosity Shop* (New York: Charles Scribner's Sons, 1939), 2:59, 73.

while in many developing countries, particularly in Latin America, urbanization races ahead of industry.

The technologies of construction and transportation have permitted tremendous concentrations of people. As many as eleven cities have over 5 million inhabitants. Even larger populations reside in several *megalopolises,* or sets of cities. "Bos-Wash"—the original megalopolis extending from Boston to Washington—contains 40 million people; "Chi-Pitts" (Chicago to Pittsburgh), 20 million; the British Midlands, 10 million; the greater Ruhr in Germany, 15 million; the Randstad (Amsterdam-Hague-Rotterdam) in the Netherlands, 8 million; Kobe-Osaka in Japan, 10 million; and greater Tokyo, 25 million.

What is the basic force behind the urbanization of society? Agglomeration—the bringing together of people and activities in cities and towns—and the opportunities thus provided. Cities are the most obvious result of society's desire to minimize the distance separating human activities. Society discovered that it was easier and cheaper to minimize the *distance* and *time* separating leaders, artisans, craftsmen, and producers in cities and to bring the needed food and resources together. People did not necessarily prefer living in cities, but it gave them an opportunity to rise socially and economically. In the cities, new ideas could be tried, labor and customers could be found, businesses were most likely to succeed, and fortunes could be made.

As centers of cultural, economic, and social control, cities are vehicles for the preservation and transmission of society's values and traditions. But as temporary homes for travelers and transients and as havens for nonconformists—social reformers, critics, idealists—cities are also a breeding ground for new life-styles, new ideas, and social change.

Probably one-third of the world's city dwellers were not born in cities. Often these newer migrants find it difficult to adjust to urban life, although they are still attracted by its opportunities. In the more developed countries, however, generations have been born in the cities and have stayed there, creating a different, urban culture.

Despite the greater freedom of urban life, cities are not totally free from external constraints. Their wealth is built on specialization and exchange—exchange not only with other cities, towns, and countries but with farming areas, on whose ability to produce a food surplus the city ultimately depends.

URBAN SETTLEMENT

Urbanization is a complex process. Urban settlement begins in especially favorable areas and gradually diffuses outward from early points of origin. The development of the urban system depends on investments in transportation, storage, production, and exchange. The theory of urban location reflects this complexity. For now, we will simply sketch general patterns of urban settlement, leaving the more technical explanation for later chapters.

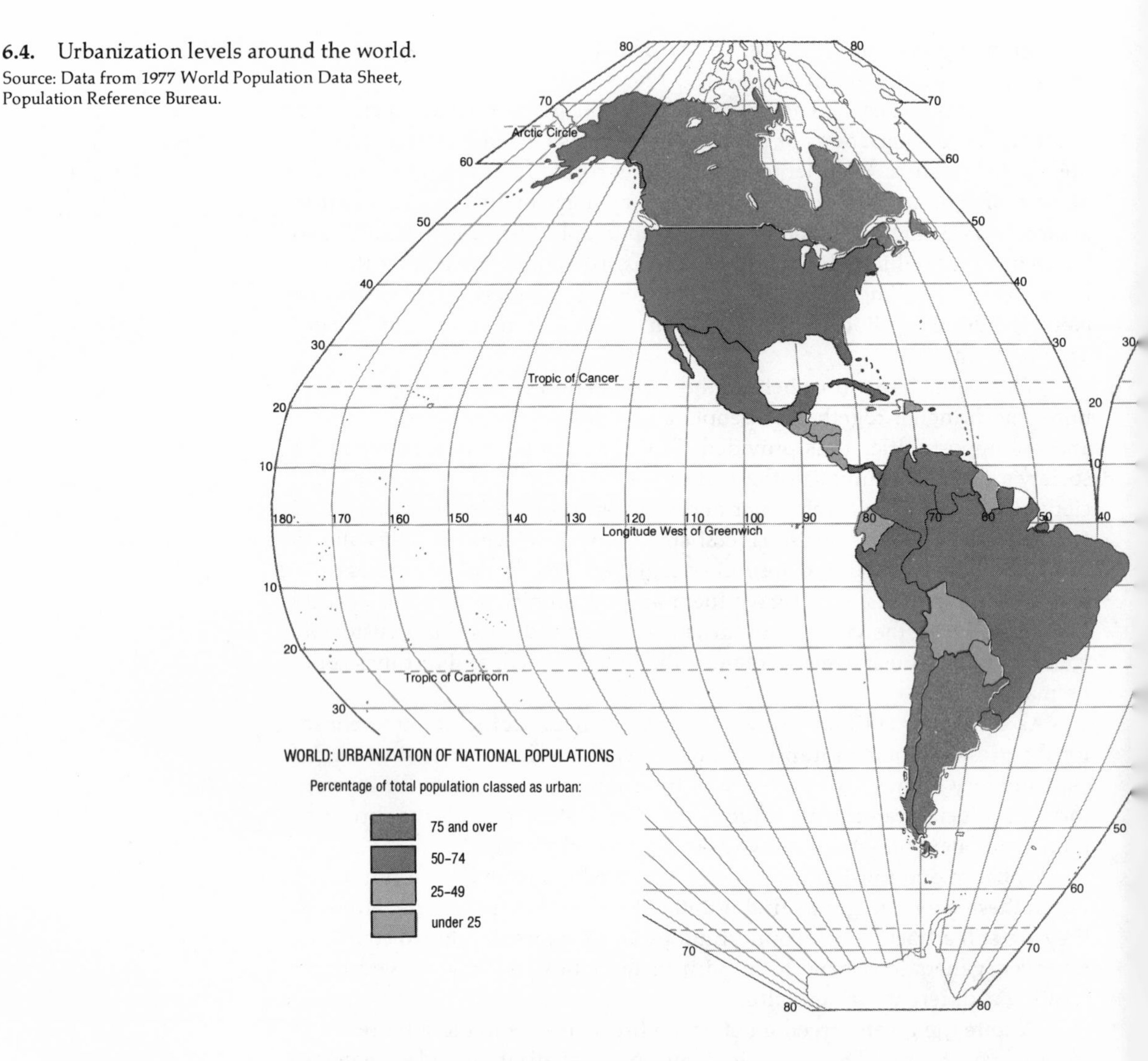

6.4. Urbanization levels around the world.
Source: Data from 1977 World Population Data Sheet, Population Reference Bureau.

Urbanization is a global experience. The same geographic principles explain the development of the urban system in all technologically advanced societies and, to a lesser extent, in less advanced societies too.

The Structure

The urban landscape is composed of places supported by central place activities (exchange, control, services) and processing, or industrial, activities. In agricultural areas, market towns have arisen to serve farmers. These towns are *central places* for the collection and distribution of farm produce and for the provision of goods and services needed by the rural population. Other kinds of central places include administrative capitals and religious centers (figure 6.5), which serve wider regions, and ports. In areas rich in mineral resources, particularly coal and iron, towns have developed as *proc-*

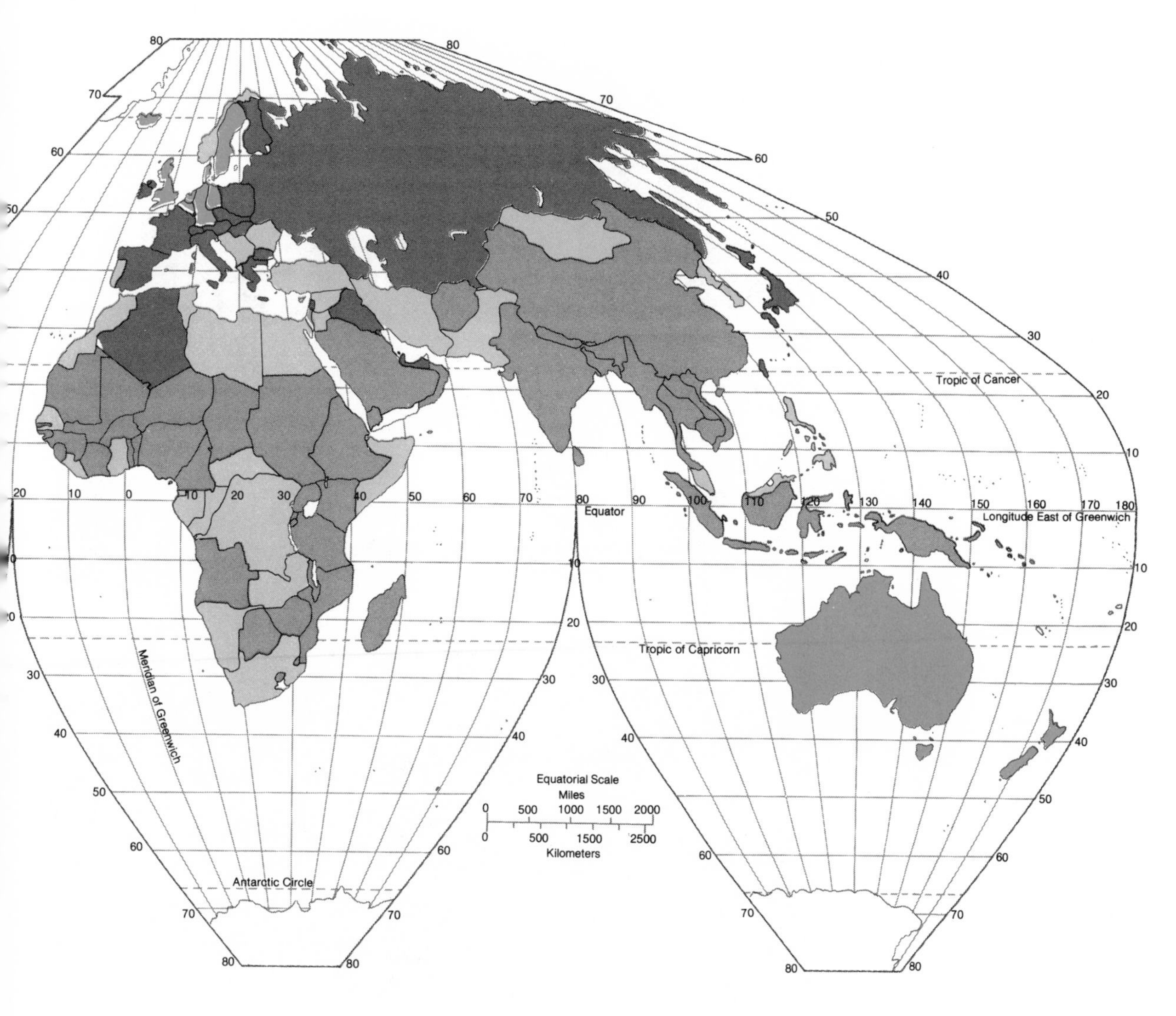

essing centers in which raw materials are assembled and made into semi-finished or finished goods.

Urban settlement often forms a fairly regular network of central places in agricultural areas that offer few nonfarming opportunities such as mining or manufacturing. Well-known examples include the central place pattern of Iowa, or of Bavaria, Germany, and the north China plain. Typically many small villages with limited services, such as schools, gas stations, and grocery stores, develop to serve the nearby rural population. Fewer but larger towns offering more specialized services, such as repair shops and clothing and appliance stores, develop to serve a larger trade area. And finally a major city with still more highly specialized services, such as wholesaling and legal and financial services, arises to serve the entire region.

Figure 6.6 shows the urban pattern that has emerged around Colum-

6.5. The holy city of Benares, India. About a million Hindus a year visit Benares to bathe in the sacred River Ganges and to worship in the city's fifteen hundred temples. Many Indians come there to die and be cremated, believing that their souls will be freed from the cycle of reincarnation (birth and rebirth). In addition to its role as a religious center, Benares is an important commercial and industrial city. Its modern name is Varanasi. (Photo from Sovfoto/Eastfoto.)

bus, in south central Ohio. Note the hierarchy of service centers—villages, towns, and cities—and the network pattern. In contrast, mineral-rich areas have a clustered and often linear pattern of mining centers and industrial towns. The linear string of settlements along the river north of Dayton is characteristic of industrial location.

Industrial towns must have some services, of course, and most large service centers also have industries. In regions where services and industries are about equally important, a complex settlement pattern often develops. This is because several different forces—historical, environmental, central place, industrial, and entrepreneurial—have influenced the growth of towns.

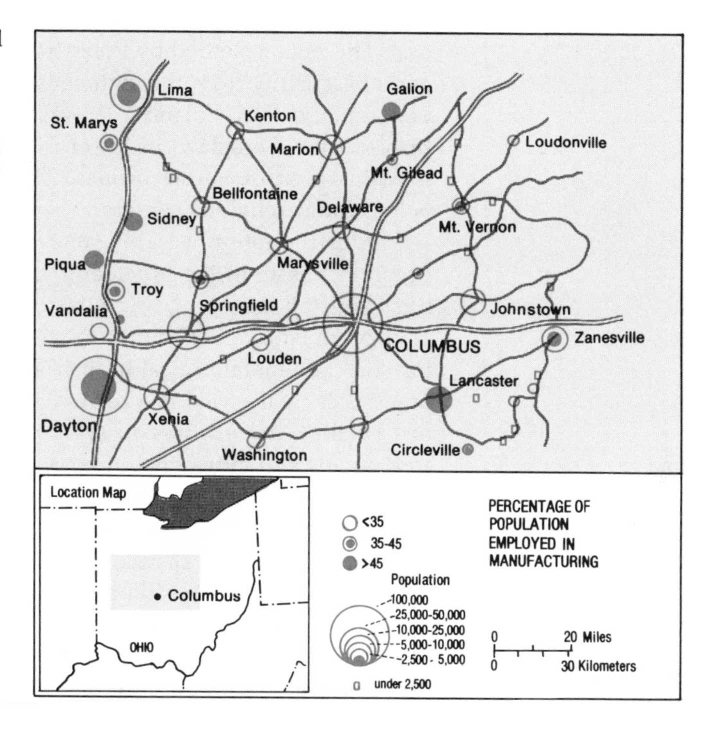

6.6. The urban pattern of south central Ohio. This local urban network is composed of central place patterns (notably around Columbus) and an industrial pattern (the string of urban places north of Dayton). Places in which less than 35 percent of the population is employed in manufacturing are primarily service centers.

The urban structure of Silesia (Slask) in Poland demonstrates this complexity. Silesia is located in the basin of the Oder River in extreme southeast Poland (figure 6.7). The lower valley near the German border is good for farming, and much of the upper valley is rich in minerals, especially

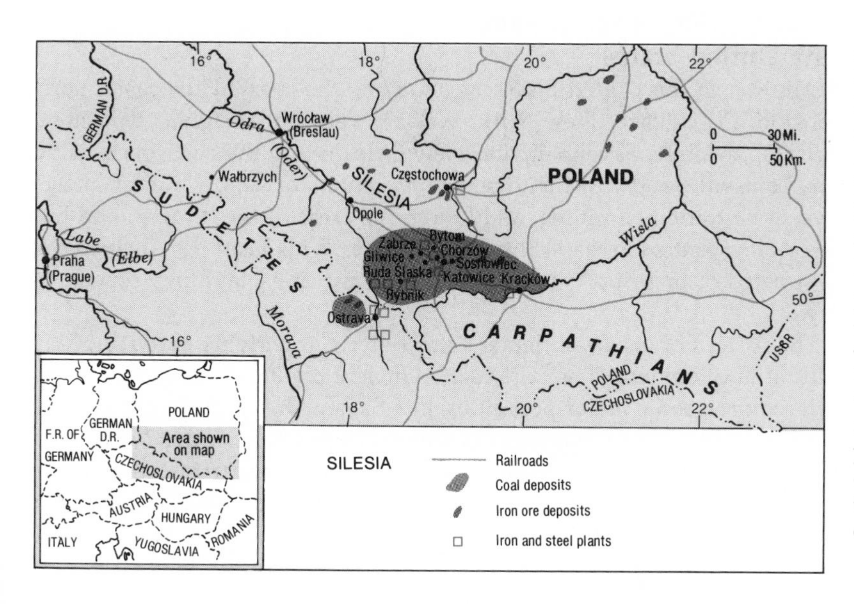

6.7. The urban pattern of Silesia, Poland. Before Silesia became industrialized in the nineteenth century, the region was dominated by three central places: Breslau (Wroclaw), Opole, and Krackow. In the nineteenth century, several industrial cities emerged: Katowice, Gliwice, Rybnik, Zabrze, Ruda Slaska, and Sosnowiec. Some of these cities, notably Katowice, also functioned as central places. Nonetheless the traditional cities have maintained their importance as central places.

coal. The region, settled by Poles from the fifth to the twelfth centuries, was part of Germany from the eighteenth century to World War II (a small portion remained in Poland). It was dominated by three medieval cities: Breslau (now Wroclaw) in lower Silesia; Opole, in upper Silesia; and Krakow, just east of Silesia proper. These were central places—markets and political and religious centers.

Industrialization, notably steel production based on local coal and iron, began in Silesia in the nineteenth century. Access to raw materials and to river and rail transport was vital to the region's development as a major steel-producing center. (Since World War II, Silesia has been Poland's leading steel and metallurgical industrial center.) The localization of coal in the upper Oder Valley led to a relative shift of population from the agricultural lower valley around Breslau (Wroclaw). There arose a cluster of coal and steel towns and cities, including Katowice, Gliwice, Rybnik, Zabrze, Ruda Slaska, and Sosnowiec. These industrial towns gradually acquired service functions, and Katowice became a fairly important central place. Yet even today the trade and service sectors of this *conurbation* (cluster of cities), which contains over 2 million persons, are less developed than those of the historic central places—Wroclaw, Opole, and Krakow. This is partly because there is no single dominant central place and little tradition of a service economy in the conurbation.

The Process

The structure of any actual urban landscape did not appear all at once but evolved over a long time. Many factors influence the location and development of central places and industrial centers. We will follow this process first in the United States and then in a less developed country, Nigeria.

Service Center Development in the United States

Colonization in the United States began along the coast. The coastal ports of Boston, Providence, New York, New Haven, Philadelphia, Baltimore, Charleston, Mobile, Savannah, and New Orleans and the river ports of St. Louis, Louisville, and Cincinnati served as centers of supply and exchange between the trapping, mining, and lumbering areas of the frontier, and Europe or older settled areas in the United States. This head start helped the best situated of these ports to become important cities, and they remain so today.

The pattern of urban settlement also reflects underlying patterns of agricultural development. The fairly self-sufficient plantation economy of the South required some major port cities, like Savannah, Charleston, and New Orleans, but hindered the growth of a network of smaller inland cities and towns. A pattern of small, closely spaced hamlets similar to the compact settlements of New England accompanied the spread of farm development west of the Appalachians. The fine, almost geometric network of small central places in many areas settled in the nineteenth century resulted in part from the Range and Township Survey System. Land was divided into townships measuring exactly six miles square. Each township was divided into thirty-six sections of one square mile, with (theoretically) one hamlet to each township. The hamlets provided the farm population with essential services, including locally run schools.

The Range and Township Survey System was adopted by the Congress of 1796, which determined the national land policy for the settlement of the U.S. frontier. Congress felt that such a compact, orderly system would be more conducive to well-organized local government than the system that prevailed in the South. There individuals purchased a warrant for a given number of acres, which they could locate anywhere in the public domain. This method led to both large and small, dispersed, loosely run units typical of plantation settlement. The Range and Township System was simpler, cheaper, and less subject to dispute than the more individualistic system of the South.

Thus many hamlets in the American frontier were deliberately planned. Some became villages offering a wider range of services. And some villages became market towns or county seats, or both, by virtue of their favorable location or the aggressiveness of local entrepreneurs.

Many larger towns were arbitrarily located by railroad executives to serve as places for the speculative sale of railroad land or as railheads, terminals, or way stations for the nation's expanding rail system. Huntington, West Virginia, for example, was founded in 1871 as a western terminal for the Chesapeake and Ohio Railroad. Collis P. Huntington, president of the railroad, selected a site on the Ohio River near the border of West Virginia, Kentucky, and Ohio and close to southern West Virginia coalfields (figure 6.8). At first the town survived primarily on the strength of its monopoly

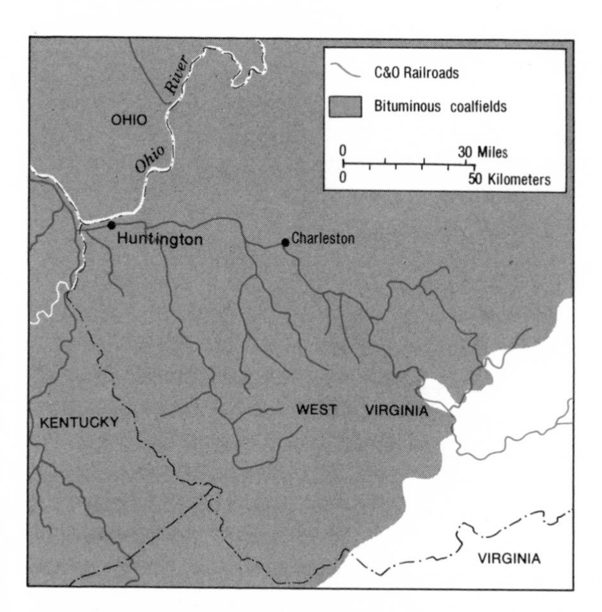

6.8. The location of Huntington, West Virginia. This planned terminal for the Chesapeake and Ohio Railroad became an important transshipment point (from water to rail and vice versa) and inland port owing to its favorable location on the Ohio River.

over all car repairs for C & O trains. But later it became the nation's largest port for inland vessels and a major transshipment point (barge to rail and vice versa) for coal and other materials. Huntington's location near the Ohio River and the West Virginia coalfields enabled it to develop beyond its simple function as a railroad town. Other town sites were determined by state agents who defined county boundaries and established county seats—typically in the center of the county.

Such planned settlement has proved surprisingly rational and efficient. Only the modern era of relatively cheap and rapid automobile travel has shaken the central place structure that accompanied American frontier settlement. Many hamlets and small towns off the highway system have become obsolete, especially after school consolidation, while larger towns and cities linked by the system have flourished.

Further up the central place hierarchy, many U.S. cities are also to some extent creatures of the railroad. Important cities often developed in places where railoads crossed natural barriers and provided access to the interior of the country. Atlanta, Georgia, for example, was founded as a railroad town in 1837. Located at a gap through the Blue Ridge Mountains, it became an important crossroads between the South Atlantic states and the Gulf South. Cities located at railroad junctions and terminal points also prospered (see figure 10.10). Even the success of the old prerailroad port cities was influenced by later railroad development. Those with good access to the interior of the mid-nineteenth-century United States benefited from the railroad. New York, Philadelphia, and Baltimore were able to steal the commercial advantage from such port cities as Charleston and New Orleans, which were comparatively remote from the agricultural Midwest and the industrial Northeast. (These cities are now growing rapidly again, as the South industrializes.)

Industrial Center Development in the United States

Early U.S. industrial settlements developed in a rather closely spaced pattern in New England and the Midwest. Most of the early small industries along the eastern seaboard depended on waterfalls to power their mills. From New England to the southern Piedmont (high, rolling land east of the Appalachian and Blue Ridge mountains), a fairly fine network of small industrial towns still exists, particularly near falls on streams and rivers (figure 6.9). Some of these towns have declined because they were unable to adjust to more modern sources of power and to shifts in consumer demand. Others are important industrial centers—for example, Trenton, New Jersey; Richmond, Virginia; and Columbia, South Carolina. Scores of small mining towns and a few large industrial cities grew up near mineral resources, especially coalfields. Such patterns prevail in eastern and western Pennsylvania, eastern Ohio (the Pittsburgh region), West Virginia, and northern Alabama (the Birmingham region).

Industrial cities with access to both water and rail transport often became **break-of-bulk ports** where cargo was divided into smaller parts, particularly for transshipment from high-volume carriers, such as barges, to lower-volume carriers, such as trains or trucks. The Great Lakes cities of Chicago, Detroit, Cleveland, and Buffalo developed facilities to handle the transshipment of grain, coal, iron ore, and manufactured goods.

Change in the Relative Importance of U.S. Cities

Most early U.S. cities and towns have at least survived, if not prospered, for economic reasons: the exchange or production of goods and the provision of services. But political factors often determined which ones succeeded and which did not. Political considerations influenced the location of land grants for railroads. Towns on the rail system fared better than those that were bypassed. Other successful cities and towns included state capitals and the national capital (whose location was determined by political compromise between conflicting northern and southern interests, as well as by its centrality in the late eighteenth-century United States), county seats, and other administrative centers. University towns also thrived.

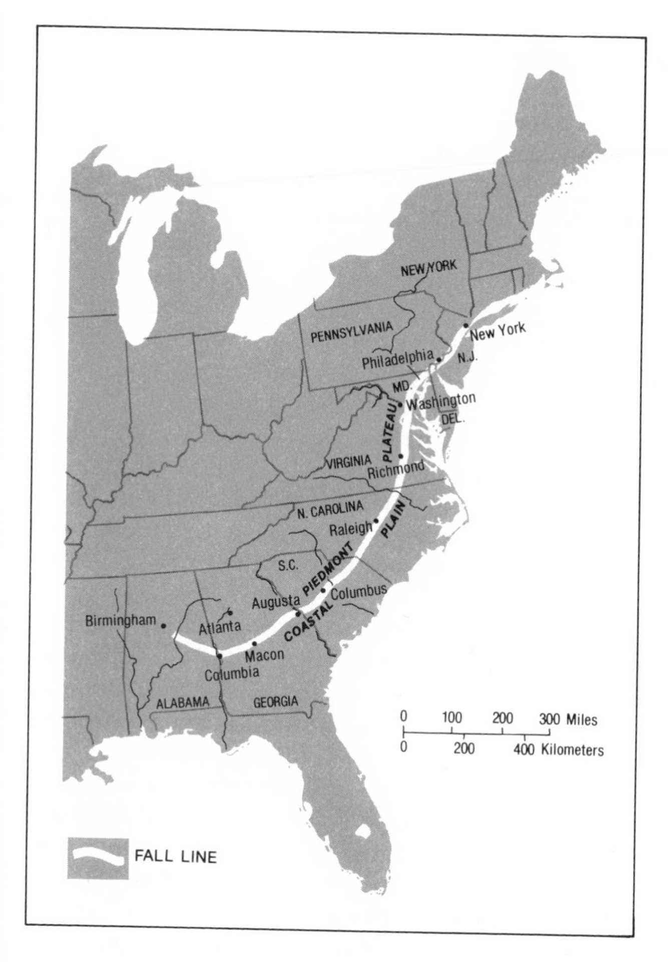

6.9. Industrial settlement in the eastern United States. Many industrial towns grew up along the fall line, where rivers flowing toward the Atlantic Ocean cut deep channels into the softer soil of the coastal plain, forming waterfalls.

The urban system was produced over the generations by people who wanted to succeed. In general, they appear to have used space fairly efficiently. Of course, historical inertia and changes in the economy have created problems for many cities. At any given moment the urban system is less than ideal. Some places are declining because they are no longer efficiently located or because they have lost their original base of people or resources. The shift from trains to trucks and automobiles, for example, has undermined the importance of many railroad towns in the United States. Depleted mines have made ghost towns of once prosperous centers in many states. On the other hand, some places are gaining from shifts in population, changes in resource use, responses to new scale economies, and the growing importance of leisure-time activities. The discovery of oil along the Gulf Coast and the development of the technology to exploit it brought new growth to Houston, Texas, Baton Rouge, Louisiana, and other cities. Sunshine, mountain slopes, beaches, nightclubs, and national parks have attracted thousands of people to resort and retirement communities in California, Florida, Arizona, and Nevada, particularly in recent years.

The map in figure 6.10 shows the growth of the urban system and changes in the relative importance of U.S. cities over time. In 1800, while New York had already achieved preeminence and Philadelphia, Baltimore, and Charleston were important places, New England contained the largest number of industrial and port cities and towns. Many of these, like Salem (6), Newport (9), Newburyport (11), Gloucester (14), and Portsmouth (15), are but suburbs or very minor cities or ports today.

Between 1800 and 1850, new industrial centers and new gateways to the West emerged. Examples of the former are Pittsburgh, Newark, Buffalo, and Rochester, and of the latter, Cincinnati, Louisville, St. Louis, and Chicago. New York, Philadelphia, Baltimore, and Albany boomed as eastern terminals of canals and the new railways. A few southern ports remained important (Charleston, Mobile, and especially New Orleans).

After 1850, when the development of the South stagnated, new railway centers, such as Chicago, Indianapolis, Minneapolis, Denver, and Kansas City, and new industrial cities, such as Detroit, Cleveland, Milwaukee, Columbus, and Toledo, became growth centers, diminishing the importance of river ports such as New Orleans, Cincinnati, Louisville, Albany, and St. Louis (gateway cities).

After 1900, New Orleans and the old gateway cities declined further. The old industrial cities—Boston, Providence, Albany, Buffalo, Rochester, Baltimore—also declined to some extent in relative standing. But new southern centers, such as Atlanta, Houston, and Dallas, and western regional capitals, such as Seattle, Portland, San Francisco, and Los Angeles, grew in importance. So did Detroit, when the automobile became popular.

After 1950, metropolitan growth shifted dramatically to the West and South. The large northeastern industrial cities (Cleveland, Pittsburgh, Buffalo, Milwaukee, Newark, Boston) and even New York, Chicago, Philadelphia, and Detroit declined relatively. But western and southern cities rose

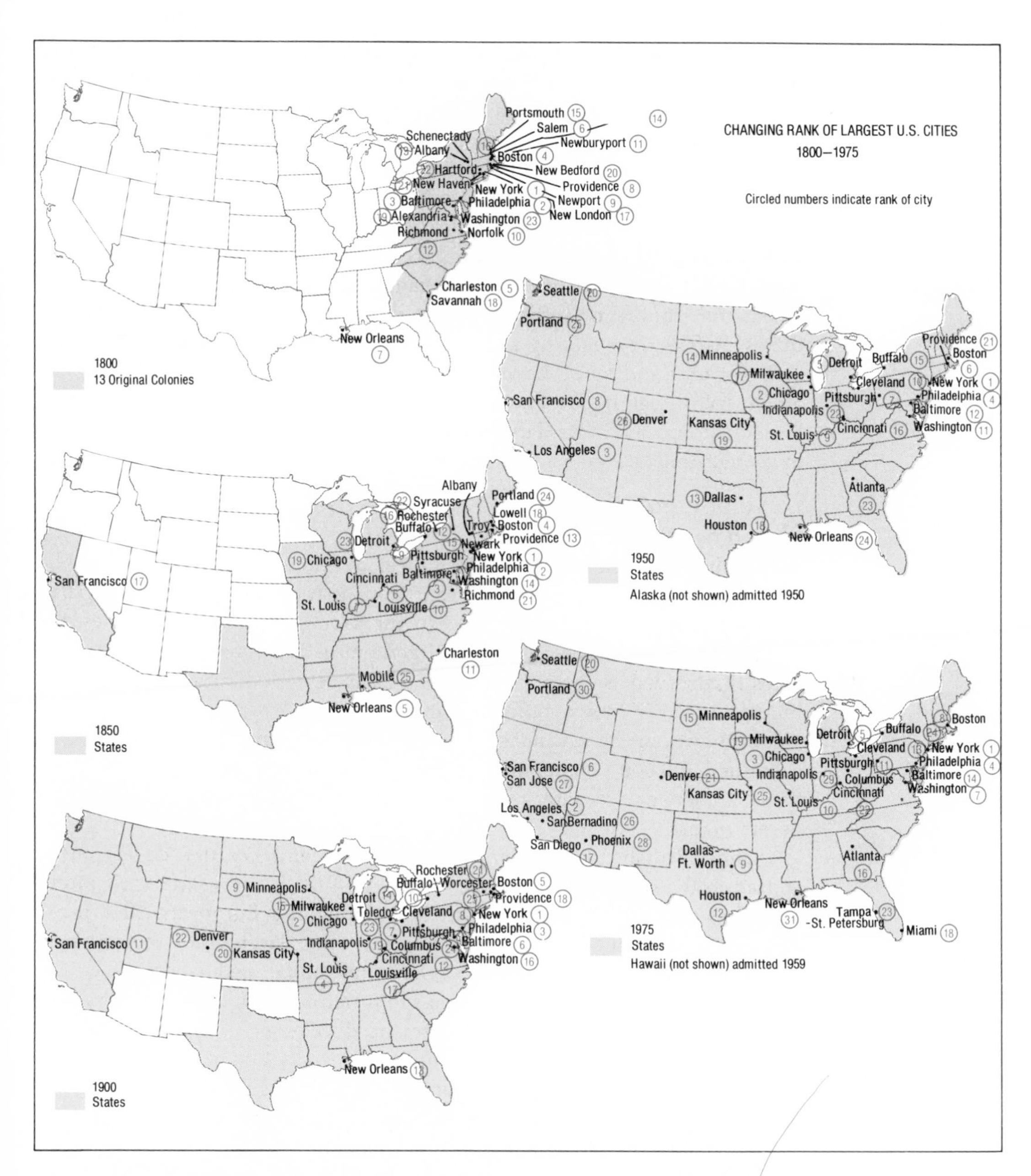

6.10. Changing rank of the largest U.S. cities, 1800 to the present. As settlement spread westward and the population and economy expanded and changed, the location and rank of major cities also changed. See text for detailed discussion.

rapidly: Seattle, San Francisco, San Diego, San José, Los Angeles, Denver, Dallas, Houston, Atlanta, Washington, D.C., Tampa, and Miami (but not old cities like New Orleans or Charleston). Yet considering that two hundred years of westward movement and internal development have passed, the continuing vitality of the older regional capitals such as Boston, New York, and Philadelphia attests to their powers of adaptation.

Fourteen of the twenty-three largest places in 1800 had lost their status by 1950. Only New York, Philadelphia, Baltimore, Boston (the top four in 1800), and Washington (then twenty-third) remained among the leaders in 1975. Many places rose quickly but soon declined in relative position—Albany, Cincinnati, St. Louis, Louisville, Buffalo, Rochester, Cleveland, Minneapolis, Milwaukee, Indianapolis—reflecting the speed with which the frontier of settlement and development moved in the United States. A few grew rapidly and maintained a high rank—notably Chicago and Detroit—and some have moved upward fairly continuously, for example, Washington, San Francisco, and, more recently, Atlanta, Los Angeles, Houston, and Dallas. Finally a few have broken into the top ranks only in the last generation—San Diego, San José, Miami, Tampa, and Phoenix.

Development of the Urban System in Nigeria

The evolution of the urban system in less developed countries such as Nigeria is in many ways quite different from that of the United States. But some parallels can be drawn between the processes in both countries.

Before European colonization, Nigeria already supported many inland cities that served as regional markets and as centers of administrative and religious control. Perhaps the most important was Sokoto, the seat of Moslem authority in the savannas, or grasslands, of the north. In 1800, Sokoto contained 180,000 people; many were farmers who commuted out to their fields. Other major centers were Kano and Katsina, on the fringes of the desert (figure 6.11).

Under British colonization in the late nineteenth century, development shifted to the southwestern coast, where the British had established cacao, oil palm, and rubber estates. Both indigenous and European cities prospered, including Ibadan, Ogbomosho, and other large cities in the cacao belt, especially Lagos. Lagos was the colonial capital, major port (cacao export center), and terminus of Nigeria's first railroad. The British built a second railroad in the east, which led to the development of a new colonial port, Port Harcourt, and of inland mining cities like Enugu and Jos. The inland extension of the railroads favored certain cities such as Kaduna, Kano, and Zaria, at the junction of the two lines. Places not served by the railroads, notably the old religious center of Sokoto, declined.

After Nigeria won independence in 1960, the larger northern tribes gradually gained political power. With the commercialization of agriculture and the beginning of industrialization, new urban development has surged in the north. Kano, in particular, has been revitalized as a northern economic and cultural capital. The rapid growth of oil production and export from the Niger delta has created another urban-industrial zone around Port Harcourt. But southwestern Nigeria is still the most urbanized and industri-

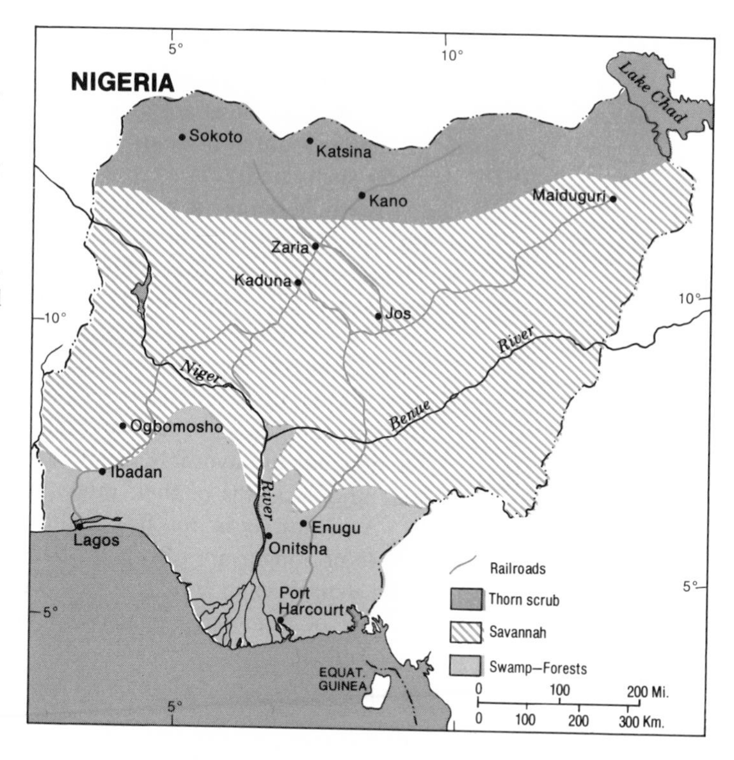

6.11. Major cities and transport systems in Nigeria. Until the British colonized Nigeria in the nineteenth century, the most important city was Sokoto, an administrative, religious, and market center. Under British rule, development shifted to port cities such as Lagos and Port Harcourt. Inland mining centers like Enugu and Jos were linked to port cities by railroad. The railroad also promoted the growth of Kaduna, Zaria, and Kano. While northern cities have benefited from modern commercialization and industrialization, southwestern Nigeria remains the most highly developed area.

alized area. (Compared to Kano and Port Harcourt, which contain about 300,000 and 180,000 people, respectively, Ogbomosho has a population of about 300,000; Ibadan, 630,000; and Lagos, 850,000.)

Thus many of the forces influencing urban location in the United States have also affected cities and towns in Nigeria. Early port cities have tended to remain important population centers (historical inertia); some cities have become more prosperous because of location on railroads (accessibility); service centers have arisen in agricultural areas, mining and industrial centers in mineral-rich areas (comparative advantage); and cities in certain regions have benefited from shifts in political power.

URBAN VARIATION

As cities and towns develop over time and across space, a complex urban landscape unfolds, consisting of places that vary in size of area and population, in political and cultural importance, in activities that support the population, and in economic role.

Size

Imagine a world in which all people and activities could be concentrated in one area and suppose that the benefits of agglomeration were unending. The result would be a single world city, where transport costs would be at a minimum. But such a city does not and cannot exist. Because of the space required for agriculture and the sporadic location of natural resources, people and activities are dispersed. Transport costs reflecting distance appear. Many different central places and processing centers are necessary under such conditions.

A whole range, or hierarchy, of urban places has evolved, from countless small hamlets to many large cities and a few enormous metropolises. The smaller places have few activities and limited linkages with other places. Usually they provide frequently needed services, such as grocery and drug stores and gas stations, for very local areas, or they exploit very local resources. Some grow into large cities by providing a wider variety of services to a more extensive area or by producing a larger volume of goods that can compete in distant markets, or both. The success of such places often depends on favorable location or the entrepreneurial skills and shrewd planning decisions of their inhabitants.

A point may be reached in the development of a city where increasing size is no longer an advantage. Transport costs begin to rise sharply and scale economies may be reversed (figure 6.12). Large cities may become incapable of organizing space and disposing of wastes efficiently. Traffic congestion, crowding, visual blight, pollution, and crime often become serious problems.

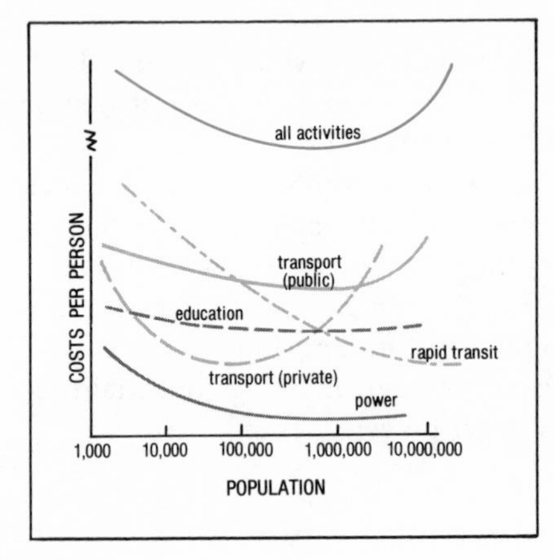

6.12. Relationship between economies of scale and urban size. Moderate-sized cities seem more efficient than very large metropolises or small cities and towns. Yet places of every size play an essential role in the national economy. Note that private transport (the automobile) is relatively inexpensive in smaller cities and public transport in larger ones. Rapid transit is the least costly form of transport in larger cities. Power costs fall until the population reaches about 500,000 and then remain fairly constant. The cost of education does not change significantly in cities with more than 50,000 people. The rising inefficiency of the largest cities seems to be due to the higher cost of private and public transportation.

Cities cannot expand indefinitely. Growth is limited not only by the problems just mentioned but by the cost in time, money, and effort involved in getting enough food, water, and raw materials and developing large enough markets to support the population. Thus a set of regional metropolitan centers can compete successfully with each other. The entire range of urban places, from smaller metropolises down to local villages, is important to the overall economy of nations. Each level in the hierarchy is able to provide the goods and services demanded by a dispersed population.

Primacy

The territorial size of a region or nation influences the number and distribution of large cities within that unit. The smaller the unit, the greater the proportion of cultural, political, and economic activity that can optimally be located in a single center. The concentration of activity in one place is known as **primacy.** Many countries that are relatively small in area, such as Great Britain, Denmark, France, and Thailand, contain one dominant city (London, Copenhagen, Paris, and Bangkok, respectively). Larger countries, such as the Soviet Union, the United States, and Canada, or those formed from the merger of several smaller states, such as Germany, Italy, and Spain, tend to have regionally rather than nationally dominant cities. The dominant cities in Italy, for example, are Milan in the north, Rome in the center, and Naples in the south.

In developing countries, growth tends to be concentrated in the capital or the most favorably located city. Examples are Nairobi, Kenya; Luanda, Angola; Buenos Aires, Argentina; and Abidjan, the Ivory Coast. The reason may be attributed to the desire to maximize development. When capital is limited, it is most efficient to concentrate economic investment (such as transport or manufacturing), government services, and educational, health, and cultural institutions and services in a single city. Still some countries try to counter the dominance of such cities by developing a new capital. Brazilia, for example, is the new capital of Brazil, and Ankara, that of Turkey; a new capital is also being proposed for Nigeria.

Relative Location

Smaller cities within a country tend to predominate in areas of extensive agriculture, such as the U.S. plains, the Canadian prairies, and the north Indian plain, and in forested or mountainous regions, such as the Appalachian and Ozark uplands, the Canadian Rocky Mountains, and the Swiss Alps. The rural activities forming the economic base of such regions cannot readily support many large population centers. Moreover the machines now used in extensive agriculture and in mining have reduced the need for human labor. Thus many rural communities are losing population, particularly those small places within thirty kilometers (eighteen miles) of larger towns or off major highways. Figure 6.13 shows the pattern of growth and decline in Saskatchewan, Canada.

Larger cities tend to be concentrated in areas of early commercial and industrial development, such as northwestern Europe and the northeastern United States. These were the first regions to become highly industrialized and densely populated. Even after industry spread to other areas, they retained their base of technically skilled people and their reputation for high standards of quality. These regions still constitute the largest markets in the

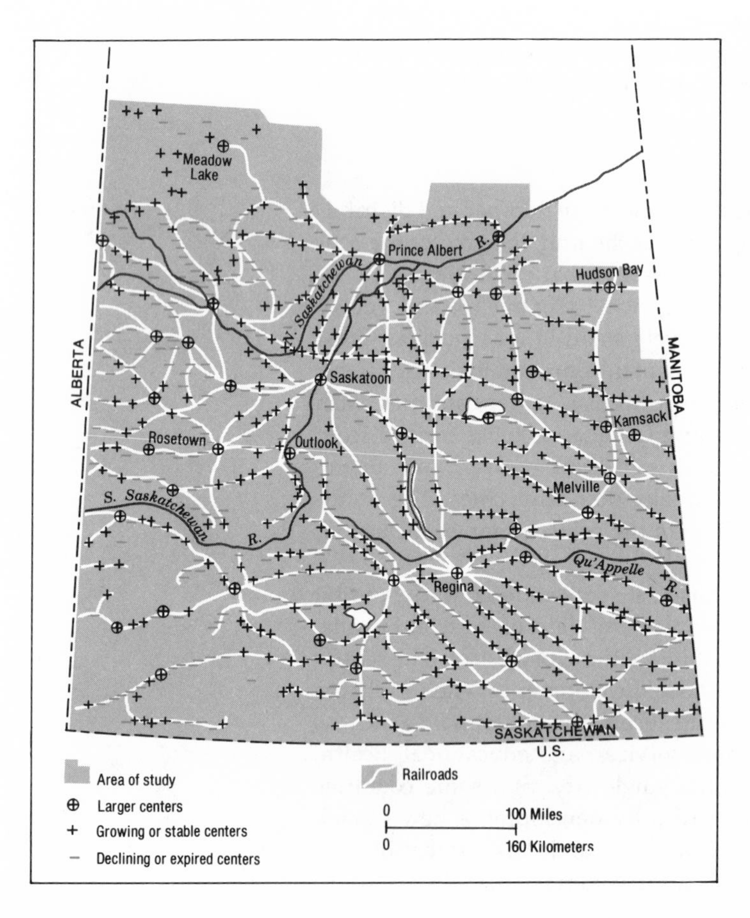

6.13. The growth and decline of trade centers in Saskatchewan, Canada, 1941–1961. What does this study reveal about the influence of accessibility on the development of trade centers?

Source: Adapted with permission from G. Hodge, "Prediction of Trade Center Viability in the Great Plains," *Papers and Proceedings of the Regional Science Association*, Vol. 15, 1962.

Western world. Large cities have also developed in highly accessible locations along coasts and rivers, on bays, and at focal points of transport networks. Some of these, like Berlin and Madrid, are not ancient cities but fairly modern ones that were purposefully developed.

In developing nations, many people still live in millions of small villages. The development of inland cities and towns, which serve this dispersed population, often differs from that of the newer coastal cities that grew up in the colonial era (seventeenth to nineteenth centuries). We have already noted the different stages in the development of inland Kano and coastal Lagos in Nigeria. A similar dichotomy has occurred in China. After 1949 the Chinese concentrated on internal development, to the benefit of inland cities such as Peking and Sian. Since 1970, however, China's renewed focus on trade and other external matters has stimulated the development of coastal cities such as Shanghai and Canton.

Optimal Size

The ideal size for a city depends on the context. Some regions are more efficiently served by a single large metropolis, others by several smaller cities.

Some activities, such as magazine publishing, require the agglomeration of many facilities or services and function best in metropolises. Others, such as textile production, do not need many auxiliary services and can profitably locate in small towns. An extremely small place, of course, cannot support even minimal services, and an extremely large place may support a variety of services most people cannot afford. The best restaurants and nightclubs in New York City or Paris, for instance, are beyond the means of the average person. Extremely large cities may suffer from congestion, pollution, and very high costs of living. Between the extremes are a variety of urban places—some large, some small—that are useful to society and belong in the ideal landscape.

Although thousands of smaller places function successfully, a population of about 250,000, in the United States, appears to be the minimum size capable of supporting a wide range of exchange and service activities and high-quality cultural, medical, and educational facilities. In giant metropolises, these advantages may be outweighed by the social and economic costs of a large population. Hence the smaller metropolises in the United States and Europe have grown particularly rapidly during the past twenty years. In developing nations, however, the largest metropolises tend to grow fastest because they still provide the highest return on capital investment.

Houston, Texas, is a good example of a fast-growing medium-sized metropolis. With a population of nearly two million, including the suburbs, it supports the University of Houston and Rice University, several theaters, concert halls, art centers, museums, sports arenas (most notably, the Astrodome), the Texas Medical Center, and the Manned Spacecraft Center. But as the leading manufacturing and industrial city in the Southwest, Houston is also experiencing some of the problems of larger metropolises, such as pollution and congestion.

In sparsely populated regions, smaller metropolises—like Salt Lake City, Utah, Boise, Idaho, and Billings, Montana—may have as important a cultural and economic role as much larger places in densely populated regions. It is therefore impossible to suggest an optimum size of cities. Evidently most of the population in Europe and North America would prefer to live in smaller cities and towns, even though the metropolis is the most efficient location for many economic activities.

Functional Class

Cities differ markedly in the specific central place or processing activities that define their character and support their populations. They are often classified according to the activities, or *functions,* that constitute a proportionately larger share of their economy compared with the average for all similar places. The pattern of urban functional classes in the United States is displayed in figure 6.14. The urban-industrial core, extending from the Northeast westward through Illinois, is dominated by specialized manufacturing cities. The periphery tends to support centers of trade. And some of the larger cities, notably regional capitals such as New York, Boston, Chicago, Atlanta, and Los Angeles, are *functionally diversified* rather than specialized: no single kind of activity employs a much higher percentage of people or has relatively greater economic value than on the national average.

Although it is generally true that large cities tend to be diversified, many emphasize one or more functions. Many cities in the Northeast, such as Pittsburgh, Buffalo, and Albany, concentrate on manufacturing. This is

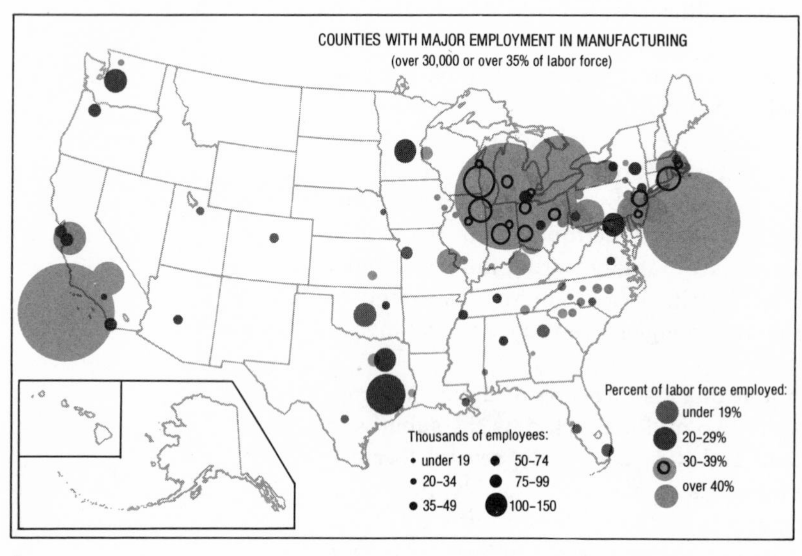

A

B

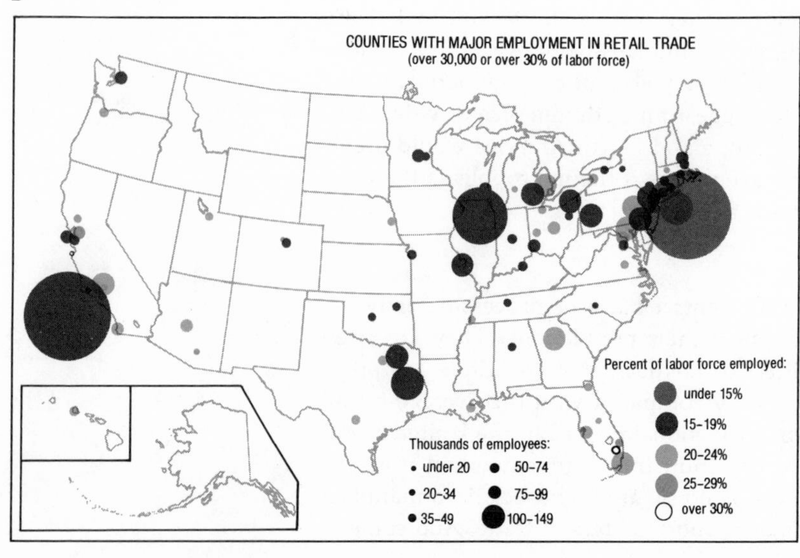

6.14. Urban functional classes in the United States. Map A: Counties with major employment in manufacturing, 1976. The dominance of the northeastern urban-industrial core in manufacturing remains impressive, but California and Texas are becoming numerically important, and a large percentage of the labor force in parts of the South is employed in manufacturing. Map B: Counties with major employment in retail trade, 1976. The distribution of retail trade activity reflects fairly closely the distribution of population and purchasing power; however, a higher than average percentage of the labor force is employed in retail trade in a few suburban and peripheral areas. Map C: Counties with major employment in transport and utilities, 1976. Although New York City is the nation's dominant transport center, a number of other coastal and river ports and regional trucking and airport centers create a somewhat peripheral pattern. Map D: Counties with major employment in finance, 1976. New York City is of overwhelming importance in finance, followed by other northeastern centers and major regional capitals.

because industry first developed in the Northeast, local resources were available, and the dense population provided a market for manufactured goods. Boston and New York have been noted especially for finance since the days of early American trade: clipper ships designed and made in Boston brought wealth to the city through the East Indies and China trade from 1790 to about 1830, and the opening of the Erie Canal in 1825 made New York City the gateway to the West and the financial center of the United

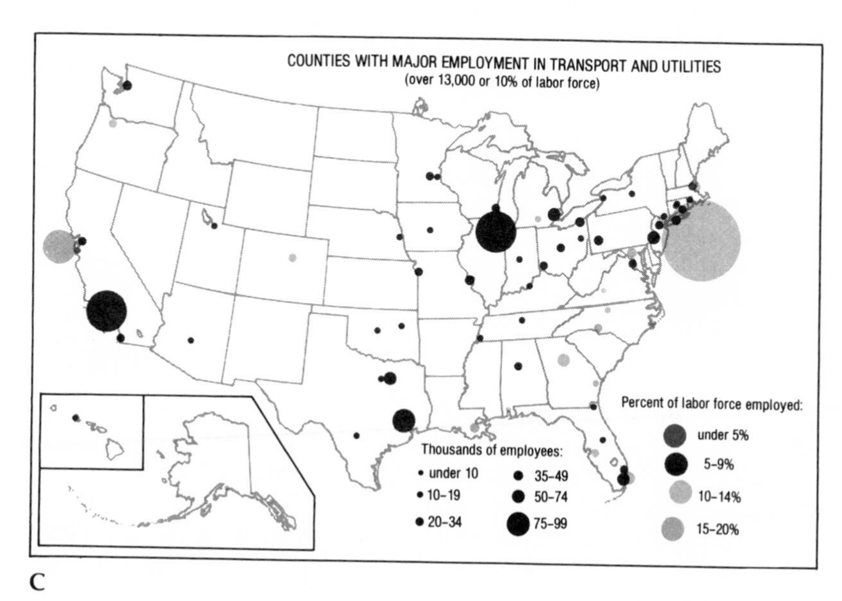

C

D

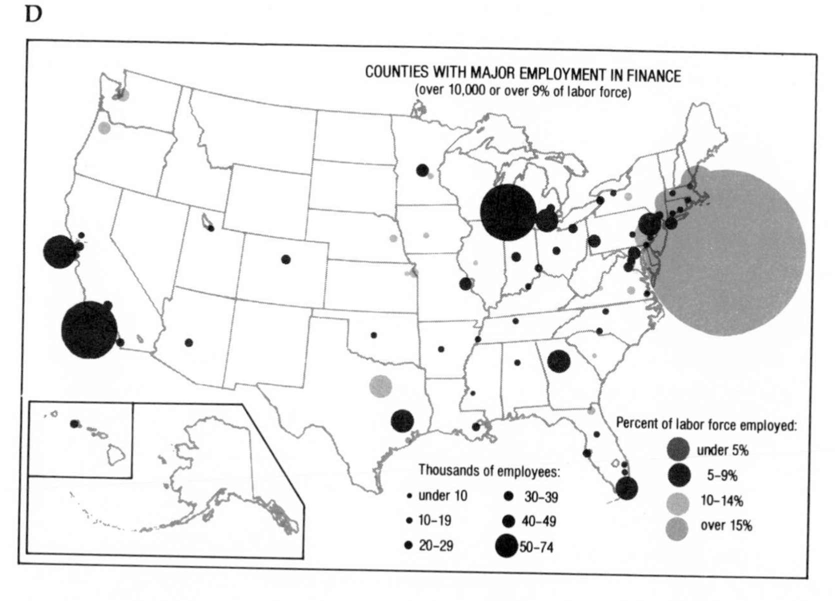

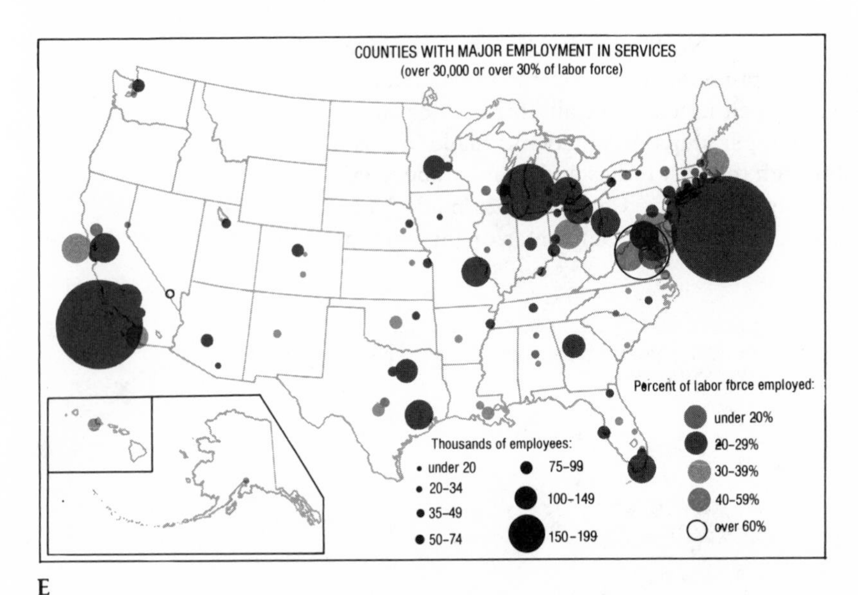

Map E: Counties with major employment in services, 1976. Services tend to be distributed according to population, but an unusually high percentage of the labor force is employed in services in Washington, D.C., state capitals, and resort centers such as Las Vegas.

E

States. The gateway cities of the plains—such as Des Moines, St. Louis, and Omaha—and the regional capitals of the South and West—Nashville, Charlotte, Richmond, Atlanta, Dallas, Portland, and Denver—specialize somewhat more in wholesale and retail trade, transportation, and services. Their location between a farming hinterland and the industrial Northeast gives them a comparative advantage for distribution.

At the small-city level, many industrial towns based on the processing of agricultural, mineral, and forest resources appear on the periphery (low-density regions far from metropolitan centers). For example, Saline, Kansas, Williston, North Dakota, and Bend, Oregon, specialize in the processing of wheat, oil, and lumber, respectively.

There are also other kinds of functional classes. Some scattered towns specialize in administration or education. These are usually state capitals or college towns, such as Lincoln, Nebraska (the administrative capital of the state), and Chapel Hill, North Carolina (home of the University of North Carolina). Many towns in the West, Appalachia, and the Gulf South specialize in mining; one is Butte, Montana. Finally a few cities specialize in transportation services and facilities, or function as resorts; Miami, Florida, does both.

Compared to the United States, cities in most developing countries are fairly diversified but tend to emphasize services and trade. They offer a wide range of government, marketing, transport, and personal services. Many of these services are provided by individuals and families outside the wage economy, for example, street vendors and peddlers. Similarly some modern manufacturing may coexist with a sizable handicraft sector, again employing many, but at rather low wages.

As a specific example, let us take another look at Nigeria (refer to figure 6.11 again). Lagos, the main transshipment point and commercial center, specializes in government and cultural services and in trade, although it is also an important industrial center. Ibadan, a diversified metropolis, is a center of trade, services, and handicrafts, as well as modern industry. Port Harcourt specializes in oil refining and other heavy industries and in foreign trade; Jos specializes in tin smelting; and Enugu in coal mining and trade. Onitsha is Nigeria's major market, or exchange center (by looking at the map, you can see why Onitsha has a comparative advantage as a market). Benin is an ancient city that still specializes in handicrafts. And Zaria and Kaduna, located at railroad junctions, are old cities that are acquiring a modern industrial character through the manufacture of textiles, bicycles, and other products.

Economic Role

More important than functional class as such is a city's role in the regional or national economy. Its *economic role* is revealed largely by the pattern and strength of its external relations with the surrounding region, nation, or world (figure 6.15). Cities specializing in central place activities usually provide services for a spatially contiguous, limited area. But those specializing in processing activities often serve the entire national or international economy. Seattle and Los Angeles, for example, manufacture aircraft, which they export to other regions and nations; Gary and Pittsburgh manufacture and export steel.

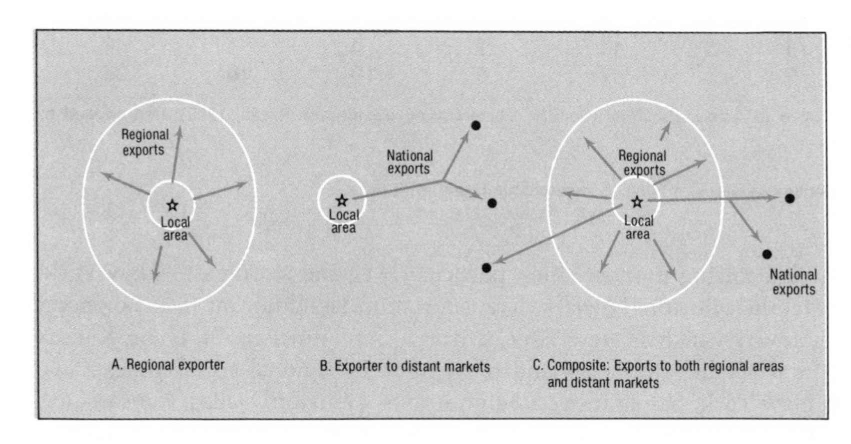

6.15. The economic role of cities and their external relationships. Stars represent cities; dots are markets. Cities emphasizing central place activities export services primarily to local markets within a ten-mile range and to regional markets within a hundred-mile range (A). Cities based on mining or industrial activities export mainly to national markets (B). Most large cities export to both regional areas and distant markets (C).

In an economic and spatial sense, exported goods and services make possible the existence of cities. Exports are the products of a city's comparative advantage. They provide the income with which the city can purchase the goods, including food, and services it does not produce for itself. Table 6.1 lists the export specialties of the largest U.S. cities. The estimated percentages of exports in each category are based on the level of employment above that needed for internal purposes. Notice the overriding importance

Table 6.1. Export Specialties of Largest U.S. Cities, 1970 (estimated percentages of exports[a] by economic sector)

City	Mfg.	Trans.	Whole. Trade	Retail Trade	Finance	Prof. Services	Admin.	Other[b]
New York	50	10	4		11	2	5	18
Los Angeles	62	3	1		4	1	3	26
Chicago	72	10	1		3		2	12
Philadelphia	75	4			2		5	14
Detroit	87	3					1	9
San Francisco	30	11	3	2	8	9	11	26
Boston	52	5	2	1	6	12	6	16
Pittsburgh	73	7		2	1	3	2	12
St. Louis	61	10	2	1	3	1	4	18
Washington		4			3	12	61	20
Cleveland	77	5	2		1	2	3	10
Baltimore	53	8		2	3	3	10	21
Newark	68	5	1		7	1	2	16
Minneapolis	43	10	7	5	6	9	4	16
Buffalo	71	7		2		4	2	14
Houston	31	11	7	5	4	2		40
Milwaukee	77	4	1	3	2	1	3	9
Seattle	45	9	5	4	5	7	4	21
Dallas	33	8	7	6	9	6	2	29
Cincinnati	58	7	2	4	3	2	3	21
Kansas City	38	15	5	5	6	3	6	22
San Diego	36	1		6	5	6	15	31
Atlanta	32	11	6	5	7	1	6	32
Miami	8	15	3	11	7	4	4	48
Denver	25	9	5	7	6	10	10	28

Source: E. Ullman et al., *The Economic Base of American Cities* (Seattle: University of Washington Press, 1975). Reproduced by permission.

[a] Export out of metropolitan area.

[b] Includes construction, personal and business services, mining, communication, and utilities.

of manufacturing in most cities, particularly in the Northeast—New York, Philadelphia, Boston, Newark. Transportation is significant in major ports and gateways, such as New York, Chicago, San Francisco, St. Louis, Kansas City, and Miami; wholesale and retail trade in regional economic capitals, as in New York, San Francisco, Minneapolis, Houston, Dallas, Kansas City, Miami, and Denver; finance in New York City and other regional capitals, and administration in Washington, D.C. Exports may be highly specialized within these broad categories—for example, transport equipment in the manufacturing sector of Detroit, Seattle, and San Diego; the U.S. Navy in San Diego; and aircraft operations and tourism in Miami.

Export activities are often referred to as **basic activities.** The population engaged in export (basic) activities creates a demand for various locally produced consumer goods and services, which are sometimes called **nonbasic activities.** Nonbasic activities include retail trade, entertainment, recreation, educational, and professional and administrative services. They provide income for much of the city's population and are necessary for high standards of living.

The same class of activities may be both export (basic) and internal (nonbasic). While most services are used by the local population, some are "exported," for example, hospital services, insurance, university education, and television. Conversely while most manufactures are usually exported, some, like printing (newspapers) and bakery products, are consumed locally.

The essential point is that the innovation of cities was a departure from the traditional notion of a family or tribe's self-sufficient relation to its land. People who lived in cities had to accept an interdependent relation with other people and areas. They had to import food and raw materials—and to export goods and services to pay for these commodities. Exportation also allowed cities to accumulate the savings necessary for advances in civilization. Marx correctly observed that the concentration of power in cities enabled them to "appropriate the surplus" from the countryside and that this was probably a necessary step in the industrialization process.

CURRENT TRENDS

Urbanization is proceeding rapidly in developing countries. But in most advanced nations, the entire historical rationale for cities is being tested: metropolitan populations are stabilizing, and rural and small-town populations are growing, a change from previous trends.

Until 1970, the concentration of people in cities, metropolises, and sets of metropolises, such as the Boston-Washington megalopolis, appeared to be the dominant trend everywhere. Small hamlets and villages in the United States, Europe, and Japan had generally been losing population, while larger cities and towns had been gaining. The larger places had become more accessible through faster, more efficient transportation and the development of the highway system. Many industries needed to locate in large cities that offered highly skilled labor pools, high-quality business services, and large markets. The population of metropolitan Chicago grew by about 36 percent between 1950 and 1970; that of Los Angeles, by 70 percent.

Since 1970, metropolitan growth in Europe and North America has slowed substantially. In the most highly developed regions, the decline of large central cities is barely offset by continuing but slower suburban growth. Smaller cities and towns and rural areas are growing once again. Many industries are decentralizing to smaller places. People who work in the city often prefer to live in the countryside. And more and more older people are retiring to smaller communities.

Despite such trends, even the most advanced nations are not yet completely urbanized. In some regions and countries, opportunities are opening for the development of new central places and industrial centers. In others, older settlements are trying to adjust to change. The urbanization process is uneven. For example, the large metropolises of the northeastern United States have reached a size where diseconomies (higher costs due to higher levels of pollution, congestion, crime, and other problems) are setting in. The new central places and industrial clusters in the South are still expanding. And urbanization has perhaps just peaked in the Midwest, where many small urban places are declining and population is concentrating into larger centers, from county seats to regional capitals such as Omaha and Minneapolis-St. Paul.

Population continues to flow into larger cities in the developing world. São Paulo, Brazil, grew by 5.4 million people from 1948 to 1975. The urbanization of most developing countries today differs from that of Europe and North America in the nineteenth and early twentieth centuries, when urbanization tended to coincide with population growth and industrialization. In developing countries, rapid population growth has preceded urbanization, and urbanization has outpaced the growth of industry and job opportunities. The countryside in Brazil, India, Indonesia, and other third world countries has become so overcrowded that millions of people have moved to the cities, which cannot fully employ their labor or provide adequate housing and services (see chapters 9 and 13).

Some scholars suggest that current trends in city growth are ultimately leading to extreme metropolitanization—the dominance of society by enormous metropolises or megalopolises, each containing between 2 million and 50 million persons, and in which 80 to 90 percent of the total population would reside (see figure 15.3). Several exist already—New York, Tokyo, Paris, Mexico City, Buenos Aires, Los Angeles-Long Beach, and London, to name only a few. Nonmetropolitan areas would be occupied by the minimal rural and small-town population needed to carry on extensive activities (agriculture, forestry, fishing, mining).

Recent data, however, do not support this view. Many of the larger metropolises in the United States and Europe are growing, but slowly. While a larger proportion of the total population may come to be concentrated in metropolitan regions, *satellite cities* and rural regions are growing the fastest. (Satellite cities are near but not part of metropolises. Trenton, Wilmington, Lancaster, and Reading, for example, are satellites of Philadelphia.) Deterred by such problems as overcrowding and high crime rates, more and more people would rather invest in or live in smaller places than in extremely large ones.

Although people and businesses have decentralized even beyond the suburbs, metropolitanization in a cultural and economic sense is still quite evident. A few large metropolitan centers dominate most of the rural, small-town, and small-city populations (see figure 6.2). Decisions made in these centers touch the lives of all people. Financial decisions made on Wall Street in New York influence the success of businesses throughout the United States; political decisions reached in Washington, D.C., affect the well-being of every citizen. Metropolitan centers dominate the worlds of art, music, drama, and communications. Faster land travel and the frequent use of air travel for business purposes have led to the consolidation of higher management, exchange, and finance in the largest metropolises. The cultural and economic dominance of the metropolis is not a particularly new phenomenon. The greatest change in the past fifty years is the concentration of power in fewer places, diminishing—at least until recently—the role of regional centers with populations of 25,000 to 250,000.

Some nations have tried to control the pattern of urbanization both to limit metropolitan growth and to develop new areas. Great Britain first attempted to do this in the 1930s. Since World War II, British government and industry have joined together in building new towns—towns that otherwise would not have existed—in nonmetropolitan areas. The Soviet Union has had a similar policy but has approached it from a different perspective. The Soviet government's control over investment has permitted it to establish a widespread pattern of newer cities and metropolises as part of a long-range planning and development scheme. Italy has diverted public and private investment into cities in its less developed south. In all cases, success has been limited. The dominant metropolises—London, Moscow, Leningrad, Milan, and Rome—have continued to attract a disproportionate share of people and activities.

CONCLUSION

Cities—particularly large metropolises like Chicago—offer an exciting mix of people, activities, and life-styles. Forty-four million people a year pass through Chicago's O'Hare International Airport. Those who come to visit the city can enjoy some of the world's best restaurants, nightclubs, theaters, orchestras, stores, and museums. They do not even have to enter a building to see works by famous artists—a Chagall mosaic, Picasso sculpture, and Calder stabile (a monumental brilliant red, slender, arching "Flamingo") adorn public plazas. After strolling through some of Chicago's distinctive neighborhoods, such as Old Town, an outstanding example of urban renewal, they can view the entire city from atop the world's tallest building, the 1,454-foot-high Sears Tower. And they can explore the self-contained 100-story John Hancock Center, where a person could live, work, shop, dine, and find entertainment without setting foot in the outside world.

Not everyone, of course, is enchanted by city life. For some, the city is a ghetto, with inadequate schools, poorly maintained housing, and dangerous neighborhoods. The middle class can escape city problems by moving to the suburbs, but the poor have few alternatives. Even welfare options in housing and job training are often limited in scope and imagination. Other people—not just the poor—also find much to dislike in the city. Although many cities are more neighborhood oriented than the suburbs, persons who come from small communities may suffer from the relative isolation and anonymity of city life. Some do not appreciate the cultural diversity that others find stimulating. The congestion, noise, speed, and intensity of life cause many people to become physically or mentally ill. And, especially in the United States, those who value individualism chafe at the regulations needed to control large numbers of people in a limited space.

Yet despite the disadvantages, most people still want to live near a city, if not directly in one. With 70 percent of the U.S. population already working and residing in cities, where the aggregate investment is colossal, we are not likely to see any radical change of pattern. In the developing world, cities are still considered, and indeed are, places of superior opportunity. Thus the process of urbanization will probably continue unabated for a long time to come.

SUGGESTED READINGS

Berry, B.J.L. *The Human Consequences of Urbanization.* New York: Macmillan, 1973.

______, and Horton, Frank. *Geographic Perspectives on Urban Systems.* Englewood Cliffs, N.J.: Prentice Hall, 1970.

Bourne, L. S., and Simmons, J. W., eds. *Systems of Cities.* New York: Oxford University Press, 1978.

Carter, H. *The Study of Urban Geography.* Garden Grove, Calif.: Arnold, 1972.

Davis, Kingsley. *World Urbanization.* Berkeley, Calif.: University of California Press, 1970.

Garner, Barry, and Yeates, M. *The North American City.* New York: Harper & Row, 1976.

Gottmann, J. *Megalopolis.* Cambridge, Mass.: MIT Press, 1964.

Johnson, J. H. *Urban Geography.* Elmsford, N.Y.: Pergamon, 1967.

King, L., and Golledge, R. *Cities, Space and Behavior.* Englewood Cliffs, N.J.: Prentice-Hall, 1978.

Murphy, R. E. *The American City.* New York: McGraw-Hill, 1974.

Pred, Allan. *City Systems in Advanced Economies: Past Growth, Present Trends and Future Developments.* New York: Wiley, 1977.

Smailes, A. E. *The Geography of Towns.* London: Hutchinson University Library, 1960.

7

The Central Place System

The Development of Central Places

The Principles of Centrality and Agglomeration

Central Place Theory

Central Place Activities

7 The Central Place System

Think for a moment about your surroundings. How far away are the nearest cities or towns? Do they differ in character and size? Can you see any order to their arrangement? It is hard to make sense of the urban landscape. Villages, towns, and cities seem to be randomly scattered. Some places appear rather similar in character, others quite different. The underlying spatial order is not readily apparent because so little of the urban landscape is visible to us, and theoretically regular patterns are modified by rivers, mountains, and other physical features. To distinguish basic patterns in the location and character of urban places, we need to use maps and theoretical models.

The theoretical emphasis of this chapter and the next may seem abstract at first, but the models we present should help you see your own landscape in a new way. You will learn that the location and social and economic roles of different places are not accidental. They are the necessary result of purposeful decisions and investments made by individuals and groups.

We have seen that the urban landscape evolved both from central place activities, such as commerce, transportation, cultural activities, and administration, and from processing or manufacturing activities. The geographic principles and patterns of central place location will be discussed in this chapter. Those of industrial location will be reserved for the next. Bear in mind that in the following pages we will explore only one of the bases for the system of cities and towns: central place functions.

THE DEVELOPMENT OF CENTRAL PLACES

The human desire for goods and services produced by others led to the development of central places. People in all societies have wanted greater material security and luxury than they could provide for themselves—more durable or elegant storage chests, for example, or softer material for clothing. To obtain such products, a place was needed where goods could be exchanged—a market, where farmers could perhaps trade surplus grain for a merchant's fine cloth.

In early subsistence societies, the level of supply and demand was not high enough to support permanent markets. Villagers could best exchange their small surpluses locally by barter. Still, some people wanted or needed products that were not locally available. The problem was how to meet small levels of demand dispersed over a large area. One solution was the itinerant peddler; another was the annual or occasional fair held in a place accessible to many communities. Widely scattered groups of people were willing to make an effort one or more times a year to exchange goods at these fairs.

As specialization, productivity, and the demand for "imported" goods increased, fairs took place more often and served more towns and villages. In Europe during the twelfth and thirteenth centuries, groups of farmers, small merchants, entertainers, and craftsmen developed *periodic markets.* (These still exist in China, Africa, and other parts of the world.) Periodic markets circulated among a set of villages, staying perhaps a day or two in each place. Eventually permanent markets were set up in the larger villages. The most accessible villages became permanent market centers. Then as now, the success of central places depended on their location. They had to be situated so as to satisfy the mutual needs of buyers and sellers at least effort. Ideally the location of central places forms a pattern that minimizes necessary travel.

Central places serve other needs besides the exchange of goods. Human societies also require direction, religious or ethical control, military protection, and social contact—needs satisfied most efficiently in central locations. Thus administrative and religious centers and fortified villages and towns sprang up throughout Europe (figure 7.1). The growth of central places also accompanied the development of settled civilizations in Asia, Africa, and South America—for example, Tombouctou, Africa, and Oaxaca, Mexico.

THE PRINCIPLES OF CENTRALITY AND AGGLOMERATION

In chapter 2, we suggested that *centrality*—location at the distance- or time-minimizing focus of an area—was basic to our understanding of human spatial order. Central place theory formalizes this idea. The other key to spatial order (and to the development of the central place system) is the principle of *agglomeration,* the grouping of people and activities for mutual benefit.

Early agricultural societies agglomerated into villages, the most efficient social and spatial form for their semicommunal way of life. A village located at the center of a community's lands minimized the total distance villagers had to walk to their fields. It was the most efficient arrangement, provided that families participated in communal activities, held some lands in common, and continually rotated the use of land holdings so that no one family monopolized the best or most accessible fields. The extent of agglomeration (size of the village) was usually limited by the lack of efficient transportation and by the inability of early agricultural economies to support large populations.

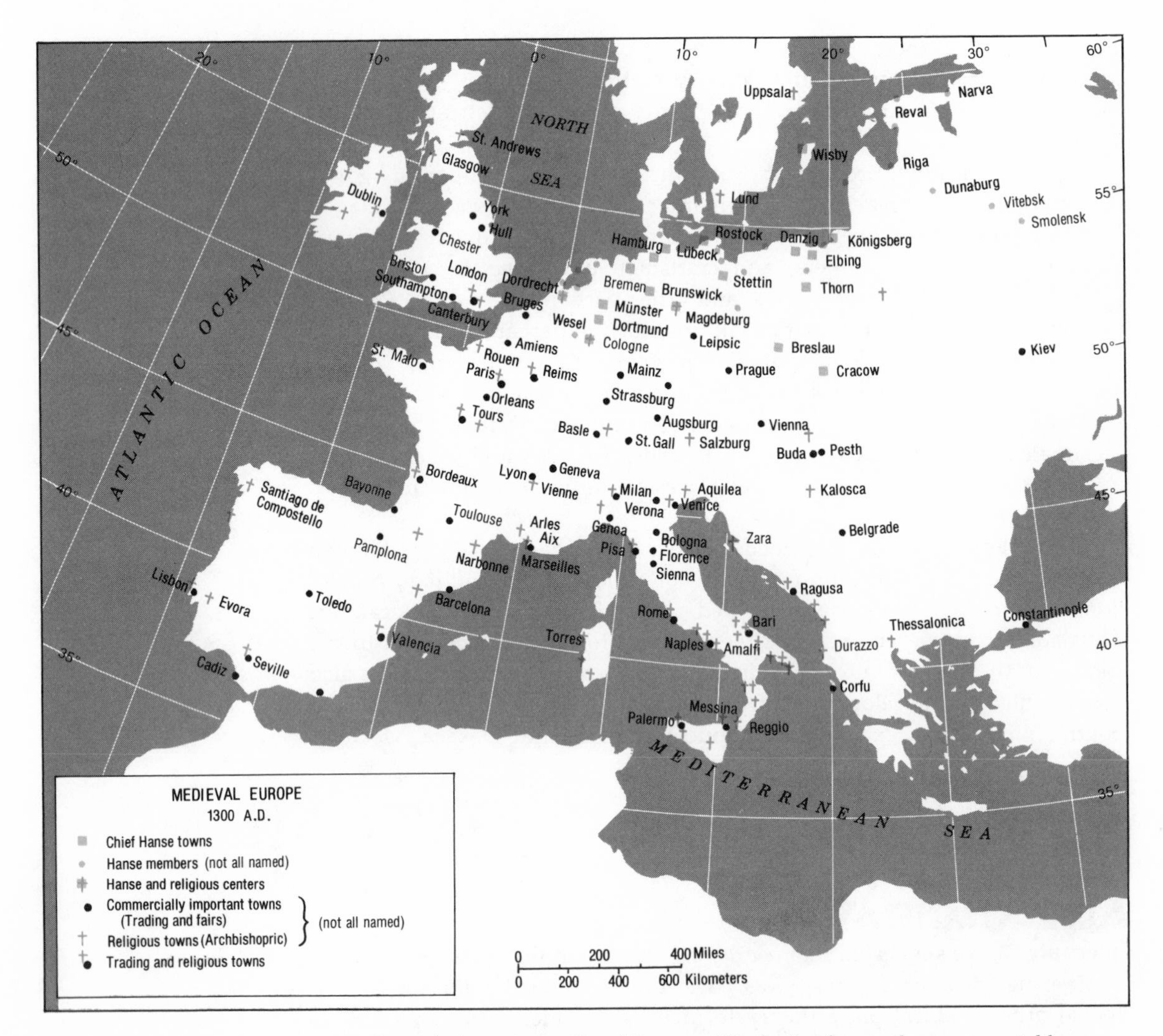

7.1. Central places in medieval Europe. "Hanse" refers to the commercial league formed by merchants in northern Germany.

Local villages often specialized in activities favored by the physical environment, such as pottery or basketweaving if clay or reeds were available. No one village could provide or support a large variety of activities. Some activities—for example, the making of specialized knife blades—might require the combined purchasing power of several villages for adequate support. It would therefore be worthwhile for some people to devote their efforts solely to those activities.

Theoretically all villagers would have to travel to a whole set of different villages to obtain the various goods and services they needed. But, if several activities were located in a central village, the total distance people would have to travel would be greatly reduced. Among a set of villages, one was probably situated more favorably than the others—perhaps on a better road or at the junction of a road and a river. This village would develop into a market town. In turn, the better located of the market towns would emerge as commercial cities, in an extension of the same distance-minimizing, trade-maximizing principle.

Today as in the past, there are many advantages to the agglomeration of activities in a town or city. A person coming to town to buy clothing is likely to see and buy other things as well. Thus the shopper saves time and effort, and retailers benefit from the additional trade. Moreover interaction between related activities, as between banks and retail stores, is easier in an agglomeration. The city, a large agglomeration, can bring together a greater number and variety of buyers, sellers, producers, consumers, and professionals than villages or towns can. But such benefits are won at a cost: as the city dominates an ever larger area, transportation time and cost from the countryside to the center inevitably rise.

The advantages of agglomeration could theoretically induce all activities to locate in a single world city. But, as we noted in chapter 6, such a city is impossible. People have dispersed across the land surface of the earth, and agriculture—a space-consuming activity essential to human survival—requires the dispersion of farms and service centers. Other needed resources, such as fossil fuels and metals, are also scattered, preventing the agglomeration of processing activities in a single place (see chapter 8). Furthermore while metropolitan activities greatly benefit from agglomeration, even the small village is a rational and efficient central place, minimizing the distance people in the local area must travel to satisfy everyday needs.

CENTRAL PLACE THEORY

Given the gradual evolution of central places, it may seem surprising that a central place order exists. But if we look at the economic and social bases for the development of central places, we will see why specific patterns are likely to emerge. The central place system reflects society's attempt to maximize spatial interaction at least cost and to minimize the distance separating related economic, social, or political activities. The principles governing the spatial order of central places apply to both free enterprise and centrally planned societies across many cultures.

Central places compete to market farm produce and to supply goods and services to the population dispersed around them. Such competition tends to create a regular pattern of central place size and location. Most individuals at one time or another must go to all the central place levels generated by this competition, from small hamlets that furnish only the most basic services to large metropolises that provide all kinds of services, in-

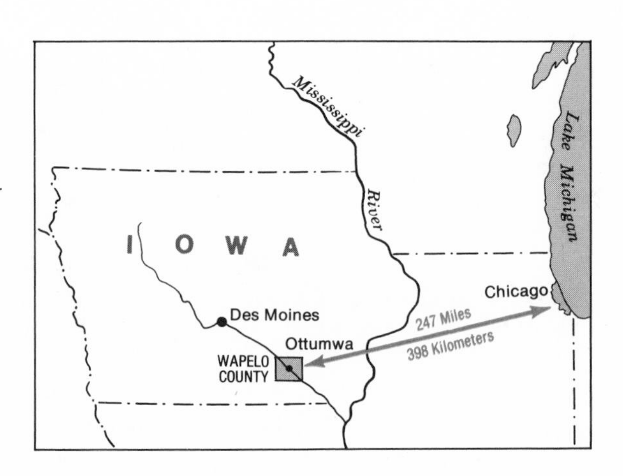

7.2. The central place hierarchy in a portion of Iowa. People living in this area can satisfy everyday needs in Ottumwa, the county seat. For more specialized services, they must go to Des Moines, the state capital. And very specialized services require a trip to Chicago, the highest level center in the Midwest.

cluding the most specialized. For example, farmers who live near Ottumwa, Iowa, will take care of most of their needs in Ottumwa, the county seat (figure 7.2). But they may go to Des Moines to look at new farm equipment, visit a specialty hospital, buy quality clothing, or attend the symphony. And they will have to go all the way to Chicago for the opera, for very specialized medical treatment, and for other activities found only in metropolitan centers.

What kind of activities a given central place can support was one of the questions explored by Walter Christaller, a German geographer who formulated classical central place theory. Christaller's theory, discussed below, enables us to determine how large an area one center can serve efficiently and what kind of activities it can best support. It also helps explain the spatial distribution of central places and the development of the central place hierarchy.

Spatial Equilibrium

To understand central place theory, it is useful to begin with the concept of **spatial equilibrium.** This term refers to the economic balance that should exist between the location of a central place (or central place activity, such as a store) and the dispersed population it serves.

The following example illustrates the principle. Suppose you wanted to open a grocery store. You know that the best location for your store would be where the greatest number of paying customers would have to travel the least to make a purchase. For a time, you may have to incur losses until the volume of sales reaches the critical level, or **threshold,** where sales exceed costs. The threshold for a grocery store might be attained with a minimum of five hundred customers or by selling to nearly all potential customers within, say, a one-kilometer radius of your store. Profits then increase as long as the revenues from new customers reached by extending the service area beyond one kilometer exceed the costs of serving them (assuming your store pays for delivery). Of course, the customer probably pays indirectly

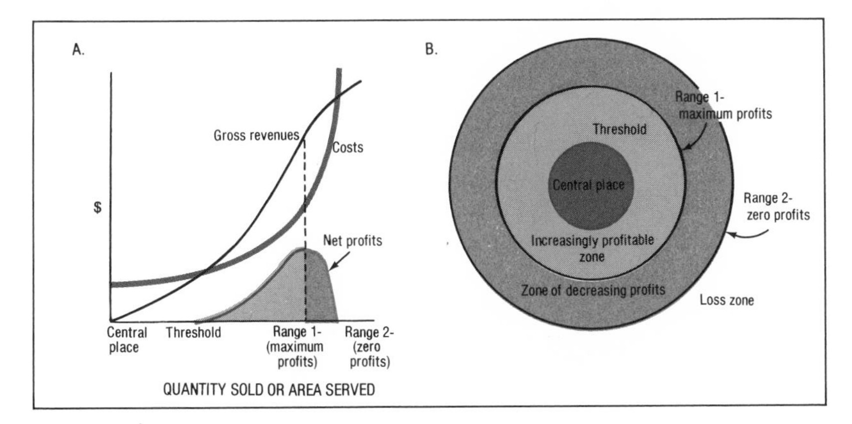

7.3. Threshold and range of a central place activity. An activity such as a fast-food outlet will typically have greater costs than revenues until a threshold volume and market area are reached (perhaps customers within a half-mile). Profits will then increase as long as revenues from more distant customers exceed the cost of serving them (if the store pays for delivery), until the point of maximum profits is reached, at range 1 (perhaps at two miles). After this, profits decline until the maximum range of sales (range 2) is reached. At this point (three miles) costs equal revenues. Presumably no rational seller would offer goods to customers beyond the maximum-profit range. If customers pay for their own transportation, the maximum range is simply the distance from which no more customers will come.

through higher prices. The maximum-profit **range** (range 1 in figure 7.3) defines the ideal, or optimal, service area size for your store. Range refers to the maximum distance over which a seller will offer a good or service or from which a customer will travel to make a purchase. Beyond that, the cost of serving additional customers will be greater than the additional revenues you make on their purchases. It would not make sense to attempt to serve those customers. A further point will be reached where your total costs will equal the income derived from providing the service (range 2 in figure 7.3), and you will no longer make a profit. If the average customer buys $10 worth of groceries and the profit margin is $1 over wholesale and operating costs, the maximum-profit range—range 1 (optimal service area size)—will be where the cost of delivering to the next farther customer just equals $1. For example, at $.50 per kilometer per $10 of groceries, this would be at two kilometers, since even at almost two kilometers, total profits will still be rising, although the margin is very small.

The optimal service area size is often determined by how quickly the demand for goods and services falls off and by how far customers are willing to travel. If customers pay for their own transportation, they will have less money to spend on purchases. For example, if customers allowed $10 for groceries but must spend $1 of it on transportation to a particular store, they will be able to buy fewer groceries than if they lived near the store. If customers are unwilling to spend more than $1 for travel, this will be the limit of effective demand.

The time limit that people are willing to allow becomes the critical factor for certain kinds of central place activities. No one wants to wait long for police and fire protection, taxis, or public transportation; no one wants to spend much time on a trip to school, to grocery stores, to the gas station, or to the hospital. Such activities must therefore be reasonably close to the population they serve. The maximum range in which an activity can profitably operate reaches its limit when demand starts to decline because of time or travel costs. This range will be small for frequently or quickly needed goods and services, large for those less commonly demanded. Beyond the maximum range, a demand for the good or service from that supplier no longer exists.

The range in which profits can be made is similar for many kinds of activities. For example, grocery stores, taverns, and gas stations need to meet only a low level of demand *(threshold)* in a fairly small area in order to survive. Their *range* (distance customers are willing to travel) is also small, though larger than the threshold area. Many central place establishments, such as variety and department stores, sell goods with varying but overlapping threshold and range requirements. The precise combination of activities that can theoretically be supported by one center is determined by the similarity of their thresholds and ranges, the population density and purchasing power of the service area, the nature of transport costs in the area, and the kinds of goods and services demanded in that area or culture.

In theory, if an activity with a given threshold is successful in a central place, all activities with lower thresholds should also succeed there. Thus if gas stations and taverns have lower thresholds—perhaps 500 customers—than grocery stores—with 1,000 customers—then centers that support grocery stores should also be able to support gas stations and taverns. It would be possible, of course, to find local gas stations and taverns in centers too small to support grocery stores or, in some small crossroads, general stores that also sell alcoholic beverages and have one or two gas pumps out front. Typical combinations of activities in different sized centers will be described later in this chapter.

The Spacing of Central Places

Central place theory is primarily concerned with the optimal spacing between service activities or centers. Consider the spacing between two competing establishments, such as your grocery store and that of a competitor. It would seem logical for both of you to locate at points where each could realize maximum profits, or serve all potential customers, without fighting over the same territory. In our example, you and your competitor would locate four kilometers apart. Customers beyond the two-kilometer radius would have to pay extra for delivery or buy less. But there is a risk when two competitors allow themselves the maximum-profit range: a third might enter the market midway between the first two, reducing all three to marginal thresholds in which the level of demand is barely sufficient to support each establishment. In our example, the stores would now be two kilometers apart and would probably have many fewer customers. Thus two competing activities or shopping centers will usually allow less than maximum space between themselves and accept a less than maximum, but still highly satisfactory, profit margin. Their location would be to the customers' advantage, since now all are within delivery range. In our example, you and your competitor might locate three and a half kilometers apart, thus discouraging a potential competitor from opening a store between you and the other store. Competition tends to result in a spatial separation that yields only minimally acceptable profits, a fact that explains why existing shopping districts fight against the establishment of new ones.

Central place theory could be considered a **spatial monopoly** theory. Each center has a competitive advantage in a given territory because of the distance factor: transportation costs are less for customers who go to the closest seller or center, but the extent of the spatial monopoly is strictly limited by competition. Referring again to the example above, if one grocery store raises its prices, its cost advantage within its spatial monopoly, or captive market, is partly eliminated. The neighboring grocery store can capture part of the first grocery store's market and either put it out of business or force it to return to its original price level. On the other hand, if the grocers collude, all may raise prices simultaneously. Such collusion appears typical of major oil company service stations.

Theoretical Hexagonal Structure

Now that you are acquainted with the concepts of the spatial equilibrium between an activity or center and its customers and the optimal spacing between an activity or center and its nearest competitor, the problem is to determine how, in theory, a set of centers is structured in space.

Assume for now the following conditions: (1) the region consists of a uniform plain (no environmental variation, such as mountains or rivers) in which population and purchasing power are evenly distributed; (2) transport costs increase with distance (the rate for each kilometer is the same, with no discount allowed for long-haul trips); but (3) potential demand does not weaken with distance. Given these conditions, imagine a set of seven central places or grocery stores, each having its own spatial monopoly. Experimental juggling of their positions will reveal that only a triangular arrangement of competing centers will, for any one center, make its six competitors equally distant (figure 7.4). This triangular "equidistant spacing" is theoretically the most efficient: the "unserved" area is minimized, and all customers are as close to a center as possible. As you can see from figure 7.4A, however, there are unserved areas outside the circles. The risk is that new competitors might enter to serve these areas. If the centers are moved a bit closer together, their service areas will overlap. And if the overlap area is divided between the centers closest to it, a hexagonal service area pattern will emerge (figure 7.4B).

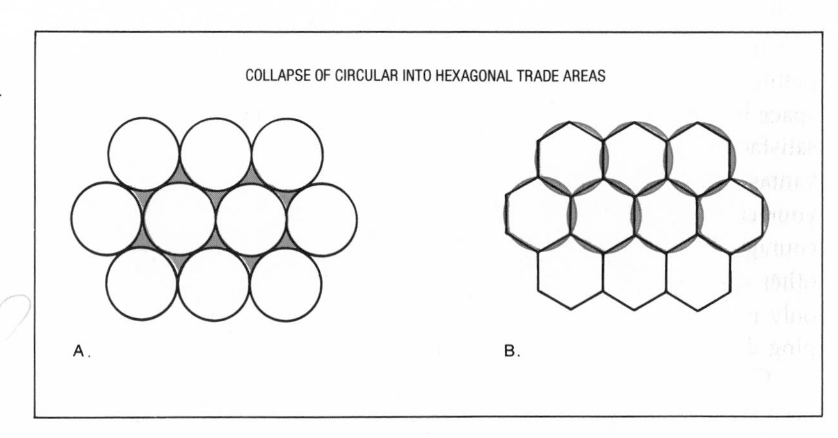

7.4. Development of hexagonal service or trade area pattern. In the circular service pattern (A), the shaded triangular areas outside the circles are unserved. To avoid the unserved areas, the circular pattern "collapses" into a hexagonal structure (B).

What is wrong?
a) Assumes too much
b) out of date

An isolated circle is a more optimal form than an isolated hexagon, but a set of circles (or a set of any other shape) is not as efficient as a set of hexagons, in the sense of packing areas together. Total distance between customers and a center is minimized, and so is the disparity between the farthest customers at the corners and those near the centers.

The Central Place Hierarchy

Ideally, then, the most commonly demanded services will be provided by the smallest centers, whose service areas are arranged in a hexagonal pattern. But how many of these will offer less common services requiring the support of larger populations, and how will these higher level places be arranged? Christaller theorized that the sizes and locations of the centers will vary according to three basic principles: marketing, transportation, and administrative.

3 principles

The Marketing Principle

The **marketing principle** minimizes travel and ensures that everyone will be as close as possible to all levels of the hierarchy. To understand how a central place hierarchy develops according to this principle, consider a set of hamlets five kilometers apart, each containing a grocery store, tavern, and gas station. Suppose that one hamlet in figure 7.5A succeeds in adding higher threshold activities, such as a drive-in movie theater, a restaurant, and a dry cleaner, and becomes the circled *B*-level place. The new village-level activities require perhaps twice the number of customers that are in a hamlet's service area. Thus the circled *B*-level place will need to sell these higher threshold activities to some customers in the service areas of the six neighboring hamlets (circled *C*-level places). None of these six hamlets can acquire village-level activities now, since some of their customers are needed to support the hamlet-turned-village. The closest centers that can compete with it—that can also add the higher threshold activities—are the boxed *A*- and *B*-level places in figure 7.5A, located beyond and exactly as far from the neighboring hamlets as the circled *B*-level place itself is.

If boundaries are drawn midway between these six higher level centers, connecting the *C*-level places, a set of larger hexagonal service areas will result. Each of the higher level centers, or villages *(B)*, is spaced nine kilometers from the next village and serves the population in one-third of the service areas of the six surrounding hamlets. Hence the total area and population served by the village (for village-level activities) is three times larger than that of the hamlet. Similarly if any of the villages acquires town-level activities (the boxed *A*-level place in figure 7.5A), a third ("town-") level service area will be created. The towns will be spaced fifteen kilometers apart and will serve an area and population nine times larger than the hamlet's and three times larger than the village's.

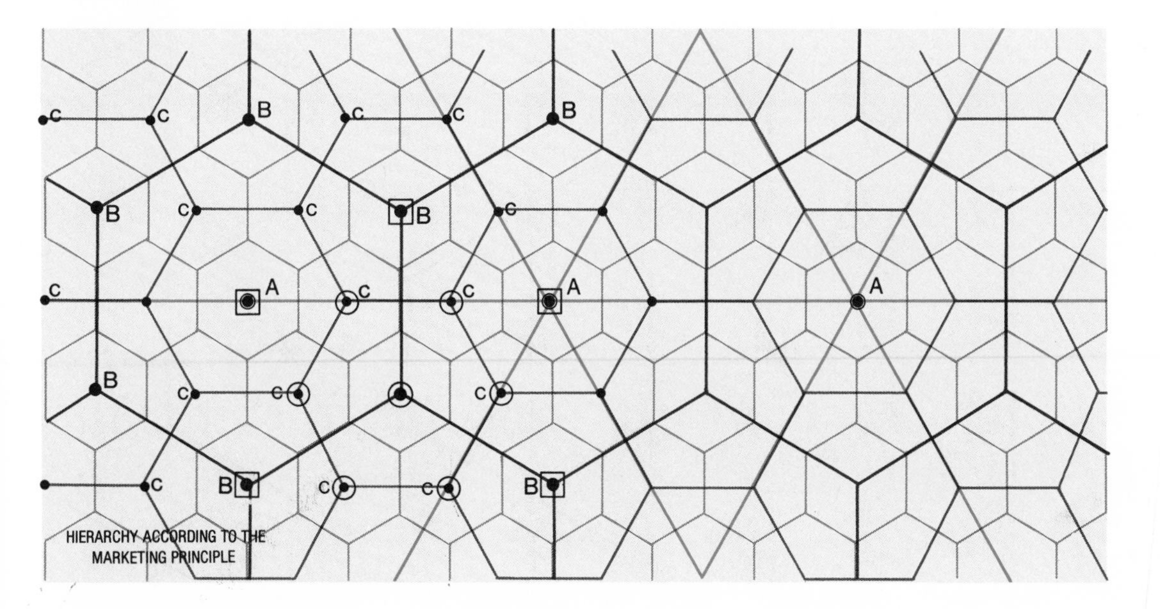

7.5. Central place hierarchies.

Diagram A: Development of the central place hierarchy according to the marketing principle. This figure shows a theoretical arrangement of towns *(A)*, villages *(B)*, and hamlets *(C)* and their hexagonal service areas. The orange lines radiating from the towns are transport links. To understand how such a pattern develops, assume that all the places started as *C*-level hamlets. One of these hamlets acquires village-level activities and becomes the circled *B*-level place in the figure (lower center of the set of places). These activities need more purchasing power than is available in the hamlet's service area: they require customers from one-third of the service areas of the six surrounding hamlets (circled *C*-level places). These hamlets cannot become villages because some of their customers are needed to support the new village, but the next hamlets out can become villages (boxed *B*-level places). In this way, village-level service areas contain three times the area of a hamlet (the original hamlet's service area and one-third of the six other hamlet's service areas). Similarly suppose that the village northeast of circled village *B* adds town- *(A-)* level activities, which require more customers than are available in the village's service area. This new town will require customers from one-third of the *B*-level service areas around it. Again, the town's service area contains three times the area of the village's.

The Transportation Principle

The **transportation principle** ensures that the hierarchy of transport routes reflects the hierarchy of central places. The marketing arrangement of central places minimizes travel but is not conducive to the most efficient transportation system. The important transport links (orange lines in figure 7.5A) between larger centers *(A)* do not pass through the next smaller centers *(B)*. For more efficient transportation, a different arrangement of central places is needed (figure 7.5B). If instead of the pattern shown in figure 7.5A,

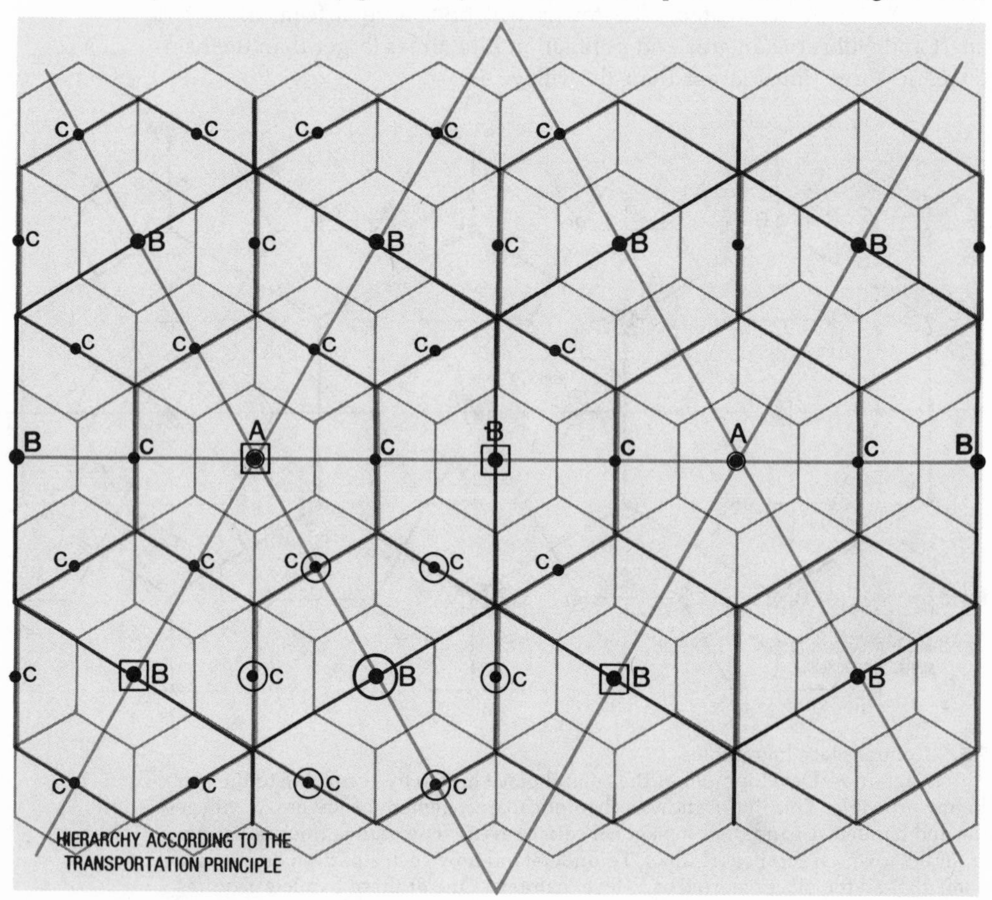

Diagram B: Development of the central place hierarchy according to the transportation principle. In this case, assume that the first hamlet mentioned above requires four times the purchasing power of its own service area in order to add *B*-level activities. If it serves half the customers in the surrounding *C*-level service areas, then the next closest places that can do the same are the boxed *B*-level places. Similarly *A*-level places (towns) require half the customers from the six surrounding *B*-level service areas. This model is said to follow the transportation principle because larger roads connecting *A*-level places pass through intermediate *B*-level places.

every other hamlet in a line (boxed *A*- and *B*-level places in figure 7.5B) were to acquire village-level functions, the service area population of the hamlets *(C)* would be shared equally by two villages. Thus the service area of each village *(B)* is four times larger than the service area of each hamlet, and the service areas of towns *(A)* are sixteen times larger than those of hamlets.

In this theoretical arrangement, transportation is more efficient because routes connecting cities pass through towns, and routes connecting towns pass through villages (the orange lines in figure 7.5B). (Customers, however, must travel a little farther to reach a center at a given level of the hierarchy than would be necessary under the marketing principle.) Historically centers of a lower order indeed tend to develop midway along a road between two centers of a next higher order, suggesting that this hierarchical structure may be the most likely one to occur.

In both the marketing and transportation patterns, the service areas of smaller centers are not completely contained within the service areas of larger centers but are divided among them. They are divided equally among three larger centers in the marketing pattern, and between two larger centers in the transportation pattern (see figures 7.5A and B). This split may seem confusing from a planning point of view, but many smaller places in a median position actually show a "torn loyalty" between two or three larger ones. For instance, people living in northeastern Connecticut shop in both New York City and Boston for highest level activities (figure 7.6). You can probably think of an example in your region.

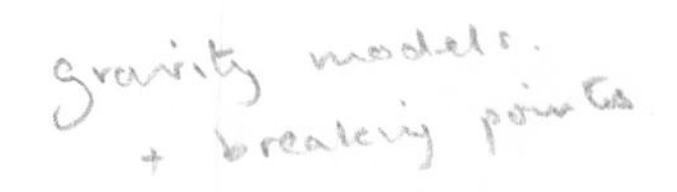

Whether or not a lower level place, such as a village, is intermediate between two or, less commonly, three higher level places, such as towns, there is a tendency toward nesting; most of the villagers may come to prefer only one of the towns rather than dividing their visits evenly between them (figure 7.7). This fact suggests that customers achieve greater convenience and time savings by having fewer destinations.

The Administrative Principle

The **administrative principle** ensures that smaller service areas will nest fully within larger ones, avoiding split loyalties. For administrative purposes, such as running a school system or governing a state, the division of smaller areas between higher level centers and areas is not practical. For example, it is more efficient to nest a primary school area entirely within the larger secondary school area. If smaller areas are not to be divided, the most logical (distance-minimizing) arrangement is for a larger center to serve the entire service area of the six surrounding centers. Under this theoretical arrangement, each service area is seven times larger than the next smaller one (see figure 7.5C). In patterns based on the administrative principle, **nesting** is automatic but the transportation system is relatively inefficient. Customers must travel farther than would be necessary if the marketing or transportation principle governed the arrangement.

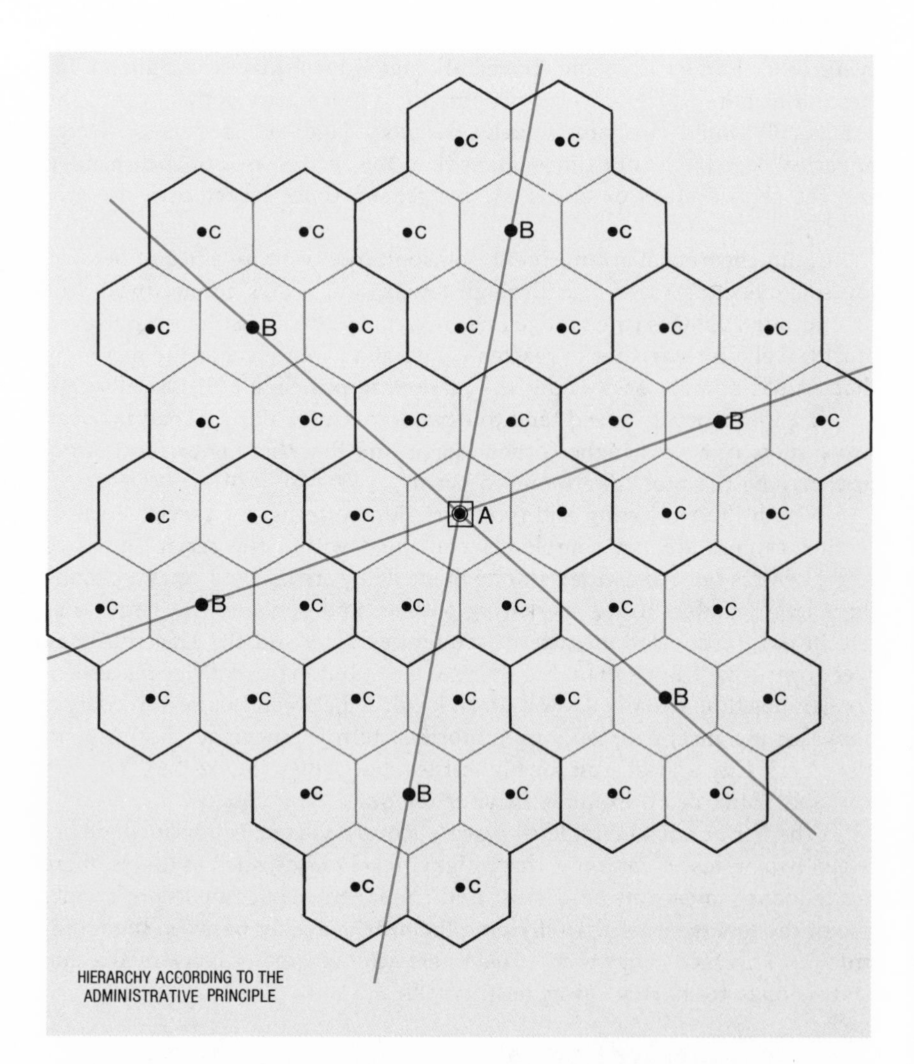

Diagram C: Development of the central place hierarchy according to the administrative principle. In the administrative arrangement, service areas of smaller centers are wholly nested within service areas of larger centers. The pattern that most efficiently meets this requirement is one in which larger areas are seven times larger than the next smaller areas.

One might expect that in an economy where regional self-sufficiency is a planning goal, as in the Soviet Union, an administrative structure would develop in which lower order centers are nested within higher ones. But in reality the Soviet pattern more closely approaches the transportation principle. A better example of administrative principle nesting is the organization of Federal Reserve districts in the United States. Seattle, Portland, Salt Lake City, and Los Angeles are nested within San Francisco's Twelfth District. (Although Los Angeles is now larger, the headquarters remain in San Francisco, far more important in 1912 when the system began.)

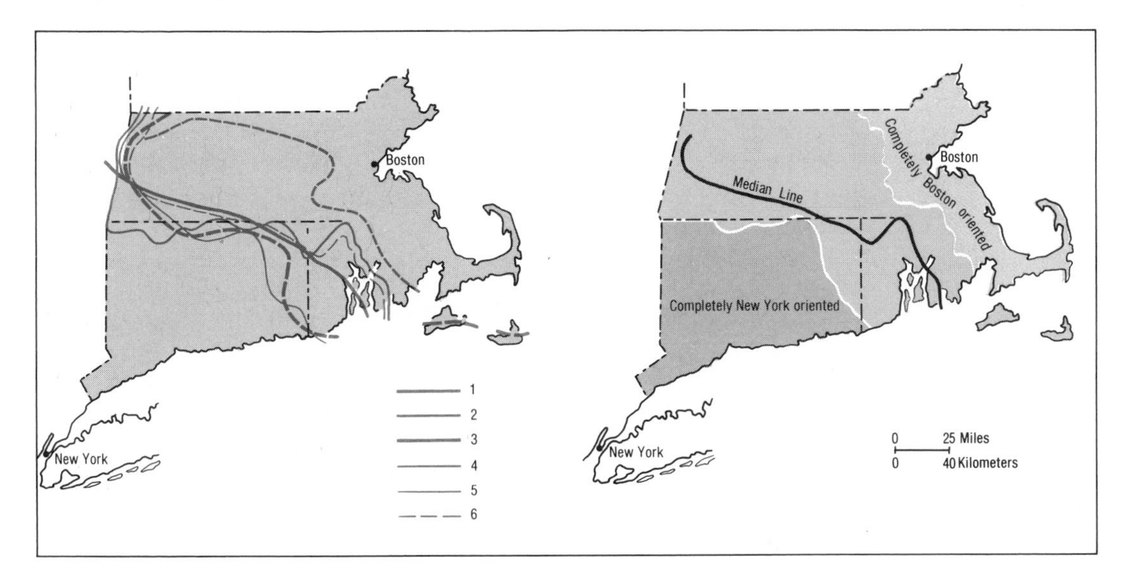

7.6. Delimitation of service area boundaries between Boston and New York. In the left-hand portion of this figure, boundaries for five activities were determined: (1) railroad passenger flows, (2) newspaper circulation, (3) telephone calls, (4) work addresses of industrial firm directors, and (5) correspondence. These defined a composite boundary (6—the median line, corresponding to the middle of the other lines, on the map to the right). As expected in theory, New York, being larger, extends its zone of dominance somewhat at the expense of Boston.

Source: Data used with permission from H. L. Green, "Hinterland Boundaries of New York City and Boston in Southern New England," *Economic Geography*, Vol. 31, 1955.

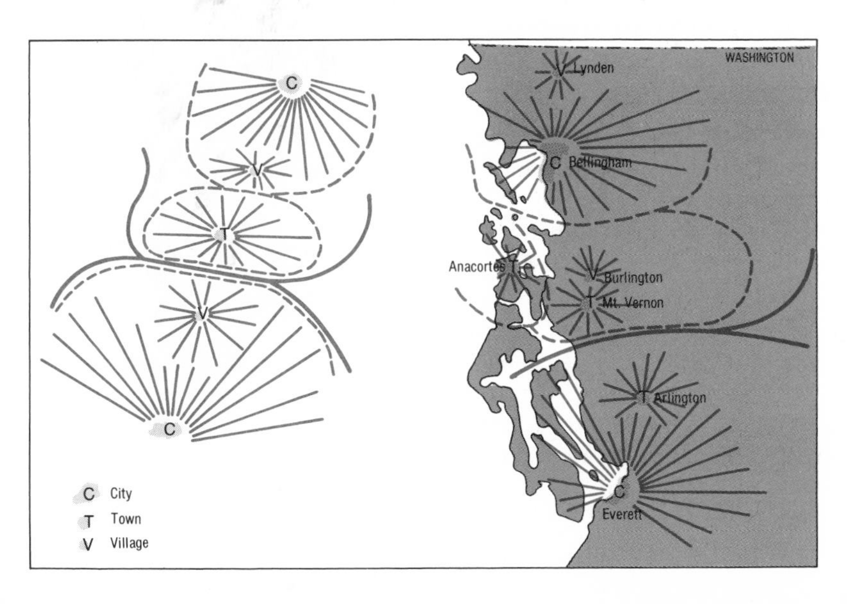

7.7. The nesting process. The diagram on the left shows that the town and its trade area, midway between the two cities, are served entirely by one city rather than being divided. This nesting process is illustrated by actual cities and towns in Washington: Mt. Vernon and the neighboring village of Burlington are served by Bellingham, leaving Arlington to Everett.

Other examples of administrative principle nesting may be found in small nations or isolated regions where a large central place dominates the entire urban system. In Denmark, France, Poland, Uruguay, and Thailand, for instance, a single ring of provincial cities surrounds the primary city. But at lower levels of the central place hierarchy, the transportation or marketing principle, or both, might govern the pattern, as is true at the hamlet and village level in Denmark.

Each of the three principles—marketing, transportation, and administrative—generates a specific theoretical pattern. Most actual landscapes are the product not of one but of all three tendencies and perhaps of other forces not mentioned, such as cultural or recreational needs. Marketing arrangements tend to develop in densely populated rural areas where transportation is limited, as in parts of China, India, and Europe. But when economic exchange, or trade, is limited and government is particularly strong, then administrative efficiency becomes more important than economic efficiency. Trade barriers and national boundaries tend to bring about an administrative arrangement, at least at the metropolitan level. And as economic efficiency gains importance, marketing or transportation patterns may begin to transcend administrative or political boundaries. Improved transportation tends to readjust central place patterns according to the transportation principle. All three principles underlie society's organization of space and often result in composite rather than theoretically pure patterns. (For an example of a composite settlement pattern, see figure 6.6.)

The central place patterns we have described are the result of economic competition. In the real world, of course, not all towns have developed in this way. As we noted in chapter 6, many were arbitrarily located by railroad executives, land developers, or governments. The fact that such towns have survived and thrived, however, suggests that the location decisions often achieved the goals of spatial efficiency.

Commonly Recognized Central Place Levels

As central places grow in size and functional variety (combination of activities), they require a larger population for support and a wider spacing between themselves and competing centers. Hamlets provide only minimal, frequently needed services, while cities and metropolises provide all the services of smaller centers plus specialized services only the highest level centers can support. Various central place levels are described below, with the approximate number of people they contain, the size of the population they serve (including their own residents and the people dispersed in the surrounding area), and the **central place functions** they typically support (figure 7.8). The data are typical for the United States; the levels, sizes, and particular mix of activities differ in other cultures.

The *hamlet* may have as many as 800 residents or as few as 5, though its population usually ranges between 200 and 600. Most hamlets serve about

2,000 persons and provide such everyday conveniences as gas stations, grocery stores, drugstores, and taverns. They often have schools and perhaps a grain elevator or other farm-related activity. Most people do not want to travel more than five minutes to reach these activities. Improved routes have reduced travel time and effort, thus increasing the distances people are willing to travel. Because larger centers with more functions have become easily accessible to motorists, many hamlets within fifteen kilometers (nine miles) of a larger town have declined. On the other hand, many survive or have been established along major highways, primarily to serve a traveling population.

The *village* contains between 800 and 4,000 persons (average population about 2,000) and serves as many as 10,000 persons. It offers a greater variety of services than the hamlet, including some that people may use only occasionally but for which demand is constant: bakeries, several churches, a restaurant or fast-food outlets, a post office, a high school, a general store, a hardware store, an auto repair shop, a farm-implement-and-feed dealer, and perhaps a doctor, dentist, and bank.

The *town* has a population range of 4,000 to 20,000 (average size 10,000) and serves 25,000 to 100,000. The typical American town is epitomized by the county seat, the traditional social and economic center of American life. Its counterpart may be found in many countries—for example, the Chinese *hsien* (district capital) or the French *préfecture* (department capital). Here are usually found a hospital, newspaper, large high school, department store, lawyers, courts, and places of entertainment. People in the surrounding region utilize the town's medical, dental, and veterinary services; attend its churches; patronize its restaurants, movie theaters, dry cleaners, car dealers, and many retail stores; and deal with its insurance or real estate agencies. Most towns are probably no more than a half-hour to an hour, or twenty-five to forty kilometers, from most of the people they serve. In the United States and Europe, the metropolis may have become the dominant influence in people's lives, but the town heritage is still strong. Most city dwellers, for example, still prefer to shop in large neighborhood, or town-sized, shopping centers.

The *small city,* containing from 20,000 to 100,000 people and serving up to a half-million, is a district capital and a distribution and communications center. The small city is usually at the bottom of the wholesale distribution ladder; it is the smallest market that can efficiently support a middle exchange between producer and retailer. It may contain a college, a department store, medical specialists, a television station, and many retail and service specialties, such as music, sporting goods, and photographic supply stores. In a sense, small cities are the outposts of metropolitan life. They are served by good roads and railroads and have fairly good air connections. They are so located that people in the area can travel to and from the center in the same day and still have time to shop and do business. In many developing countries, these may be the small provincial capitals that will later emerge as metropolises. Small cities like Kano and Port Harcourt in Ni-

A

B

7.8. The central place hierarchy. This series of photos shows the increasing variety, specialization, and complexity of activities in the central place hierarchy, from a Vermont hamlet (A) to a small Massachusetts town (B) and city (C), to downtown Boston (D). (Photo A by Josef Muench; photos B and C by Clemens Kalischer; photo D by Mike Mazzaschi from Stock, Boston.)

C

D

geria, for example, are perhaps emerging metropolises. (To take an example from the United States, as recently as 1940 only 65,000 persons lived in Phoenix, Arizona. Now it has mushroomed to 1.2 million people.)

The *large city*—a regional, state, or provincial economic capital—has a population of 100,000 to 500,000 and serves from 1 million to 3 million. Large cities are more self-contained than smaller ones. They support a wide variety of local manufacturers and have competing television stations, extensive business services, a major airport, specialized hospitals, and perhaps a university. Wholesale trade and finance may exceed local retail trade in volume. Large cities are usually the highest level center required by a population. Most people can find nearly all the goods and services they need in centers of this size and complexity.

The *metropolis,* with 500,000 or more residents, serves between 2 million and 30 million people. Dominating the modern economy and culture, the metropolis is where major economic decisions are made, most manufacturing activities are concentrated, and most innovations are developed. It is a center for communications, education, and the arts, and provides communication, trade, and cultural links to other nations. The principal ties between the United States and Western Europe, for example, are between New York City and London, Paris, and Rome. People living a thousand miles from the metropolis may rarely, if ever, set foot in its streets, but they still feel its influence through the agricultural commodities market, the banking structure, mail-order houses, wholesale-retail relationships, and other metropolitan functions. In the United States, for example, many corn belt farmers support the Chicago sports teams, read the Chicago newspapers, and study the city's commodity prices and prime interest rates.

Consider the structure of central places in your region. A person who lives in the state of Washington, for example, might observe the following pattern:

Level	*Population*	*Function*
Hamlet (Washtucna)	**325**	**Crossroads; small service center in wheat area**
Village (Ritzville)	**2,000**	**Service center in wheat area**
Town (Moses Lake)	**12,000**	**Main service center of Columbia basin irrigation area**
Small city (Walla Walla)	**35,000**	**Larger service center in wheat area; college town, etc.**
Large city (Spokane)	**285,000**	**Capital of the "inland empire"**
Metropolis (Seattle)	**1,400,000**	**Leading commercial and industrial center of U.S. Northwest**

Table 7.1. Central Place System in a Modified Transportation Pattern

Level	Type of Place	Size Range[a] (population)	Number of Places[b] A	B	Spacing Range[c] (km)
1	**Hamlet**	**5–800**	**3072**	**12,288**	**3–10**
2	**Village**	**800–4,000**	**768**	**3,072**	**6–20**
3	**Town**	**4,000–20,000**	**192**	**768**	**12–40**
4	**Small city**	**20,000–100,000**	**48**	**192**	**24–80**
5	**Large city**	**100,000–500,000**	**12**	**48**	**50–160**
6	**Metropolis**	**500,000–2.5 million**	**3**	**12**	**100–320**
7	**World metropolis**	**2.5–15 million**		**1**	**640–1,280**

Note: The higher level places are five times the size of the next lower level.
[a] Size range: Ideal sizes for a national economy depend on nation's total population and degree of urbanization.
[b] A: A nation of perhaps 50 million people. B: A nation of perhaps 200 million people.
[c] Spacing range: Spacing depends largely on density and thus varies within and between economies.

The theoretical principles discussed earlier have an important bearing on the actual size, quantity, and spacing of centers in the hierarchy. Table 7.1 approximates these figures for a central place system structured after the transportation principle. (Some geographers might add a seventh level—a *world metropolis,* or *megalopolis,* with 2.5 to 15 million persons.)

Central Places in Non-Western Societies

The central place system occurs in non-Western as well as Western societies. Villages, towns, and cities serving rural China constitute a well-organized system in which the market structure tends to parallel the rise in social, administrative, and (before the Cultural Revolution) religious complexity. Villages may be served only by itinerant peddlers or periodic markets. Towns have *standard markets* for the purchase and exchange of farm produce and everyday goods, as well as restaurants, tea shops, and social and political clubs. Small cities tend to have *intermediate markets,* with a greater variety of foods, clothing, furniture, and services—and better health facilities and secondary schools, despite the current egalitarian emphasis of Chinese society. Both marketing and transportation arrangements have been identified on the Chinese landscape, the former tending to prevail where topography is slightly more rugged and transport routes are more limited.

The central place structure in India is similar to the one in China. Again, examples of marketing and administrative patterns have been identified. Table 7.2 lists typical activities found at lower levels of the hierarchy.

Table 7.2. Activities and Their Thresholds in Agricultural Regions of India

Activity	*Number of Establishments*	*Thresholds*	*Number of Places*
Highest level	200		1 (city)
College		25,000	
High level	120		2 (large towns)
Hospital		12,000	
Secondary school		6,000	
Medical supplies		5,500	
Utilities		5,300	
Engineering goods		8,000	
Medium level	35		8 (towns)
Laundry		4,700	
Auto and tractor repair		4,600	
Bank		4,300	
Medium-low level	7		7 (large villages)
Jeweler		2,600	
General store		2,000	
Carpenter		2,000	
Clothing		1,700	
Flour milling		1,400	
Cycle repair		1,400	
Low level	2		12 (villages)
Tailors		90	
Primary school		80	
Grocery		270	

Source: R. P. Misra, *Regional Development Planning in India* (Delhi: Vikas, 1974). Reproduced by permission.

Theory and Reality

Central place theory has been widely criticized because neither research nor maps show the pure patterns of development associated with the marketing, transportation, and administrative principles. Not only is the geometry of the real landscape irregular, but central place sizes are inconsistent, population density is uneven, rivers and mountains distort the theoretically homogeneous plain, and mining and manufacturing towns seem to intrude in the landscape at random. What then is left of central place theory?

First of all, it is naive to seek perfect geometrical patterns in the real world. Moreover central place theory as formulated is incomplete. Many nonservice functions, such as mining and manufacturing, help to create and support urban centers. The real contribution made by the theory is that it has revealed general tendencies in human spatial behavior and has explained why clear variations exist in the size, quantity, functional complexity, and spacing of urban places. The test of its validity lies not in the existence of pure geometric forms but in the answers to the following questions:

1. *Do entrepreneurs seek to serve the potential customers of an area and to carve out somewhat monopolistic trade (service) areas for themselves?* Certainly they do. This is the basic planning or decision-making problem that confronts the developer of a prospective shopping center, for example.
2. *Do similar kinds of central place activities tend to be regularly spaced?* Experimental research shows evidence that the distance between places offering a similar variety of activities does not vary significantly within physically and culturally homogeneous regions. Think about the spacing of planned shopping centers in the metropolis nearest you, or about the spacing of a well-known fast-food chain's outlets in your area.
3. *Do individuals tend to minimize the distance they travel to satisfy their needs and desires?* Many researchers have observed that individuals tend to minimize the distance they travel, though for various reasons, they will often go beyond the closest opportunity.
4. *Does the central place hierarchy exist in real life?* That is, does an individual use different central place levels to obtain different kinds of goods and services? We touched on this point earlier but will elaborate on it below.

From our own shopping behavior, each of us is aware of the central place hierarchy. We often go to the local store or gas station, travel less frequently to a larger town or shopping center, and only occasionally use the downtown shopping district of a large city (unless we live nearby). Yet geographers and others have debated heatedly whether a central place hierarchy actually does exist in view of the observed continuum of urban sizes. If all urban places in the United States, for example, were ranked according to population, a fairly continuous size gradation would be seen. The scale would not be neatly divided into central places with 50,000 people; 250,000; 1,250,000, and so on. Where in the hierarchy would real places of 100,000 fit in? With the level 4 centers (small cities) or with the level 5 centers (large cities)?

The range of urban sizes implies the existence of a central place hierarchy. But we cannot expect to observe a clear hierarchical division for at least the following three reasons. First, almost half the support of urban places comes from activities other than central place services (such as manufacturing), the proportion varying widely from place to place. The center with a population of 100,000 could have half its labor force in manufacturing for export from the region. Thus the number of people involved in central place services actually approximates 50,000. Second, the density and relative purchasing power of the population dispersed around central places varies greatly. Centers of about 40,000 in the sparsely populated western United States are widely spaced and offer the kinds of services (though perhaps less profitably) found only in places of about 60,000 to 100,000 in the densely populated Northeast. Third, entrepreneurs and others often make mistakes. If a person decides to establish an intermediate-level function, such as a movie theater, in a lower level center (hamlet), it may not survive. The net effect of these and other factors is to cause a great deal of variation around any theoretically expected central place size, resulting in a continuum rather than a strict hierarchy of central places.

One way to determine whether a hierarchy exists is to examine an area within a range of distances actually traveled by people. If we select a range within which most trips to central places are made, we will discover that there is a discernible hierarchy of size despite varying industrialization and other factors. Look at your own county or city. Where do you usually buy gas or milk and bread? Where do you visit the dentist or see a movie? And where do you go with friends on a special occasion or shop for unusual gifts? Chances are, you use the gamut of central place levels at one time or another during the year. The hierarchical structure of central places is also suggested by the pattern of wholesale trade activities in the United States (figure 7.9).

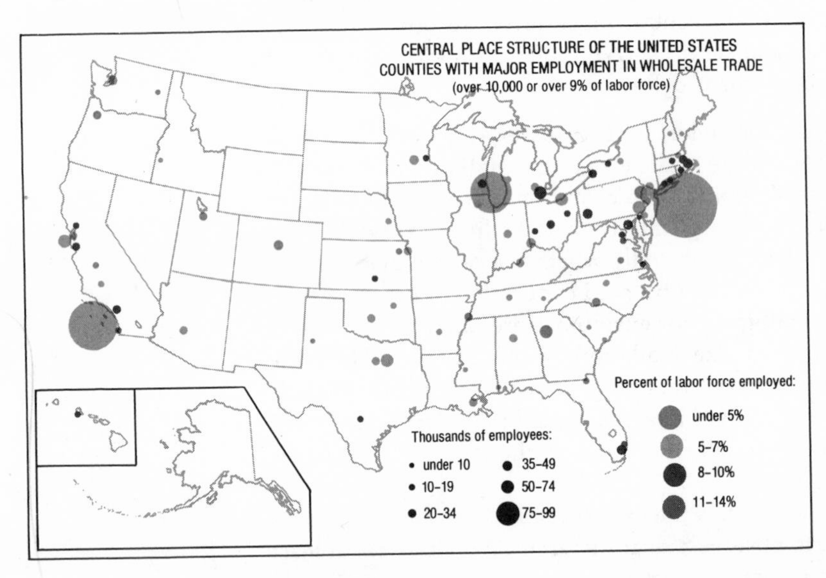

7.9. Central place structure of the United States: counties with major employment in wholesale trade, 1978. The hierarchical ranking of cities in the U.S. economy is fairly well indicated by the absolute and relative importance of wholesale trade employment.

Modifications

In recent years geographic researchers have found evidence confirming several elements of central place theory: variation in the thresholds and ranges of central place functions, the mutually repellent force of competing central places, spatially monopolistic zones, and the hierarchical structure of central places. But the fact that pure central place landscapes cannot be observed has led researchers to propose modifications of central place theory:

1 The uniform-plain premise has been relaxed so that physical and cultural variation may be taken into account. An example

is the variation in population density resulting from rugged terrain or a particular settlement pattern, such as the Range and Township System in the U.S. West.

2 The premise that central places develop simultaneously over limitless space has been relaxed so that theories for the gradual spread of development from areas of early settlement could be admitted. An example is the diffusion of development in the United States from the eastern seaboard (see figure 3.8).

3 The premise that people always behave rationally and act on accurate, complete information has been relaxed, and the effects of irrational behavior and error on ideal patterns have been analyzed. An example is the overlapping of service area patterns that occurs when people shop beyond the closest opportunity.

4 Geographers have recognized that higher level centers dominate service areas that are larger than predicted by central place theory. For example, people living closer to Peoria, Illinois, than to Chicago still shop in Chicago for goods and services also available in Peoria.

5 Geographers have recognized that central place theory conflicts with the agricultural gradient theory; both theories should be modified. For example, according to central place theory, population is fairly evenly distributed throughout the region; but according to the agricultural gradient theory, population density decreases with distance from very large central places.

6 Geographers (including Christaller) have recognized that nonservice activities, notably manufacturing, contribute to the growth of central places. For example, Chicago's development was due to manufacturing as well as service activities.

Each of these modifications is discussed in greater detail below.

Spatial Variation

Variation in the physical environment, and therefore in rural population density, affects the size of theoretically hexagonal service areas and modifies their shape. Central places in rugged and sparsely populated regions have large, irregular service areas both locally and over the entire region. For example, towns strung along Florida's southeastern coast are prevented from forming more typical central place patterns by the almost uninhabitable swampy Everglades to the west.

In local areas, rugged topography and irregular transport routes have limited the growth of the central place system to hamlets and villages containing few activities. Small sets of lower level settlements, known as **dispersed cities,** are commonly found in mining or logging areas, as in parts of the Appalachian United States, the Ruhr, upper Silesia in Poland, and the Donbas in the Ukraine. Such conditions result in longer travel time, reduced consumption of goods and services, and less demand for higher level activities.

Even if physical conditions are fairly homogeneous, the quality of transportation between larger centers is better than that between smaller ones. As a result, corridors of greater development appear along major railways and other routes, producing sectors of greater and lesser population density. Settlement clings to the Hudson-Mohawk corridor in New York State and the Trans-Siberian railway in the Soviet Union.

Before the twentieth century and the development of the train, truck, and automobile, port cities were the most important centers of exchange. Hence in coastal countries or countries with major rivers, the central place system is tipped toward, or heavily dominated by, ports. The port function of importing, exporting, and transshipping goods is a special kind of central place activity: the "hinterland" is usually asymmetric (the port is on one edge), and the "foreland" is a disparate set of import sources and export destinations. Most of the world's leading central places—metropolises such as New York, London, and Tokyo with over 5 million persons—are ports. The main exceptions are Paris, Moscow, Peking, and Mexico City, which were centers of historic land-based empires.

Finally spatial variation in the form of particular settlement patterns also affects the central place system. In much of the United States, for example, the system developed in the context of a rectilinear land survey. As a result, transport routes tend to run north-south and east-west, farm settlements are often square, and service areas are rectilinear rather than circular or hexagonal (figure 7.10).

Gradual Spread of Development

Admitting the gradual rather than simultaneous development of central places introduces two more complications:

1. Changes occur over time in the nature and price of goods and services, in population density and purchasing power, and in the quality and cost of transportation. Hence the factors determining threshold, range, and profitability are also subject to change.
2. It is possible that the entire central place system develops gradually as settlement spreads across the land, accompanied by changes in settlement pattern. An example is the shift from the closely spaced communities of New England to the more widely separated settlements of Montana, Nebraska, Utah, and other western states about two hundred years later.

The passage of time may have a significant effect on central place patterns. The central place hierarchy gradually develops as population and purchasing power increase and as agriculture becomes more commercialized. In parts of China and India, the hierarchy has developed mainly as a result of long-term population growth, without an accompanying rise in per capita

7.10. Rectilinear spatial pattern. This Kansas farmland exhibits the checkerboard pattern of roads and fields typical of areas settled under the Range and Township Survey System. (Photo by Grant Heilman.)

income (purchasing power): existing centers have taken on higher level functions, while farming villages have become lower level centers with some nonfarming activities. Similarly improvements in transportation—a new railway or superhighway, for example—can affect the central place hierarchy. Centers that are too close together or that are not served by the new transport link may decline in importance.

In the United States and other countries where settlement has spread gradually from its origin, more recent central places are often better adapted to modern modes of transportation and are thus spaced farther apart than comparable places in older areas (see figure 10.10).

Nonoptimal Behavior

People usually do not gather enough information or care enough about costs and profit margins to behave as location theorists wish they would. Few decision makers carefully study relevant spatial and economic factors before choosing a location. But constructing a theory as if people did behave rationally is valuable because it depicts the state toward which society

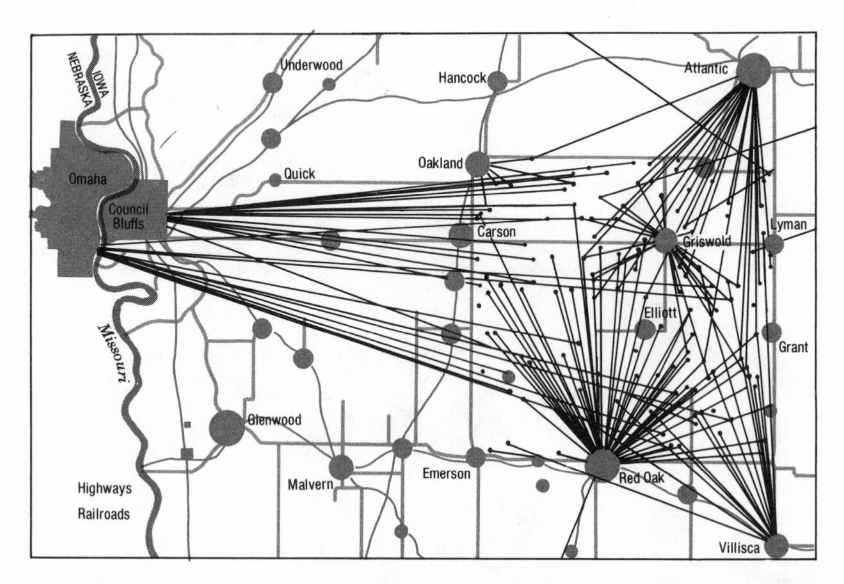

7.11. Overlap of trade areas. "Desire lines" connect rural customers (small dots) to central places. Note that the various centers do not monopolize all nearby customers. Trade areas overlap, indicated by the pattern of desire lines.

Source: Adapted with permission from B. J. L. Berry, H. G. Barnum, and R. Tenant, "Retail Location and Consumer Behavior," *Papers and Proceedings of the Regional Science Association,* Vol. 9, 1962.

moves, however imperfectly. When such **nonoptimal behavior** and uncertainty are taken into account, the following things happen:

1. The optimal location for a central place in the hexagonal structure will vary.
2. The kinds of goods and services offered by central places will vary. Some sellers will not take advantage of their opportunities, and others will try to offer goods whose thresholds cannot be met.
3. Customers will not perceive small differences in distance, or they will consider noneconomic factors such as prestige or loyalty to be more important than efficiency. The result is that theoretically clear-cut service areas will overlap (figure 7.11).

The Domination of Higher Level Centers

Central place theory defines service areas geometrically. Thus for a given level such as towns, all service areas are considered to be identically sized hexagons, even if some of the areas contain larger centers offering higher level activities. Studies of actual shopping behavior indicate that larger centers selling greater quantities of goods, owing to their larger internal market, will have a competitive advantage over smaller neighboring centers offering fewer of the same kinds of goods. Apart from the fact that larger centers are better known, economies of scale should enable sellers in those centers to enlarge their trade area through competitive pricing. The smaller neighboring centers will then sell fewer goods. Larger centers alter the basic central place structure by tending to make nearby intermediate-level centers drop to a lower rank. The smaller centers may adjust, however, by such marketing strategies as longer hours, personalized service, or delivery. Within the city, smaller centers, or business districts, may adjust by offering more convenient and cheaper parking.

Population Distribution

Central place theory originally assumed that settlement occurs on a uniform plain in which population is evenly distributed. But this premise conflicts with the gradient theory of agricultural location. You recall from chapter 5 that land rents and rural population decline with distance from large markets. As population density decreases, a larger service area is needed to support central place activities. Under such conditions, it is impossible to maintain ideal hexagonal areas. Moreover the density gradient often affects central place size and the goods and services offered at a given level. Salt Lake City, for example, is located in a sparsely populated region and has a much smaller population than cities with comparable functions in the densely populated Northeast.

An attempt to preserve hexagonal service areas while adjusting to the density and rent gradient produces distorted central place patterns, although they may in fact be quite regular and conform to the principle of competition for spatial monopoly. Figure 7.12 projects a possible surface on which a central place system is developed as geometrically faithfully as possible over a constantly changing density gradient.

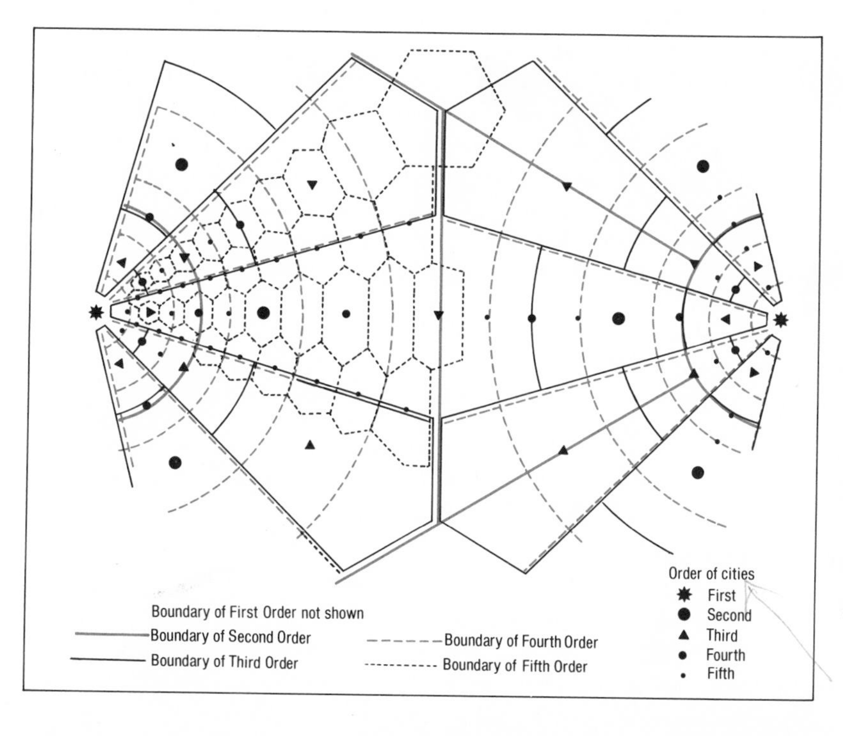

7.12. Theoretical gradient-hierarchy landscape. This figure represents a possible theoretical structure derived by combining rent gradient and central place theories. Smaller central places (dots) arise to serve the area between two large agricultural markets (stars). The hexagonal shape of the smallest trade area (for clarity, not all are shown) is maintained, but the trade areas (dashed lines) become larger and the centers farther apart with distance from the major centers (stars). Compare with figure 7.5, showing the usual central place patterns.

Role of Manufacturing Activities

Finally central place theory excludes basic town-building activities, which produce goods and services for export rather than for the local market. However, processing and other export activities directly affect the central place system. Even if settlements originate as manufacturing towns, they will inevitably need and acquire central place functions. Since industrial settlement tends to be clustered, not diffuse, the central place structure will diverge from the theoretical patterns based on uniform population density.

Central Place Theory and Planning

Central place theory has practical value for both private and public planners. People use it to plan the location of warehouses, stores, and shopping centers, as well as health facilities, schools, and other public services—and settlement itself. Manufacturing concerns often apply central place principles in planning the location of regional distribution centers. So do fast-food chains in planning the growth of their systems. Private and public concerns usually join in deciding where to locate large, planned shopping centers, which commonly develop into entirely new central places (the formation of new central places within a metropolis is discussed in chapter 9). Planners in the Netherlands have put central place theory to work on a larger scale. They have used it in the layout of new service centers for polders reclaimed from the Zuider Zee. On a smaller scale, the service centers in the Columbia River basin irrigation project were planned according to these principles.

The value of locating services at more than one central place level has proved particularly fruitful in practice. Many public health planners, for example, allow for several neighborhood health centers in addition to a few large, specialized facilities.

Developing countries are using central place principles to allocate public investment for schools, hospitals, markets, radio stations, roads, and the like, which will create or greatly influence the central place system. An interesting contrast may be drawn between the location of services in developing countries with socialist economies and in market economies. In the former, services are dispersed though fewer, reflecting an emphasis on equal access, while in the latter, facilities and stores are located where demand is greatest.

CENTRAL PLACE ACTIVITIES

Many central place activities involve the distribution of goods after they have been produced (in contrast to raw materials). Most obvious are wholesale and retail trade, but also included are local transportation and storage services, various business services—such as advertising, consulting, copying, and equipment rental—and repair and maintenance services. Other activities relate to the construction and maintenance of the built environment—buildings, housing, roads, public and private utilities, and parks. Real estate services are required for the exchange of property.

A large and growing class of central place activities provides services to individuals and groups: laundries and dry cleaners, beauty and barber shops, funeral services, fire and police protection, recreation and entertainment, libraries, health and educational services, welfare services, nonprofit institutions such as churches, communications (radio, television, newspapers), insurance and banking, and public administration. Table 7.3 summarizes central place activities in the United States.

In contrast to the manufacturing activities discussed in the next chapter, services form a hierarchy in terms of their level of specialization and scale. You are probably aware of the hierarchy from local movie theaters to metropolitan opera houses; from local playfields to municipal sports stadiums; from local libraries to the Library of Congress; from local newspaper advertisements to metropolitan advertisement agencies; from neighborhood churches to city cathedrals; from small airfields to international airports; from local branch banks to international banks; and from local school districts to national government.

Market economies in both less developed and highly developed countries tend to support a vast array of central place activities, while socialist economies—perhaps because of the historic leadership of the Soviet Union—tend to have weakly developed service sectors. Central place services dominate employment in market economies, and industrial or agricultural production is dominant in socialist economies.

Retail Trade

Retail trade (we include here not only the selling, repair, and maintenance of goods but also restaurants, taverns, and the like) is the most important central place activity in terms of economic value. The three main factors in the location of retail establishments are accessibility, agglomeration, and spatial monopoly. Operators of retail outlets usually choose sites where their stores can be seen or easily found and will attract customers. Retailers try to maximize their accessibility by locating as close as possible to a center on a transportation network. Lower level activities, such as groceries, taverns, and gas stations, are often located at the central crossroads of a hamlet or village or at highway intersections in cities. Higher level activities, such as department stores and specialty shops, are concentrated in larger centers at the intersection of more important streets or freeways.

Retail stores tend to agglomerate in groups where they help each other; a customer at one store may also be attracted by the goods in another store. A typical high-value-per-square-foot cluster includes clothing, jewelry, music, stationery, candy, and shoe stores. A typical low-value-per-square-foot combination includes gas stations, automobile dealers, automobile repairs, restaurants, and discount stores. Each retail group will try to locate as far as

Table 7.3. Central Place Activities in the United States and Orting, Washington

Activity	Number of Establishments United States	Orting	U.S. Employment	Establishments per 10,000 U.S. Population
Services	2,860,000	19	26,400,000	130
Social	1,780,000	13	19,250,000	81
Education	120,000	2	5,000,000	5.5
Government	250,000	3	5,000,000	11
Police, fire	45,000	1	750,000	2
Administration	120,000	1	2,900,000	5.5
Postal	32,000	1	700,000	1.5
Health	350,000	2	5,000,000	16
Physicians	205,000	1	{1,000,000	9
Dentists	125,000	0		5.7
Hospitals	7,200	0	2,900,000	0.3
Personal	510,000	2	1,500,000	23
Barbers, etc.	280,000	1	590,000	12.7
Laundry	97,000	0	525,000	4.4
Other	550,000	4	2,300,000	25
Churches	330,000	2	1,200,000	15
Nonprofit	200,000	2	1,080,000	9
Legal	150,000	0	450,000	6.8
Economic	1,080,000	6	7,150,000	49
Finance	430,000	3	4,200,000	19.5
Banks	30,000	1	1,900,000	1.5
Insurance	200,000	1	1,400,000	9
Real estate	200,000	1	900,000	9
Business services	330,000	1	1,980,000	15
Repairs	320,000	2	970,000	14.5
Auto	170,000	2	550,000	7.7
Trade	2,445,000	14	17,850,000	111
Retail	1,930,000	14	12,800,000	88
Grocery	270,000	3	1,950,000	12
Auto	132,000	0	1,150,000	6
Department stores	76,000	0	1,150,000	3.5
Gasoline	226,000	3	820,000	1
Restaurants	400,000	4	3,300,000	18
Clothing	130,000	0	850,000	6
Wholesale	370,000	0	4,200,000	17
Entertainment	145,000	0	850,000	6.6
Communication and utilities	160,000	1	2,215,000	7.3
Telephone	30,000	0	990,000	1.4
Media	18,000	1	150,000	.8
Utilities	70,000	0	735,000	3.2
Transport	40,000	0	280,000	1.8
All	5,465,000	34	46,465,000	250

Sources: 1972 Censuses of Retail and Wholesale Trade, Selected Services, and U.S. Statistical Abstract, and 1970 U.S. Census. Many of the figures are approximate because of limited data on sole proprietorships and unincorporated businesses.

Note: For comparison with the entire United States, figures are given for the village of Orting, Washington (pop. 2,000).

possible from its competition to ensure each retailer a fairly secure home market, or spatial monopoly.

Older planned shopping centers and business districts were built around one department store, resulting in a local spatial monopoly. This arrangement avoided the more competitive conditions of the downtown district. But partly because of intense competition among shopping centers, the newer ones have more department stores and thus greater internal competition.

Although most retail stores are clustered at the convergence of transport routes, a growing number are oriented toward highway traffic. Examples include motels, auto-trailer sales and services, rent-all dealers, fast-food restaurants, and most gas stations. Such services need ample customer parking space but cannot readily afford the high rents of commercial shopping centers.

Trade Areas

The **trade area** (area from which most customers originate) of a store, business district, or shopping center is defined by customer shopping trips. Pure central place theory outlines trade areas absolutely and geometrically. Of course, individuals usually do not act on perfect information, nor do they always behave in an economically rational manner. They do not always shop at the nearest store; perhaps a distant shopping center is more attractive than one located more conveniently, or the service may be better. Realistically, then, we should look for a field or pattern of shopping trips about the store or center. The extent of the field is limited by the distance to competing centers of the same size or larger. Its intensity reflects the concentration of customers in the area. Trade areas, or shopping fields, identified by asking people where they usually shop, are found to overlap (see figure 7.11).

Multipurpose Trips

Multipurpose trips help explain both the apparently inefficient overlap of trade areas and the efficient behavior of customers who, by combining several errands in one trip, save time and minimize the total distance they travel. People frequently make short single- or double-purpose trips, but they can accomplish more on longer multipurpose trips.

Many shopping trips involve travel to more than one retail center and often other activities, such as visits to the doctor, to a museum, or to a government office. Customers might not be able to find what they want at the first center, or they might know that another center offers a different selection of goods. A multipurpose trip to two centers may still save travel time and distance over several single-purpose trips to one closer center.

Multipurpose trips provide one of the theoretical bases for the agglomeration of activities at fewer, larger centers. Agglomeration makes it easier for retailers to maximize the availability of their services to customers. Not only do shopping centers give people the satisfaction of planned multipurpose trips, but the proximity of various stores often helps business by inducing unplanned or impulse shopping.

Wholesale Trade

Wholesale trade is intermediate between production and retail trade. Wholesalers buy goods in bulk from the producer and package them into smaller lots, which they then sell and distribute to retailers. A food wholesaler, for example, will buy a truckload of carrots from a farmer, package them into one-pound bags, and sell the bags by the crate to supermarkets.

Wholesaling is both spatially and economically efficient. Producers who specialize in one or two commodities for very wide distribution often cannot afford to spend the time or the money on labor needed to handle countless small orders from retailers. Retailers who stock an immense variety of items do not have time to order small quantities of goods from countless producers. And dividing large shipments into smaller lots would be a bookkeeping nightmare for both producers and retailers and would result in extremely high distribution costs. (The large mail-order retail firm does not really bypass the wholesalers; it is itself a major wholesaler.)

Wholesaling also eliminates expensive long-haul shipments of small lots from producers to retailers. Instead wholesalers in regional centers can realize economies of scale on large lots shipped from dispersed producers. At the same time, retailers benefit by receiving goods much faster from a regional center than from scattered, distant producers. Moreover wholesalers and distributors are sensitive to regional needs, whereas national manufacturers may wish to avoid responding directly to regional variations in demand with respect to quantity, style, packaging, and so on.

Wholesale trade is concentrated in large central places. The critical factor is not only the size of the place but its relative location (see figure 7.9). In the United States, wholesaling thrives both in regional centers (such as Portland, Salt Lake City, San Francisco, and Atlanta) and in the metropolitan centers bordering both the northeastern core manufacturing region and the midwestern agricultural periphery (figure 7.13). However, New York City is by far the dominant wholesale center of the nation.

Services

Retail and wholesale trade have long been important central place activities. In the modern economy, however, services have become a major growth sector. More and more people in affluent, high technology societies are earning their living from such services as finance, consulting, law, education, and health. Only a few of these activities will be discussed here.

Finance

Financial and related services—banking, real estate, and insurance—proliferate in most advanced economies. Multipurpose trips often include banking, real estate, or insurance transactions. Hence these services are found,

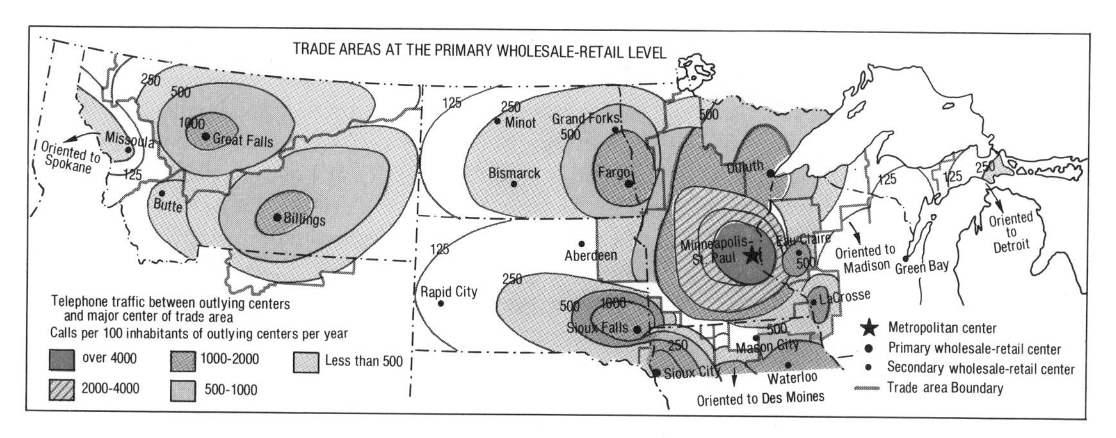

7.13. Major wholesale-retail trade centers in the U.S. upper Midwest. Telephone traffic is a fairly good indicator of wholesale-retail trade area boundaries. (Note that many trade areas cross state lines.) The location of Minneapolis–St. Paul between Great Lakes industrial centers and western agricultural centers has contributed to its importance as a wholesale-retail trade center.

Source: Adapted with permission from J. Borchert and R. Adams, "Trade Centers and Trade Areas in the Upper Midwest," Urban Report 3, Upper Midwest Economic Study, Minneapolis, 1963.

along with retailing, in nearly all towns and in district shopping centers within cities. Higher level, more specialized services, such as commercial and industrial banking and investment, are located in the central business district of larger cities, where important customers can be readily accommodated and interoffice business conducted rapidly.

Real estate firms and local insurance outlets are low-threshold establishments that can enter the market with relative ease. Since much of the business is conducted in private homes, the locational preferences of real estate and insurance agencies are not very strong. Most agencies lease space in low-rent stores on less important highways.

Schools

Schools tend to follow the administrative principle of spacing. One higher level school (a large high school) may draw its students from several lower level schools (junior highs). A centrally located neighborhood school may be convenient for students, but a policy conflict often arises over whether to provide good local accessibility or to realize scale economies in teaching staff, facilities, and materials with larger, more widely spaced regional schools. Regional schools, however, entail higher busing costs for taxpayers and greater time costs for students. The conflict between proponents of neighborhood and of regional schools is aggravated by the use of the larger schools and busing for purposes of racial integration.

Health Services

Health services range from local physicians and dentists to clinics, community hospitals, and highly specialized teaching hospitals. Such services are a good example of the central place hierarchy and spatial regularity that result from competition for secure trade areas. Most villages support doctors and dentists but are too small for clinics and hospitals. The nearest hospital is usually located in a higher level center—a town or small city. Larger cities and metropolises contain hundreds of private doctors and dentists, support several hospitals and clinics, and may feature at least one or two specialized teaching hospitals. The location of health services, like schools, is controversial. On the one hand, practitioners and managers prefer to agglomerate in specialized centers to reduce costs and the travel of professionals. On the other hand, patients and some health planners prefer to decentralize the services so that local needs may be met more directly. Local income levels, however, are so uneven that not all communities could afford to support local health services—a reminder that purchasing power, as well as population density and environment, varies in the real world and affects spatial patterns.

CONCLUSION

Central place theory helps us recognize and understand the spatial order of the urban landscape. We can better appreciate why a city like Chicago contains huge wholesale houses, provides medical students with highly specialized facilities, and is a major marketplace for the nation; why it is located relatively far from other metropolitan centers like New York and Los Angeles; why it is surrounded by many villages, towns, and cities; and why such a large agglomeration may attract some people and activities but drive others to smaller places where pressures may be less wearing and the competition less severe.

SUGGESTED READINGS

Berry, B.J.L. *Geography of Market Centers and Retail Distribution.* Englewood Cliffs, N.J.: Prentice-Hall, 1967.

______, and Pred, Allan. *Central Place Studies: A Bibliography.* Philadelphia: Regional Science Research Institute, 1965.

Borchert, J. R. *Trade Centers and Trade Areas in the Upper Midwest.* Upper Midwest Economics Study. Minneapolis, Minn.: University of Minnesota Press, 1963.

Christaller, Walter. *Central Places in Southern Germany.* Translated by C. W. Baskin. Englewood Cliffs, N.J.: Prentice-Hall, 1966.

Davies, Rosser L. *Marketing Geography.* Corbridge, England: Retailing and Planning Associates, 1976.

Doxiadis, C. *Ekistics.* New York: Oxford University Press, 1968.

Everson, J. A., and Fitzgerald, B. P. *Settlement Patterns.* London: Longmans, 1966.

Scott, P. *Retailing.* London: Hutchinson University Library, 1970.

Vance, J. E. *The Merchant's World: A Geography of Wholesaling.* Englewood Cliffs, N.J.: Prentice-Hall, 1970.

8

Industrial Location and Urban Systems

8 Industrial Location and Urban Systems

The next time you pass through an industrial area, try to figure out why that steel mill or electronics firm or shoe factory is located in that site. Where does the plant get its raw materials? Where does it distribute its products? The industrial landscape, like the central place landscape, makes sense when you analyze it. There is a reason for the location of a particular industry in a particular site. Someone gave careful thought to that location at one time. In this chapter we will explore the rationale behind industrial location. We will also study the process of industrialization and the patterns that underlie the industrial landscape from the Atlantic seaboard of the United States to the Inland Sea of Japan.

INDUSTRIALIZATION

Several changes occurred in the landscape as a result of the industrial revolution, which began in Great Britain during the last half of the eighteenth century. The resource that had traditionally attracted settlement—good farming land—became secondary to coal, timber, iron ore, and other fuels and raw materials. New towns sprang up in agriculturally unproductive but minerally rich areas. Large population concentrations formed in the English Midlands, the Ruhr district in Germany, and other regions where large coal and iron ore deposits were discovered. Urban settlement patterns shifted from the dispersed arrangement of central places to clusters of industrial towns.

By the early nineteenth century, industrialization had spread from Great Britain to the rest of Europe and North America. In most countries it began in a few core areas: in the capital and other large cities and in areas containing water power, coal resources, or surplus farm labor. Over time, manufacturing became increasingly elaborated (altered from original raw materials) and more concentrated in metropolitan areas.

As more and more people followed industry into the cities, the cities became less and less fit for people. Industrial and municipal wastes fouled rivers and coastlines, and a thick haze of smoke, soot, and fumes poisoned the air. During the nineteenth century and well into the twentieth, thousands of people died from diseases caused by pollution in the factories and slums of industrial cities and mining towns.

Yet industrial cities continued to attract throngs of migrants and new enterprises. Although industry endangered human health and degraded the environment, it still enabled people to escape the often worse poverty, filth, and overcrowding of the countryside. Children of factory and mine workers had a chance to be educated and to rise socially and economically. Eighteenth- and nineteenth-century critics condemned cities and industry, looking with romantic nostalgia on the beauty and simplicity of rural life. But for the mass of the peasants, poverty was severe, disease endemic, and opportunities almost nonexistent.

Despite the grim conditions of early industrial cities, industrialization has unquestionably raised levels of human well-being. Through the development of machinery, better storage, and other innovations, it has increased agricultural productivity and improved rural conditions. Industrialization has helped absorb surplus labor from the countryside, made possible better housing, and, perhaps more important, provided the basis for the gradual elimination of poverty and the growth of services through the rising productivity of labor. Indeed it is because of the high efficiency of manufacturing that society can enjoy a rich variety and abundance of goods and services. This is why towns, counties, and nations endeavor to attract new industries to their areas. Several regions in the United States, for example, have been trying to entice Japanese and European automobile companies to locate assembly plants in this country. And oil-producing nations are insisting that more refining and petrochemical plants be located on their soil.

In the United States, the industrial core was and still is concentrated in the Northeast, the area of earliest development. But industrialization spread as the distribution of population, agricultural production, and mineral exploitation gradually shifted west and south. Typically **resource-oriented industries,** such as food processing and furniture, and industries that could use less skilled, lower cost labor, such as textiles, first filtered down from the core-area cities. During the twentieth century, such industries have been replaced in the older cities by manufacturers of machinery and other more elaborated goods. In the 1920s the shift of textile production to Virginia and the Carolinas extended the industrial core southward. In the 1960s lower wage industries, such as textiles, furniture, and appliances, began to spread into the Tennessee Valley, the Ozarks, and the northern plains, where industrialization continues to grow.

Manufacturing is declining in relative importance in the older industrial areas of Western Europe and the United States, but it has been progressing rapidly since World War II in Eastern Europe and the Soviet Union, and in China, Taiwan, Korea, Japan, India, South Africa, and Brazil. The other countries of Asia, Africa, and Latin America are now beginning to industrialize. Socialist countries such as the Soviet Union, China, and various nations in Eastern Europe tend to stress the long-term, independent (bloc) development of heavy industry for internal use. In contrast, other countries such as Japan and Brazil are repeating the European and American pattern of capitalist expansion. And developing countries such as Singapore, Hong Kong, and Korea are emphasizing the manufacture of light industrial goods for trade with the developed countries.

If we look carefully at the past two centuries of industrialization and urbanization and at the development of countries around the world today, it becomes apparent that industrialization has been and continues to be **the** driving force behind the growth of cities and even behind the modernization and commercialization of agriculture. Central place services to an agricultural hinterland are not sufficient in themselves to generate a high level of urbanization or a high standard of living. It is true that cities can and do expand even in the absence of industrialization, as in many third world countries today. But these countries realize that their cities will be unable to provide adequate services and opportunities without industrialization.

Modern economies are too complex to permit us to identify the relative contribution of central place and industrial activities. But in a country that is 75 percent urban, such as the United States or West Germany, 25 percent of the urbanization is probably a result of providing services for the rural population, and the other 50 percent is probably due to activities unrelated to serving the rural population—to industrial activities and services oriented to the urban population.

FACTORS AND PRINCIPLES OF INDUSTRIAL LOCATION

Most of our discussion will concern the factors and principles of industrial location in market economies. Industrial location in centrally planned societies will be treated separately.

If industrial raw materials and fuels were available everywhere, producers would probably locate at the precise central place that would provide the most profitable market for their goods. Needed supplies could be obtained locally, and central place patterns would dominate the landscape. In both industrial location and central place theory, a primary objective is to minimize transport costs. But there are two important differences between the theories. First, in industrial location, most resources are scattered unevenly in the physical environment (see figure 14.2). The cost of transporting raw materials is often so high that firms will locate near their resources rather than at the market. Hence the central place pattern is broken. Second, in many industries production costs, including both investment in plant and machinery and operating costs, are greater than transport costs. Thus spatial patterns become more complex. In contrast, customers in central place theory are assumed to be evenly distributed, and transport costs for finished products or for services are the most important factor.

A look beneath the surface will reveal that many factors besides the cost of marketing products influence industrial location. Consider the industries in Chicago. To determine why the electrical goods industry is prominent there, you must first identify the industry's general orientation. You probably know that the manufacture of radios, television sets, and other electrical goods requires substantial technically skilled labor; therefore the industry is likely to be labor oriented. Chicago is a high-level central place that contains both the schools needed to train skilled workers and the opportunities for recreation, shopping, and so on that would appeal to the workers and their families. Moreover the city itself is a large market for electrical goods and has excellent transport connections for exporting products to other regions. The industry can also benefit from agglomerating with other industries and services located in Chicago. Needed materials, such as fabricated metal, are produced in the area, and capital, advertising, and marketing services are readily available. Thus Chicago is an excellent location for the electrical goods industry. But the ideal location for an industry can change. As the production of electrical goods has become more routinized and skill level requirements have fallen, television and radio manufacturers, for example, have found it more profitable to shift assembly operations to lower cost areas such as Taiwan, Korea, and Puerto Rico. Other industries could be analyzed in much the same way.

Despite the obvious complexity and apparent irregularity of the industrial landscape, the underlying spatial order tends to be logical, predictable, and universal. The theory of industrial location will be discussed later in this chapter in order to demonstrate the principles of this spatial order. Meanwhile we will examine some of the factors that affect industrial location.

The Locations of Resources and of Markets

Two important variables influencing the location of industries are the locations of resources and of markets. Raw materials must be obtained from specific points within very large areas. Markets may be either diffused throughout a region or nation or concentrated in a few urban centers. Often resources and markets are widely separated, and the manufacturer must overcome the distance as efficiently as possible.

If the finished product costs more to transport than its raw materials or component parts (because of bulk or weight gain), the producer will probably locate near a major market to save on distribution costs. Examples of **market-oriented industries** include construction, bakeries, newspaper publishing, and beer and soft drinks. If one raw material is much heavier or bulkier than the other materials or the finished product, the producer is likely to locate near the source of that material to save on procurement or assembly costs. Examples of resource-oriented industries include food and lumber processing, canning and freezing, sugar beet refining, sawmills and pulp mills, smelting and refining. And for some industries, both market and resource locations are critical. Examples include meat packers, creameries, flour mills, and feed mills. Such industries often locate midway between their markets and resources.

Transportation Cost and Quality

Assembling raw or semifinished materials and distributing the product to market are processes that entail specific time and transport costs. These costs vary with the type of carrier (rail, barge, truck), the difficulty of handling the goods, and the distance. It is not simply the unit cost of transportation that is critical but the quality of the transport system itself—for example, the frequency of service or the delivery time. Improvements in transportation, such as larger vessels, have lowered the costs of importing some kinds of raw material. Hence firms that have traditionally located near raw material sources may instead choose to locate near large ports. The steel industry, which now often uses imported iron ore, has expanded from the Great Lakes region to coastal cities like Philadelphia and Baltimore (figure 8.1). In European countries such as France, the Netherlands, Italy, and Great Britain, many new steel mill complexes have also been built in coastal sites.

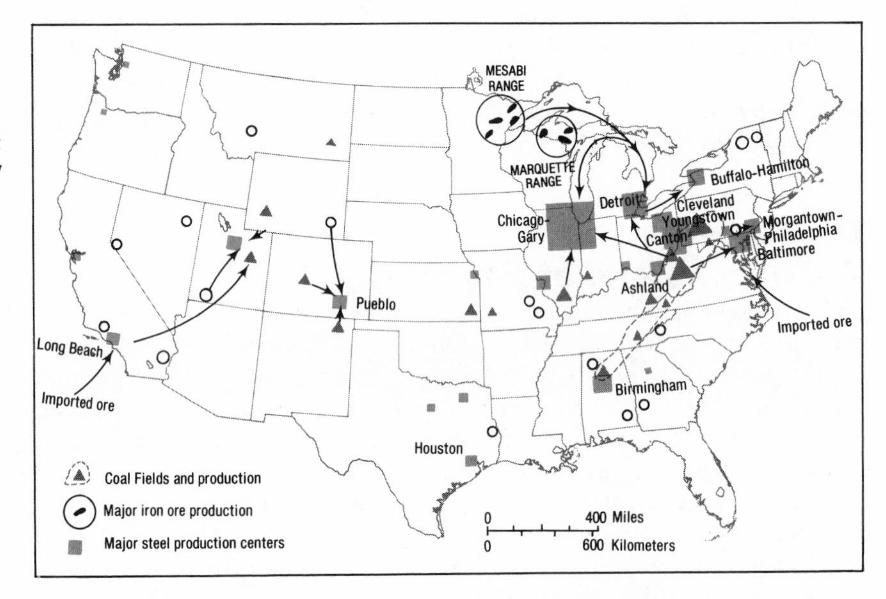

8.1. Relative location of iron ore, coal, and steel production in the United States, 1970. Major steel production centers exhibit a mixed orientation to raw materials and markets.

Production Costs

Industrial location would be much simpler if the costs of transporting raw materials and finished products were the only variables to consider. But production costs also vary in different places and affect location decisions. These costs include the price and quality (productivity) of land and labor; the purchase and interest cost and the longevity of machinery; corporate and inventory taxes; construction costs; and the cost of maintenance, heating, and air-conditioning. High land costs and taxes in central cities often induce industries to move out to suburban locations. Variations in taxes and restrictions on industries may lead to regional industrial shifts. In both Europe and the United States, lagging regions may offer low tax rates or less regulation to attract industries.

For some industries the influence of labor costs and quality is greater than that of transport costs. Labor costs and quality vary from place to place. Some areas have highly skilled labor, and some have an abundance or scarcity of labor, skilled or not. Labor costs for the producer are reflected in these spatial variations. The pool of relatively highly skilled labor in New England, along with other factors such as nearby research centers, has attracted part of the electronics industry to Massachusetts. The pool of relatively low-skilled, low-cost labor in the rural South has attracted textile and other labor-intensive industries to Virginia and the Carolinas. On a more global scale, labor-intensive industries are shifting from high-cost areas like Sweden, Germany, and the United States to developing countries like Korea, India, and Mexico.

Scale (Internal) Economies

Scale economies refer to the expected reduction in average unit production cost when more units of a product are manufactured. As the scale of production, or volume of output, increases, production efficiency tends to improve. The minimum costs of development, new machinery, plant, and overhead are spread over more units. Hence cost per unit produced diminishes; labor productivity rises; more internal specialization is possible; supplies can be purchased and transported at lower bulk rates; machinery and specialized personnel can be more fully utilized; and inventories of raw materials and products can be kept relatively low. These result in scale economies.

Scale economies are not gained indefinitely, however. **Internal diseconomies** may result from traffic congestion within the plant, plant reconstruction costs, labor shortages, pollution-abatement costs, in-plant organizational problems, or input supply problems such as fuel or material shortages. Most important from a spatial perspective, the lower unit costs must be balanced against probable increases in **external diseconomies,** that is, in assembly and distribution costs. The need to tap more distant resources and markets results in higher transport costs for the firm.

If internal diseconomies do not occur and if the firm sells its products to a market in a single location, average costs will continue to decline as the scale of production increases. But because most markets are in fact located at several points and are spatially separated from each other, a greater scale of production will eventually force a firm to sell to more distant markets and to utilize more remote suppliers. When a beer manufacturer moves into a new metropolitan market without establishing branch plants, the consequent rise in transport costs may outweigh the scale economies from higher production in the home plant.

Scale of production is thus limited by the size of regional or national markets and by external diseconomies—the rising costs of transporting supplies and products over longer distances. Industries that incur high transport costs tend to seek protected market areas and are likely to form spatial monopolies or oligopolies, where only a few firms attempt to compete in a given market. The steel and cement industries in the United States are examples of spatial oligopolies.

In some industries, such as aircraft and machinery of many kinds, transport costs play only a minor role in determining the optimal scale of production. For these industries, no significant spatial monopoly is possible, and firms must share virtually the entire national or international market. When transport costs are low, scale is usually limited by the firm's competitive share of the total market rather than by external diseconomies. Prices are typically rather uniform, and plants will seek locations where production costs are minimized. Hence they will tend to cluster for the sake of achieving agglomeration economies. The importance of scale economies is revealed by the success of the European Economic Community (EEC) (see page 311 and the growth of multinational corporations seeking to tap the combined markets of more countries.

Demand

The success of industries in a market economy depends on the willingness and ability of consumers to pay a price that will more than cover production costs. Demand encourages the development of new products and industries. Conversely a lack of demand for certain products and industries leads to their decline. Changes in demand drastically affect the fortunes of places and regions. Short-run fluctuations in aircraft and ship orders, for example, have hurt some local economies, and long-run declines in demand, as for a given firm's automobiles, may cause severe problems.

Because demand changes fairly rapidly in consumer-oriented economies, industries must be able to adapt their technology and spatial behavior in response to these changes. Automobile firms, for example, can vary the mix of models produced at various assembly plants according to regional demand. Demand for specific brands also affects the competitive strength of a firm. By exploiting brand loyalty or name familiarity, a firm may be able to enlarge its market area far beyond the typical range for its goods. A recent example is the success of Coors beer in the United States. The beer is produced in Colorado and distributed in eleven western states. Although Coors has no branch plants (not even in California, its largest market), its reputation for high-quality beer has made it competitive with national breweries.

Capital

Industrial activities often require large outlays of capital for plant and equipment. The greatest amounts of capital are generated in well-established industrial locations. Hence new industries are attracted to existing industrial centers such as Chicago, Detroit, Los Angeles, London, Paris, and Milan. Capital may theoretically be mobile, but investors tend to be interested in known areas and accepted lines of activity. Moreover investors and prospective producers in a capitalist economy seek fairly short-term profits. Such constraints discourage the building of new plants in undeveloped regions and thus contribute to the disparity in regional income levels. This fact helps to explain the relative poverty of the rural Deep South and the relative prosperity of the Great Lakes region. Regions or countries in which capital formation is limited, as in most developing countries, will tend to emphasize labor-intensive industries, while more highly developed countries tend to emphasize capital-intensive industries.

Agglomeration (External) Economies

Many producers locate near related activities to take advantage of agglomeration economies. Agglomeration may occur among firms that share the same resources or market or that have strong **interindustry linkages.** An agglomeration of related firms is called an **industrial complex.** A fabricated metals and machinery producer may locate near an iron and steel plant to save on assembly costs. And they may be joined by a chemical plant, which utilizes by-products of the coke used in steel production, thus completing this typical industrial complex. Such agglomerations are found in the Pittsburgh and Chicago regions in the United States, in the Ruhr district of West Germany, the Donbas in the Soviet Union, and the Inland Sea area of Japan (figure 8.2).

Textile and apparel manufacturers would also appear to form a compatible partnership. But textile plants are usually separate from apparel plants because their labor and marketing requirements are much different. Similarly alumina processing, aluminum refining, and aluminum fabrication may be spatially separated because of differences in energy, labor, and transport requirements.

Agglomeration of related firms may increase the aggregate, or total, transport costs incurred in assembling raw materials and marketing finished products. But agglomeration is advantageous as long as savings in transfers between firms, the sharing of business services, the presence of a skilled labor force, and other benefits more than offset the higher initial assembly and final distribution costs.

Even unrelated firms may enjoy indirect benefits from agglomeration, such as the availability of capital, business and communication services, transport facilities, or a large labor pool. In Seattle, a mail-order department store, a fancy-goods department store, and an aircraft firm have shared a Chicago freight-forwarding firm to get bulk train-load rates for shipping all kinds of goods to Chicago. Another incentive for agglomeration is the possibility of sharing a common technology. Computer firms and digital watch firms both use semiconductors and are therefore both likely to locate near the same semiconductor firm. And light industries that employ mostly women, such as the candy industry, are often attracted to areas where heavy, male-oriented industries such as steel predominate. Garment-sewing factories in Appalachian mining areas are another example. Such complementary agglomeration is due to the large, originally untapped female labor pool available in these areas.

8.2. Inland Sea, Japan. Raw materials and semifinished or finished products flow into and out of Kobe, Japan's leading seaport. Barges crowding the wharves house some eight hundred families who transport cargo between freighters and the port. Petrochemical plants, steel mills, textile plants, and shipyards agglomerate in this heavily industrialized region of Japan. (Photo from Wide World.)

Competing producers may agglomerate if transport costs are low and brand loyalty is high for their product. In this case, agglomeration advantages include the presence of a skilled labor force and the spin-off (or stealing) of ideas and personnel from competitors. In addition, prospective customers can compare different brands more easily when several producers are located in one area. This applies particularly to the Paris, London, and New York fashion markets, to jewelry manufacturers, and to some concentrations of furniture makers and machine-tool makers.

Perhaps most important, agglomeration economies in the form of large and diverse market and labor pools make fertile ground for new and innovative firms and products, which, if successful in gaining a larger market, may shift routine production facilities to lower cost regions.

Substitution

To some degree, the factors of production—capital, labor, various resources, transport services—may be substituted for one another or used in different combinations in the effort to maximize profits. In nearly every industry, some kinds of materials can be substituted for others. Either coal, oil, gas, or atomic energy can be used to produce electric power, and either cardboard, plastics, paper, or wood can be used for containers. The material selected depends on its relative price and availability. In the steel industry, scrap iron can be substituted to some extent for pig iron (figure 8.3). Since scrap from junked cars is readily available in large cities, the industry can locate either at its market (the city) or near iron ore deposits. When producers have a choice between two materials, one located in a remote area and the other near a major market, they will probably choose the latter and locate near the market. All producers are basically market oriented, since all—even resource-oriented ones—need a market for their product.

A firm may substitute among the mix of goods produced in its plants, usually in response to changes in demand. An automobile firm, for example, may direct its plants to shift their emphasis from larger cars to smaller ones.

Capital and labor may be substituted for one another to some extent. An investment in machinery will reduce labor requirements and costs. In areas where labor is scarce or expensive, as in Scandinavia or Switzerland, capital investment is a logical alternative. If machinery cannot be substituted for labor, the industry will be limited to locations where labor costs are lower or workers are uniquely skilled. American radio and television manufacturers have been attracted to countries with lower wage labor, such as Mexico and Korea.

A substitution, or trade-off, may be made between transport and production costs. Aluminum processors may be willing to locate in remote areas in order to take advantage of the cheaper electric power available far from metropolitan markets. Indeed the abundant hydroelectric power of the Pacific Northwest has attracted many aluminum processors, despite higher costs incurred for transporting raw alumina from Gulf ports and for shipping processed aluminum to final markets. In general, when producers

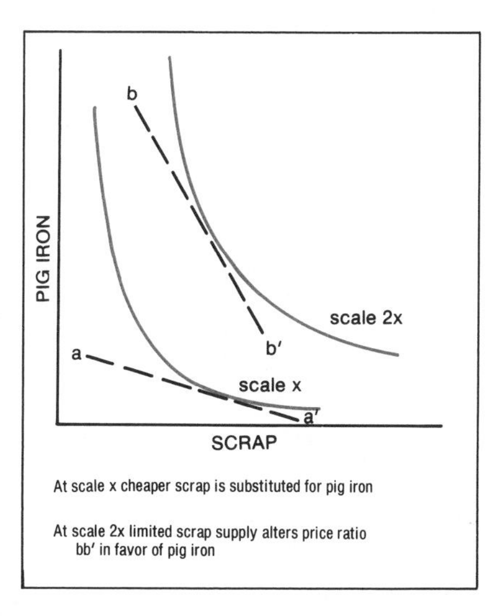

8.3. Input substitution in steel production. Solid lines show the combinations of pig iron and scrap that will produce a constant amount of steel. The slope of the dashed lines indicates the ratio of pig iron price to scrap price: line *aa′* shows that scrap is cheaper if not too much is needed and the scale of steel production is limited, while *bb′* shows that pig iron is cheaper if more inputs are needed for larger-scale production.

move to a location that will reduce production costs (power, labor, or taxes), they will have to pay higher transport costs.

One other fairly common substitution is between economies of scale and transportation outlays. Large-scale producers can substitute scale economies for higher transport costs and greater **market penetration;** they can enlarge their markets by seeking more distant customers and paying for the increased cost of transportation with the savings gained from high-volume production. For example, large metropolitan bakeries can ship goods to distant places and still sell at lower prices than charged by small local bakeries. Smaller, higher cost producers may successfully compete by accepting a smaller market and therefore paying less for transportation, especially if their plants are far from their competition, or if, like many local bakeries, they maintain higher quality. As in central place theory, spatial separation allows each plant to maintain a spatial monopoly over some of its customers. Thus a firm may find it more profitable to maintain both large and small plants in different regions.

Organization

The location decision rarely concerns a single plant in isolation. Rather the plant may be one of many operated by a firm, which may produce a variety of goods. The firm, as a competitive organization, also seeks to protect and strengthen itself. It may be vertically integrated, incorporating within its organization all manufacturing functions, from the mining of raw materials to the assembling of finished products. The Ford Motor Company, for example, owns coal and iron ore mines, ships, steel plants, and parts plants, as well as engine and assembly plants. Such ownership integration tends to favor locational clustering for reasons of efficient product and communication flows within the firm, departing from what might be the best location for an independent producer. Other firms have a horizontal structure: they produce a variety of goods. Some petroleum companies, for example, now control coal mines and gas and uranium sources as well. In both vertical and horizontal integration, the firm is trying to reduce competition and uncertainty and to protect itself from fluctuations in costs and prices of individual products. Integration is often accomplished by merger with or acquisition of other firms. This may result in the closure of some formerly competing plants and a concentration of production in favored areas.

Large firms, particularly in high technology and energy-related industries, have become multinational, operating plants in several countries. This arrangement gives the firm great flexibility, permitting it, for example, to locate the production of labor-intensive components in countries where labor costs are low and to place headquarters and research and development in desired metropolitan environments. Some critics fear that these giant firms control too many resources and account for so much production and international trade that they can unfairly influence prices, suppress innovation, and even dominate smaller governments.

INDUSTRIAL LOCATION THEORY

The preceding discussion of factors and principles of industrial location demonstrates the rather bewildering complexity of decisions facing producers when choosing a location. Nonetheless a fairly simple theory can be formulated to explain and predict much industrial location. The theory will provide general principles for understanding the industrial landscape.

Classical industrial location theory was formulated by Alfred Weber, a German economist. His theory may be summarized as follows: The best location for a plant is usually at the point where transport costs are at a minimum. This location, however, can deviate to take advantage of sites with low labor costs or of agglomeration in larger cities.

Weber developed the now-famous concept of the *location triangle.* Given two necessary raw material sources and a market, or a single raw material source and two markets, the optimum location for a plant is at the transport-minimizing point within that triangle (figure 8.4). But if the cost of shipping a unit of one raw material or product exceeds that of the others combined, the optimum location will be at that terminal point (source or market).

Industrial location theory has interested many scholars, who have brought it to a highly sophisticated level. We will describe only some of the simpler additions to the theory. Economist E. Hoover observed that the presence of *tapered transport costs*—lower rates per mile for longer distances—strongly favors location at markets or resources (figure 8.5). But if there are many markets and resource points for a particular industry, a cen-

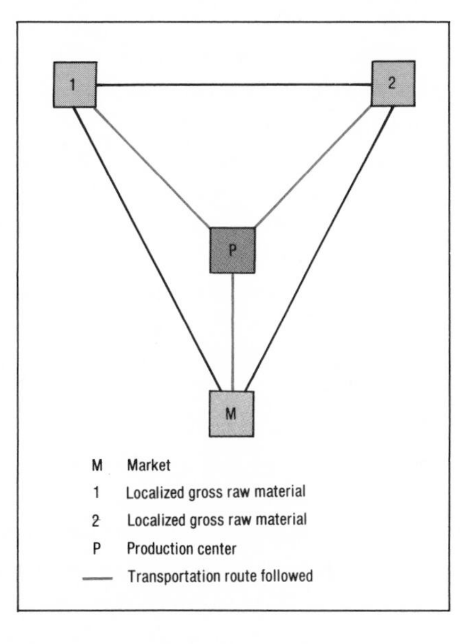

8.4. The location triangle. Two raw material sources *(1* and *2)* and one market *(M)* form a triangle. The ideal location for production *(P)*, where total costs are minimized, is within the triangle closest to the location that entails the highest transport costs. If the cost of shipping the finished product to market is highest, then the best location would be relatively close to market, as in this diagram.

Source: Redrawn with permission from R. J. Sampson and M. T. Farris, *Domestic Transportation,* 3/ed. Copyright © 1975 by Houghton Mifflin Company.

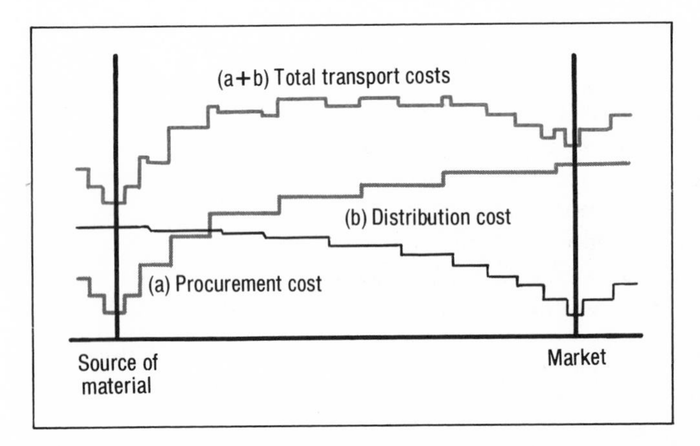

8.5. Realistic orientation to market or material end point. In this example, the cost of procuring raw materials (a), is greater than the cost of distributing finished products to market (b). But because of the typical transport-cost structure in which short- and long-distance shipments are charged rates under actual costs and are in effect subsidized by middle-distance shipments, a market location is better than any intermediate location beyond the immediate vicinity of the raw material.

Source: Redrawn with permission from Edgar M. Hoover, *Location of Economic Activity* (New York: McGraw-Hill, 1948).

tral transportation site—a transshipment point like Buffalo or New York, for example—might be best. Other scholars have studied the possibility and effects of substituting between capital and labor or other factors of production, and the effects of increasing scale. Smaller plants, for example, may prosper at more dispersed locations, while larger plants may need more central metropolitan locations. In his book, *Industrial Location,* D. Smith introduced the idea of the *spatial margin* (or zone) *of profitability:* the area within which the plant can make a profit. He suggested that plants may still make satisfactory profits even if they depart from the optimum site (decision makers may have many different reasons for doing so).

Finally scholars are paying increasing attention to the effects of uncertainty on industrial location. Not only do producers lack perfect information on probable costs of land, labor, capital, resources, and future prices, but they know even less about the behavior of competitors or about possible future regulation, tax changes, and the like. Government intervention is a constant concern of private firms operating in developing countries. As a result, producers make location decisions in the face of uncertainty and take considerable risks in each major investment or change of product. Given this uncertainty, Smith's zone of profitability may make more sense than a precise optimum location, which may be optimum for only a short time. The geographic impact of uncertainty, then, is both to lead to departure from optimum locations and to favor known rather than potential places and markets.

Patterns of Industrial Location

At least four spatial patterns, which we will refer to as *A, B, C,* and *D,* are derived from the principles of industrial location. When transport costs are more important than production costs, two very different kinds of location occur: resource oriented and market oriented. A clustered, resource-oriented pattern *(A)* develops when the cost of transporting raw materials is the dominant factor. In figure 8.6, each plant in such an arrangement has its own spatial monopoly over a resource area. When the cost of transporting the finished product to market predominates, a regular central place market-oriented pattern develops (pattern *B*—figure 8.7).

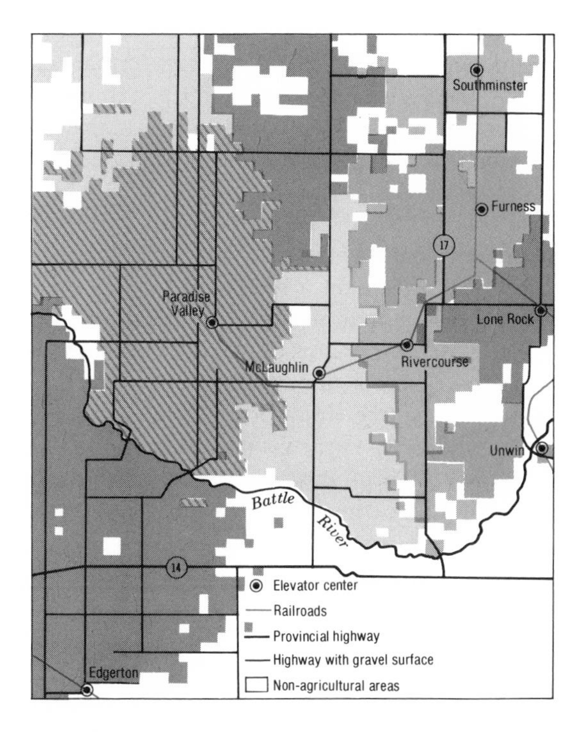

8.6. Supply areas of grain elevators in a portion of Alberta and Saskatchewan, Canada. This map illustrates spatial pattern *A*: when the cost of transporting raw materials (grain in this example) is the critical factor, plants (grain elevators) tend to locate near their resources. Each grain elevator center in this region has a local monopoly over its supply area.

Source: Adapted with permission from Edgar C. Conkling and Maurice H. Yeates, *Man's Economic Environment* (New York: McGraw-Hill, 1976).

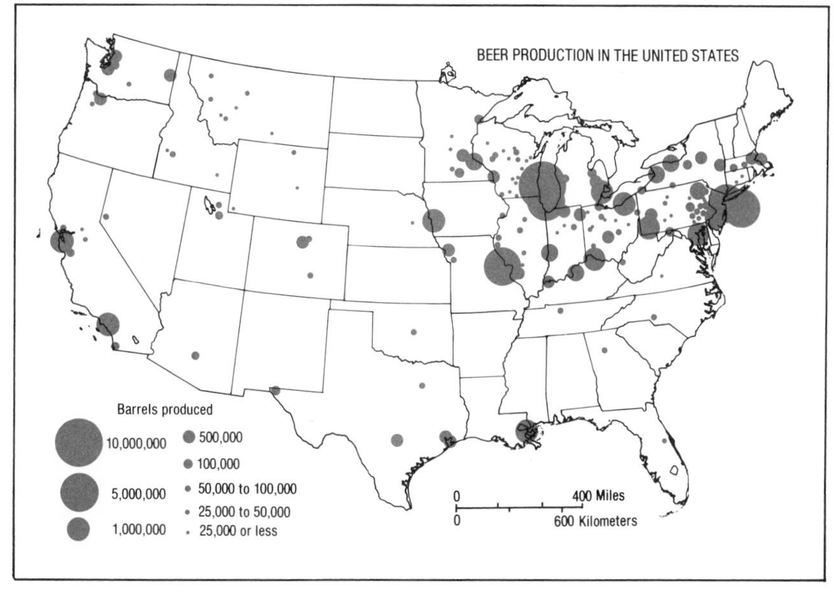

8.7. Beer production in the United States. This map illustrates spatial pattern *B*: when the cost of transporting finished products (beer in this example) is the critical factor, plants (breweries) tend to locate near their markets. Note the fairly low productivity in the South, for historical-cultural reasons, and the very great importance of Milwaukee and St. Louis, the locations of the industry's leaders.

Of course, transport costs are not always the controlling factor in industrial location. If production requires a highly competent labor force, the industry will tend to locate in the higher order central places that attract skilled and talented people. The type *B* spatial pattern that results in this case is due to labor immobility or availability or other agglomeration advantages rather than to distribution costs.

If production costs are the key factor, the industry will locate in areas where low-wage labor is available, energy resources are plentiful and inexpensive, or taxes are relatively low. Each firm will attempt to gain a spatial monopoly over a small supply area—in this case, of labor or energy. The spatial pattern formed by such industries might look like pattern *A*.

For some industries, transport costs for both raw materials and finished goods are important. A transshipment point is often the most favorable location for such industries. They may try to gain a spatial monopoly over their suppliers, their markets, or both. The pattern *(C)* formed by industries whose location is influenced by both markets and resources may be considered a compromise between patterns *A* and *B* (figure 8.8).

Finally some industries are relatively free from locational constraints, especially those with high-value products, such as stereo components, where transport costs are much lower than production costs. Such industries tend to locate freely in any region that contains a sufficient number of suppliers and markets and labor of sufficient quality, forming the fourth pattern, *D* (figure 8.9). Once established, however, these industries may be tied to local labor supplies and business services.

Optimum Scale of Production for the Single Plant

Industrial location theory has a very practical value: it enables firms to find not only a balance, or spatial equilibrium, between sources and markets, but also an optimum scale of production for their plants (figure 8.10). As an example, consider a plant, such as a cement manufacturer, that serves many customers—that is, a firm whose markets are spatially diffuse. Because of scale economies, the costs of producing cement keep falling with increasing volume of production. It would seem profitable for the cement maker to increase production continuously as the market expands and more customers become available. However, transport costs increase as a result of serving more distant customers and gradually outweigh the savings from lower unit production costs. Thus an optimum scale of production can be found (scale *B* in figure 8.10), in which sales revenues exceed production and transport costs to permit maximum profits for the manufacturer. In other cases, there may be more than one profitable level of production, for example, if the producer's markets are few, large, and unevenly distributed.

If customers pay a price that includes transport costs, then demand will fall with distance from the plant. If the number of potential customers declines rapidly with distance, then the policy of charging for delivery will probably yield maximum profits. Appliances, for example, may be delivered free within the urban area, but delivery charges may be imposed for sales outside the area. If the number of potential customers increases with distance, plant management may find it profitable to quote a uniform delivered price to all customers. This is true for nationally advertised products

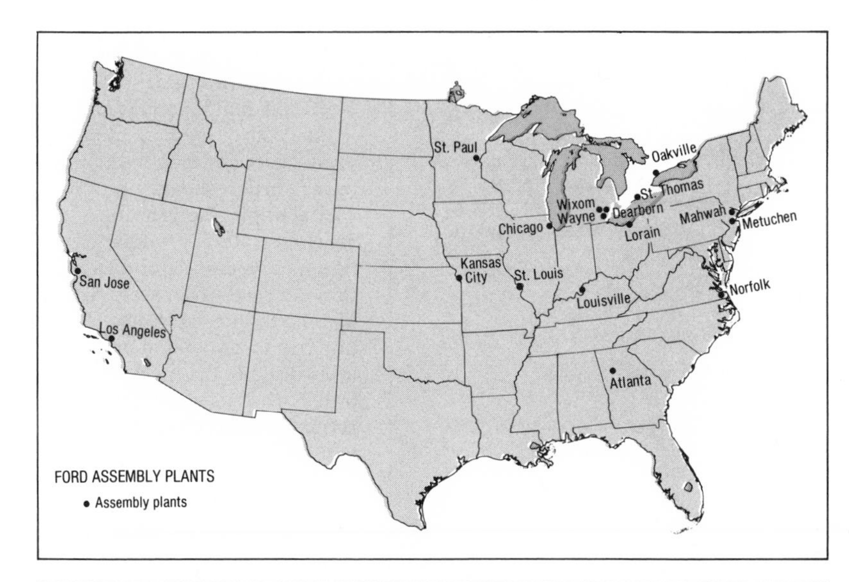

8.8. Ford assembly plants in the United States. This map illustrates pattern *C*: when the costs of transporting both raw materials and finished goods (steel and automobiles in this example) are critical factors, plants tend to locate near suppliers, markets, or both. The pattern of plants reflects concentration around Detroit (a major supply area as well as a transshipment point) and decentralization to major regional markets and strategic gateway cities.

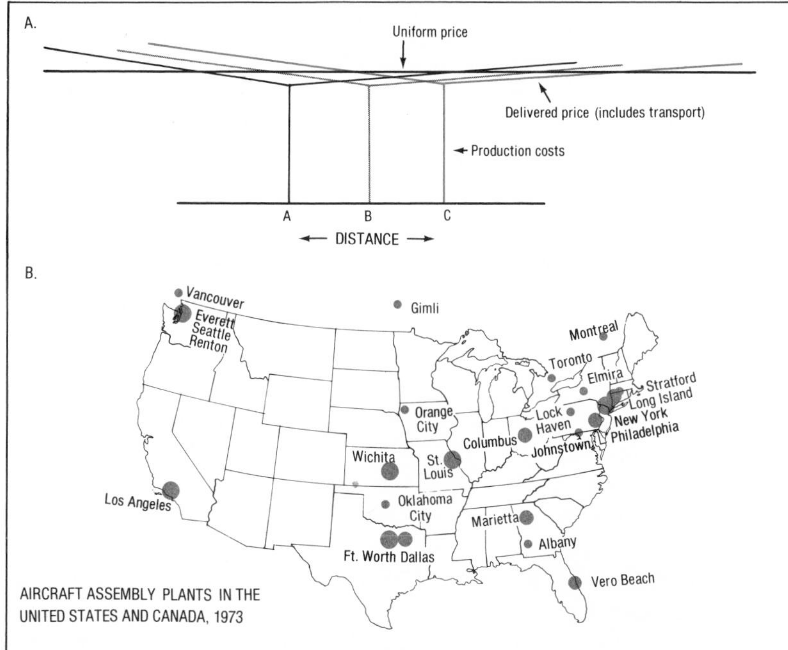

8.9. Shared markets: aircraft assembly, 1973. This map illustrates pattern *D*: when costs of delivery are low and delivery prices virtually the same, even widely separated clusters of firms, as at *A*, *B*, or *C*, can compete over a wide area. The firms will prefer to quote a uniform price and will not attempt to capture a local market. Aircraft assembly is an example of such location and pricing.

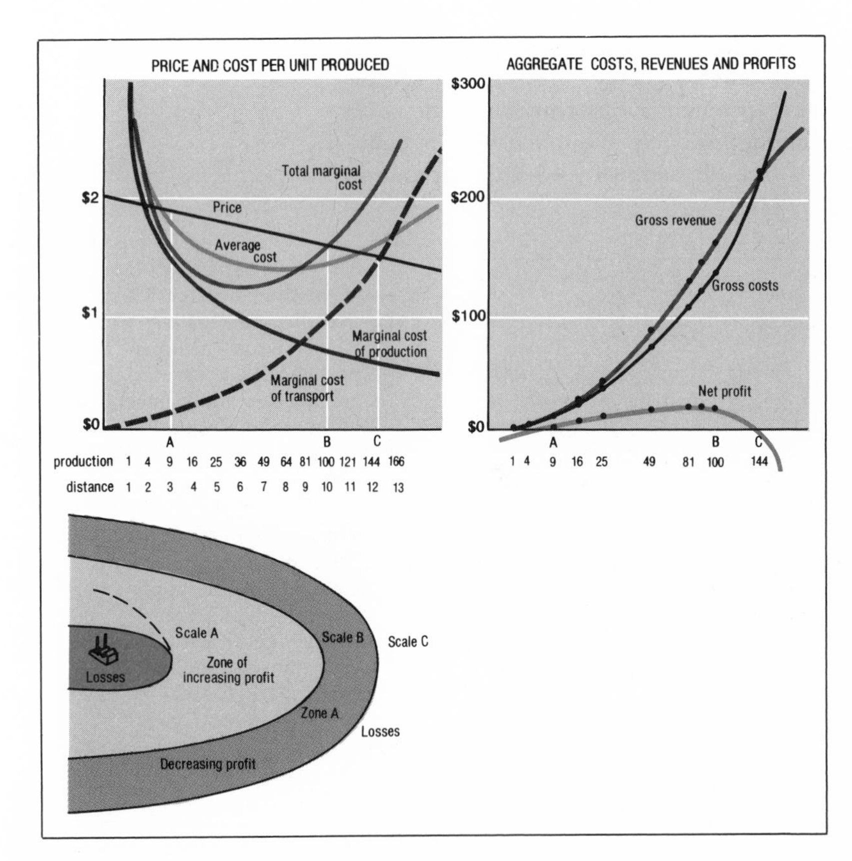

8.10. Economies of scale and their spatial implications. The graphs in the upper portion of the figure show the relationships between scale of production, distance, costs, and profitability for each unit produced by a manufacturer and for total (aggregate) production. (Marginal costs in the left-hand graph refer to the costs of the next unit produced.) The map in the lower portion shows the spatial implications of increasing scale of production. Profits rise as production increases from scale *A* to scale *B*. But as production increases beyond optimum scale *B*, the higher transport costs incurred for serving a larger, more distant range of customers begin to erode profits until, at scale *C*, costs exceed revenues and losses are incurred.

like cameras and calculators. Even though nearby customers will in effect be discriminated against, the plant can attract a larger number of customers and realize greater economies of scale.

Optimum Location for Multiple Plants

When there is only one large producer of a given commodity, as in the example above, economies of scale are possible until transport costs exceed a certain level. By admitting more than one plant into the landscape, we can greatly reduce transport costs and thus raise profitability. But multiplant location is more complex.

Let us consider optimum multiplant location under the simplest circumstances. Imagine a set of customers located along a highway. If there is only one plant serving these customers, the best location for the plant would be near the middle customer. If there are two plants, however, the problem becomes more complicated. A famous example provided by economist H. Hotelling illustrates the equilibrium solution when two units are involved.

Two competing ice cream vendors will probably find it safest to agglomerate by setting up their stands in the center of a beach. Each vendor may dominate the customers on his own side of the beach, or the two may share the whole market. If the two vendors are not competing (they both work for the same firm), the optimum solution would be to locate at two points one-quarter of the total distance from each end of the beach. In this way, the aggregate distance traveled by customers would be minimized. If the vendors are competing, however, this solution might not work. The vendors would be tempted to move toward each other in an effort to capture a larger territorial area of the market. The risk here, of course, is that the customers at each end of the beach may be lost altogether and more competition—new vendors—will locate at the edges of the market area. To avoid this risk, both must accept an equal share of the market, or one must be content with a smaller, protected share of the market while the other dominates a larger share. The vendor with the larger share then locates at a point that minimizes the aggregate distance customers in his area must travel and at the same time avoids unserved territory where new competition might locate.

In this example, each of the two vendors tries to establish a spatial monopoly in order to gain the competitive advantage in a portion of the market. Industries also tend to locate in a spatially monopolistic pattern when the costs of distributing finished products are much greater than the costs of obtaining raw materials or when customers or suppliers are spatially diffuse (figure 8.11). The *spatial penetration* (farthest distance to which the plant can distribute goods and still make a profit) of any one plant is checked by the ability of the plant's competitors to obtain supplies and distribute finished products to markets closer to them at a lower delivered price. This spatially monopolistic pattern is most evident on a local or regional scale with respect to the location of dairies, newspaper companies, bakeries, and producers of other goods for which demand is limited in time and space.

On a national scale, the territorial size of most advanced economies is so great that there are many markets and raw material sources for virtually every industry. For some industries, such as publishing, transport costs become significant only on a national scale, even though some newspapers and magazines have regionally produced editions. For other industries, such as automobile assembly and steel, regional markets are large enough and transport costs from any one location great enough to warrant some dispersal of plants and to permit at least a limited spatial monopoly (see figure 8.8).

For industries with relatively low transport costs, location near particular raw material sources or large markets is less critical, and firms may seek regions where their production costs will be lower—that is, where they can save on labor costs and realize other agglomeration benefits (figure 8.12). Frequently this strategy results in clusters of competing plants sharing large regional or national markets, such as manufacturers of machinery and communication equipment. If plants in a particular industry do not benefit from

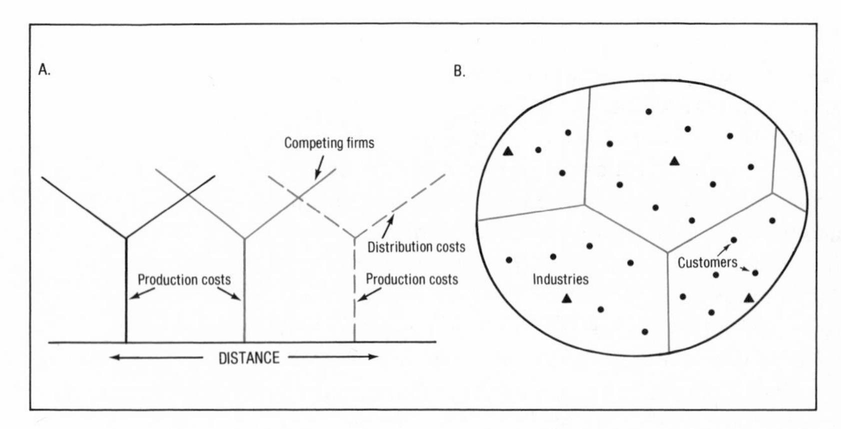

8.11. Spacing and competition among plants with high distribution costs. Diagram A shows that competing plants tend to disperse when costs of transporting finished goods to market are high. Rapidly increasing delivered prices limit the spatial extent of each plant's market (the plant cannot afford to deliver far or their scattered customers are not willing to travel far). New competitors will try to find a poorly served portion of the market and obtain a spatial monopoly over it. Diagram B shows four dispersed plants, each with a spatial monopoly over customers.

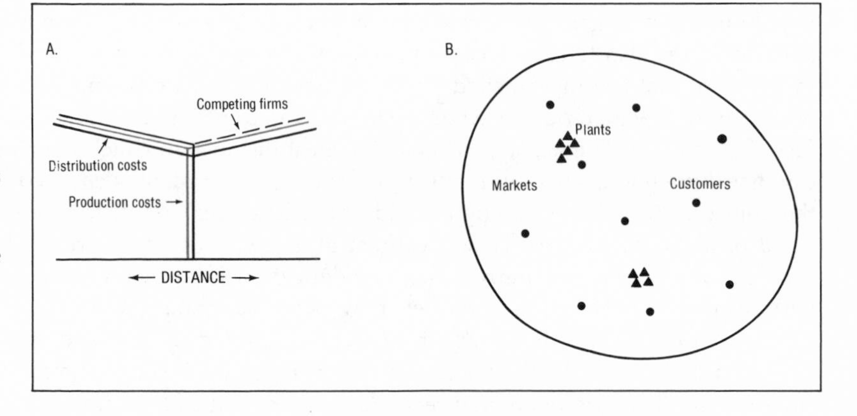

8.12. Spacing and competition among plants with low distribution costs. Diagram A shows that competing plants tend to agglomerate when costs of transporting finished goods to market are low and production costs relatively high. The plants can realize extra savings by agglomerating in an area with low-cost labor or other advantages. Diagram B shows two sets of agglomerated plants (customers do not necessarily order from the nearest plant).

agglomeration, they will disperse so that each has a monopoly over a small market area and shares a wider, overlapping market area with the other plants.

Agglomeration economies may induce a firm with several branch plants to consolidate production at a single location. Assume that production can occur at a set of low-cost sites. Additional transport costs would be incurred if all production were carried out at one location, because goods would have to be distributed over greater distances. But, as Weber pointed out, if agglomeration economies exceed the added transport costs, then all the plants should be located at the least-transport-point within the area. Thus a brewery may find it desirable to close small branch plants and consolidate production at one modernized large plant.

Spatial Adjustment

Changes in material and labor costs, technology, demand, and transport routes and rates may cut into the profitability of a plant at a particular site. But because of substantial investment in an existing plant, the producer is more likely to modify the plant's operation than to abandon it and build elsewhere. Major production and managerial changes at existing plants are in fact far more common than are changes in location. If substitution and other forms of adaptive behavior are limited or impossible, however, the plant may have to close. The older the plant, the lower the investment, and the less restrictive the material and labor requirements, the greater the willingness of an industry to move. From the 1920s through the 1950s, many New England textile producers with obsolete buildings and equipment found it more advantageous to move away than to renovate. Such relocation modifies the landscape by discouraging migration and growth in regions abandoned by major industries and stimulating the development of regions to which industry shifts.

Different kinds of changes call for different forms of **spatial adjustment.** Increases in the cost of raw materials may be offset by importing foreign materials of a higher quality or lower cost or by using alternative materials, such as plastic instead of wood. Increases in the relative cost of labor may be met by increased automation where feasible. Some electronics firms in Chicago now purchase more labor-intensive components from overseas subsidiaries.

Radical changes in technology may make traditional manufacturing processes obsolete and force a plant to shut down unless the location remains favorable for the new process as well. In the United States, steel companies are closing some older plants with higher costs, older technologies, or plants that are in central locations, such as Chicago or Youngstown, Ohio, and that are unable to meet new pollution standards.

Changes in demand can often be met by altering the mix of products, although this is limited by the capability of existing machinery. One interesting example is the shift from milk and butter to yogurt products in U.S. and European dairies.

Firms can also adjust by increasing the demand for their products. Often such change is rather sudden, as with new aircraft orders. Critical decisions include whether to increase production in existing plants, to build new factories, or to subcontract part of the increase to other firms. On-site expansion may be difficult because of lack of space, inadequate nearby sources of materials or labor, or environmental restrictions.

Some firms adjust to competition and uncertainty by acquisition of or merger with suppliers and competitors or by diversification into products likely to even out fluctuations.

Reductions in transport rates, permitting a rise in the scale of production, have weakened the competitive position of many smaller producers. But some manage to survive by shifting to custom production or by providing special services, like fast delivery. If such spatial adjustments can be made, the landscape will remain relatively unchanged.

Industrial Location in Centrally Planned Societies

The location of industries in centrally planned societies is subject to the same factors of costs, productivity, and profitability that affect industrial location in market economies. In practice, methods for choosing industrial sites are probably almost identical to those used by private firms. Yet theoretically at least, some differences can be expected. If the state is the general owner of all industries, it is likely to accept a longer period of loss while waiting for a new unit to become productive. Thus industries have a better chance of developing in peripheral areas. One might also assume that peripheral areas would develop faster and more extensively than similar regions in market economies because of the egalitarian goal of investing capital more evenly across the nation. Soviet experience partly supports these expectations; new industrial regions, as in western Siberia, have indeed been developed. But industry still tends to concentrate in traditional areas.

An economic system based on the concept of benefiting society as a whole rather than rewarding individual enterprise would tend to encourage the formation of more efficient spatial patterns, particularly if investment capital were limited. Plants producing similar goods are not likely to compete for the same market because there are many fewer brands in a centrally planned society. That is, the choice of automobiles, vacuum cleaners, and other products is much more limited. But the absence of a competitive market that determines the prices of raw materials, labor, and finished products according to supply and demand (however imperfectly) often appears to result in nonoptimal allocation of resources, especially capital; production of goods for which there is little demand; and inadequate production of goods and services for which there may be great latent demand. If it were not for the policy of regional self-sufficiency common in many centrally planned societies, the absence of competition would probably lead to fewer but larger plants and greater concentration and specialization. In capitalist societies, in contrast, competition can spawn too many plants producing the same kinds of goods.

The impact of large-scale industrial planning errors in socialist societies may be worse than that of many small failures in capitalist societies. For example, in the Great Leap Forward—China's 1959 industrialization program—planners attempted to apply the technology and organization of handicraft production to the manufacture of major industrial goods such as steel. The results were disastrous: there was no uniformity of production, raw materials were wasted, transportation facilities became severely congested, and productivity lagged far below expectations. After 1961 greater standardization and modernization returned to some sectors, such as oil refining and steel. But until 1977 the Chinese continued to emphasize small-scale local industries, both to utilize more fully the nation's large rural labor force and to curb urbanization and reduce rural-urban differences. Only since 1977 have the Chinese made a new commitment to the development of modern urban industries as well.

In reality, patterns of industrial location are remarkably similar among capitalist and socialist countries, perhaps because in the more developed countries the basic structure was already established in a presocialist period. Possibly patterns in newly industrializing socialist countries like Tanzania or Angola will differ more obviously.

Theory and Reality

The theory of industrial location as outlined in this chapter does not take into account all the variables that affect location. One reason for the discrepancy between theory and reality is that decision makers lack the perfect information needed to make the optimum location choice. As a result, nonoptimal decisions may be the rule rather than the exception. Moreover human perceptions are often influenced by personal attitudes. A person who loves the seashore might perceive a coastal region to be the most favorable location for a firm, despite evidence to the contrary. Then, too, many producers are willing to remain in only moderately successful locations as long as they make at least some profit.

Decision makers often temper a profit-maximizing goal with a noneconomic one. They may be sentimentally attached to a particular place or region or to a group of employees; they may prefer rural to urban settings when the latter objectively would be better, or vice versa; they may seek greater volume or visibility rather than profit; they may pursue growth for its own sake; they may be harassed by and fearful of competitors; or they may feel that they can succeed despite the odds. It is also not unusual for a government to subsidize or nationalize a major plant or firm for political economic reasons, such as to prevent a rise in unemployment or to avoid price increases. Thus money-losing shipyards and other kinds of factories in many countries have been kept alive by a shift to public ownership or subsidy; the price of such staples as bread or rice in many countries is kept artificially low; and, very commonly, uncompetitive industries have been saved by the imposition of tariffs against more efficient foreign producers.

Most decisions in the real world are imperfect. But persons choosing an industrial location usually consider most of the factors discussed earlier and other factors as well, using a time-honored, comparative-cost accounting approach. The criteria for selecting the best location in an economically rational manner include the following: Is the location central to customers and related activities? Are production costs minimized at that site? Can maximum profits be achieved there? Possible sites are narrowed to a few by general considerations of competition for markets, transport position, and linkages to other industries. For each possible location, the expected total costs and revenues (allowing for the adjustment behavior of competitors, to some degree), variations in the raw materials and products, and other factors are compared in detail. While mathematically less sophisticated than some theoretical techniques, these practical methods are behaviorally sophisticated in that the reaction of competitors is estimated. Also practical industrial location decisions can and do take into account the kinds of personal and psychological constraints that would be inappropriate for a general theory of industrial location. Although industrial location theory, like central place theory, cannot account for all that we observe, it does help explain and predict overall patterns of industry and the behavior of many firms, as we will see in the following summary of industrial systems.

INDUSTRIAL SYSTEMS

Modern industrial societies produce countless products. The location of plants and the marketing of their products illustrate the principles we have developed. A brief summary of the major branches of manufacturing may be helpful in relating location theory to the real world.

Most countries have fairly similar patterns of demand for goods, and the larger, more developed countries have similar patterns of production, but there are variations. Nearly all countries produce a variety of food products, some textiles and clothing, some books, magazines, newspapers, and other printed matter, metals, furniture, lumber, and the like. Developing countries may also produce labor-intensive electronic components, toys, textiles, apparel, and other goods for export to more developed countries. Among developed countries, the larger ones tend to have a full range of industries, including chemicals, steel, motor vehicles, and machinery, although only a few have significant aircraft or computer sectors. The smaller developed countries tend to specialize in just a few sectors—for example, the Netherlands specializes in petrochemicals, jewelry, and electrical goods and machinery, and Sweden in steel, tools, paper, and glassware. Socialist economies generally emphasize heavy industry, such as steel and machinery, whereas market economies stress consumer goods.

We will classify industries according to their basic spatial pattern and economic orientation: transport (resource or market), complex (resource and market), production cost and agglomeration, and interindustry orientation.

Transport-Oriented Systems

Resource-Oriented Industries

Resource-oriented industries depend on raw materials that are spatially limited and fairly expensive to transport because of low value or high waste per unit. They tend to favor location at or near resources. Major industries in this group include parts of the primary metals sector, such as smelting and refining, food and wood processing, and some chemicals.

Smelting and Refining. A number of metals—iron, copper, zinc, lead and aluminum among them—are essential to many industries. The primary production processes involved in obtaining these metals are the smelting and refining of ores. The poorer the ore, the lower the metal content and the more likely that processing will be necessary at the mine. When the proportion of metal to valueless material approaches 50 percent, as with iron ore and some bauxite, little treatment is needed. But in many other ores, the proportion of metal is as low as 1 percent. Shipping such ore is not feasible. Smelting—a processing step in which the metal is extracted by heat produced with coal, petroleum, or gas—is indicated whenever transport costs for the raw ore exceed the combined costs of smelting and fuel and metal transport. Smelting, an example of extreme resource orientation, is the kind of industry most commonly found in remote, sparsely populated, and often rugged areas that contain the demanded ore. For example, copper is smelted near copper mines in Arizona, Utah, and Montana in the United States and in Chile, Rhodesia, and Zaire.

Aluminum. Power-intensive industries such as final aluminum, magnesium, and titanium refining represent a special kind of resource orientation. They require huge amounts of electricity, the largest and most variable cost factor. To produce one ton of aluminum, 16,000 to 18,000 kilowatt hours of electricity are needed (the average home uses about 5,000 kwh per year). Optimal location for such industries is in regions where electricity is abundant and relatively cheap. Quebec, Norway, the Pacific Northwest, and Ghana are favored because of their large hydroelectric resources. The high value per unit weight of aluminum offsets the extremely high shipment costs for intermediate raw materials (mainly alumina) and finished aluminum.

Food and Lumber Processing. The largest group of resource-oriented industries includes food and lumber processors. Many of the plants are located in less densely populated areas, owing to the space needs of farming and the peripheral location of many of the best lumber resources. Agricultural products and lumber are for the most part bulky and perishable and are therefore costly to handle and ship. Processing increases their **transferability** by eliminating excess bulk and weight and by reducing perishability.

Canning and freezing plants that process riper, less transferable food such as fruits, vegetables, sugar beets, and fish tend to locate as close as possible to their suppliers, often in remote and sparsely populated areas. Output per plant tends to be rather low because of high assembly costs. In the United States this industry is particularly important in California, Florida, Hawaii, Oregon, Washington, Michigan, and other areas specializing in fruit and vegetable production. In Europe increasing numbers of canning and freezing plants are appearing in the specialty fruit and vegetable areas of countries such as Spain, Italy, Greece, and Bulgaria. Other resource-oriented food processors include wineries (located near vineyards) and cigarette manufacturers (located near tobacco farms). But because of high value, these industries are sometimes located in metropolitan areas.

Meat packers, creameries, flour mills, and feed mills are basically resource oriented but must also consider their final markets. Hence the suppliers are located as close to market as competition permits, and the processing plants are often located midway between the suppliers and the market. In the United States, gateway cities between the agricultural and urban-industrial regions of the nation are the preferred locations for these plants. Meat packing, for example, is concentrated in the string of cities bordering the western edge of the corn belt, from Fargo, North Dakota, to Fort Worth, Texas (see figure 5.12).

Creameries, especially for cheese and butter, are similarly located in the midst of milk-producing areas in these gateway cities. Flour and feed mills may be found in grain-producing areas throughout the world, but they tend to concentrate in larger places with superior transport facilities—for example, in Minneapolis, Kansas City, and Buffalo (reached by Great Lakes transport).

Lumber processors—sawmills, veneer-plywood makers, shingle mills, woodpulp mills, and turpentine and rosin plants—are resource oriented. Their products undergo significant weight and bulk loss through processing. Though not notably responsive to scale economies, sawmills and other lumber processors are very responsive to transport costs and thus tend to be small and dispersed. Depletion of local resources, however, restricts the useful life of the mills. Thus an intermediate-sized processing scale, utilizing a larger supply area, is becoming a more profitable solution. Pulp mills and makers of cheaper grades of paper not dependent on scrap are also located near lumber resources. But since the plants are very large, they must also be located at favorable transport points. The lumber-processing industry is important in many sparsely settled regions in the U.S. Pacific Northwest and South, in Canada, Scandinavia, the Soviet Union, and, increasingly, in tropical countries.

Market-Oriented Industries

Industries whose distribution costs are high tend to locate at their market, usually a higher order central place. Many finished products are less transferable than their component parts because of weight, bulk, or perishability gained through manufacturing; thus they cost more to distribute.

Soft-drink bottlers and others in the beverage industry are market oriented (see figure 8.7). The principle ingredient of soft drinks, water, is available in most places, and the syrup base has little bulk. The final bottled product, however, has poor transferability because of its gain in bulk. Hence most bottling plants are located at their markets. Similar industries include fluid-milk bottling, perishable bakery goods, ice manufacturing, and local newspapers, for which demand is limited in time and space. The distribution of fresh-baked goods virtually reproduces the pattern of cities, especially in Europe. Much of the construction industry is also market oriented, particularly residential construction and the manufacture of construction materials—sand, gravel, brick, concrete, and concrete block.

Distribution costs for many industries are greater than assembly costs, but not so great as to limit delivery to small areas. In an economy where distances between markets may be vast, as in the United States or the Soviet Union, distribution costs may be high enough to permit the existence of somewhat protected regional markets (limited spatial monopolies). Some industries have therefore opened branch plants in regional centers. While much of the automobile industry is highly concentrated—in the United States, for example, production is centered in Detroit—assembly plants are dispersed in the large cities of major regions (see figure 8.8). The same is true of much of the steel, cement, petroleum refining, food processing, and furniture industries, among others. The broad industrialization of California demonstrates that most kinds of manufacturing can succeed if the regional market is large enough and far enough from the more traditional centers of industry.

Industries Oriented to a Spatial Complex of Resources and Markets

For some industries—generally those in which transport costs vary more than production costs—both resources and markets influence location. The use of several resources and markets may complicate the problem of location. The industry may locate at or near any one of the resources or markets, or it may choose a site between them. Rarely are the transport costs for a single kind of raw material or product so great that it alone determines location. The problem is reduced to finding the point where aggregate costs are minimized.

Iron and Steel. The iron and steel industry is a particularly vital industry with linkages to the fabricated metal, machinery, hardware, and construction industries and to the transport and military sectors of the economy. The industry is spatially and technically complex. It uses a mix of raw materials that allows for some substitution possibilities, its processing technology can be varied, its products are highly diversified, and it sells to a wide variety of markets.

The material requirements for the blast-furnace production of primary pig iron are coking coal, iron ore, and limestone. Scrap can be partially or largely substituted for pig iron in the production of steel. Moreover coking coal is not essential for steel production, and different kinds of power sources may be utilized. Modern techniques have reduced fuel requirements substantially. Hence the location of finished steel production is freer than that of pig iron, although the advantages of integrated production with iron are great enough to warrant agglomeration between the two.

Iron and steel producers have traditionally located near coalfields, particularly those closest to markets; but a more complex orientation has evolved during the past few decades. Coal requirements have dropped to one-third the volume required in the 1830s; demand has risen for goods and materials that have been processed and thus have higher distribution costs; more scrap is available; and older iron ore deposits have become depleted.

When most iron ore in the United States came from Minnesota's Mesabi Range, the major iron and steel centers were the Great Lakes port cities located fairly close to coalfields—Cleveland, Chicago, Detroit, and Youngstown. But in recent years, steel production has expanded into coastal markets like Philadelphia and Baltimore because of the rising demand for steel products in metropolitan markets, the plentiful supply of scrap now available, and the greater dependence on imported ore. Steel centers have also developed in Utah, Colorado, Texas, California, Washington, and Oregon. Demand has risen high enough to permit production in these new areas at a scale competitive with the older, more distant centers.

Integrated plants producing both pig iron and finished steel products have the advantage of centralized management and lower fuel and transport costs, but the separate production of steel may also be profitable. Steel alone is produced in regional centers like Houston, Los Angeles, San Francisco, and Seattle, which lack readily available coal and are far from large integrated basic producers. Regional steel mills may purchase pig iron from other producers, but they can also utilize a large quantity of scrap. One of their chief advantages is much faster delivery to regional markets.

The steel industry forms similar but not identical patterns in other parts of the world. In Europe the industry was traditionally located near major coalfields—as in Wales and near Newcastle and other coal areas in Great Britain; in the Ruhr, Germany; in Silesia (Slask), Poland; and in the Donbas and Kuzbas, the Soviet Union—or near major iron ore fields, as in French Lorraine or the Krivoi Rog and southern Urals in the Soviet Union. The same shift to major markets and coastal locations has occurred in Western Europe. Most newer steel mills are located in or near ports on the North Sea or the Mediterranean, where coal, often from the United States, and iron ore can be imported. Perhaps because of a less developed consumer goods sector, the industry in the Soviet Union and Eastern Europe is still rather resource oriented.

Petroleum Refining and Petrochemicals. The location of petroleum refineries and petroleum-based chemical plants is simpler than that of the iron and steel industry. Nodal points—the oil field, the market, and certain intervening transit points—are the most favorable sites. (**Nodes** refer to terminals or intersections on a transport network.) The processing of crude petroleum is efficient, resulting in very little weight or volume loss. The slightly higher cost of handling and shipping refined products, which are more varied and volatile than the crude, does not outweigh by too much the cost of handling and shipping the slightly greater bulk of crude petroleum. Hence the advantages of locating near markets are not overwhelming. Given the use of multiproduct pipelines and tankers, field refineries producing a few products in sufficiently large volume can compete with those located at markets. Yet the greatest refinery capacity is at markets, a fact that has long disturbed oil-producing areas, some of which require construction of refineries at the resource.

In Europe and increasingly in the United States, refining and petrochemical plants tend to locate in coastal ports with good access to the interior, as along the U.S. Gulf Coast. This shift reflects the greater dependence on imported petroleum. In the Soviet Union, production is oriented to petroleum resources in such areas as the Caucasus, the Urals, and western Siberia and to the needs of other industries rather than to metropolitan consumers. Petroleum and related industries are gradually increasing in the major oil-producing nations as they begin to demand more home processing in an effort to industrialize.

Electric Power. Generating electricity from steam involves high costs for transporting fuel and for transmitting electricity. In the past, transmission costs have been higher than the costs of transporting the fuel. Thus large power plants are usually located near major markets. But technological improvements in high-voltage, long-distance transmission as well as in fuel efficiency now permit coal-, gas-, petroleum-, and nuclear-powered stations to be installed at the source of the fuel, several hundred miles from final markets. The use of Rocky Mountain coal, once a major power source for U.S. railroads, is rapidly rising again as huge power plants in remote areas such as New Mexico and Montana are now feeding power into regional grids.

Production Cost–Oriented and Agglomeration-Oriented Systems

Industries Oriented to Production Cost Savings

Spatial variations in the productivity, cost, skills, and availability of labor strongly influence the location of some industries. The use of semiskilled and skilled labor constitutes a major part of the technology and adds much of the value of products in such industries as electronics and printing and publishing. For others, labor costs may be the critical factor if transport costs for raw materials and products are relatively low. For still other industries, labor orientation may be a matter of choice rather than of necessity. Textile manufacturers, for instance, may prefer to use lower wage labor rather than automate their technology.

Textiles. Textile manufacturing has typically been a labor-intensive industry, with a traditionally high ratio of women to men. Competition for labor from nearby industries has posed some problems, but textile mills have found it possible to survive as long as workers willing to accept lower wages are available. The textile industry has been affected by changes in the competitive position of other industries and of labor. A brief sketch of the industry's history in the United States illustrates this relationship.

During the 1800s, textiles were made in New England, where capital, labor, and water power were readily available. As the nation's economy developed, machinery, shipbuilding, and other industries began to bid up the price of male labor. In the early 1900s, unions were organized and welfare legislation was passed, restricting the hours and conditions under which employees could work. Work stoppages put additional pressure on employers to raise wages and improve working conditions. Confronted by these changes and demands, many textile mill operators moved to small towns clustered in the South, where conditions of surplus labor were comparable to those in New England a century earlier. Other advantages of locating in the South were lower taxes, cheaper services and utilities, the absence of unions, better access to cotton supplies, and fewer competing industries.

Some segments of the textile industry, however, have not moved from

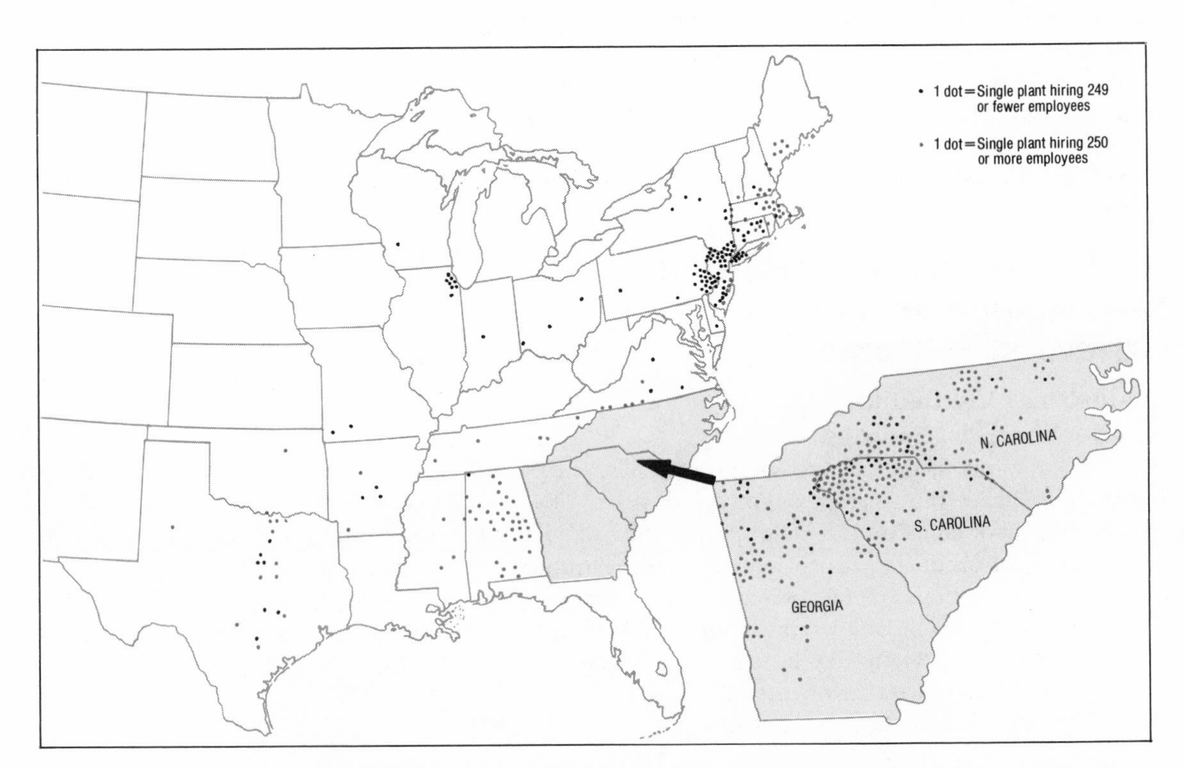

8.13. Location of textile manufacturing in the United States. While many textile mills have relocated in small towns clustered in the South and West, some segments of the industry have remained in traditional northeastern locations.

Source: Adapted with permission from Richard S. Thoman, *Geography of Economic Activity* (New York: McGraw-Hill, 1962).

the Northeast (figure 8.13). These include finishing, finer woolens, pattern printing, and some knitwear mills, which either require more highly skilled labor or have continued to employ immigrant and minority workers at lower wages. Those industries that have moved in many cases have created unusual new urban settlement patterns, particularly in the North Carolina area (figure 8.14).

Not only has much of the textile industry left the northeastern United States for the U.S. South, but many firms have also left the country. Because textiles require much labor but not too much capital and can be produced successfully at moderate scales, the industry is basic to the early industrialization of less developed countries. In Europe textiles are declining in the wealthier countries, and production is moving to southern Europe and overseas to such developing countries as India, Korea, and Hong Kong.

Furniture. The location of the furniture industry is a good example of the influence exerted by markets, resources, and lower labor or processing costs. The location of the early furniture industry in the United States was oriented to the labor supply and hardwood forests of southern New England and the Michigan-Ohio region. Like the textile industry, however, many furniture makers later moved South. Competition for labor in the North and the superior hardwood forests of the South induced the shift to the Carolinas and other southern states. The traditional northern centers,

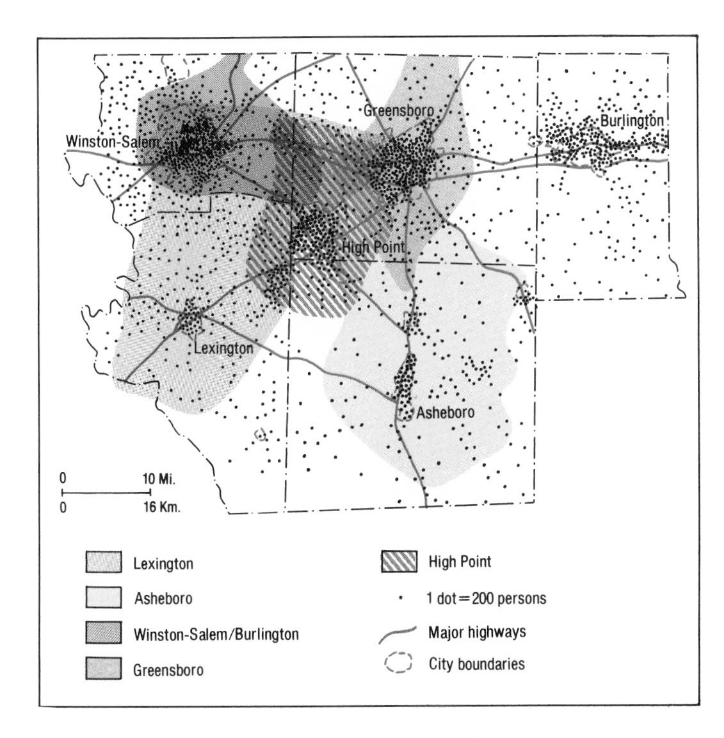

8.14. The North Carolina Piedmont dispersed city. The unusually dispersed pattern of urban settlement (only half the population of 800,000 lives in the larger cities) directly reflects the dispersed pattern of the location of textiles, furniture, and related industries.
Source: Redrawn with permission from Charles R. Hayes, *The Dispersed City,* Research Paper No. 173, 1976, University of Chicago, Department of Geography.

such as Grand Rapids, Michigan, adapted by producing higher quality furniture; they survived because customers were willing to absorb the higher labor costs.

Electronics and Appliances. Certain mass-produced electronic goods made with a standardized technology—radio, television, electrical parts, and small appliances—are labor intensive. The electronics industry has gradually migrated from New York and other traditional northeastern centers to smaller cities with lower wage labor just outside the industrial core (in Iowa, Kentucky, Virginia, and Minnesota) and to depressed areas with surplus labor, such as New England in the 1950s. These industries, too, have partly shifted and are growing rapidly in Japan, Korea, Singapore, and other less developed countries.

Other Industries. A few industries that have high labor requirements and pay moderate wages have not migrated but have remained in traditional locations, usually because of a specialized labor pool. In the United States, cutlery, brassware, silverware, jewelry, fine stationery, and watches are all manufactured in cities and towns where the industries originated. Similarly traditional patterns occur in Europe, but specialized production, such as the Swedish glass industry or the Swiss watch industry, is often concentrated in smaller cities and towns.

Industries Oriented to Agglomeration Savings

Industries with low transport costs and strong technical linkages to resource- or market-oriented industries may locate near their suppliers or their customers to take advantage of agglomeration economies. Such *secondary,* or dependent, *location* is characteristic of plants using the by-products of other industries and of plants supplying parts for larger industries. Chemical plants, for example, utilize the by-products of iron and steel mills and of petroleum refineries. We have already mentioned the industrial complexes that contain both iron and steel and chemical plants. Container manufacturers often locate near food processors and other industries that they supply. Firms making specialized machinery typically locate in areas where their products are used: drilling equipment is made in Los Angeles and Houston; wood-processing equipment in the Pacific Northwest, eastern Canada, and Scandinavia; aircraft components are produced in major airframe centers, such as Los Angeles and Seattle; and the huge automobile parts industry is located in a wide belt around Detroit or near major automobile manufacturing centers in Europe and Japan. To some degree, production of high-quality paper is located fairly close to major publishing centers like New York, London, Paris, and Moscow.

Industries such as apparel and publishing tend to be oriented to the agglomeration advantages of large metropolises and their suburbs. These industries have fairly high labor costs, and their products usually command high market prices and have high value added through manufacturing. Transport costs do not significantly influence location, and markets may be somewhat diffuse. But labor is an important variable. Employees with the desired qualifications often prefer living in metropolitan areas. Thus even if production costs were much lower in smaller cities and towns, apparel and publishing would still have to locate in higher level centers.

Apparel. The apparel industry has traditionally consisted of small-scale enterprises with a limited range of products and a small number of employees. The industry's extreme concentration in metropolitan areas reflects historical inertia as well as the attraction of agglomeration benefits. In the United States, nearly half of all clothing is made in New York City, mostly within a small section of Manhattan. In Europe, the centers of the apparel industry are Paris, London, Rome, and Moscow. These metropolises contain large pools of skilled workers willing to accept moderate wages, especially women and recent immigrants. In addition, the availability of capital, transportation facilities, communications, advertising services, and other marketing and distribution services in the metropolis provide important agglomeration advantages. The apparel industry requires much publicity, which is rapidly diffused from metropolitan centers, and depends strongly on the dictates of fashion, which is influenced by the tastes of metropolitan fashion setters. But very standardized and mass-produced segments of the apparel industry, such as blue-jean manufacturing, have in part accompanied the shift of textiles to lower labor cost regions. And some specialized branches, such as sportswear and outdoor clothing, are thriving in the western United States and other new centers

Printing and Publishing. The location of the printing and publishing industry (with the exception of local newspapers) is also oriented to such cultural capitals as New York, Boston, London, Paris, and Moscow. By centralizing in a metropolis, the industry can take advantage of skilled labor pools and communications facilities.

In the United States, publishers of trade books and high-quality periodicals requiring centralized or cooperative distribution are concentrated in New York. Publishers of high-volume material, such as Bibles, telephone books, mail-order catalogs, and mass-circulation magazines, are concentrated in more central locations, notably Chicago. However the current trend toward much higher mailing rates for books and magazines may force the industry to decentralize.

Instruments. Plants producing scientific instruments, drugs and medical supplies, and cameras and film are oriented both to the agglomeration advantages of large markets and to traditional centers. For example, photographic equipment plants are located in Rochester, New York, and Jena and Stuttgart, Germany. These industries have shifted to some extent to areas of lower cost labor overseas.

Interindustry-Oriented Systems

The largest group of industries in technologically advanced societies comprises machinery and machine tools, transportation and communication equipment, appliances, and other elaborated goods (made from simpler manufactured products). Characterized by strong interindustry linkages, these industries use a variety of semifinished materials, such as fabricated steel, and distribute their products to a rather wide market composed mainly of other industries. Because they are not tied to a given resource or market by transport costs, time constraints, or labor costs, they enjoy a limited degree of locational freedom. The optimum location is usually within the set of resources and markets contained within the urban-industrial core of the economy. In the long run, location decisions are influenced by transport position, quality of the labor force, agglomeration benefits, and relationships to other industries and to competitors. The complexity of the industrial location decision reflects the complexity of advanced society as a whole.

Fabricated Metals. The fabricated metals sector is intermediate between makers of steel, copper, aluminum, and other metals and makers of final products containing these metals. It includes producers of cans, hardware, plumbing and heating equipment, and wire. The industry tends to be located near its largest users, for example, near automobile manufacturers or steel mills.

Motor Vehicles. The automobile is the highest value industrial product and the largest single consumer of steel, rubber, and glass in the United States and other advanced countries. The ideal location for automobile makers would be at the center of the transportation and industrial network. Thus Chicago would probably now be the best location in the United States, but for historical reasons Detroit is the center of the industry. The Detroit area was an optimal location fifty years ago, when the automobile industry was first developed and the largest market was still along the Atlantic seaboard. Detroit had many advantages: automobile makers could draw on the experience of Detroit's renowned carriage and boat-engine manufacturers, and they could also utilize the city's skilled labor force and its nearby lumber resources. Detroit bordered two markets, the wealthy East and the developing West. Moreover the city and its environs were the home of Henry Ford and some other early automobile manufacturers. These shrewd entrepreneurs successfully defeated competitors in New York, Philadelphia, and Chicago. The decision to standardize and mass produce the automobile was perhaps the key to their success. Once the industry was well established, Detroit's agglomeration advantages and experience overshadowed any transportation or labor advantages other cities may have had.

Today Detroit remains favorably located with respect to suppliers, though not quite so favorably with respect to final markets. Population shifts and the rising cost of final distribution have resulted in a partial dispersal of body-making and final-assembly operations to serve distinct regional markets. Related industries are not far from Detroit. Parts suppliers are scattered throughout Michigan, Indiana, and Ohio. Glass production is concentrated in the nearby Toledo, Ohio, area, and rubber tires in the Akron, Ohio, area. To extend our example of interindustry orientation, the rubber industry, which requires large amounts of coal, is located near the eastern Ohio coalfields and is also reasonably close to Detroit.

Similar patterns are found in Europe and Japan. So capital intensive and complex is the motor vehicle industry that the basic design remains limited to only a few developed countries. However assembly plants for cars and trucks have been dispersed to many parts of the world—for example, to South Africa, Nigeria, and Brazil.

Machinery. Machinery, both general and electrical, is essential to the growth of productivity in advanced societies. A large quantity of steel and other basic materials is consumed in the manufacture of machinery, and the value added through manufacturing is high. Many different industries depend on machinery, from primary industries such as smelting to the most consumer-oriented ones such as frozen foods. Thus it is difficult to speak of machine producers as a group. Their markets are diffused throughout the urban-industrial core. Distribution costs for finished products are higher than assembly costs for components. But the fact that each plant sells to many customers suggests that the best location is one that is central to all the markets rather than simply the largest market.

The production of electrical machinery, including communication and copying equipment, predominates on the Atlantic seaboard; nonelectrical machinery oriented to the construction and automobile industry is concentrated in the center of the industrial core; and agricultural equipment and food-processing machinery are concentrated at the western end of the industrial core. Machinery production is heavily represented in almost all northeastern metropolises, and it dominates the economies of many smaller and intermediate-sized cities as well.

Subdistrict specialties have evolved from early inventions and spinoffs. These include the machine-tool industry centered in Cincinnati, turbines and generators in Milwaukee and Chicago, textile machinery in New England, paper machinery in Wisconsin and New York, mining equipment in Minneapolis, and printing machinery in New York.

Freedom of Location

True freedom of location for industry does not exist, although a few small specialty manufacturers may locate in remote areas and survive because of the uniqueness of their brands or because of strong employee loyalty. Large military-industrial firms, such as aircraft, ordnance, and missile and rocket development, may also locate in peripheral sites partly for security reasons and partly because of the federal government's half-conscious desire to spread the wealth to less developed regions or to aid politically powerful but economically weak peripheral regions.

The aircraft industry comes close to being genuinely footloose, enjoying a moderate degree of locational freedom (see figure 8.9). At first, aircraft manufacturing was located near the spruce forests of Michigan and the Pacific Northwest, which provided lightweight but strong construction material for early planes, and near large cities. The locational preferences of the industry's pioneers and the specifications of World War I contracts determined some of the important centers: Los Angeles, Seattle, St. Louis, and Hartford. California's favorable climate for experimentation and pleasure flying was another incentive for the Los Angeles location. During World War II, the government created centers in Wichita, Kansas; Marietta, Georgia; and Fort Worth, Texas. Although transport costs for the aircraft industry are negligible, the need for a large, skilled labor force suggests that the established centers will remain dominant and that metropolitan areas, with their extensive airport facilities, might become sites for future expansion.

Classification of Manufacturing in the United States

Table 8.1 summarizes the major branches of manufacturing in the United States; the pattern is typical of most advanced industrial countries. The machinery and transportation sectors, and the metal and other industries upon which they are based, are numerically dominant. Note the large number of establishments, low value added per establishment (value added to goods, or returns to labor and management, in the process of manufacturing), and small number of employees in publishing and lumber and wood. In contrast, metals, electrical machinery, and labor-intensive textiles have a much larger average number of employees, and primary metals and transportation equipment have higher value added per establishment. Textiles and leather have low value added per employee compared to chemicals, transport equipment, and petroleum refining. Interestingly handicrafts are significant in terms of employment even in the United States and would be relatively far more important in most developing nations. (Although handicraft production is increasing in the United States, few facts are available.)

INDUSTRIAL REGIONS

At a global scale, the major industrial regions comprise Europe (both East and West), the United States, Canada, and Japan. These regions combined produce about 90 percent of the world's industrial output.

The pattern becomes more complex when we consider industrial regions at the national and continental scale. Three principal kinds of industrial regions can be identified: regions of heavy industry, emphasizing

Table 8.1. Manufacturing in the United States, 1974

Industry	*Employment*	*Establishments*	*Value Added ($ millions)*	*Average Number of Employees per Establishment*	*Value Added per Establishment*	*Value Added per Employee*
Resource oriented						
Food, tobacco	1,860,000	29,000	$ 43,000	64	$1,483,000	$23,120
Lumber, wood	626,000	34,000	12,500	18	368,000	19,970
Resource and market oriented						
Paper	706,000	6,000	15,000	118	2,500,000	21,246
Chemicals	1,057,000	11,500	36,000	89	3,139,000	34,055
Petroleum refining	199,000	2,100	8,000	95	3,810,000	40,200
Stone, glass	690,000	16,000	14,000	43	875,000	20,290
Primary metals	1,344,000	7,000	29,000	192	4,143,000	21,580
Agglomeration oriented						
Publishing	1,112,000	42,000	22,000	26	524,000	16,784
Apparel	1,365,000	24,000	15,000	56	625,000	10,900
Labor-cost oriented						
Textiles	988,000	7,300	13,000	137	1,806,000	13,158
Furniture	517,000	9,500	6,800	54	716,000	13,153
Leather (shoes)	300,000	3,300	3,000	91	909,000	10,000
Interindustry oriented						
Rubber and plastic	676,000	9,250	13,500	73	1,460,000	19,970
Fabricated metals	1,505,000	30,000	31,000	50	1,033,000	22,263
Machinery	2,218,000	41,000	45,000	54	1,098,000	20,290
Electrical Machinery	2,030,000	12,500	35,000	162	2,800,000	17,241
Transportation equip.	1,821,000	14,400	59,500	126	4,132,000	32,674
Other	986,000	31,000	13,000	32	420,000	13,000
Handicrafts	2,000,000	1,000,000	[a]	[a]	[a]	[a]
All	20,000,000	330,000	414,000	61	1,212,000	20,500

[a] Data not available.
Source: U.S. Census of Manufactures, 1974.

mining, iron and steel, chemicals, and machinery; regions of light industry, emphasizing labor-oriented activities such as textile production; and regions of diversified industry, containing a complex mix of activities clustered in and around great metropolitan centers and smaller national economic capitals.

Major regions of heavy industry include the Great Lakes cities from Milwaukee to Buffalo, the Atlantic coastal ports from Boston to Baltimore, and the Texas-Louisiana Gulf Coast region in the United States; the Lancashire, southern Wales, Newcastle, and Glasgow areas in the United Kingdom; the Ruhr in West Germany; the Rotterdam-Antwerp area in the Netherlands; Silesia in Poland; the Donbas, Kuzbas, and southern Urals districts in the Soviet Union; the southern Inland Sea area in Japan; southern Manchuria in China; the Calcutta and Bombay districts in India; and the Johannesburg area in South Africa.

Major regions of light industry include the U.S. Southeast and smaller towns in the Northeast; Manchester and other areas in the United Kingdom; many small towns and cities in France, Germany, and Italy; smaller cities in central Asia and Japan and around Moscow and Leningrad in the Soviet Union; and larger cities in several developing countries.

Major regions of diversified industry include the leading world metropolises, notably New York, Chicago, Los Angeles, London, Paris, Milan, Tokyo, Moscow, Leningrad, São Paulo, and Buenos Aires. At a smaller scale, diversified industry is also concentrated in the economic capitals of many smaller and developing countries—for example, in Lagos, Nigeria; Nairobi, Kenya; Bogotá and Medellin, Colombia; Athens, Greece; Bangkok, Thailand; and Istanbul and Ankara, Turkey.

An overall structure of world industry is discernible. Despite national boundaries and the division between socialist and capitalist economies, the level of industrial interdependence among countries is high and is growing. Large multinational corporations have succeeded in part in overcoming political and economic barriers and have further internationalized the world economy.

The three core industrial regions of Europe, North America, and Japan dominate the higher technology sectors of machinery, transportation equipment, computers, instruments, and the like, while the developing, peripheral nations are favored for the increasingly decentralized resource and labor-intensive sectors.

CONCLUSION

To summarize our discussion of industrial location, we will make some very general observations. First, industrial activities, like central place activities, help to create cities and towns. So responsive are industrial activities to economies of scale that they will almost inevitably promote some development in the surrounding region. Even small factories are likely to spawn small urban places.

In addition, industrial activities tend to be geographically more concentrated than is the general population. Firms seek the largest, densest markets in order to save on distribution costs, to achieve scale economies, and to maximize profits. Market-oriented firms reinforce the central place structure. Many firms seek the agglomeration benefits of large metropolises and concentrate in these highest level centers. Others disperse to smaller cities and towns to save on production costs—land, labor, taxes, and the like. And some firms locate near their raw material sources, which may or may not be near existing central places. Such firms often create new towns outside the central place structure.

If raw materials were found everywhere and production costs did not vary from place to place, the theory of industrial location could be reduced to central place theory. A regular pattern of dispersed central places would be the most efficient and profitable arrangement for both industrial and central place activities. But because raw materials are unevenly distributed and costly to transport and because labor and other production costs do vary spatially, industrial patterns are not as consistent as central place patterns.

We have emphasized the spatial and economic factors that influence industrial location. While these are fundamental, environmental factors are also becoming significant. Before constructing a plant in a new site, many U.S. firms are now required to prepare an environmental impact statement, a detailed assessment of all potential effects of the project on the environment, which must be approved by state or federal agencies. Recently construction on a nuclear power plant in Seabrook, New Hampshire, was halted because the Environmental Protection Agency reversed its approval of the plant's environmental impact statement. In the Soviet Union, new pulp mills are being directed away from sensitive environments like Lake Baikal. Such constraints will become more common as environmental awareness and concern increase in high technology societies. Ironically our quality of life may deteriorate because of the very activities that provide material well-being in the first place. The tension between environmental concern and the need for jobs may erupt into serious political conflict.

Urban geography is concerned not only with the patterns and problems of the central place and industrial landscapes but also with the organization of individual human settlements. In the next chapter we will examine the microlandscape of the village, town, and city.

SUGGESTED READINGS

Alexandersson, G. *A Geography of Manufacturing.* Englewood Cliffs, N.J.: Prentice-Hall, 1967.

______. *The Industrial Structure of American Cities.* Lincoln, Neb.: University of Nebraska Press, 1956.

Estall, R. C., and Buchanan, R. *Industrial Activity and Economic Geography.* London: Hutchinson University Library, 1961.

Fuchs, V. R. *Changes in the Location of Manufacturing in the U.S. Since 1929.* New Haven, Conn.: Yale University Press, 1962.

Greenhut, M. *Plant Location in Theory and Practice.* Chapel Hill, N.C.: University of North Carolina, 1956.

Hamilton, F. I., ed. *Spatial Perspectives in Industrial Organization and Decision-Making.* New York: Wiley, 1974.

Karaska, G., and Bramhall, D. *Locational Analysis for Manufacturing.* Cambridge, Mass.: MIT Press, 1969.

Miller, E. W. *A Geography of Manufacturing.* Englewood Cliffs, N.J.: Prentice-Hall, 1962.

Smith, David. *Industrial Location.* New York: Wiley, 1971.

Weber, A. *Theory of the Location of Industries.* Translated by C. J. Friedrich. Chicago: University of Chicago Press, 1929.

9

The Micro-geography of Human Settlements

Urban Structure

Urban Growth

9 The Microgeography of Human Settlements

One of the most fascinating views seen from the air is the panorama of a city unfolding below. In an approach to a large American city, signs of human habitation gradually emerge from a landscape of fields and forests. A few scattered houses appear, surrounded by trees and spacious lawns; then more homes, usually in rows or clusters. Soon suburban schools, shopping centers, parking lots, and industrial parks come into view. The built-up landscape intensifies into a dense mat of city blocks laced by a web of roads, highways, and bridges. Skyscrapers rise from the heart of the city, circled by a wide band of apartment buildings, housing projects, row houses, and parks. High-rise buildings and shopping districts sprout up at intersections and along major highways radiating out from the central core. Farther from the center, the intensity of activity and density of settlement diminishes. More and more space separates buildings and houses. Around the fringe, the city fades into the rural landscape, with a few outlying suburbs and shopping centers linked to the city by expressways and railroads. Then fields and forests once again dominate the landscape.

This generalized picture reflects the basic structure of nearly all large cities today. But if you had flown over a non-Western city—Cairo, Egypt, for instance—you would have noticed some differences: fewer skyscrapers but many prominent mosques; less land devoted to parks, streets, and parking; more widespread high-density settlement (considering the low city profile); and a sharper dividing line between the built-up area and the surrounding farmland.

In this chapter we will examine the internal structure (microgeography) of urban places, particularly the large city. Many of the principles governing the location of central place and agricultural activities also govern the location of urban activities. Thus cities are, in a sense, a smaller version of the larger human landscape.

URBAN STRUCTURE

We will define as urban an area that contains an agglomeration of people and in which most of the land has been altered and developed for commercial, residential, and industrial use. It is often difficult to determine where the countryside ends and the town or city begins. There is a gradual shift, or *gradient*, of intensity, from relatively little use and sparse population in the remote rural hinterland to extremely intensive use and dense population toward the central core of the city (figure 9.1). The urban area is the more intensive portion of the entire gradient.

There are some similarities between the gradient pattern of urban ac-

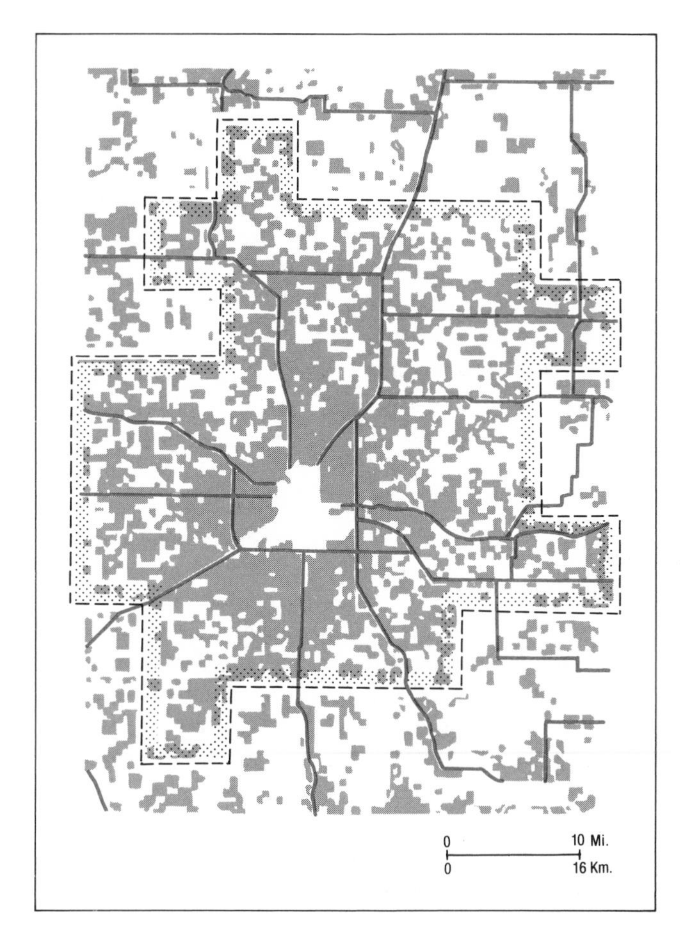

9.1. The urban-rural gradient, Grand Rapids, Michigan. Predominantly nonfarm (urban) areas, in gray, include residences of commuters and extend far beyond city limits. There is no sharp boundary between urban and rural land use; the intensity of use and density of population diminish toward the periphery.

Source: Redrawn by permission from *The Professional Geographer* of the Association of American Geographers, Vol. 14, 1962, H. Stafford, Jr.

tivities and that of agricultural activities (chapter 5). Nothing may seem less related than the location of dairy farms in the countryside and of jewelry stores in the metropolis, but the same competitive forces are at work in each case. The dairy farmer competes for a location as close to urban activities as possible (some dairy farms are located even within urban areas); the jeweler competes for a location at the peak of the urban intensity gradient, where pedestrian traffic is greatest. Both activities require central locations, and both can afford to compete for such locations.

The gradient pattern of land-use intensity and population in the city developed gradually as a result of competition for access to the center. Some activities, such as large hotels and theaters, must be near the center to succeed. But others, such as golf courses and single-family homes, can locate farther away, reflecting the willingness of people to travel farther for more space and less expensive land.

Just as an agricultural zone of given intensity might be used for different crops as a result of environmental or cultural forces, so might an urban zone of given intensity or value be divided into sectors of different land use or social and economic class. Cities often have separate industrial and wholesale sectors, for example, and distinct lower-, middle-, and upper-income residential sectors.

The gradient and sector theories of urban structure, which will be examined more thoroughly below, can be considered extensions of agricultural location theory. Similarly the location of urban commercial centers, discussed later, tends to follow central place theory. The existence of several centers has given rise to a composite theory of urban structure: the multiple-nuclei theory. The gradient and central place structure of the city are the outcome of society's pursuit of economic goals: maximizing the productivity of land and the level of interaction. But the sectoral differences result from society's pursuit of noneconomic, social, and political goals: maintaining or improving status. The noneconomic goals modify the economic ones and thus distort theoretically regular patterns of urban location.

Structure of the Village and Town

A brief look at the structure of the village and town will provide a useful introduction to the more complex organization of the large city. Figure 9.2 shows the two common village forms: (1) linear, or "row," characteristic of settlement in rugged areas or along dikes and river banks; and (2) circular and compact, characteristic of settlement in areas of level topography or where defense was essential (many clustered villages and towns, particularly in Europe, were originally walled for defense).

Villages are typically located at crossroads. In some areas, such as the American West and much of Latin America, roads form a regular grid, reflecting land division patterns, but in most places an irregular road network prevails. Commercial activities and churches are strung out along the main roads. A small grid or network of streets fills in some of the area between the main roads, but settlement usually extends much farther out along the latter. Even in the village, richer and poorer sectors develop, often separated

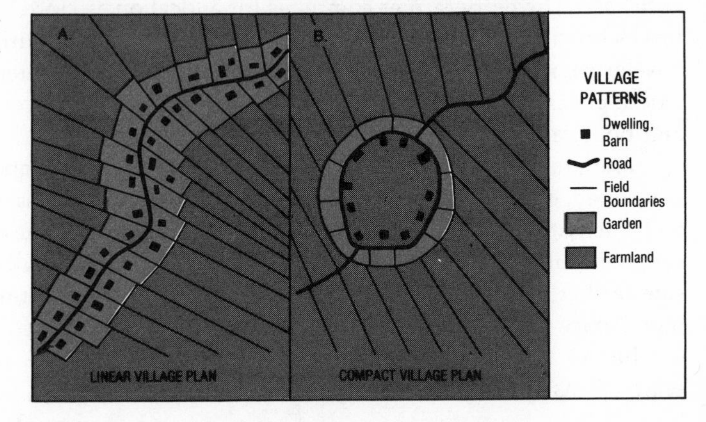

9.2. Typical village patterns: the linear village, or "strassendorf," and the circular, compact village.

Source: Redrawn from H. J. deBlij, *Human Geography: Culture, Society, and Space* (New York: John Wiley & Sons, 1977). Reprinted by permission of John Wiley & Sons, Inc.

by the main road. Most villages, especially in less developed countries, are closely linked to rural activities, and many are occupied by farm families. Indeed one of the reasons for the linear village pattern is that some farmland may extend outward from the farmers' houses and barns.

Some villages evolve into towns, with more complex internal structures. Suppose that a village has acquired an industry and has gained in commercial importance, becoming a town of perhaps 5,000 persons. Roads to other villages are added or improved. Many homes along the main roads have been replaced by stores or converted to commercial use. A central business district or "main street," forms, with a few blocks of commercial and public activities paralleling the intersection of the main roads. One end, adjoining the more affluent part of town, has the more expensive shops and more prestigious churches; the other end, adjoining the poorer section, contains other churches and some less attractive activities, such as auto repair shops and taverns. A small shopping center may have developed out on the main road where the town extends the farthest.

The town's industry is probably located on the main road in the poorer sector, where land costs are lower or where land is available or zoned for industrial development. Lower-income housing has been built nearby, perhaps in a major plat, or subdivision. At the other end of town, the growing upper-middle class has also added a subdivision, perhaps on land situated somewhat above the town. The concentration of jobs from the commercial core to the factory has created a demand for nearby housing, especially for single young people, the elderly, and others who cannot afford single-family homes or cars. Pressure increases on close-in single-family homes, and many are converted to or replaced by apartments. Thus a clear gradient pattern of land values and residential density has developed.

Gradient Theory

You can probably distinguish the different zones in your city or town. Industries, stores, and residences are not randomly scattered but are located in certain areas. Most cities, in fact, are fairly well organized. They tend to form an intensity gradient of land use, land value, and population that ranges from highly intensive use, high land value, and dense population at the center to less intensive use, lower land value, and sparser population at the fringe (figure 9.3). One reason for the orderly location of urban activities is their differing need for access to the center and their differing ability to pay for it. As distance from the center increases, accessibility and land values (rent) decline, and units of land tend to become larger.

Activities that have the greatest need for access and can afford to pay high rents for small parcels of land compete for the limited space in the central business district (CBD). (See photo A in figure 9.4.) Examples are hotels, department and specialty stores, banks, office buildings, theaters, and restaurants. Next in line come activities that also need a central location but require more space and cannot afford the high rents of the CBD. (See photo B in figure 9.4.) These include warehouses and wholesalers, some

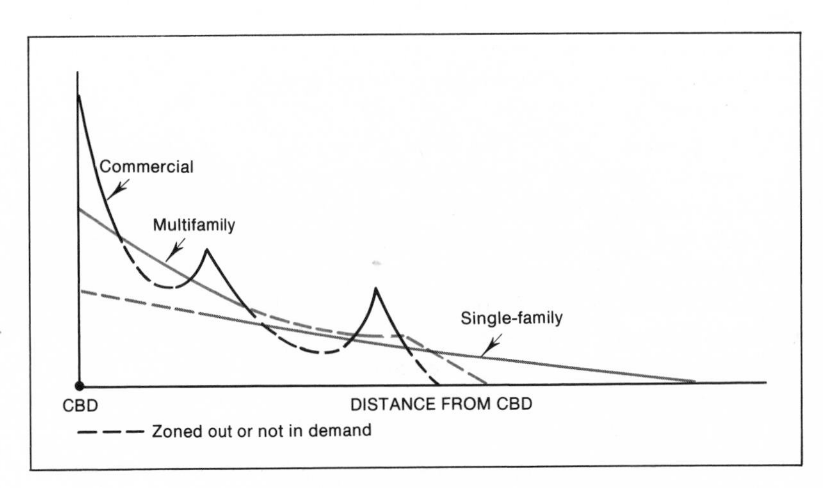

9.3. A theoretical urban intensity gradient. Competition for access to the CBD results in an orderly pattern of land use, ranging from commercial to multifamily and single-family residential use. Note the similarity between this model and the agricultural location model presented in figure 5.3. The effect of additional commercial centers is not shown in this simple model.

light industries (such as apparel workshops), hospitals, museums, and new-car showrooms. The upper stories of business establishments in this area are often used for apartments. This mixed commercial, industrial, and residential lower intensity zone tends to be followed by two kinds of residential zones: first, a more intensive apartment and multifamily-home zone (photo C in figure 9.4), and second, a less intensive single-family-home zone (photo D in figure 9.4).

Models of the Gradient Structure

The city's intensity gradient can be thought of as a series of concentric rings, similar to the classical model of agricultural location. The simplest model of urban gradient structure consists of only four rings, or zones: the retail core, a mixed commercial-industrial zone, an apartment and multi-family-home zone, and a single-family-home zone (diagram A in figure 9.5). This gradient occurs because commercial uses require access to many customers and employees and thus compete for the most accessible land; industrial-wholesale activities require access to many employees and to other businesses and therefore seek the next most central land. Within the residential zones, which are much larger than the commercial and industrial ones, density falls because the householder buys both land and transport for a given budget: some people are willing to pay much for a little space near the center, gaining greater access to their jobs and downtown amenities, while others prefer to pay less for more space toward the edge and to spend more on transportation.

The activities within each zone vary somewhat in character and intensity; thus a more realistic model of the gradient structure of large cities would include eight (or more) zones:

1 *Central business district*—occupied by commercial and governmental activities that require high visibility and access and that consume relatively little space.

2 *Mixed commercial-industrial-residential zone*—occupied by activities (and people) that need a moderate amount of space and require proximity to activities in the CBD.
3 *High-rise apartment or tenement zone*—occupied in some areas by upper- and middle-income people who want to be near downtown amenities and in other areas by a very low-income and transient population in old residential hotels and deteriorated apartments. Most residents of this zone are young adults, older people whose children have grown up, people who do not drive, and the transient population, from the richest to the poorest. Small groceries and other shops may occupy the ground floors of buildings that line the main streets.
4 *Multifamily-home zone*—occupied by lower- and middle-income families and often by groups of students sharing rent. These zones are very extensive in generally high-density cities, such as New York, Tokyo, London, and Paris; they also predominate in Eastern European cities, where detached homes are less common. In lower density cities, as in the American West and South, apartments tend to be restricted to small areas and to major highways.
5 *Outlying business districts*—occupied by commercial activities similar to, but usually less specialized than, those in the CBD. Most of these districts are located at major intersections and along well-traveled highways.
6 *Single-family-home zones*—occupied by all socioeconomic classes but usually dominated by middle-income people. These zones occupy the largest area of the built-up city. Lot sizes tend to increase, and the age of houses decreases, with distance from the city center.
7 *Peripheral manufacturing, wholesale-trade, and shopping center zone*—occupied by industrial and commercial activities similar to those in the central city but requiring more space and better external connections or serving outer city and suburban populations. These zones reflect the increasing decentralization of activities in response to lack of space, pollution, congestion, high taxes and rents, and other problems in the city center.
8 *Suburban home zone*—occupied largely by middle- or upper-income families willing to trade off proximity to downtown amenities for private property, open space, and neighborhood homogeneity. Low-income suburbs may form around peripheral industries.

The Influence of Physical Environmental Factors

If ability to compete for access to the city center were the only factor governing the location of urban activities, then the actual pattern of land use

A

B

C

9.4. The urban gradient: Montreal, Canada. This series shows the decline in intensity of activities and density of population outward from the CBD in photo A to the mixed commercial-industrial-residential zone in photo B, the multifamily-home zone in photo C, and the single-family-home zone in photo D. (All photos from NFB Phototheque; photo A by André Sima; photo B by D. Bancroft; photo C by Gabor Szelasi; photo D by N. Gregoire.)

D

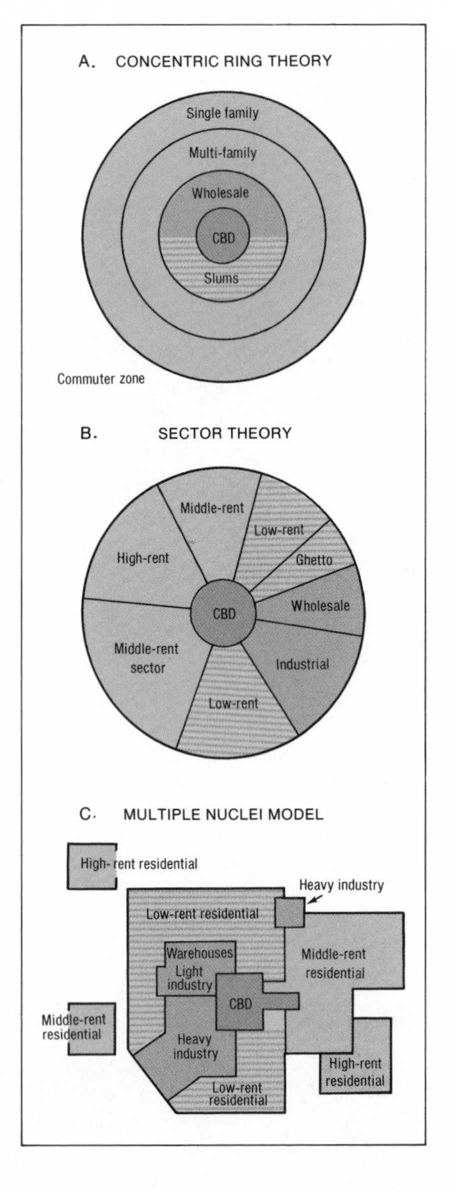

9.5. Three theoretical models of urban structure. Diagram A: the simplest model, based on competition for access to the CBD, consists of four concentric rings: the retail core (CBD), the commercial-industrial-residential zone, the multifamily-home zone, and the single-family-home zone. The retail core is, of course, at the most intensive point of the urban gradient. Diagram B: the sector model reflects the incompatibility of various land uses and socioeconomic class. The city is divided into sectors, which are often divided by transport routes radiating from the CBD. Diagram C: the multiple-nuclei model, with several commercial centers, attempts to depict a more realistic combination of patterns.

would approach the theoretical one outlined earlier. But improved transportation has weakened the importance of centrality and has strengthened the influence of environmental factors on location. Many industries and stores have moved out of the central city to the suburbs. For such activities, level terrain that can accommodate spacious parking lots and one-story buildings is a more attractive feature than centrality. And gently rolling land, especially land with a view, is often preferred as a residential site even if it is fairly remote from the central city.

Less directly, the location of commercial activities and the layout of road systems are also influenced by terrain. Shops and services catering to motorists often spring up along arterial roads. In cities built on irregular terrain, many of these roads are constructed along valleys and easy ridges, where grading problems are minimized. Yet some cultures, including the American, prefer a grid pattern and impose it even where difficult, as in hilly San Francisco. In contrast, street patterns in most African and Asian cities tend to be irregular and mazelike even in flat terrain.

Microclimatic features, such as a prevailing wind, high humidity, and frequent instances of air inversion, also affect the location of urban activities. More desirable land windward to industrial sites is often used for high-income housing. Low-income housing may be interspersed haphazardly among factories and wholesale stores, reflecting earlier times when workers had to live within walking distance of their jobs.

Conflict inevitably arises over environmentally favorable sites. Various interest groups, for instance, compete for the lakefront in Chicago: residents of apartment buildings along the lake consider it their front yard; the public wants to preserve the lakefront for beaches and recreational activities; industries are attracted by easy access to Great Lakes transportation; port facilities have developed along the lakefront; and possibly a large airport will be built there too. Many waterfront areas desirable for residential use have been taken by industry. But in some areas—Boston is one—old waterfront warehouses and factories are now being converted to fashionable shops and restaurants and luxury condominiums.

Sector Theory

Even if the physical environment did not modify the location of urban activities, we would still be unable to find perfectly concentric rings, each composed of activities with similar needs for access to the center, in the real world. This is because the economic goal of maximizing the productivity of land is greatly modified by another goal: the desire to maintain or improve one's social and economic status. Many people identify with a particular social group or economic class and fear that contact with other social groups (perceived as inferior) or with industry will depress the value of their property and lower their social status or security. Such attitudes generate feelings of incompatibility and protectiveness. The structure of the city clearly demonstrates that people are willing to sacrifice economic optimization for social reasons. For example, a middle-class Spanish-speaking family living in a poor area may be able to afford better housing. But the family may prefer to remain near other Spanish-speaking people for the cultural support provided by the neighborhood.

Activities of similar intensity—light industry and apartment buildings, for example, or different ethnic groups—may compete for land in the same zone but prefer not to locate near each other. Such incompatibility tends to divide the city like a pie into sectors of different land uses, ethnicity, or socioeconomic class (see diagram B in figure 9.5). The boundary between sectors often consists of a transport route radiating from the city center or an industrial zone or major park.

Each sector has its own intensity gradient. Historically the upper income groups have been able to bid for and maintain the most environmentally desirable sectors or zones; middle-income groups strive to locate near the upper-income sectors; and lower-income groups and heavy industries are relegated to the less desirable areas.

Within the upper-income sector, the gradient shifts from the most intensive activities in the CBD to the least intensive at the fringe. Fashionable shops are typically located downtown, followed by expensive high-rise apartment buildings and townhouses, and then by high-income residential suburbs that may contain small shopping areas and some light industry, such as research laboratories utilizing professional labor, but more likely private golf courses and country clubs.

Within the lower-income sector, the gradient shifts from inexpensive shops located near factories and wholesale houses, to older homes subdivided into rooms and apartments, to smaller single-family homes, and finally to lower-income, mixed residential and industrial suburbs. Ethnic and racial neighborhoods tend to divide the high-intensity zone within the lower-income sector. But a racial ghetto may become a separate sector with its own intensity gradient after enough of its residents earn higher incomes and can afford better housing. In black ghettos in American cities, for example, low-income blacks tend to live in multifamily tenements or public housing near the city center, and middle-income blacks tend to live in apartments and single-family homes farther out (figure 9.6).

The larger middle-income sectors (most cities have more than one) begin in the apartment zone beyond the CBD and extend through ever newer single-family housing out to the rural fringe. Some major light industries may appear, as well as office complexes associated with large suburban shopping centers.

Urban sectors can be clearly observed in Chicago, where the topography is generally level and approaches the homogeneous plain assumed in theoretical models. Chicago's North Side, which parallels Lake Michigan, is considered the most desirable residential area. The South and West sides are heavily industrial, with a preponderance of low-income housing. Development to the east, of course, is blocked by the lake.

The upper-income sector rises north from the Loop (Chicago's CBD), beginning with the Magnificent Mile of exclusive shops, modern high-rise office and apartment buildings, and expensive restaurants along Michigan Avenue. The Magnificent Mile terminates in yet another glittering zone known as the Gold Coast—a long strip of luxury apartment buildings facing the lake. This predominantly white, upper-income residential zone extends fifty kilometers into North Shore suburbs, such as Evanston, Winnetka, and Lake Forest.

The sector running south from the Loop is quite different. The Near South Side had originally been as fashionable as the Near North, but gradu-

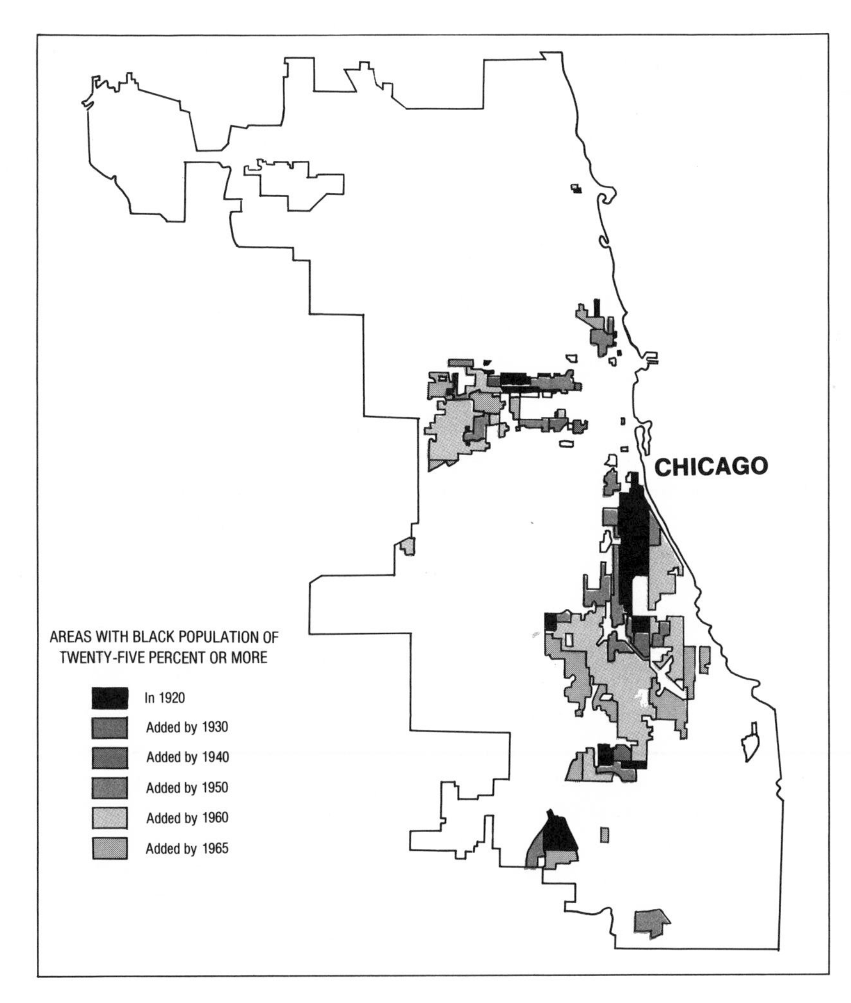

9.6. Expansion of the black residential area, Chicago, 1920–1965. The development of the black ghetto in Chicago is a good example of inner-city concentration followed by wedgelike expansion primarily into middle-income areas. Since 1965 the West Side ghetto has expanded farther westward, and the South Side ghetto both southwestward and southeastward.

ally its wealthier residents moved to the suburbs. The former mansions of the rich were subdivided into low-rent rooms and apartments to accommodate the immigrants who poured into Chicago from Ireland, Germany, and Scandinavia during the mid-1800s. In time, the fortunes of the immigrants improved, and they moved to better neighborhoods in the central city or to single-family homes on the outskirts. The deteriorating tenements they left behind were then occupied by blacks, who migrated in large numbers from the South during and after World War II.

Race proved far more formidable a barrier than language or religion ever was for the immigrants from abroad. Unlike the immigrants, who became assimilated into white American society, the blacks were virtually trapped in the ghetto. Poverty, overcrowding, and inadequate maintenance blighted much of Chicago's South Side. In recent years, however, some South Side neighborhoods have been redeveloped, and modern housing projects now replace older slums. The poor, who cannot afford the new middle-income housing, have been forced into other neighborhoods. Eventually poverty, crowding, and neglect may turn these neighborhoods into new slums.

Most of the suburbs south of Chicago are mixed residential and industrial. Some, such as Phoenix, Robbins, and East Chicago Heights, are predominantly black and low income. The industrial character of the southern sector culminates in the steel mills of Gary, Indiana.

Many western zones are predominantly middle class, extending from the more ethnic inner city through older suburbs like Cicero, and now out to and beyond formerly rural service centers or industrial satellites like Aurora and Joliet. The suburban middle-class zone contains a great variety of light industry and large, planned shopping centers.

Sectorization occurs not only by social and economic class but also by land use. One or more wedge-shaped sectors bordered by major access railroads may be devoted almost entirely to industrial activities. The segregation of industrial from residential and commercial areas results largely from zoning laws and personal preference, although low-income housing may be mixed with factories in older neighborhoods.

If it were not for the perceived incompatibility between different racial, religious, and income groups, some sectorization would disappear. But given the strength and persistence of social cohesion and prejudice, a *joint gradient-sector model* simulates urban structure far more accurately than the gradient model alone. One advantage to the sector theory is that it provides a simple solution to the problem of urban growth. Increased demand for space can be met by extending the wedge outward. Indeed most forms of intraurban migration, such as movements to the suburbs, seem to take place within a sector. Families that reside in Chicago's Near North Side, for example, tend to move to North Shore suburbs. And even if social discrimination were eliminated, wedge-shaped development would still be fostered by the radial, center-oriented structure of the urban transport network.

The Central Place Structure of the City

To expand our model of urban structure, we will impose on the overall gradient-sector system one more element: the *central place hierarchical structure.* Urban commercial centers, like central places, form a hierarchy consisting of several levels:

1. *The central business district,* containing high-level activities, such as department stores and specialty shops, hotels, museums, banks, theaters, and administrative offices;
2. *Major outlying centers,* also containing high-level activities—branch department stores and banks, for instance—as well as a wide range of lower-level activities;

3 *District centers,* containing intermediate-level activities, such as hardware and clothing stores, supermarkets, movie theaters;
4 *Neighborhood centers,* containing lower-level activities, such as convenience-goods stores and dry cleaners;
5 *Local establishments,* primarily gas stations, restaurants, taverns, and grocery stores.

With respect to activities, population served, and relative size of market area, these levels are roughly equivalent to (1) larger cities, (2) small cities, (3) towns, (4) villages, and (5) hamlets.

The location of urban commercial centers follows central place structure and at the same time both reflects and modifies the intensity gradient. The emergence of outlying commercial centers as a city increases in size is a natural economic process. Theoretically when the total market area served by the dominant center (the CBD) becomes large enough to support seven centers (the CBD and six smaller centers), six competing centers will arise (figure 9.7). The CBD will contain the highest-level activities because it can draw on the entire urban market for support. Hence the administrative principle (see p. 210) generally applies to the CBD with respect to the larger outlying business districts. Lower-level centers typically follow the transportation principle (p. 209), with smaller centers appearing approximately midway between larger ones.

In our study of the central place system, we found that central places of the same level tend to be spaced fairly evenly apart, particularly when the surrounding population is evenly dispersed. But because people in suburban areas live much farther apart than do people in the inner city, the spacing between urban commercial centers of the same level increases with distance from the CBD; at lower suburban densities, the centers need a larger market area to reach their threshold of support.

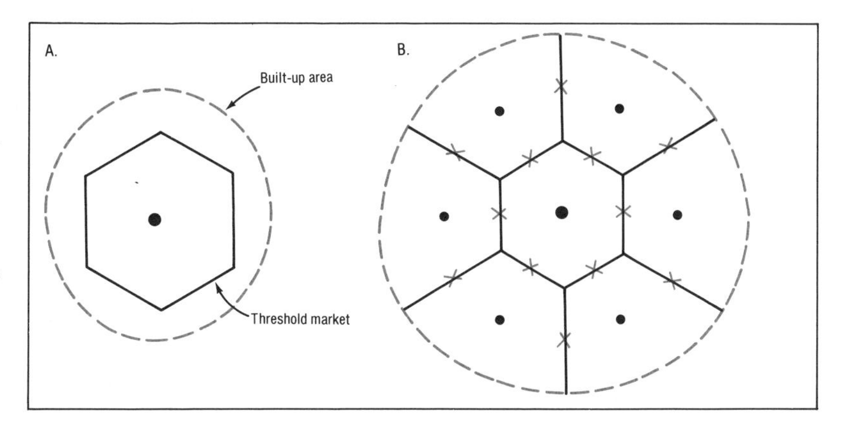

9.7. Development of the commercial-center hierarchy. When the built-up area of the city exceeds the threshold necessary to support one center, outlying centers (small dots) will arise. The original center (large dot) continues to dominate the entire market area for higher-level goods and services. Increasing population density or affluence (purchasing power) may induce the development of more closely spaced neighborhood centers midway between the larger ones.

Urban trade areas overlap to a greater extent than do trade areas in a larger region (figure 9.8). Most urban commercial centers are not more than a mile or so apart, and the goods and services they offer may be only subtly differentiated. Thus centers that have a spatial monopoly over a wide trade area are rare. Internal urban mobility also contributes to the overlap of trade areas: people usually live and work in different areas, and even after moving to a different part of the city, many people maintain former relationships with doctors, dentists, bankers, and the like.

A system of roads and highways connects major outlying centers, first to the CBD and then to each other. The urban gradient is modified by the effect of these roads and centers, which produce ridges and peaks of intensity (figure 9.9). Commercial establishments, especially those oriented to motorists, develop along the more traveled roads, raising land values and intensity. In the typical American city, where automobile ownership is high but public transit poor, these major roads are coming to support an increasing portion of total commercial activity, including department stores and other activities usually found in business districts.

Outlying shopping centers have an even more pronounced effect on the gradient, punctuating it with areas of greater intensity. As figure 9.10 shows, many retail stores have followed the middle class to the suburbs, reducing the commercial concentration and intensity of the CBD and increasing the density of people and stores in outlying districts. Indeed so great has been the decentralization of employment and most notably of high-income population in some cities during the last decade that the CBD has been reduced to merely one of several centers in a multicentered city (see diagram C in figure 9.5). Los Angeles is perhaps the most notable example (see figure 12.4); Houston is another. In many cases the multiple centers are the result of a gradual merger of originally distinct places into one metropolis, for example, Los Angeles–Long Beach, Minneapolis–St. Paul, and Albany–Schenectady–Troy, to mention but a few in the United States.

The early outlying business districts, oriented to a streetcar-riding population, have been supplanted to varying degrees by the rise of planned shopping centers oriented to an automobile-driving population. The shopping centers are usually farther from the CBD in the post–World War II suburbs of American and European cities; the older business districts in the inner cities were built up before World War II (and much earlier in Europe). The postwar planned shopping centers have supplanted most traditional business districts in importance largely because of their superior microgeography—the agglomeration of a large number and variety of stores within a tight pedestrian mall, surrounded by adequate and usually free parking.

The commercial structure of the large city differs from the structure of central place systems in two important respects. First, sectoral division by class and race leads to a differentiation in the size, quality, and specialization of districts and neighborhood shopping centers. Commercial centers

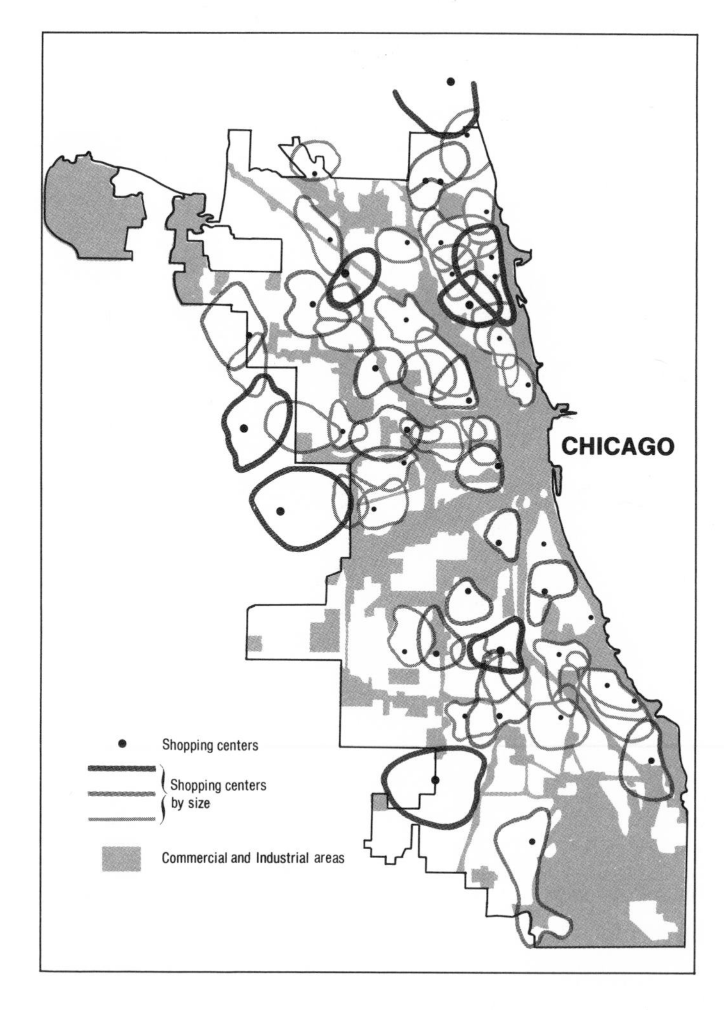

9.8. Overlapping trade areas around major Chicago shopping centers. Width of lines indicates size of shopping centers (dots). Shaded areas are industrial and commercial districts.

Source: Reprinted by permission of the University of Chicago, Department of Geography, from B. J. L. Berry, *Commercial Structure and Commercial Blight: Retail Patterns and Processes in the City of Chicago*, 1963.

may contain ethnic stores and institutions, or the stores may reflect the preferences and purchasing power of richer or poorer neighborhoods. Second, the large size of the metropolitan market permits another agglomeration benefit: the ability of certain centers to specialize in a narrow range of goods and services, such as antiques, furniture, automobiles, health services, and the like.

In summary, the central place structure of the city modifies the intensity gradient and is in turn modified by urban sectorization: the hierarchical structure weakens in industrial and low-income residential areas where purchasing power is lower, and it intensifies around outlying shopping centers. A more complete model of urban structure, then, would include an intensity gradient modified by a hierarchy of commercial centers and divided by sectors of different land uses and socioeconomic class.

9.9. Effect of arterials and outlying commercial centers on the urban gradient. This theoretical model shows the ridges and peaks of intensity that form along major arterials and at points of greater accessibility (commercial centers).

Source: Reprinted by permission of the University of Chicago, Department of Geography, from B. J. L. Berry, *Commercial Structure and Commercial Blight: Retail Patterns and Processes in the City of Chicago,* 1963.

Urban Structure in Non-Western Cities

Many of the elements described above are also found in non-Western cities but with some important variations. The non-Western city has a gradient pattern of intensity, land value, and population, but it is less pronounced and somewhat different from that of the typical Western city. The city center is less developed vertically (fewer skyscrapers) and is dominated by religious and administrative structures rather than by commercial ones. Housing in the core is likely to be occupied by the upper classes for reasons of prestige. This is radically different from the typical pattern in most modern Western cities (although in some, like Paris, the rich live in the center, and in many others the wealthy are redeveloping some central-city neighborhoods). No transition zone of industry and commerce is likely to be found in the non-Western city. Instead there is a densely populated residential zone where shops and workshops (in which people live) are interspersed. Lower density single-family houses—often row houses—extend outward. At the outskirts, more and more of these dwellings include intensive gardens and some livestock. But overall the density of most non-West-

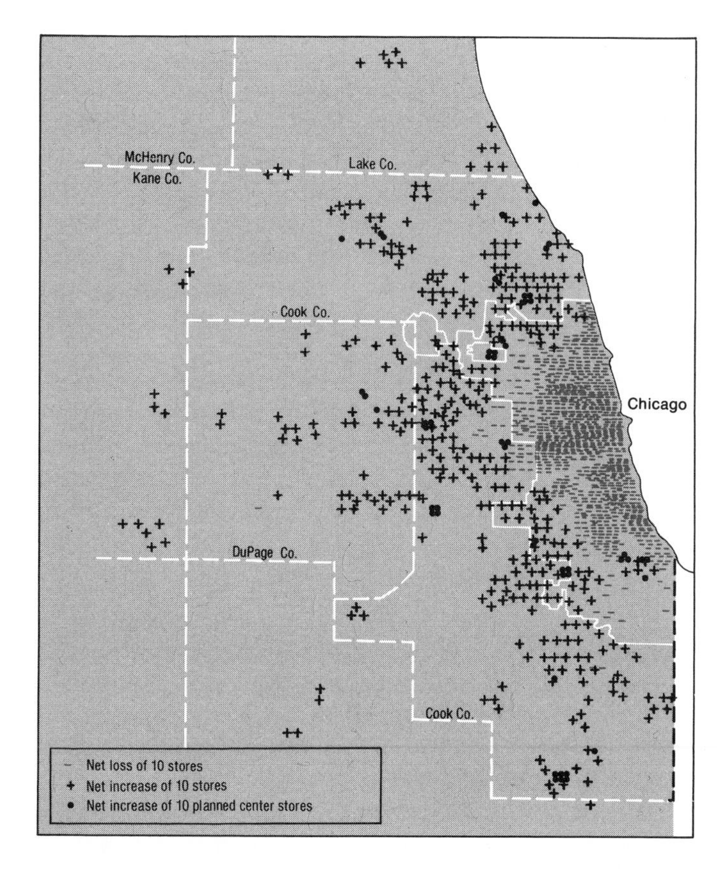

9.10. Changes in Chicago retail establishments. Retail stores have followed the migration of population to the suburbs. Since 1962, when this map was prepared, the suburbanization of retail stores has continued at a rapid pace.

Source: Reprinted by permission of the University of Chicago, Department of Geography, from B. J. L. Berry, *Commercial Structure and Commercial Blight: Retail Patterns and Processes in the City of Chicago*, 1963.

ern cities, despite the low profile of buildings, is far higher than that of most Western cities because houses are much smaller, families are larger, and streets are narrower.

While the non-Western city has a less obvious gradient pattern, the social differentiation may be even sharper. The city tends to be divided into several quarters, each with its own market (commercial core) and distinctive cultural institutions. Special areas or streets devoted to particular occupations and handicrafts are common. Each major part of the city tends to contain all income classes, often living in close proximity. However as in the Western city, the status and prosperity of the quarters differ markedly. For example, Malaysian cities typically have a more affluent European area, a middle-class Chinese area, and a poorer Malay area.

The commercial structure of the non-Western city is again similar to but less regular and hierarchical than that of the Western city. The central market, or commercial area, is usually next to the political and religious center, and smaller markets appear in other quarters. Shops line the major roads in non-Western cities, too, and the agglomeration of merchants and specialized vendors, particularly craftsmen such as silversmiths, is highly developed. Many cities in developing countries have a dual commercial structure: a Western-style CBD, imposed during the colonial era in the European quarter, often at the terminus of a new railway or highway, and a traditional market-temple core, a short distance from the CBD.

To illustrate non-Western urban structure, let us return to Cairo, the capital and dominant metropolis of Egypt, which contains about 6 million people in an area of but 200 square kilometers and has ten times the density of the average American city. In some parts, density is as high as 1,000 persons per acre. About 18 percent of Egypt's population lives in Cairo, and 30 percent of its industry and commerce is concentrated there. Old Cairo, founded in the tenth century, was built on high ground above the eastern bank of the Nile, across from the ancient Memphis and Giza pyramids. It is a traditional, crowded city, divided into Arab, Turkish, and Levant (eastern Mediterranean) quarters and still containing old, narrow craft- or occupation-specific alleys with nothing but coppersmiths, pottery makers, and other craftsmen. New Cairo developed mainly after 1863 and the coming of the railroad. It spread north and west from the old city across the alluvial plain to the other side of the Nile, Many of its main streets are framed by modern high-rise buildings, like those of Western cities. But off these streets winds a tortuous network of alleys lined with older mud-bricked houses and domed mosques, preserving the traditional character of old Cairo.

Neighborhood Structure

Many city dwellers may be dimly aware of their city's intensity gradient, its land use and socioeconomic sectors, and its hierarchy of shopping centers. But the element they most clearly perceive is their own neighborhood, those small regions of the city shaped by the forces that create both the gradient and the sectorization of population. Neighborhoods rarely are legal entities, but they have a distinctive character rooted in the community's social, economic, and ethnic composition.

Before World War II, when mobility was more limited, most social and economic activities took place in the neighborhood (although the principal wage earner may have worked elsewhere). Neighborhood identification was strong. After World War II, explosive urban growth and high mobility appeared to have destroyed the meaningfulness of the neighborhood, at least in the American city. But, whether because of dissatisfaction with big government, or to protect the investment in one's home, or from the human need for community, a remarkable resurgence of community feeling and organization has occurred.

In the non-Western city, the neighborhood remains strong. This is commonly the result of a shared culture, perhaps even a legal association with a territory, defined by participation in the church, temple, or mosque. Northern Ireland and Lebanon provide current examples of the sensitivity and cohesion of religious neighborhoods.

Political Structure

We have discussed the city as if it were a single entity—and it is a single economic, functional unit. But it typically consists of many political units spawned in the process of urban growth. Political fragmentation also results from the attempt of various social and economic groups to maintain or enhance their status, a goal they perceive to be helped by legal separation.

Growing metropolises not only annexed land at the rural fringe but often spread far beyond the reach of the central core, encompassing former rural service centers and even spilling over into adjoining counties and states (figure 9.11). Many other communities outside the central city incorporated themselves as independent suburbs in order to control their own future, frequently by zoning out undesirable land uses or social classes.

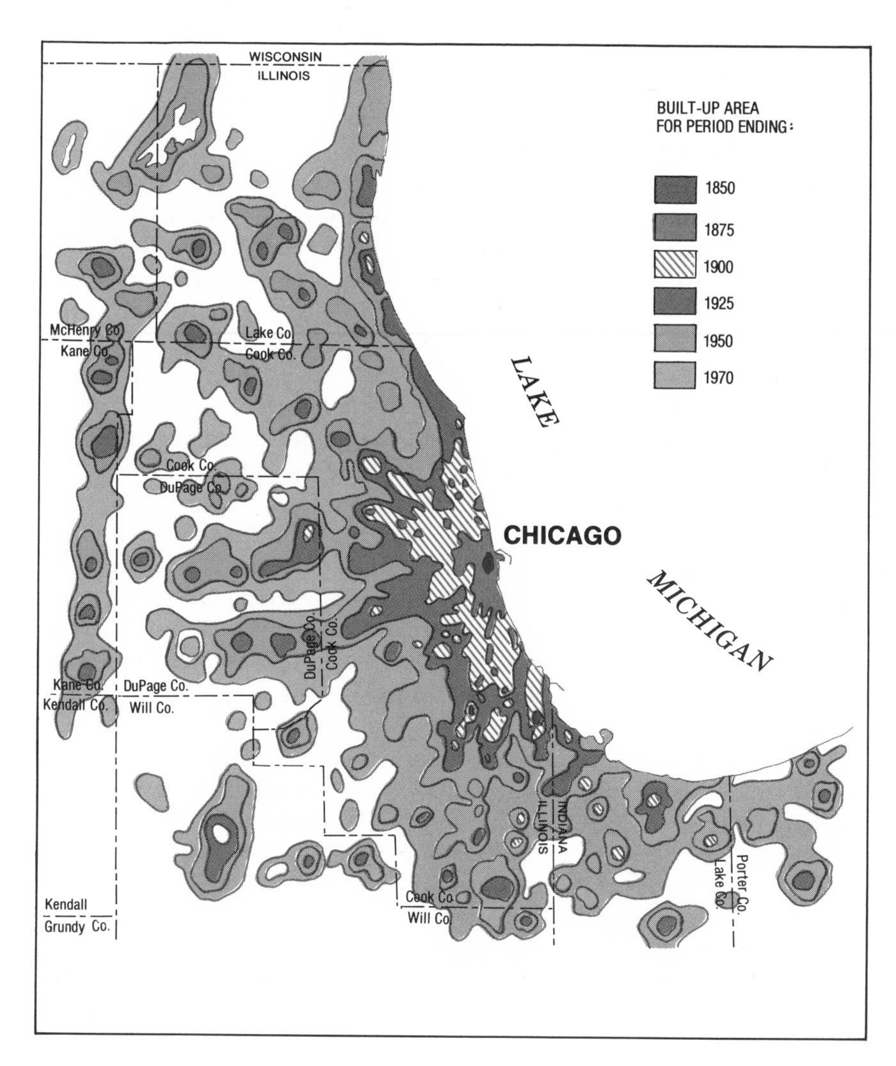

9.11. The expansion of Chicago. The Chicago metropolitan area expanded rapidly in the 1950s, when 20 percent of the city's population moved to the suburbs.

Source: Redrawn by permission of the University of Chicago Center for Urban Studies.

As a result, the typical metropolis in many parts of the world includes dozens of separate political units (see figure 13.2). Metropolitan Chicago contains eight counties, two of which are in Indiana; six major urban areas besides Chicago proper (Gary, Hammond, East Chicago, Waukegan, Joliet, and Aurora); numerous suburbs; and myriad special districts. The last provide services not authorized to cities and counties, such as schools, water and sewers, fire protection, airports, and transportation. Problems associated with such political complexity, including the dichotomy between the declining, poorer central city and the growing, richer suburbs, are discussed in chapter 13. Once established, political units take on such an aura of tradition that it is extremely difficult to create a single unit out of the real metropolitan area. Thus virtually all large metropolises, Western and non-Western, in socialist and market economies, contain a large number of separate suburbs.

Transportation and Urban Structure

Transport routes have historically determined the pattern of urban development (figure 9.12). Before the advent of modern modes of transportation, most people had to live within walking distance of their jobs. Residences were crowded around the CBD, at the convergence of the most important roads, or near industries on major roads and railways. The wealthy often lived farther from the center, but not more than a horse-and-buggy ride away.

With the introduction of public transportation (horse-drawn trolleys) in the 1830s, the more affluent population began to spread outward along trolley lines. As cities increased in size, demand rose for better transportation to the center. The horse-drawn trolley was succeeded by the cable streetcar, then the electric streetcar, the subway, and the commuter and elevated railroads, which increased the accessibility of the CBD and intensified land use along the affected routes. Branches of development began to extend outward along these routes in starlike fashion. At first, public transportation was used mainly by the rich. But gradually costs fell, permitting the suburbanization of the middle classes and leaving the poor concentrated in the inner city.

When the automobile and motor bus began to compete with the streetcar and train, existing routes proved inadequate to the task of handling this additional traffic. The CBD became so severely congested that commerce in the central core suffered—and competing centers prospered. While in the preautomobile era most homes and businesses were located near streetcar and railway lines, the automobile permitted the dispersal of population and economic activities. The starlike pattern that formerly characterized cities blurred, and the importance of the CBD waned further.

Efforts to improve access to the CBD and to revitalize the central city have led to the development of rapid transit and freeway systems. Rapid transit trains were intended to concentrate activities in the CBD and at outer stations (figure 9.13). This has occurred in places where automobiles are scarce, as in the Soviet Union, but apparently too few people in the United States use the system to bring about these results. Rapid transit systems would be efficient if people were willing to give up the use of their cars for commuting and to live in high-density apartment zones near transit lines. But many prefer to drive to work from low-density suburbs. Traffic congestion and pollution in the CBD have worsened, not improved, and the profitability of transit systems has steadily declined in areas where cars now dominate, as in the United States, Canada, Western Europe, and Japan. Freeway systems of high-speed, limited-access highways have contributed to the decentralization of people and activities by increasing the accessibility of peripheral areas. Now many people commute from the city to jobs in the suburbs. Limited access highways have also aggravated the congestion

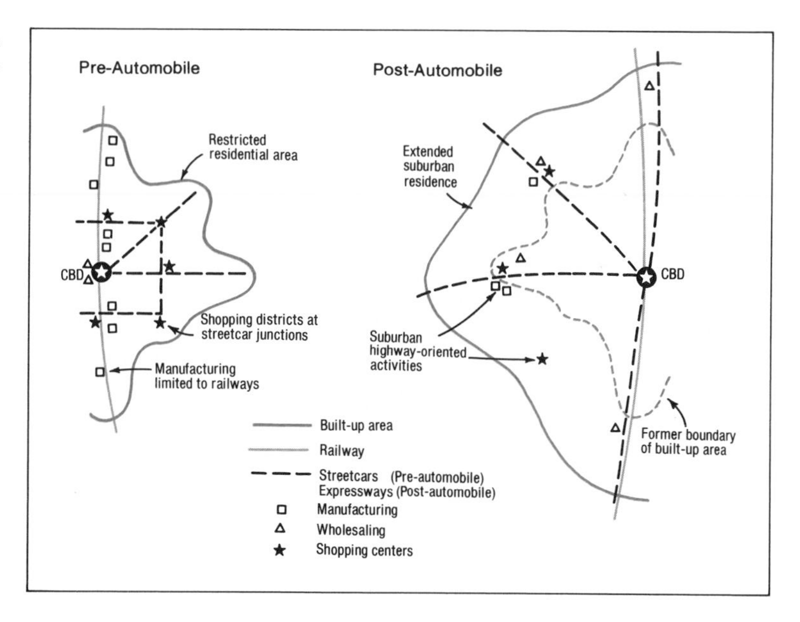

9.12. The pattern of urban development before and after the automobile. In the preautomobile era, the growth of the residential zone was restricted to areas served by railways and streetcars. Economic activities were concentrated in or near the CBD and along transport lines. The automobile has extended the residential zone and permitted manufacturing and wholesaling activities to disperse to the suburbs.

and hastened the decline of the CBD by funneling too many cars into local roads better suited to the streetcar era.

The only existing form of public transportation flexible enough to reach a widely dispersed population is the bus. But because buses have to compete with automobiles on congested roads, they are often frustratingly slow. Until a more practical form of rapid transit is developed—one with a reasonably extensive network—urban transport problems will continue.

In non-Western cities, automobiles are less common but congestion is

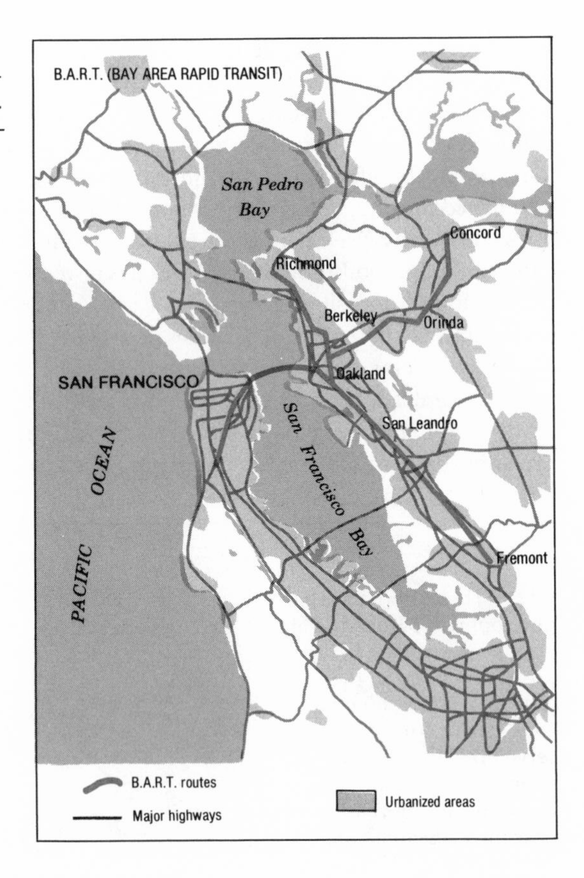

9.13. Bay Area Rapid Transit (B.A.R.T.). Fast, efficient, and convenient—for those who live near the line. Most people still prefer to live in low-density suburbs, even at the cost of a long drive to work.

still severe, primarily because most cities have only a few, narrow, through roads. The street network typically consists of small lanes connecting squares or markets. Modern traffic—cars, buses, trucks—must share the roads with dense pedestrian traffic and often with livestock.

URBAN GROWTH

Urban structure is never static; cities are always growing and changing, a process that intrigues geographers. Cities grow in two ways: first, by expanding around their periphery, and second, by intensifying their existing areas. The outer limits of the city expand as more and more families demand additional, newer homes (see figure 9.11). Intensity of land use at the center rises as competition for accessibility increases. Because space is so limited in the central core, growth is upward as well as outward. Growth in cities may for a short time be accommodated by expansion alone, or, if development is curtailed by zoning laws, physical barriers, or severe land shortage, by intensification alone. Over the long run, however, both processes will normally occur.

During the past few decades many metropolises have expanded more rapidly than they have intensified, and overall population densities have declined. This phenomenon is due partly to the decentralization of employment, which itself is a result of lack of space in the center and the shift to dependence on cars and highways, and partly to increasing affluence, which permitted more of the population to realize its dream of a suburban house and garden. In many American cities the decline in population density is also due to the flight of the white middle class to the suburbs. High crime rates in the central city, pollution, traffic congestion, high taxes and rents, inadequate services, crippling municipal-employee strikes, controversy over the busing of schoolchildren for desegregation purposes, and fear of racial integration have driven many families out to the suburbs. In addition, the construction of highways through cities and of parking facilities in the central core has lowered population density by preempting large tracts of land formerly used for residences.

The fact that cities are rapidly expanding at the edge does not necessarily diminish the process of intensification at the center. The pageant of demolition and construction that one may observe in most central cities attests to the changes that are taking place through intensification. As demand for central locations rises, older houses and apartment buildings around the CBD are taken over for commercial use or are demolished and replaced by taller structures, parking lots, or garages. All over the world, this process conflicts with the preservation of historic buildings and homes.

The apartment zone is pushed outward into the single-family-home zone, particularly along arterial roads. New apartment buildings replace older single-family homes in the more desirable middle- and upper-income neighborhoods, while in less desirable areas older homes are subdivided into small apartments. In this process of intensification, the more affluent residents of older neighborhoods in the central city tend to move out to the suburbs, and middle-income residents tend to move to new, more spacious housing near the edge of the city. Their institutions and enterprises (churches, clinics, shops) often follow them out, and the existing buildings are converted to new uses, reflecting the differing needs and preferences of the succeeding populations. Some churches in central cities, for example, have been converted to theaters.

The urban periphery comprises a fairly wide zone of transition, ranging from solidly urbanized areas to more widely separated homes, subdivisions, and factories. Because far more land is ultimately available for subdivision or industry than is demanded, the specific location of such areas is somewhat random. Much land on the periphery of cities is in speculative nonuse: it has been sold by the original owner, usually a farmer, but has not yet been subdivided or built up.

Most of us think of the outer suburbs as the domain of the middle- and upper-income classes. But in fact many lower-income people live on the fringe, some in older farmhouses, others in trailer courts. In America, the poor who live in such locations often do so by preference; they do not like city living. But in much of the world, the poor live on the edge by necessity, creating quasi-legal squatter settlements. Only the upper- and middle-income classes can afford the relatively high inner-city costs.

The patchwork, or sprawl, pattern at the edge is characteristic mainly of U.S. and Canadian cities. Many criticize the wasteful use of land and energy, the loss of productive farmland, and the large expenditures for road construction and maintenance, utilities, and fire and police protection. But North American society is willing to pay these costs. In Europe, stronger land-use planning and higher land costs minimize the amount of sprawl.

Urban Growth in Market Societies

The process of urban growth varies in different kinds of society. In societies where central planning is minimal, as it is in most of North America, urban growth may be described as a process of diffusion and replacement, or *succession,* involving a great many private and public land use decisions. The process unfolds as follows:

1. The spread of the built-up area, itself a diffusion of development, is expressed mainly by the extension of roads and single-family homes, many of which are occupied by families with school-age children. These new households have been established in part by young families who moved from slightly older homes farther in or who settled for a few years in the apartment zone, close in or along major roads. Also included are families fleeing the arrival of other cultural groups and fearing loss of status, as well as families whose incomes have risen and who now seek a higher status area or more space. In this way, the older housing stock filters down to the expanding lower income population. A wave crest of very active building moves outward. But because of a largely unregulated land market, the outer edge of development is not sharp. Filling in continues in already settled areas, and some widely spaced homes and subdivisions sprawl well out into surrounding farmland and woodland. As part of the filling-in process, schools, parks, and shopping areas are installed or further developed.
2. The core commercial area spreads at the same time into the surrounding older residential area, first converting, then replacing the structures. This edge, too, is uneven, reflecting private and individual land use decisions.
3. The growth in central employment makes it profitable to convert older single-family homes into multifamily housing, particularly in lower income sectors, or to replace them with such housing. The multifamily-home zone thus spreads outward from the growing commercial core, primarily along arterial

roads served by public transit. In American cities this apartment zone is now mainly occupied by an elderly population or a very mobile, transient young population without children. The zone tends to expand as the population structure changes in this direction. In both American and European cities, the trend to later marriage and few children and the higher costs of gasoline and of suburban homes have induced the affluent young to rehabilitate some inner city housing, particularly in historic areas. In this case, it is the poorer population that is displaced to other older housing. The portion of single-family homes in Europe and elsewhere is much lower and chiefly limited to upper- and middle-income classes. Most families, even with children, occupy the apartment zone, much of which may consist of fairly high-quality public housing. Consequently there is less segregation by age, or life-cycle stage, than in American cities.

4 In many cities the original buildings, commercial or residential, gradually become obsolete or deteriorate, and replacement or renewal often occurs, again beginning at the core and spreading outward.

Increasingly the market governing urban change is regulated through planning—that is, through zoning laws, usually designed to separate presumably incompatible uses such as housing and industry; through policies on the rehabilitation or replacement of housing stock; and through controls over the extension of new roads, sewers, transit routes, and water service.

Urban Growth in Centrally Planned Societies

The pattern of urban growth in centrally planned societies differs somewhat from its counterpart in market societies. In the absence of a free land market, government planners carefully control the conversion of farmland to urban use and of one form of land use to another. Housing in some centrally planned societies is mostly public and consists of large apartment complexes, complete with shops, services, and even employment opportunities. Rather than an extensive zone of single houses interspersed with vacant farmland at the urban fringe, the larger and denser housing estates may end abruptly at occupied farms. The more affluent or privileged may have a suburban house or cottage in addition to a centrally located apartment. Compared to cities in private enterprise societies, there is a better balance between residences and employment within sectors or zones, but shops are far more limited in number and variety. Population density tends to be rather high, not because land is scarce, but because of conscious planning and the attempt to minimize costs of providing transportation and other services. Despite the lack of a free land market, central accessibility is still valued, and a weak gradient occurs. The CBD is even more dominant with respect to shopping and services than it is in private enterprise societies.

Urban Growth in Third World Societies

Urban growth tends to be very different in many developing countries where society still consists of a small elite, a small middle class, and a majority of poor. The wealthy often maintain residences in the urban core. (This also occurs to a limited degree in Western cities like New York, Paris, or London. However, all but the very wealthy tend to be forced out by the pressure of conversion to commercial or office use.) The lower and middle classes occupy an extensive area of housing in the city itself. And the poorest and most recent city dwellers tend to occupy squatter settlements outside the city proper. These semiurban, semirural slums may lack even minimal services, such as water, electricity, and sewerage (figure 9.14). Gradually such settlements may become permanently incorporated within the city, and some services may be added, often through the ingenuity of the residents themselves. (This pattern was common in the American South before this century, when most blacks lived in slums outside cities. But it has gradually given way to the dominant North American pattern in which the poor supplant the wealthy in the urban core, while the latter move to

9.14. Slum housing on the outskirts of Rio de Janeiro, Brazil. (Photo from OAS.)

desirable suburban locations.) Very rapid urbanization in developing countries is now transforming this inherited structure, as high-rise office and apartment buildings replace older housing, and suburban housing estates are built. Population growth often exceeds the rate of new employment. This results both in overcrowding, as the unemployed coming to the city move in with employed relations, and in the overdevelopment of the service sector—the reverse of the pattern in centrally planned societies.

CONCLUSION

The microgeography of the city parallels and reflects the complexity of the human landscape as a whole. We have seen that the gradient and sector patterns of the city are extensions of agricultural location theory and that the hierarchy of urban commercial centers follows in many respects the principles of central place theory. Noneconomic motivations related to maintaining social distinctiveness and status, however, exercise a great influence. The spatial organization of the city is as logical and predictable—in theory—as the patterns we have examined in other chapters. And it is as responsive in reality to the influence of human behavior and the physical environment. Another element that modifies the spatial structure not only of the city but of the human landscape as a whole is spatial interaction: the movement of goods, people, and ideas. In the next two chapters, we will explore the patterns, processes, and principles of these movements.

SUGGESTED READINGS

Alonso, William. *Location and Land Use.* Cambridge, Mass: Harvard University Press, 1964.

Berry, B.J.L., and Kasarda, J. *Contemporary Urban Ecology.* New York: Macmillan, 1977.

Bourne, Larry, ed. *Internal Structure of the City.* New York: Oxford University Press, 1971.

Chinitz, Benjamin, ed. *City and Suburb.* Englewood Cliffs, N.J.: Prentice-Hall, 1965.

Clark, C. *Population Growth and Land Use.* London: Macmillan, 1967.

Downs, Anthony. *Opening Up the Suburbs.* New Haven, Conn.: Yale University Press, 1973.

Hawley, Amos. *The Changing Shape of Metropolitan America.* Glenbrook, Ill.: Free Press, 1956.

Hoover, E. M., and Vernon, R. *Anatomy of a Metropolis.* Cambridge, Mass.: Harvard University Press, 1959.

Hoyt, Homer. *Structure and Growth of Residential Neighborhoods.* Washington, D.C.: Federal Housing Administration, 1939.

McGee, T. G. *The Southeast Asian City.* London: Bell, 1967.

Muth, Richard. *Cities and Housing.* Chicago: University of Chicago Press, 1969.

10

Trade and Transportation

10 Trade and Transportation

Every day hundreds of trucks pull into and out of Charlotte, North Carolina, one of the major trucking centers of the United States. The goods they haul are products of different regions: textiles from the Carolinas, electrical goods from Chicago, machinery from New York and New Jersey, meat and grain from Iowa and Kansas. This interregional movement of goods reflects the location of specialized agricultural and industrial production in the United States. Most of the production in advanced societies enters into trade, with the exception of grain fed to a farmer's own livestock and the at-mine generation of electricity from coal (the electricity, of course, enters into trade).

Now consider another trade and transportation scene, this time in the ancient city of Tombouctou, in the small African republic of Mali. Situated on the Niger River as it penetrates the southern Sahara, Tombouctou has long been a center for the exchange of goods between the wet tropical coast and the Sahara and even the Mediterranean (figure 10.1). Tombouctou's strategic transport position helped it become an important city of the great Mali empire from the eleventh to the fifteenth century; it is now the second largest city in Mali. Riverboats carry livestock and wheat from northern Africa through Tombouctou southward to Niger, Nigeria, and Guinea, and camel caravans slowly bring bananas, rice, and cocoa from those regions northward to Algeria, although here, too, the truck is becoming important.

The development of transportation between centers of production and markets in Northwest Africa and in the United States have permitted trade to develop and enabled regions to specialize in the kinds of production for which they have a comparative advantage (figure 10.2). Trade centers like Tombouctou and Charlotte have prospered, and so have many of the regions that produce surpluses for exchange.

While trade and other forms of interaction reinforce the **spatial structure** of places, they can also modify it. For example, a new road linking a formerly inaccessible region to a market center may stimulate trade and production, leading to the growth of towns in that region and to higher levels of well-being for the population.

Location and interaction go hand in hand. Together they express the spatial order of society. The flow of goods, people, and ideas between places gives life to the patterns we have studied and in fact helps create them. It was Chicago's ability to specialize in certain activities and to trade with other specialized places—to forward grain and livestock from the West to the industrial East and to send manufactured goods into the rural hinterland and smaller cities of its trade area—that enabled it to rise as a great commercial and industrial center. Today ocean travel via the St. Law-

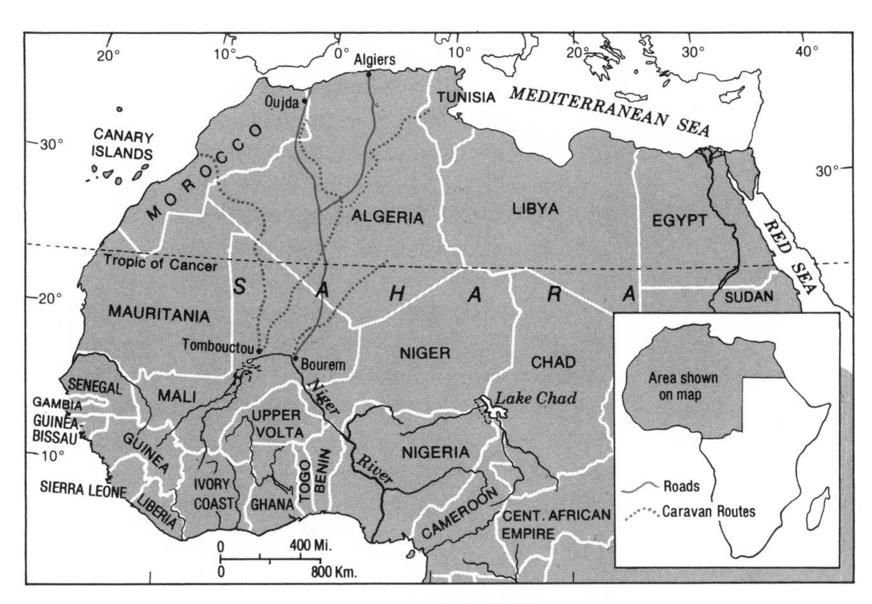

10.1. Northwest Africa: the location of Tombouctou. Favorable location on the Niger River, with access to the tropical coast of Northwest Africa and caravan routes to the north, contributed to Tombouctou's development as an important trade center. (The new road, however, goes north from Bourem, 200 miles to the east on the Niger.)

rence Seaway, air travel, and modern telecommunications link Chicago economically and culturally to places around the world. All these forms of interaction help knit the world into a more interdependent whole.

Although we have discussed both location and interaction in previous chapters, our emphasis has been on the way society structures its space. Now we will focus on how society organizes its movements through space. After briefly describing various facets of the movement of goods, we will tie these data to theories and models of transportation. These topics belong primarily to the domain of economic geography but are essential to a more complete understanding of the human landscape. In chapter 11, we will examine the movement of people and ideas.

TRADE

Trade and Transportation

Because of the critical role of transportation in the development of trade, it would be helpful to review briefly some milestones in the history of transportation.

The first seafaring ships are believed to have been developed by the ancient Phoenicians in the eastern Mediterranean during the second millennium B.C. These ships permitted both the long-distance movement of people and goods and the maintenance of long-distance supply lines. For the first time, humans gained the freedom to grow in numbers beyond the food capacity of the local area.

Later during the first few centuries of the Roman Empire, investment in a widespread network of roads, supplemented by existing sea lanes, made

A

10.2. Movement of goods in low and high technology societies. Trade flows may generally be smaller and slower in less developed societies than in more advanced ones, but the principles governing trade are the same. Goods move from regions of surplus production to regions of deficit supply. In photo A, wood and bowls are transported by boat to market in Segou, Mali. In photo B, containers of freight await transfer to ships in the port of Brooklyn, New York. (Containerization is revolutionizing ocean transport by reducing port time and handling costs.) (Photo A from United Nations; photo B by Ellis Herwig from Stock, Boston.)

B

possible the unity of the far-flung empire. When that road network deteriorated after the fall of Rome in the fifth century, the social and economic fabric of Europe also decayed.

Similarly the Chinese empire depended upon road, canal, and ocean transportation for internal unity and external trade. Most notably between the late tenth and mid-fifteenth century, the Chinese dominated exploration and trade from India to Japan.

Beginning in the eleventh century with the rise of the Hanseatic cities of the North Sea and Baltic and of Italian city-states like Venice and Genoa, sea trade again developed in Europe. Improvements in ship design and navigation instruments promoted the voyages of discovery, the establishment of European colonies, and the growth of long-distance trade. Transport improvements also helped to revive commerce and industry. (This period of expansion fostered the development of the major port cities still dominant today in Europe and its former colonies.)

Rivers and, in the eighteenth century, canals became principal arteries of commerce. Sailing ships, whose capacity and speed increased greatly after 1700, were efficient and effective. Land transportation, on the other hand, remained poor until the nineteenth century, when the steam engine revolutionized the movement of people and goods. Of course, existing modes of transportation had been improved from antiquity on, but the substitution of mechanical energy for animal power was an extraordinary breakthrough. The railroad drastically reduced the cost and increased the speed of land transportation. The expansion of the rail network accelerated the development of interior resources and farmlands, particularly in Central Europe, Russia, and North America (figure 10.3).

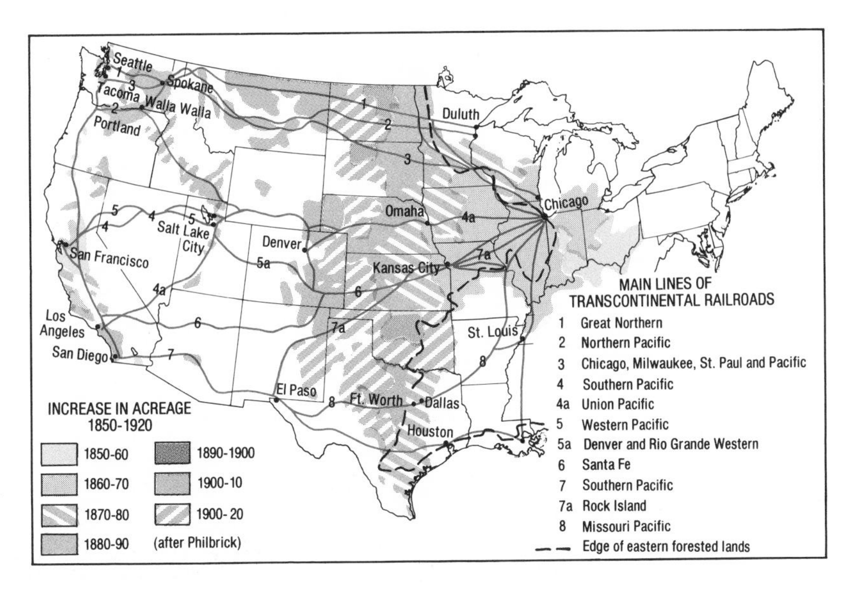

10.3. Transportation and land development. As railroads branched westward in the last half of the nineteenth century and the early twentieth, new lands were opened to agricultural settlement and development. The railroads encouraged westward migration and enabled farmers to ship goods rapidly and relatively cheaply to northeastern urban markets via Chicago and other distribution centers.

Source: Adapted from Allen K. Philbrick, *This Human World*. Copyright © 1963 John Wiley & Sons. Reprinted by permission of John Wiley & Sons, Inc.

More recently, the automobile and motor truck have increased human mobility. Pipelines have provided relatively cheap transportation for petroleum and other products. The highway system has reduced rural-urban cultural distinctions and has opened a wider, more diffuse area to settlement, development, and trade. The astounding progress of air transportation during the past few decades has shrunk the barrier of distance, at least for the affluent, and has in effect created one world (figure 10.4). Today few locations, however remote, are more than a day's trip from anywhere else.

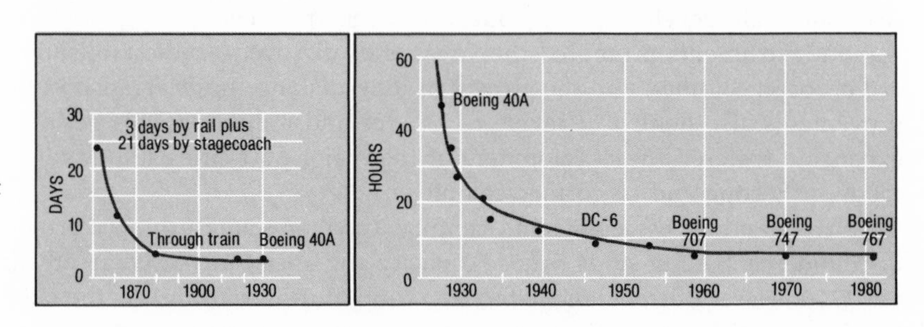

10.4. Reduction in the time-cost of distance. This diagram shows changes in the amount of time needed to cross the U.S. continent since 1850, using the fastest means of transport available. It provides a good example of the increasing interconnectedness of high technology societies.

Principles of Trade

Improvements in transportation have enabled societies to engage in higher levels of trade. But trade does not develop entirely spontaneously. The flow of goods, like the location of human activities, is governed by certain principles, which may be summarized as follows.

Trade can occur profitably if the difference between the cost of producing a good in one area and the market price in another will at least cover transportation costs. That is, trade can flow from areas of surplus production, where prices are lower, to areas of deficit supply, where prices are higher, as long as transportation costs do not exceed this margin. Germany, for example, has a comparative advantage for the production of steel and other heavy industrial goods. It produces a surplus of steel, which costs less to produce and is priced lower in Germany than in many other places. Similarly Denmark has a comparative advantage for the production of butter and other foods. These two nations will trade if there is a *complementarity of supply and demand* (an excess demand for food products in Germany and a demand for steel in Denmark) and if the cost of transporting the butter or steel is less than the difference between the market price of the good and production costs.

If conditions of *comparative advantage, complementarity,* and *transferability* (a transport facility at an affordable cost) permit trade, the importing region benefits because it will enjoy lower prices and higher consumption than would be possible in the absence of trade. For example, we may feel that we pay Arab nations a high price for the oil we import, but we would be paying a far higher price and consuming less if it were not for this trade. This principle underlies the operation of models used to predict the optimal pattern of production, consumption, and trade for commodities among regions or countries. *Optimal,* as in the location theories previously discussed, means the maximum complementary shipment of goods from surplus to deficit areas at the minimum cost of transport.

Ideally these economic forces govern patterns of national and international trade. But international flows may be affected by historical inertia, cultural preferences, customs duties, tariffs, and quotas. Great Britain maintains preferential trade agreements with its former colonies, as does France with its former territories. The United States sets lower tariffs on goods from certain countries and maintains a trade embargo against nearby Cuba, which, despite the distance, trades heavily with the Soviet Union and Eastern Europe. And the United States and other countries impose quotas on imported cane sugar by country.

Trade barriers may be established to discourage excessive consumption of luxuries or to help a country's balance of payments (balance between the value of imports received and the value of exports shipped). A developing nation that wants to protect its infant steel industry may impose high duties on imported steel, restraining the flow from other steel-producing nations. On the other hand, a mature industrial nation may reimpose tariffs to protect labor-intensive industries like textiles and toys against competition from newer industries with lower production costs in developing countries.

As a nation gains confidence in the strength of its own industry, it may enter into tariff-reduction agreements with its principal trade partners. The small but highly developed nations of Western Europe recognized that the large internal markets and resources of the United States and the Soviet Union and the absence of trade barriers within each nation permitted greater specialization, agglomeration, and economies of scale. To gain similar advantages, France, West Germany, Italy, Belgium, the Netherlands, and Luxembourg formed the European Economic Community (EEC), or Common Market, in 1958. The EEC almost completely abolished internal barriers to flows of goods, capital, and labor. The result was a market increase in trade, income, and economic efficiency. Each nation in the European Community could specialize in products for which it had a comparative advantage (figure 10.5). So successful was this integrated economic community that Eire, the United Kingdom, and Denmark later elected to join it. (International trade, in the context of world economic development, will be discussed further in chapter 13.)

Trade and Distance

The flow of goods is highly responsive to the friction of distance. There is a rather regular decay in the value and volume of trade as distance increases (figure 10.6). Local (short-distance) movements of goods represent a large proportion of the total volume and cost of movements, even in countries with highly developed transport networks, because local suppliers and consumers are widely dispersed. In less developed countries, goods tend to

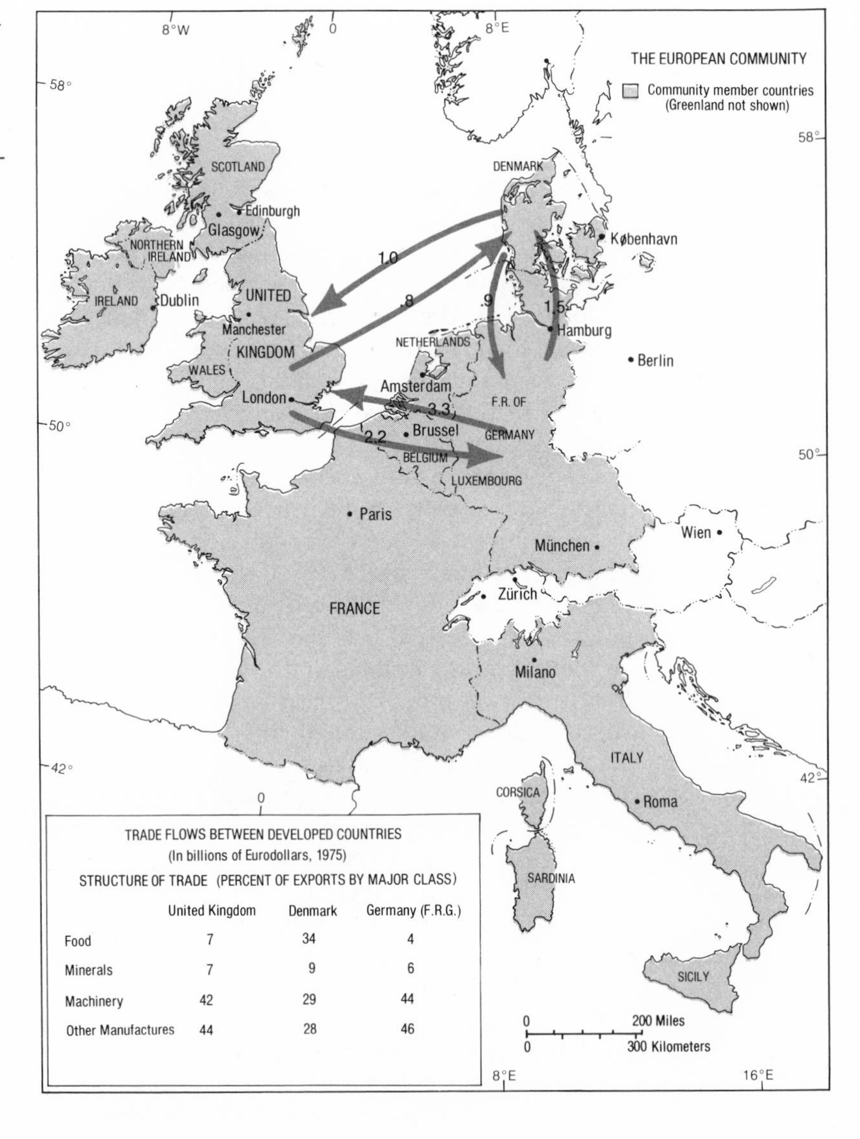

	United Kingdom	Denmark	Germany (F.R.G.)
Food	7	34	4
Minerals	7	9	6
Machinery	42	29	44
Other Manufactures	44	28	46

10.5. The European Community: trade flows and specialized production. Trade flows among developed countries are large and are dominated by manufactured goods, but flows of food and raw materials are still important.

Source: Base map redrawn with permission from the Commission of European Communities, Brussels, Belgium.

move even shorter distances—much more food, for example, is consumed very near the place of production.

Intermediate-distance regional and interregional flows reflect the specialized structure of an economy. In the United States, industrial production is relatively concentrated in the northeastern quarter's urban-industrial core, and agricultural production in the Midwest, Far West, and South. Large-volume flows of agricultural and industrial goods within and among regions in the United States form a skeletal pattern comprised of relatively few major routes (see figures 10.7 and 10.8). International flows principally reflect comparative advantage among nations, as expressed by the special-

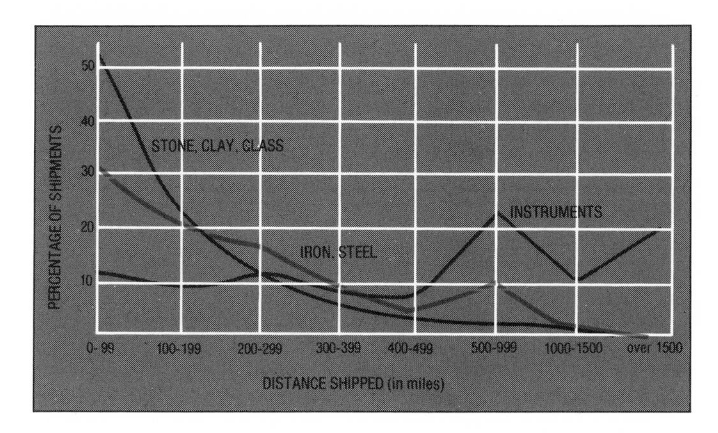

10.6. Distance traveled by selected commodities, United States, 1965. Trade volume usually falls with distance, although some peaks may occur at greater distances, where they coincide with major industrial markets. In general, the lower the value per pound, the shorter the average distance shipped.

ization of production in national economies. Less developed nations tend to specialize in food and raw material production, and advanced nations in manufactured goods.

The Flow of Raw Materials

The flow of raw materials to manufacturing centers is essentially a process of spatial concentration: goods are funneled into fewer large urban places from many scattered units of rural production, such as farms, mines, and logging operations (figure 10.7). Within a region, raw materials are brought to many smaller towns and then delivered into the regional center. Within a nation, the flow of raw materials is chiefly from the peripheral regions to metropolitan centers. Consider the flow of raw materials in and from Montana: lumber cut over wide regions in western and southern Montana is brought to dozens of sawmills and to a smaller number of pulp and paper and other mills in the state for shipment to California and the Midwest; wheat from central Montana is brought to dozens of elevators in small towns for shipment to Portland or Seattle for exporting, and east to Minneapolis and Milwaukee; animals are shipped live to Pacific Coast and Midwest feeder operations or meat packers; copper is smelted in Butte and Anaconda before being shipped to industries in California and the East; and coal is exported mainly in the form of electricity, entering the power grids of the Pacific Northwest and the upper Midwest.

International raw material flows still reflect old colonial trading patterns: resources from less developed nations are sold to processors in advanced nations. The principal raw materials include petroleum products, agricultural products (tropical fruits, sugar, rubber, fibers, wheat, meat, livestock, coffee, tea), and minerals. Of course, some highly developed nations are also major exporters of agricultural products and minerals. Canada exports petroleum, wood, wheat and other grains, and iron ore; Sweden exports iron ore and lumber; and the United States exports agricultural products and coal.

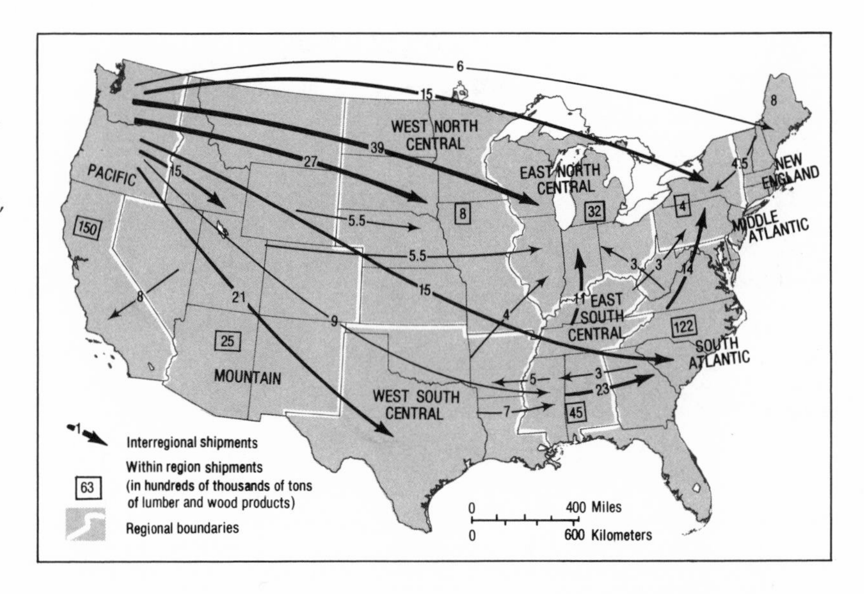

10.7. Lumber shipments, United States, 1963. Most lumber shipments flow centrally toward the northeastern urban-industrial core. In this case, one region, the Pacific Northwest, is an unusually dominant source.

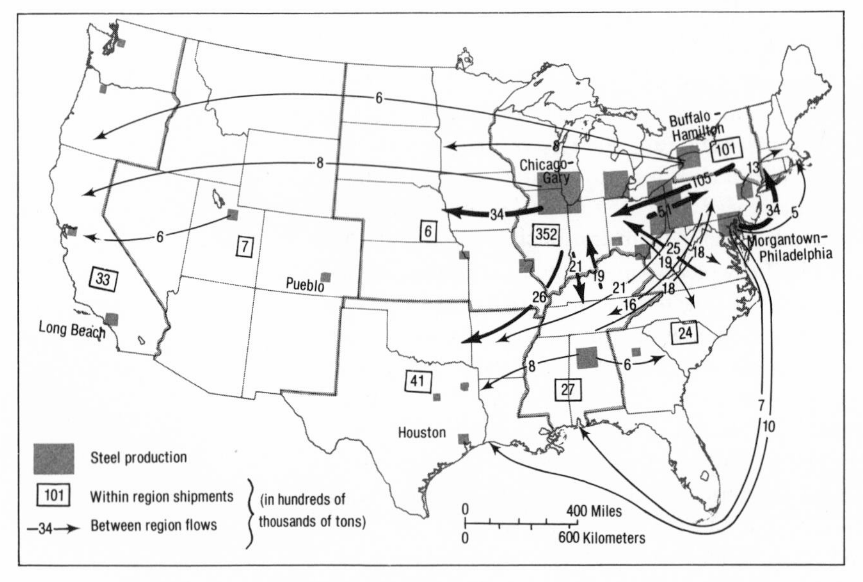

10.8. Steel shipments, United States, 1968. Most flows of industrial products such as steel are shorter than those of agricultural products and move within or out of the industrial core (East North Central and Middle Atlantic regions).

International trade in sugar and petroleum well illustrates the principles of comparative advantage and complementarity—as modified by political factors. Cane sugar is produced in and exported from many tropical and subtropical countries with a comparative advantage for sugar production because of good soil, sufficient moisture, long growing season, plentiful labor, and access to shipping (table 10.1). Sugar is demanded by most developed countries in the temperate zones, although some, like France and West Germany, meet their needs internally through beet sugar production (protected by tariffs and price supports). Because periodically there is a ca-

Table 10.1. Sugar Production and Trade, Selected Countries, 1976, (millions of tons)

	Production	Exports	Imports
World	96	22	23
Brazil	8.3	2.5	0
Cuba	6.4	5.5	0
Australia	3.8	2	0
India	6.1	1.1	0
Philippines	3.2	2	0
United States	5.6	0.5	5.2
Mexico	3.0	0	0
Soviet Union	10.1	0	3.2
Dominican Republic	1.2	1.0	0
Canada	0	0	0.9
Japan	0	0	2.8
United Kingdom	0	0	2
Mauritius	0.8	0.7	

Source: *United Nations Statistical Yearbook*, 1977.

pacity to produce more sugar than necessary to meet demand, the risk of overproduction and low prices is great. Consequently producing and consuming nations set quotas on exports and imports through the International Sugar Agreement.

The actual pattern of trade, though mutually beneficial, differs from the theoretical one in which optimizing buyers and sellers operate freely. The U.S. trade embargo prevents Cuba from selling to its neighbor, resulting in greater sugar sales to Eastern Europe and the Soviet Union. The United States tends to distribute its purchases to countries that accept U.S. private investment in sugar estates, such as the Philippines and the Dominican Republic. Similarly the United Kingdom purchases much of its sugar from former colonies like Mauritius and Australia, where private investments are maintained. Countries that depend on sugar exports as a major, if not only, source of income are greatly affected by fluctuations in the price of sugar. In 1973, for example, sugar sold for ten cents a pound; in 1974, twenty-five cents; in 1975, twenty cents; and in 1976, fourteen cents.

The other commodity we have selected, petroleum, is the dominant raw material in international trade both in volume and in value. Its price has risen more than 400 percent since 1972, to about thirteen dollars per barrel in 1977, as a result of decisions made by the marketing cartel OPEC (Organization of Petroleum Exporting Countries). Established in 1960, OPEC is an association of thirteen nations that depend on petroleum exports for most of their income. The member nations are Iran, Iraq, Kuwait, Saudi Arabia, Qatar, United Arab Emirates, Libya, Algeria, Nigeria, Gabon, Indonesia, Ecuador, and Venezuela. OPEC regulates the supply and price of its member nations' petroleum.

Table 10.2 provides data on the production and trade of petroleum in 1975. The location of surplus production and exports is fairly simply re-

Table 10.2. Petroleum Production and Trade, 1975

Country	Production (millions of barrels)	Exports (billions)	Country	Imports (billions)
World	20,000	$110.0 [a]	World	$110
U.S.S.R.	3,600	5.2		
United States	3,000		United States	21
Saudi Arabia	2,600	24.0		
Iran	2,000	18.5	Japan	20
Iraq	900 [a]	8.3	France	10
Nigeria	850	7.5	Italy	9
Kuwait	800	7.0	United Kingdom	8
Libya	700	6.5	West Germany	8
Venezuela	650	6.0	Netherlands	5
Indonesia	500	4.3	Spain	3.5
Algeria	400	3.6	Canada	3
Canada	650	3.0	Brazil	3
China	—	—		

Source: *United Nations Statistical Yearbook*, 1976.
[a] Approximate figure.

lated to currently known and exploited deposits. The largest exporters are in the Middle East and North Africa (Saudi Arabia, Iran, Iraq, Kuwait, Algeria, Libya), followed by Venezuela, the Soviet Union, Indonesia, Nigeria, and Canada. So vital is petroleum that it heads the import list of almost all other nations, from the poorest to the richest, with the greatest volume being shipped to the major industrial nations: the United States (despite its own large production), Japan, West Germany, France, Italy, and the United Kingdom (expected to become a net exporter, however, in the 1980s upon exploitation of its North Sea fields).

The pattern of petroleum flows is again rational but politically influenced. Flows to the United States and Western Europe are from countries whose fields were developed by Western multinational corporations. (These companies are still responsible for most exploration and development, even though many governments have used their revenues to nationalize the industries.) Flows also reflect to some degree old colonial ties or current spheres of influence, particularly those from Venezuela and Canada to the United States, from Libya to Italy, from Algeria to France, and from the Soviet Union to Eastern Europe and Cuba.

The Flow of Manufactured Goods

Most of the value of trade within and between advanced nations is in manufactured goods. In contrast to the flow of raw materials shown in figure 10.7, manufactured goods within the United States and other advanced nations flow most intensively from the industrial core to the periphery (figure 10.8). The heaviest volume of trade occurs within the core itself. Flows of manufactured goods occur between related activities, as between plate glass manufacturers in Toledo and automobile makers in Detroit. In addition, fairly large flows of goods such as appliances and automobiles move down the urban hierarchy, from manufacturers in metropolitan centers to wholesalers in regional centers, to retailers in smaller cities and towns, and finally to widely dispersed consumers.

In international trade, manufactured goods from highly industrialized nations are in effect exchanged for raw materials from less developed nations or for specialized manufactures of other more developed nations. The United States, for example, is the dominant producer and exporter of computers and passenger aircraft to most of the world, except that sales of advanced models to the Soviet Union and its allies are restricted for national security purposes. In return, the United States purchases petroleum, iron ore, tropical foods, and other raw materials from developing countries, and automobiles, ships, and a wide variety of industrial machinery and consumer goods from other more developed countries. Similarly Japan—the third largest industrial nation—exports a variety of manufactured goods (steel, automobiles, motorcycles, cameras, computers, and so forth) and imports virtually all its industrial raw materials and much of its food from both less developed and developed countries, particularly the United States, Canada, and Australia. Besides the simple industrial goods–raw materials exchange between developed and less developed nations, the less developed nations may themselves import some raw materials and export labor-intensive manufactured goods—India, Taiwan, and South Korea, for example, import fibers and export textiles.

The Pattern of World Trade

These various flows of goods form patterns reflecting the intensity of trade relationships. The commodity flows linking nation to nation create a striking pattern composed of an inner web of intensive ties generally oriented toward Western Europe, with smaller webs around lesser nodes (centers) and weak links at the periphery. The largest flows occur within Western Europe and consist primarily of manufactured goods. Most of the other large flows that move to and from Europe consist mainly of raw materials. This pattern is exemplified in figure 10.9 by flows of grain and oil. As you can see from figure 10.9, the United States, Japan, and the Soviet Union rival Western Europe as centers of trade.

The value of goods traded among countries amounted to one trillion dollars in 1976—at least 20 percent of the world's national product—illustrating the significance of comparative advantage and specialization despite the desire of various nations to become self-sufficient. The importance of trade is greatest for smaller countries, rich or poor, with a limited variety of resources or industries (exports represent 80 percent of the GNP of Saudi Arabia and 35 percent of Denmark's). Trade is relatively less important in the largest countries, such as the United States, China, and the Soviet Union (in the United States, exports represent about 9 percent of GNP).

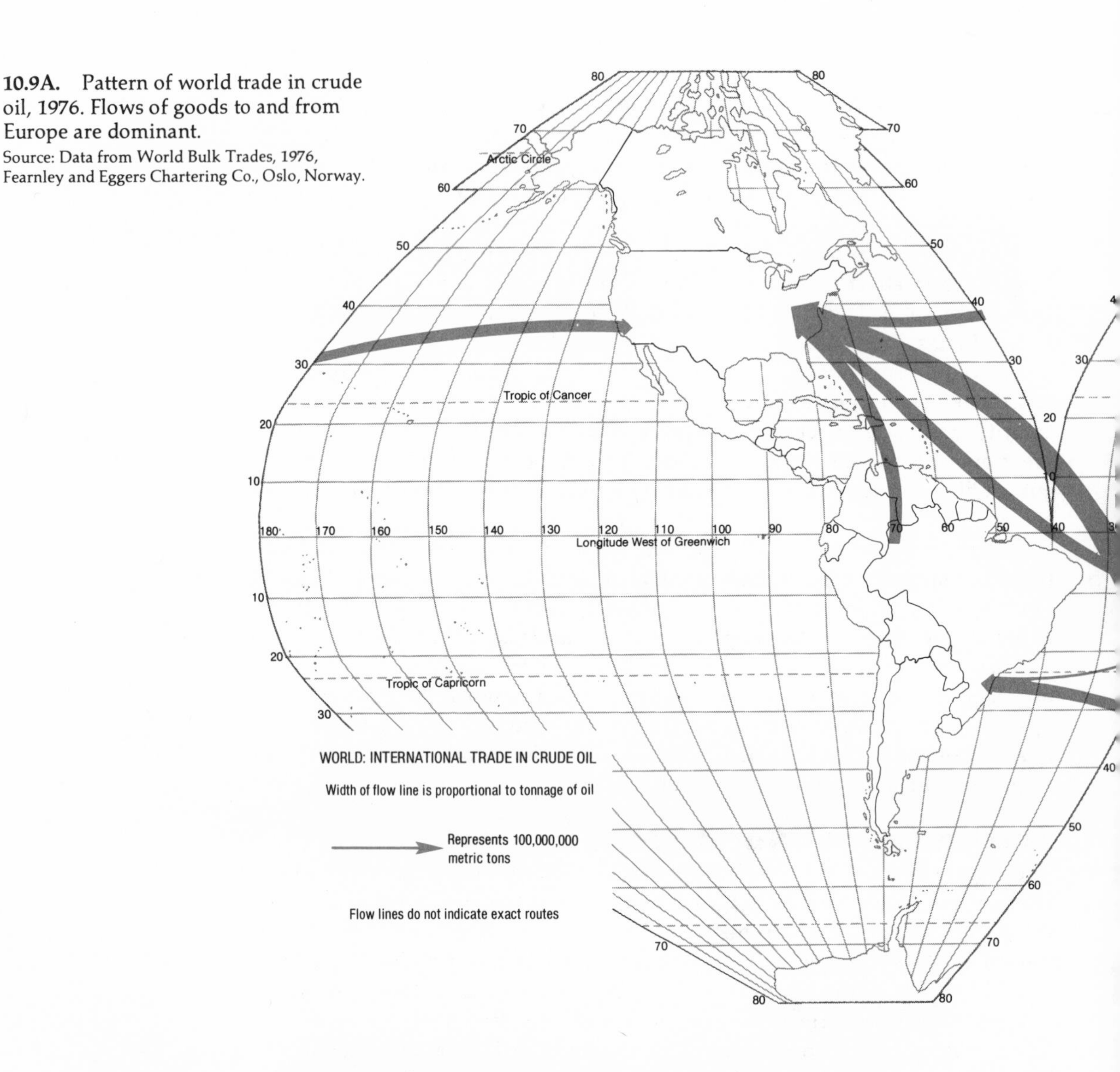

10.9A. Pattern of world trade in crude oil, 1976. Flows of goods to and from Europe are dominant.
Source: Data from World Bulk Trades, 1976, Fearnley and Eggers Chartering Co., Oslo, Norway.

The Pattern of Internal U.S. Trade

The pattern of trade within the United States is similar to the pattern of world trade. An intensive core of trade exists in the Northeast, with some competing centers in the South and West and sparse center-oriented flows from the periphery. Compared to world patterns, the U.S. economy is much more tightly interconnected by trade. This is because the U.S. economy is territorially smaller, the quality of the transport system in the United States is generally superior, disparities in economic development are much less pronounced, and there are almost no internal political-economic barriers to trade (with the exception of alcoholic beverages).

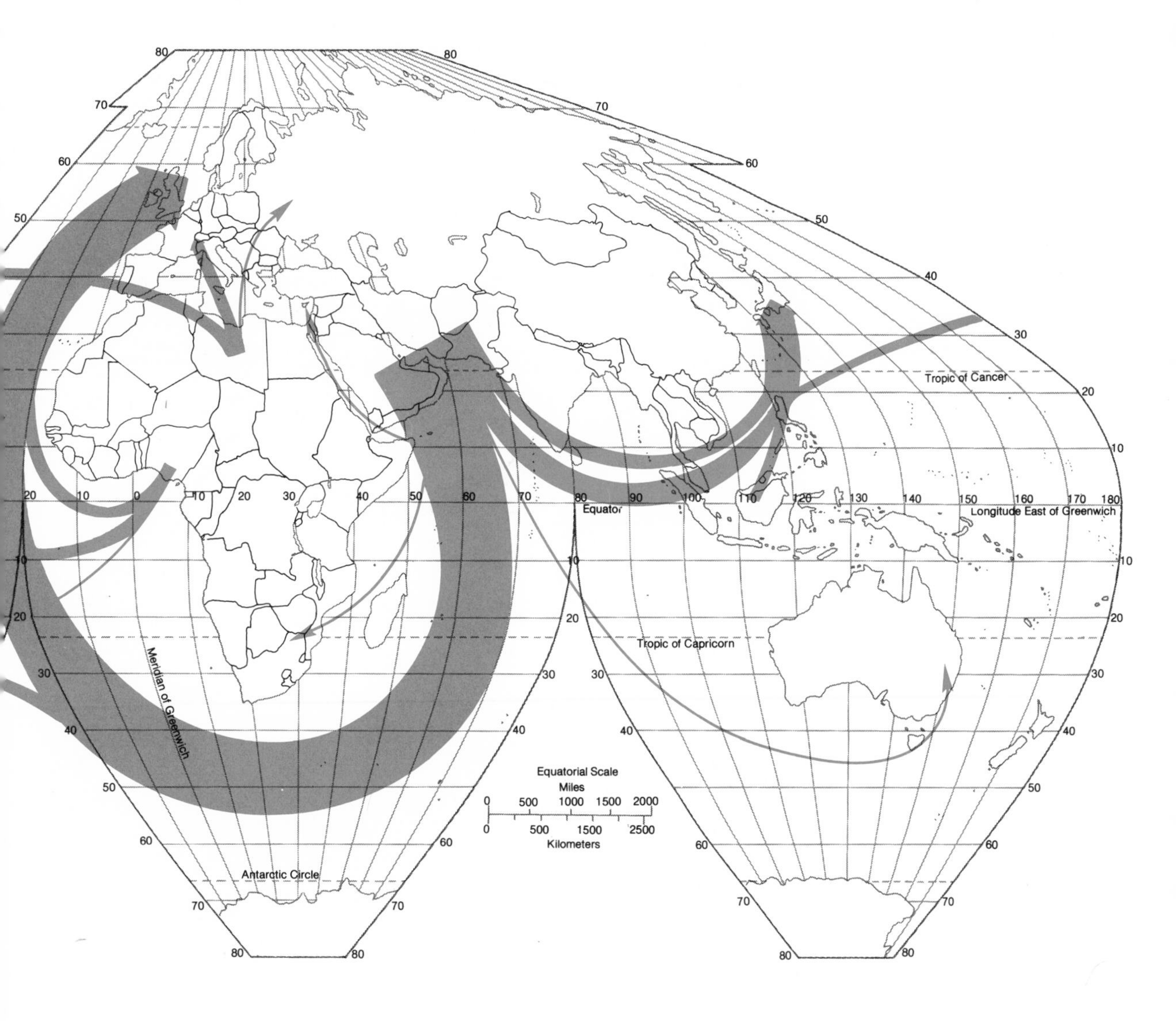

The northeastern urban-industrial core is dominated by internal movements of manufactured goods, though local movements of raw materials, especially coal, are also significant. Food and other bulk products are shipped from the periphery to the center, while the reverse flow consists largely of machinery and other finished products and consumer goods. Competing centers have arisen in California and, to a lesser extent, in the Gulf Southwest.

This pattern is well illustrated by exports and imports of Washington State (table 10.3). The total value of the state's U.S. and foreign exports, $9.2 billion, represents 36 percent of the state product and permits Washington to import the great variety of goods not produced in its specialized economy. The table shows that Washington basically exports raw materials

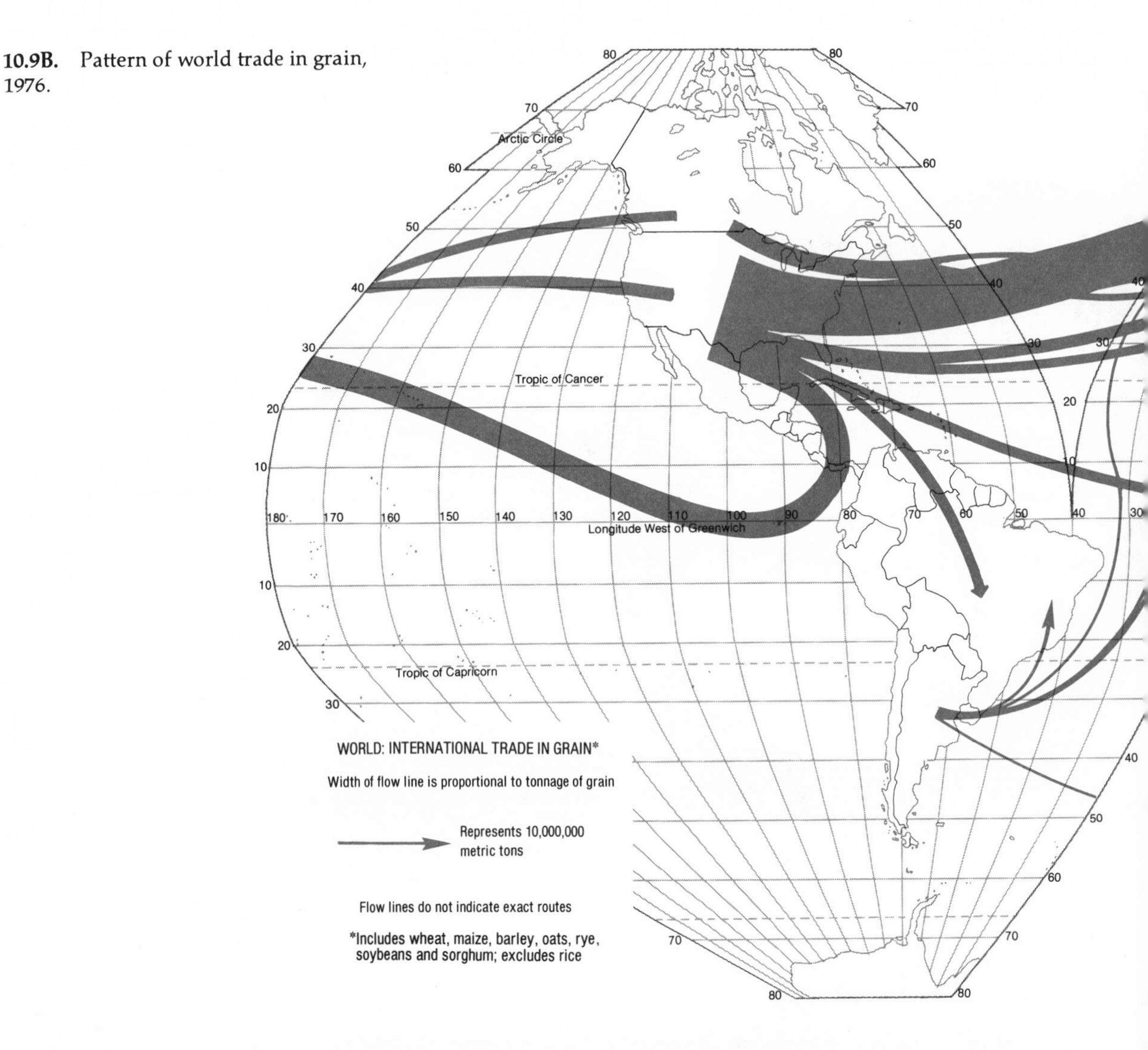

10.9B. Pattern of world trade in grain, 1976.

(wood products and grain) for which it has a local comparative advantage. It also enjoys major exports of manufactures, particularly transportation equipment (Boeing aircraft, heavy trucks, railway cars, ships), and depends on imports of meat and other foods from California and the Midwest.

TRANSPORTATION

A country's level of internal trade depends critically on the quality of its transport system. In a country where roads are few and unpaved and movement is slow, goods are exchanged only on a limited scale, and most necessities must be produced in the local area. When transportation improves, it becomes profitable for regions to specialize in products for which they have a comparative advantage and to trade their specialties for goods—including

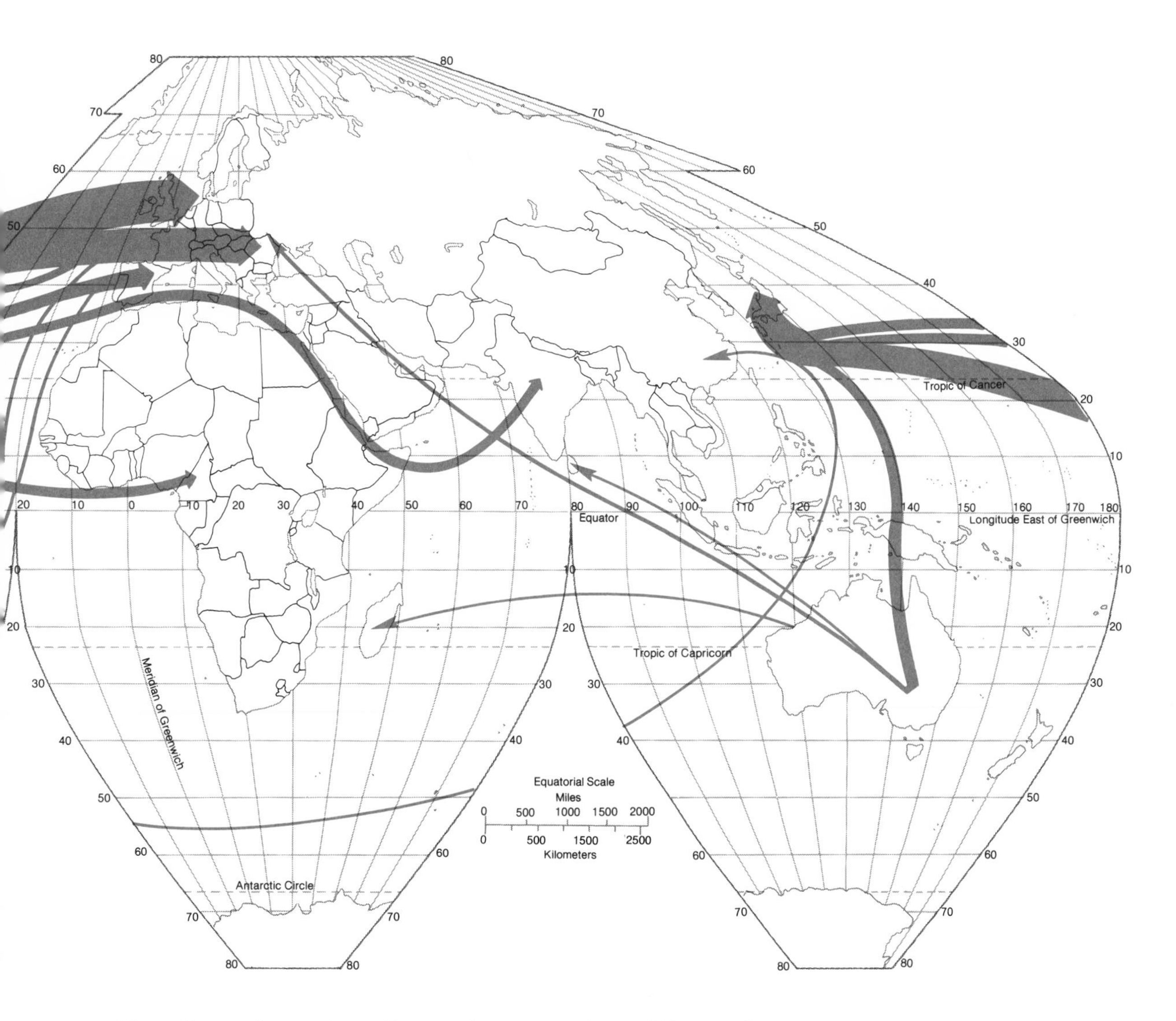

necessities—from other regions. Farmers in most regions of France, for example, could grow their own wheat, but many make a much better living by growing grapes, a more valuable crop, and importing wheat from other regions. Higher standards of living and greater productivity are achieved through specialization and trade, which are made possible by improvements in transportation.

The Development of Transport Networks

Transport networks, constructed primarily to facilitate the movement of goods and people between places, reflect the distribution of population (figures 10.10 and 10.11). But in the long run they tend to modify location. The rail

Table 10.3. Exports from Washington State, 1972

	United States (millions)	*Foreign (millions)*
Agriculture	$ 277	$ 260
Food products	782	72
Forest products	1,779	359
Metals	762	332
Fabricated metals	338	48
Aerospace	653	740
Other manufacturing	902	52
Services	1,831	390
Totals	7,294 [a]	1,952 [a]

Source: William B. Beyers and P. Bourque, *1972 Washington Input-Output Study* (Seattle: University of Washington, 1977).

[a] The total value of goods and services produced was $25,894 million.

system in the United States originated in the Northeast, the area of earliest settlement and development; it then extended west to tap the rich resources of the interior. Settlement and development followed this westward movement (see figure 10.3). Similarly the interstate highway system has stimulated the growth of settlements and the success of activities strung along its arteries, while many small hamlets at rural crossroads have disappeared. In developing countries the construction of each new road is accompanied, or even anticipated, by the settlement of new land and the growth of commerce.

Theoretically the development of transport networks follows the same principles that govern location in market economies: the aim is to maximize the utility of places and the level of interaction (trade) at least cost or effort. Building a link between two places is costly and will usually be undertaken only if, directly or indirectly, the cost of construction is repaid by increased productivity after interaction is established. The French, for example, are planning to build new roads through the interior of French Guiana, South America, to utilize the colony's timber resources for paper production. The new roads will penetrate the interior and will connect paper mills to coastal towns and to Brazil's Trans-Amazon Highway. The French expect that profits from new sales will more than pay for this costly development scheme. Although it may take years to repay construction costs, particularly in developing countries, road construction may nonetheless be justified for reasons of military security and national unity.

The goal in planning a transport route is usually to construct the shortest and least expensive path between two places. Ideally this path is as straight a line as possible. But it may be more practical to deviate if greater gains in revenue or productivity can be realized by linking intervening places. The increased cost of transportation, congestion, and time loss resulting from the deviation will be offset by the additional revenues. Topo-

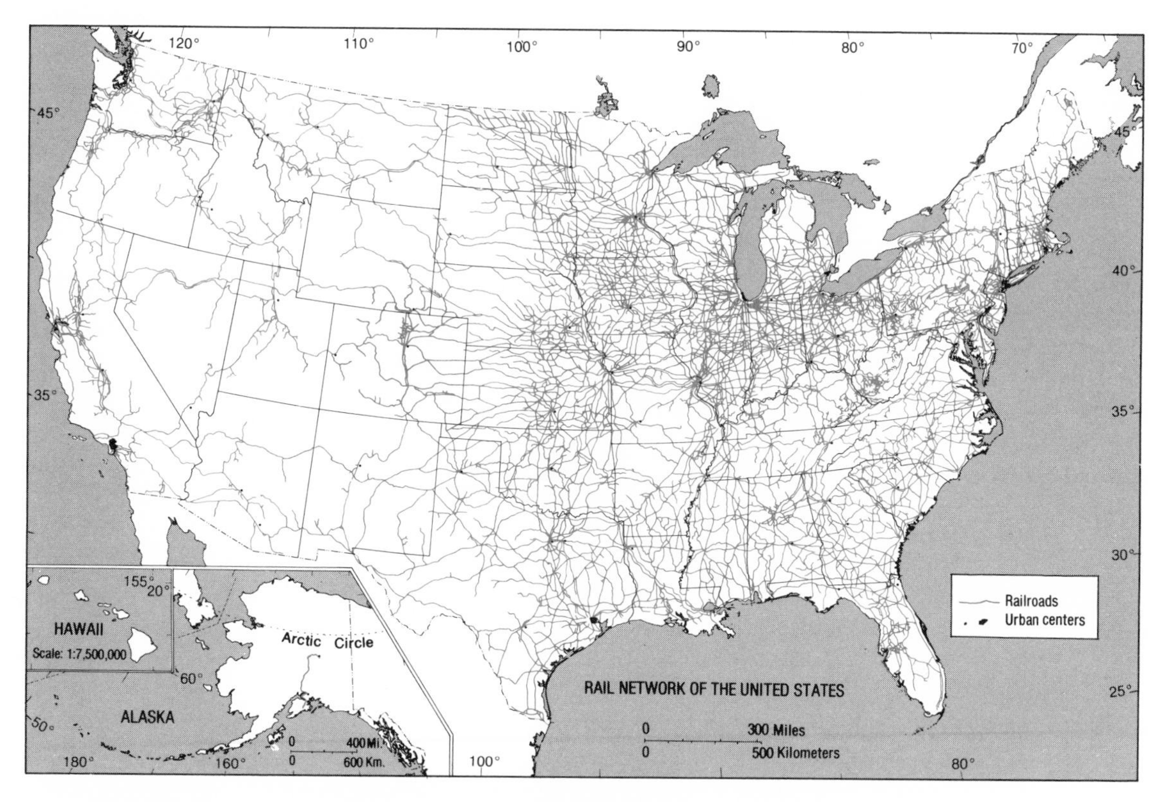

10.10. Rail network, United States, 1965. Note the fineness of the net in the agricultural Midwest and industrial Northeast and the sparser pattern in mountainous regions (such as Appalachia) and the low-density West.

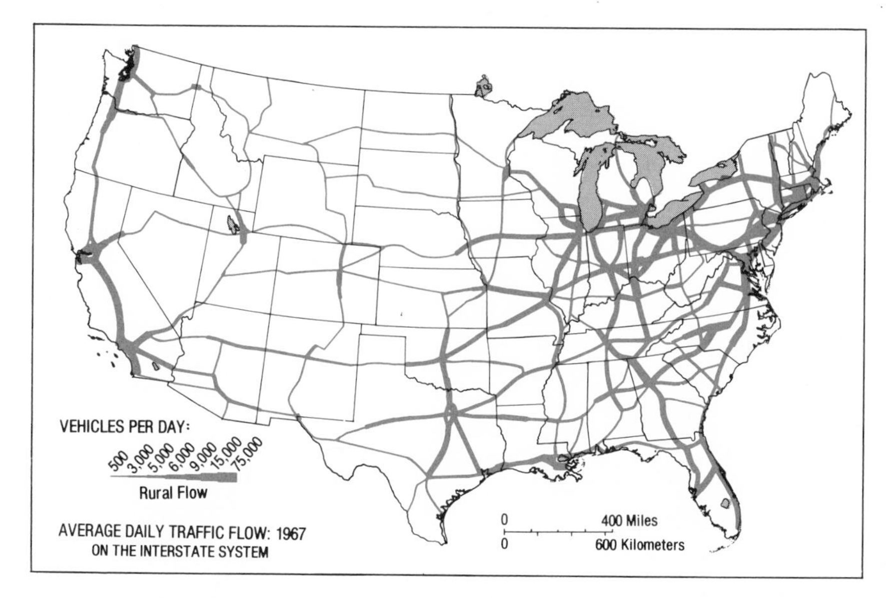

10.11. Highway traffic, United States, 1969–1970. Only flows on primary roads are shown. The flows closely reflect the distribution of urban population.

graphical variation may also justify deviation: a longer road in gentle terrain is often cheaper to build and maintain than a shorter road in rugged terrain (figure 10.12).

The development of transport networks presented in figure 10.13 is a sequence that occurs in many developing countries. In the early stages, when relatively self-sufficient local economies predominate, the transport network usually consists of partially joined local roads and possibly a few military highways, which link the capital to the provinces. With the onset of industrialization, penetration lines are built from the initial point of development (capital city or port) to peripheral areas that supply agricultural products and other resources. Processing activities may diffuse outward, and local feeder railways or access roads may be constructed for transporting resources. In time, transport links expand and connect national centers. The pattern of development then proceeds from many origins. Even before adequate routes cover the entire territory, intensive development near the point of origin makes it necessary to replace the early penetration lines with higher quality, faster links. Figure 10.14 shows the unfolding of such a pattern in Brazil.

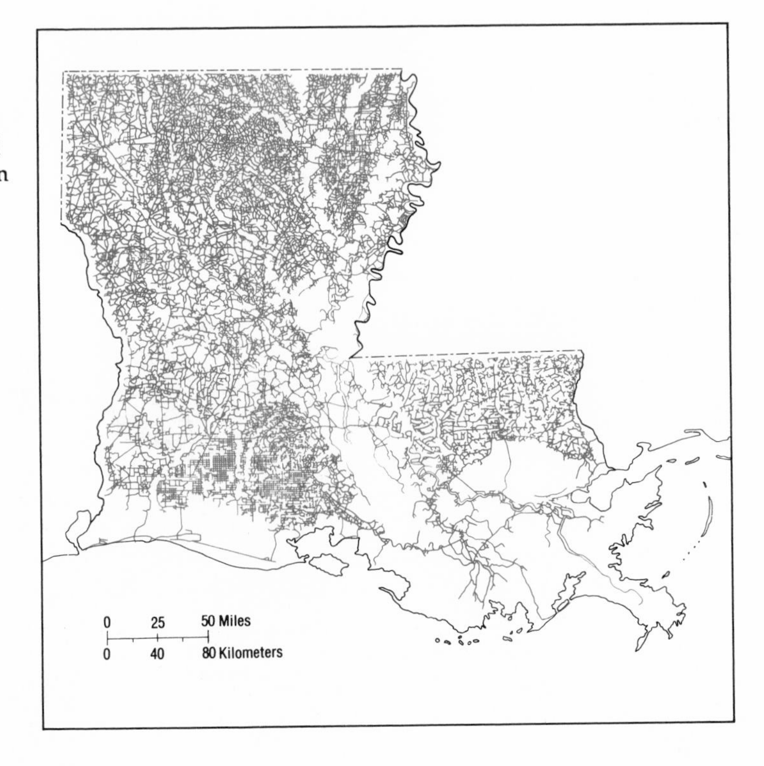

10.12. Impact of variations in local topography on transport routes, Louisiana, 1977. The road network is rectangular in level areas (southwestern portion of Louisiana), more complex in rolling areas, and absent from marshy areas along the coast and in the Mississippi Delta, where construction is difficult.

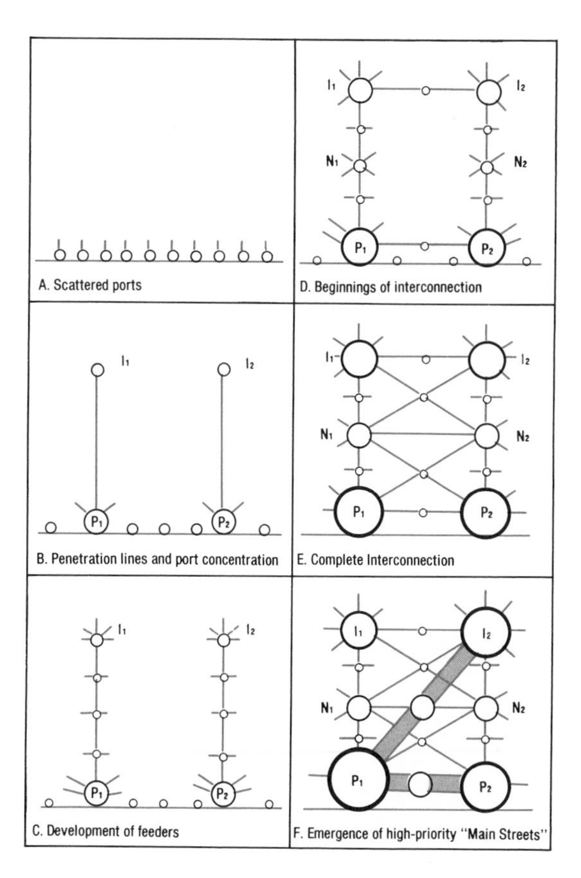

10.13. Development of transport networks. This figure depicts a typical sequence of development from initial small, scattered ports to limited penetration lines to inland centers, *I*, which results in the consolidation of port activity in two places, P_1 and P_2. Then the development of local access feeders and beginnings of interconnections lead to the growth of additional nodes of trade activity, N_1 and N_2. Finally, complete interconnection occurs, and high-quality direct routes emerge between the most highly interrelated places.

Source: Redrawn with permission from E. J. Taaffe, R. L. Morrill, and P. R. Gould, "Transport Expansion in Underdeveloped Countries: A Comparative Analysis," *The Geographical Review*, vol. 53, 1963.

Competition and Complementarity among Transport Modes

We have already given many examples of the various transport modes, but it would be useful to summarize the advantages and disadvantages of the major carriers.

1. *Ships and barges.* Ships and barges enjoy the lowest costs per volume or weight, especially for bulk commodities, because of their large capacity and special bulk-handling methods. They are preferred for transporting such commodities as sand and gravel, ores, and grain in areas where major waterways exist—for example, the Great Lakes and Mississippi-Ohio rivers in the United States and the river and canal systems in Europe. The principal disadvantages are that ships and barges are slow and waterways limited in extent.

10.14. Development of the transport network in Brazil. Historically Brazil had two transport systems: a road and rail net in the southeast and the Amazon River system. New highways are now connecting these systems.

2 *Railroads.* Railroads, with the next lowest costs, have a fairly large capacity for transporting bulk commodities where waterways are not available. Most agricultural and mineral products, as well as many manufactured products, are shipped by train. Railroads are especially competitive for fairly long-distance movements between major origins and destinations, as for transcontinental shipments of wood or farm products. They are not, however, for short hauls, small-sized shipments, or rapid delivery.

3 *Pipelines.* Once constructed, pipelines have very low labor costs and are most efficient for transporting bulk liquids or gases, such as petroleum and natural gas. Currently the railroads are fighting the possibility of shipping coal in slurry form by pipeline. The disadvantages are that they are very slow and not useful for shipping most solid material.

4 *Trucks.* Although trucks are becoming larger, their capacity is still much lower than that of trains, and their costs are higher because of labor requirements. But the highway network is more widespread, trucks can deliver door-to-door, and goods do not have to be transshipped (changed to another transport mode). Thus truck transportation is faster than rail and is very competitive with it, particularly for local and intermediate-range shipments and for volumes that constitute less than a full rail car.

5 *Airlines.* Despite their higher costs relative to land transport systems, airlines are carrying increasingly large volumes of goods. They are used especially for high-value products and when rapid shipment is desired.

Table 10.4 summarizes the share of these carriers in the movement of goods in the United States, and figure 10.15 shows the distances for which carriers of different capacity are most competitive.

The fiercest competition may occur among carriers or facilities of one kind. Airlines, for example, compete for traffic shares of high-density routes, and ports also compete for traffic. Ideally such competition leads to rate reductions and increased interaction.

Different transport modes cooperate at times. The development of freight containerization has encouraged close relations between ocean shippers and trucking firms or railways. The shippers, for example, may arrange through-shipment from Europe to Japan, utilizing U.S. railroads as a "land bridge." Similarly, railways may carry truck trailers "piggyback" on long hauls.

Theoretical Transport Networks

The major location theories have little to say about the nature of any transport network. The availability of transportation with equal ease and cost in any direction is implicit. But all the theories assume that movement occurs along the least-cost route and that a rather complete transport network exists, together with competing modes of transportation. Each theory suggests a specific transport pattern.

Agricultural Location Networks

Agricultural location theory involves minimizing the cost of transporting produce to a central market. The desire to keep transport costs from be-

Table 10.4. Goods Carried by Mode, United States, 1975

Mode	*Percentage of Ton Miles*	*Revenue, 1974 (billions)*
Ships and barges	17	n.a.
Railroads	36	$17.5
Pipelines	23	1.0
Trucks	23	16.8
Airlines	0.2	0.9

Source: *Statistical Abstract of the United States*, 1977.

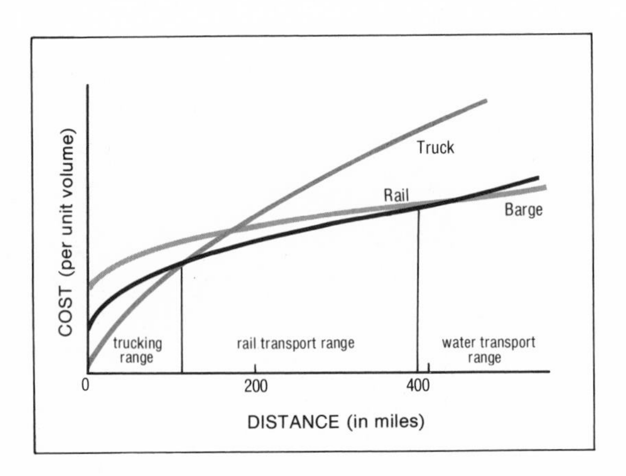

10.15. Carrier competition. The capacity of a carrier is generally related to the distances for which it is most competitive: relatively low-capacity trucking is preferred for short to moderate hauls; higher-capacity rail transport for moderate to long hauls; and high-capacity water transport for long hauls of bulk commodities.

coming unprofitably high results in competitive bidding for land closer to market, creating the agricultural gradient. The gradient aspect of agricultural location theory suggests a transport network that is completely center oriented, radiating outward from the market (figure 10.16). Decreasing fineness of the net reflects the increasing size of farms and the greater distance between them. No cross-links are required, since all movement is directed toward one market. The ideal network provides complete access to all farms and minimizes the mileage of costlier, high-capacity routes.

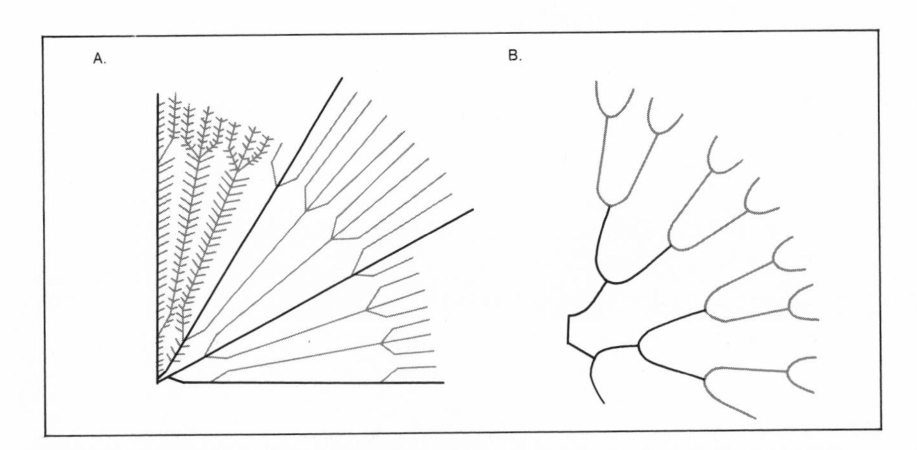

10.16. Transport networks in agricultural location theory. *A* and *B* represent two theoretical networks designed to move produce to a central market. *B* extends to points closest to individual farmers at the least distance to market. That is, the *B* network reflects a decentralized hierarchical pattern of marketing and supply, where the local producer ships to a local center, these locally collected goods are shipped to a district center, and so on. *A* may be more realistic because of human preference for straighter, more direct roads. This pattern reflects a more centralized pattern, whereby goods are shipped to a main transport link, then on directly to a more major market.

Central Place Networks

Central place theory involves minimizing the cost of distributing goods to dispersed consumers and of collecting goods from dispersed producers. The theory suggests a network oriented toward many centers on several levels (figure 10.17). Major routes serve and connect the smaller centers. The transport pattern mirrors the central place pattern. In pure theory, settlement is evenly distributed and the transport pattern strictly regular, except that the volume of movement and the quality of the route reflect the level of the interconnected centers. The central place pattern based on the transport principle (shown in figure 7.5B) is the most efficient one because the routes connecting two larger centers pass through centers of the next lower level. The mileage of high-capacity, costlier routes is minimized. Efficiency is raised by concentrating large-volume transport flows on major routes and restricting local collection and delivery to the smaller routes.

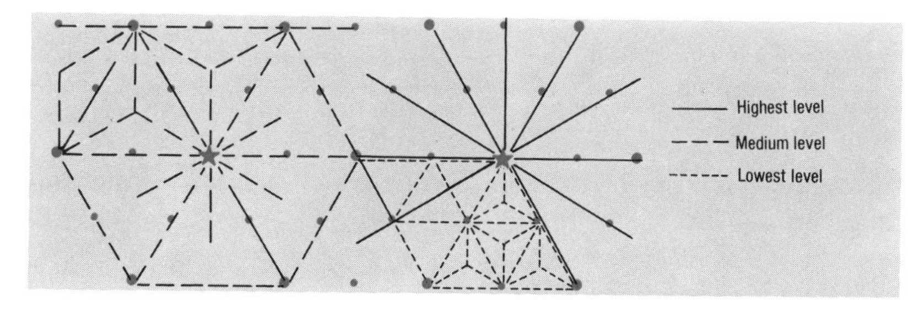

10.17. Transport networks in central place theory. Flows in a central place system mirror the geometric arrangement of places. The more important routes and flows converge on the highest level place (star) from equally large places outside the area (not shown). Medium-quality routes and medium-volume flows serve and connect medium-level places (large dots). And local routes and small flows serve the smallest, lowest level places. (More important routes carry local traffic too, of course.)

In reality, agricultural gradients and central place systems overlap and must be considered together. The theoretical network for the joint surface suggests a possible composite transport network (figure 10.18). The larger branches of the net radiate from the main central market. These branches are cross-linked at the intersections, where smaller centers are located, and local and regional routes are oriented toward the nearest center.

Industrial Location Networks

Industrial location theory involves minimizing the costs of procuring raw materials and of shipping finished products, although the importance of transport costs as a locational factor varies widely. Some industries, such as lumber and food processing, collect their materials from local areas, giving rise to a large number of low-volume transport flows. But most industries have important linkages with other industries (farm-equipment manufacturers with steel producers, for example), giving rise to a pattern of fewer, larger flows. Specific patterns of industrial location might not alter the basic structure shown in figure 10.18, but they would tend to destroy its symmetry by creating regional variations in density. For example, clusters of interrelated plants (iron and steel, fabricated metals, chemicals) would require

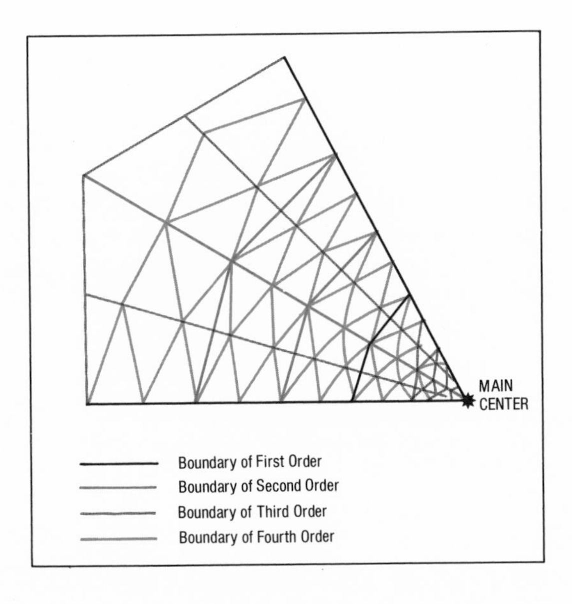

10.18. Theoretical transport network on a joint gradient-hierarchy surface. This model shows a transport network serving a landscape in which a central place system is imposed on an agricultural (and population) gradient.

strong cross-links, which would not normally appear in the composite transport network.

Theory and Reality

Transport networks in the real world are much less regular than theory would suggest. Most networks have not been set down according to a master plan but have developed gradually from areas of early settlement. (The U.S. interstate highway system is an important exception. Also transport development in newer centrally planned economies is following a long-range scheme.)

Many segments of the preinterstate highway transportation network in the United States are obsolete, reflecting the limitations of an earlier technology. In the nineteenth century, there were many small farms and hamlets. The need for local access resulted in a network of roads and railways that, by today's standards, had too many small, local-access routes. Connections between places twisted and turned with property lines and local topography. While the capacity of the network was suitable for the small volume of flows at that time, it is inadequate today. Of course, not all of our transport problems can be blamed on the past. Some arise from poor planning or from other priorities. The demand of motorists for more and better roads, for example, can lead to overinvestment in highway building and maintenance and underinvestment in public transportation or private railroad maintenance.

Investment in transport routes has a great impact on trade and economic development. The Soviet Union has an imbalanced investment in production compared to transportation. Its neglect of local access roads, in particular, has created distribution problems, impeding the most productive use of agricultural land. In the United States, the interstate highway system has contributed to the decline of some smaller places that were bypassed and to the growth of others linked by the system. Any investment in transport routes will affect the relative prosperity of centers served or bypassed, modifying patterns of location.

CONCLUSION

The flow of goods coursing through the spatial structure of society may be compared to the flow of blood through the circulatory system of the human body. Just as the circulatory system may be described either in terms of process (flow of blood, energy exchange) or in terms of structure (heart, veins, arteries), the spatial order of society may be described in terms of process (movement of goods, journey to work) or in terms of structure (location of specialized production, transport routes). Both process (movement, interaction) and structure together constitute the spatial order of society. In the next chapter we will continue our study of spatial process by analyzing the movement of people and ideas.

SUGGESTED READINGS

Bird, J. *Seaports and Sea Terminals.* London: Hutchinson University Library, 1971.

Chisholm, M., and O'Sullivan, P. *Freight Flows and Spatial Aspects of the British Economy.* Cambridge, England: At the Press, 1973.

Daggett, S. *Principles of Inland Transportation.* New York: Harper & Row, 1967.

Haggett, P., and Chorley, R. *Network Analysis in Geography.* New York: St. Martin's Press, 1969.

Hay, Alan. *Transport for the Space Economy.* London: Macmillan, 1973.

Eliot Hurst, M. *Transportation Geography.* New York: McGraw-Hill, 1974.

Kansky, K. *The Structure of Transport Networks.* Department of Geography Research Paper 84. Chicago: University of Chicago, 1963.

O'Dell, A. C., and Richards, P. S. *Railways and Geography.* London: Hutchinson University Library, 1971.

Ohlin, Bertil. *Interregional and International Trade,* rev. ed. Cambridge, Mass.: Harvard University Press, 1967.

Sealy, K. R. *A Geography of Air Transport.* London: Hutchinson University Library, 1966.

Taaffe, E., and Gauthier, H. *A Geography of Transportation.* Englewood Cliffs, N.J.: Prentice-Hall, 1973.

Thoman, R. S., and Conkling, E. C. *Geography of International Trade.* Englewood Cliffs, N.J.: Prentice-Hall, 1967.

Ullman, Edward. *American Commodity Flow.* Seattle, Wash.: University of Washington Press, 1957.

11

The Movement of People and Ideas

11 The Movement of People and Ideas

> The cars of the migrant people crawled out of the side roads onto the great cross-country highway, and they took the migrant way to the West. In the daylight they scuttled like bugs to the westward; and as the dark caught them, they clustered like bugs near to shelter and to water. And because they were lonely and perplexed, because they had all come from a place of sadness and worry and defeat, and because they were all going to a new mysterious place, they huddled together; they talked together; they shared their lives, their food, and the things they hoped for in the new country. *

John Steinbeck's account of the migration of rural Oklahoma families to California during the Depression and Dust Bowl period of the 1930s is only one of the many dramatic examples of human movement. Of interest to geographers is a whole range of movement—from the simple journey to work and shopping trip to the great mass migrations of human history. In this chapter we will analyze various kinds of human movement and describe the patterns they form. We will also examine the spread of ideas, innovations, and other phenomena across the landscape.

THE MOVEMENT OF PEOPLE

Individuals develop an identity with place through their home and their work; but most also find it both necessary and rewarding to spend much of their time on the move. Human movements fall into three categories: temporary, transient, and permanent (migration).

Temporary Movements

Persons in advanced societies have a high potential for interaction with other people and places. Rapid and efficient means of transportation and communication are at their disposal, and average incomes are high. Students in the United States, for example, can drive or take public transportation to school, part-time jobs, stores, and places of entertainment; they can fly to distant places for a vacation; and they can communicate instantaneously with friends by telephone. Persons in less developed societies may devote as much time and effort to interacting with other people and places, but the limitations of the transport and communication systems restrict the number and distances of their movements.

* John Steinbeck, *The Grapes of Wrath* (New York: Bantam Books, 1946), p. 171.

The movements described above are *temporary:* they entail a trip from a place of origin (home) to one or more destinations and back. Ideally one should be able to carry out temporary movements with minimum cost and effort. If you spend many hours driving to and from work or shopping and doing other errands, either the structure of locations is inefficient—for example, work places are located too far from residential areas—or you have not arranged your travel efficiently. One reason for inefficient movement is that people may not know where closer shopping and other opportunities are located. Another is that some may avoid certain areas because of fear or because they prefer other, more distant places.

An important characteristic of temporary movements is that they tend to reflect and enliven, rather than define, the structure of locations. However they can also lead to change; overloads of flows on the structure provide a reason to modify it. Congestion on a certain highway, for example, may eventually lead to improvements or to new highway construction.

In all countries, from the least to the most developed, the major kinds of temporary movements include the journey to work; trips to shops, services, government offices, and schools; visits to friends and relatives; and recreation trips.

Journey to Work. The journey to work is so important that it largely dictates the structure of improvements to road systems and the nature of public transit systems. The value of time spent driving or riding to work is recognized as very great (costs may comprise more than 10 percent of the typical worker's income) and has been used to justify the construction of expensive freeway and mass transit systems designed to speed the journey. Meeting the concentration of demand at morning and evening rush hours leaves the system with more capacity than it needs for much of the day.

Traditionally the urban journey to work was from the residential periphery to the industrial-commercial core. Greater affluence and decentralization of employment since World War II in most North American and European metropolises have led to a more diffuse pattern of work trips. There are many industrial and commercial centers of employment, both centrally and peripherally located. The larger and more specialized the center, the more widespread its **laborshed** area from which workers commute (figure 11.1).

As people become more affluent, they no longer consider proximity to work as the highest priority in residential location. Most take advantage of improved roads or transit to seek yet more desirable residential areas and are willing on average to spend half an hour or more commuting to work. (Many people commute to city jobs from as many as fifty kilometers away.) This trend has made it more difficult for mass transit to compete with the convenience of the private car.

Trips to Shops and Services. Compared with the journey to work, trips to shops, services, and government offices are made by more members of households and are spread over more of the day. Studies of these flows

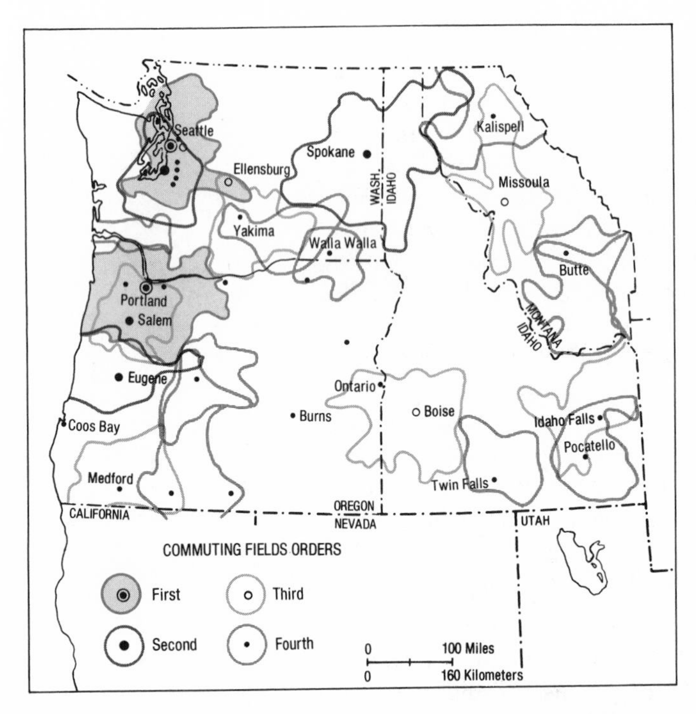

11.1. Commuting fields for four orders, or levels, of central places in the Pacific Northwest. Commuting patterns reveal the shape and extent of nodal regions, or trade areas of particular centers. Commuting fields are very large in this low-density region.

Source: Adapted with permission from R. Preston, "The Structure of Central Place Systems," *Economic Geography*, Vol. 47, 1971.

indicate that people tend to minimize distance traveled more than they do in the journey to work. The center selected, however, is not necessarily the closest nor is one center used exclusively.

Visits to Friends and Relatives. Visits to friends and relatives are still more dispersed in time and space, yet they also tend to be contained within a neighborhood. After one has moved, long trips to visit friends in the former neighborhood are gradually reduced and are replaced by trips within the new community. Because the residence pattern of friends tends to be dispersed, these trips are even more likely to be by private car.

Recreation Trips. Some trips for recreational purposes are oriented toward activities concentrated in a central area, such as a downtown entertainment district, while others are oriented toward space-consuming activities in the periphery, such as golf and racing. Most recreation trips are limited to distances people can return from in a day.

Longer recreation trips, typically lasting from one to three weeks, are made during vacations from work or school. Such tourist trips are becoming increasingly common in affluent societies. Many people from small towns visit large cities, particularly metropolitan centers. Another portion of the tourist trade is found in scenic or recreational sites such as mountains, beaches, and lakes. The choice is usually based on the degree of difference from home. In the United States, for example, people who live in northern cities often take winter vacations in Florida, southern California, Hawaii, or the deserts and mountains of the West. In Europe, Spain and Portugal benefit from similar environmental differences.

Tourist travel is geographically significant in that it fosters the development of scenic but otherwise economically limited areas and contributes to the spread of information about places and opportunities. The latter may encourage permanent movements and help create a more unified culture.

Patterns of trips, or personal activity fields, are illustrated in figures 11.2 and 11.3. About 80 percent of all trips begin or end at home. Outside of the journey to work and some recreation trips, most are made within a community or familiar sector of the city (except for when one maintains ties—with relatives, physicians, or favorite taverns, for example—after moving).

Temporary Movements in Less Advanced Countries

In less developed, mainly rural societies, the same kinds of movements are important, but they may take a different form. The journey to work may consist of two or more trips per day by one or more family members from a village to the family fields. If the village is near an urban center, one family member may commute to work, perhaps returning to the village only once or twice a week. Trips to shops and services are likely to end at periodic or town markets. On the other hand, in countries at an intermediate level of development, where few people have private automobiles or refrigerators, more trips will be by public transit (as in the journey to work) or by foot (particularly for shopping, which may be a daily activity). Trips to visit

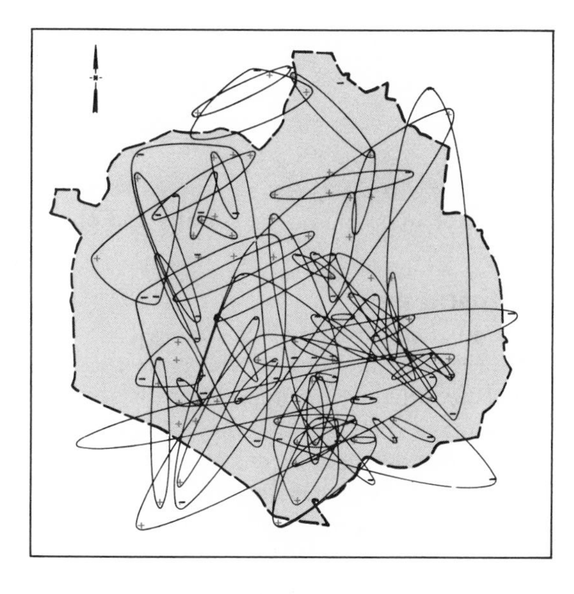

11.2. Individual travel patterns. This diagram shows the daily patterns of a sample of individual trip origins (+) and destinations (−) within Durham, North Carolina. The data are from a 1963 study. Longer-distance commuting or other trips into and out of the city are not included.

Source: Reprinted with permission from Stuart Chapin Jr., *Urban Land Use Planning,* 2d ed. (Urbana, Ill.: University of Illinois Press, 1965). ©1965 by the Board of Trustees of The University of Illinois.

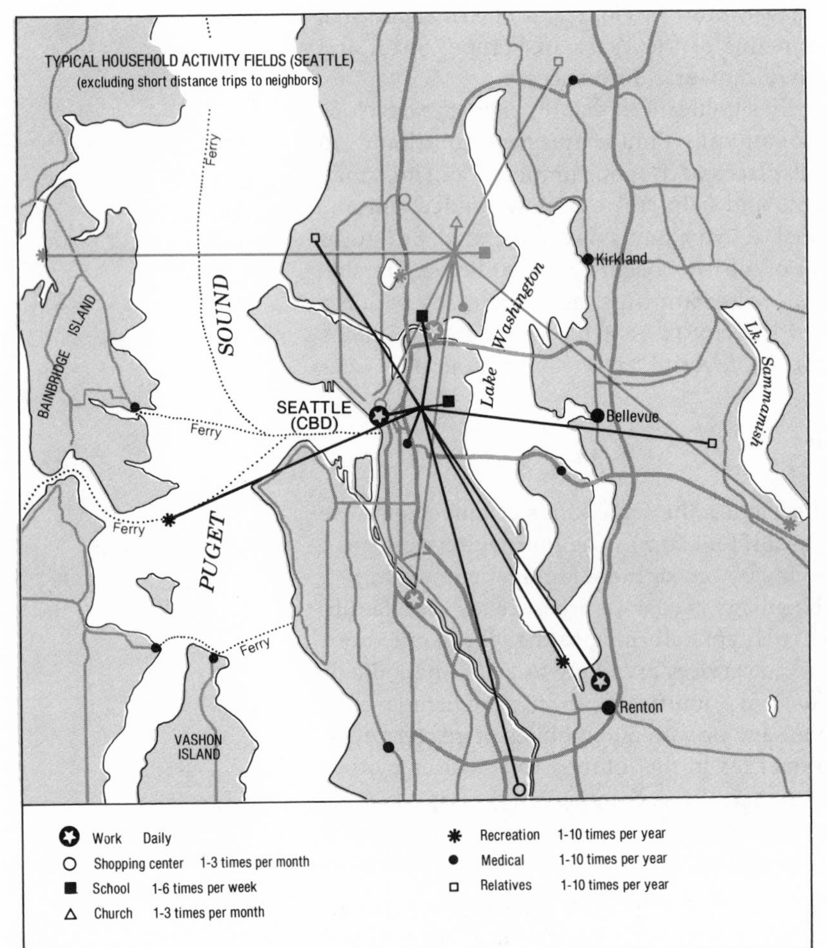

11.3. Typical household activity fields. Lines connect two households to a hypothetical distribution of major activities. Most real activity fields will have a more complex travel pattern because of visits to more shopping areas, relatives, or recreation destinations.

friends and relatives are very important in many less developed societies, and they often last for longer periods than is common in socially and physically more mobile societies.

Temporary Movements of Businesses

Businesses are less flexible than individuals in their spatial behavior. Someone who is spending a great deal of time and money on temporary trips can usually make a spatial adjustment by reducing the number or distances of trips. But businesses cannot easily make such changes; they have to maintain a certain volume of interbusiness contacts in order to survive. Publishers need to interact quickly with advertisers, wholesalers with retailers, and apparel makers with designers. Activities that require the greatest volume of mutual contacts and that can least tolerate time lost in interacting tend to cluster in central locations or in locations central to their purposes.

Examples are the gathering of physicians, laboratories, and medical supply stores around hospitals, or of bonding establishments around courts.

Communications

If personal and business interaction could be achieved only through personal contact, the complexity of the national economy could not be maintained. But of course many kinds of interaction can be carried on indirectly through previously established communication networks. The president of the United States can deliver his state of the union message to all citizens by television; corporation executives can control subordinates in branch offices through the mail or by telephone; and goods and money can be transferred by mail. Communication systems, like transport networks, require a tremendous investment in society's effort to overcome the friction of distance. Communications save time and transport costs by substituting for the temporary movement of people and effectively relaying their decisions, demands, and feelings. In high technology societies, the volume of communications—mainly by telephone, mail, radio, and television—is extremely large and greatly reduces the amount of physical travel. This fact has led some scholars to predict that communications will increasingly substitute for travel or at least reduce the proportion of business travel in favor of recreation trips. (Communications, however, are not a perfect substitute for personal contact, as demonstrated by successful firms and friendships.)

Communications are even more instrumental than transport networks in creating a national culture. Newspapers and especially radio and television are very effective in integrating a society economically and culturally—so effective, in fact, that they may be responsible in part for the erosion of regional cultural variation in such matters as dialect, food, and business activities. Many less developed nations, realizing this power of communications, wish to restrict the entry of Western television, newspapers, and magazines into their countries.

Transient Movements

Transient movements involve a temporary change of residence and may result in a permanent move. They are made for such purposes as attending college, fulfilling military or job requirements, visiting resort areas, and taking seasonal jobs. College students constitute a large transient group in the United States; roughly two million may be away from home at any one time. Most students attend schools in their home state, but many are attracted to colleges in very different environments. Transient movements to college often end as permanent movements, not so much to educational centers as to larger cities where white-collar jobs are more likely to be found.

Military requirements also lead to large-scale population shifts. Traditionally military bases have been in frontier locations, presumably for reasons of defense or for the protection of frontier settlements. Over the years in the United States, millions of Americans have moved temporarily to military bases in the South and West, and many of these movements have become permanent. The phenomenal postwar growth of California is partly due to the influence of servicemen and women who induced friends back home to see this "earthly paradise" for themselves. In less developed countries, too, military service can introduce new opportunities and expectations to young people, possibly leading to migration from home or innovation at home.

Increasingly more retired persons lead a partially transient life, spending time at two or more residences, often mobile homes. Many of these movements result in a shift of permanent residence to a resort or other environmentally attractive area.

Seasonal labor migration is an important phenomenon in many developed and developing countries. In the latter, as you recall from chapter 4, temporary migration or circulation between villages and plantations, mines, or jobs in town aids the process of modernization. Gradually the temporary migrations tend to become permanent. The government of South Africa, however, has preferred to keep black workers transient in order to maintain a legal basis for apartheid.

Transiency raises a serious social problem in the case of migrant farm workers, who move from crop to crop in response to seasonal demands (figure 11.4). A lack of education and skills prevents them from holding stable, permanent jobs. Because they are continually on the move, it is difficult for migrants to obtain the education, health care, and political representation needed to maintain a decent level of living. Various programs have been mounted to organize migrant workers, but the only lasting solution, if one is desired, is to ensure that other kinds of out-of-season jobs are available in the same areas or to end the practice of hiring migrants for farm work altogether.

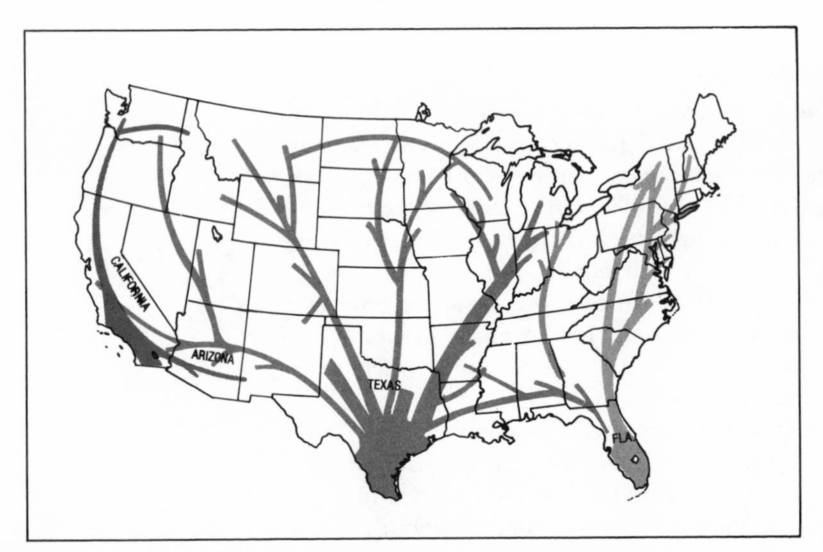

11.4. Major flows of migrant workers, United States. The flows originate in Texas, California, Arizona, and Florida, where migrant farm laborers work during the winter. In spring and summer the laborers move northward, following planting and harvesting demands.

Another form of transient labor migration is that of *guest workers,* less skilled workers from less advanced countries such as Spain, Italy, Yugoslavia, Turkey, and Algeria who are invited to work primarily on construction jobs in such countries as Switzerland, Germany, Sweden, and France. There are inevitable problems of adjustment to the different cultures for both guests and hosts. But the labor migration does benefit the developing country in that the workers send home a fairly significant proportion of their salaries and acquire skills and innovations that will be useful when they return home.

Permanent Movements: Migration

Permanent movements merit special attention, for the migration of people is a major instrument of change in society and in the evolution of the landscape. Throughout human existence people have migrated in search of better opportunities, at times pushed by the pressure of overpopulation at home and at times pulled by the hope of future rewards somewhere else. War and persecution have forced millions to flee their homeland (table 11.1), and slavery in the colonial period involved the forced migration of about 10 million Africans.

Several great world population movements were shown in figure 3.6. Some of the flows are *centrifugal,* spreading out from centers of civilization into new areas (usually already occupied). Examples include the European colonization of North and South America, Australia, and South Africa; the spread of the Chinese peoples; the westward movement of the Bedouins and Americans; and the eastward movement of the Russians. Other flows

Table 11.1. Selected European Ethnic Groups Involved in Forced Permanent Migrations, 1920–1955

Ethnic Group	*Number of Persons Forced to Migrate*	*As an Approximate Percentage of Total Ethnic Group at Time of Migration*
Germans	15,800,000 [a]	20
Poles	6,300,000 [b]	24
Jews	6,000,000	67
Czechs and Slovaks	2,000,000 [b]	16
Greeks	1,250,000 [b]	18
Latvians	310,000 [b]	30
Estonians	235,000 [b]	25

Source: After Table 3.5 (p.75) in *The European Culture Area* by Terry G. Jordan. Copyright © 1973 by Terry G. Jordan. By permission of Harper & Row, Publishers, Inc.

Note: All persons expelled, exchanged, evacuated, or imprisoned who were never able to return to their homeland, including those who perished.

[a] Excludes Jews and all ethnic Germans resident in the Soviet Union within the pre–World War II borders.

[b] Excludes Jews.

are *centripetal,* from peripheral to more developed or promising areas, exemplified by many tribal movements into Europe and China, by the recent flows of blacks into South Africa, and by the major rural-urban migrations. And still other flows are simply migrations in search of better opportunities elsewhere—examples are the original spread of humanity, the early migration of blacks southward in Africa and of Polynesians into the Pacific, and the migration of Brazilians toward the western frontier or to the developed core of their own country today.

While the volume of migration is impressive, people are far less mobile than goods. Human inertia is difficult to overcome, and many people are willing to stay home and forgo potentially greater benefits because they fear change, are unaware of better opportunities, or consider noneconomic satisfactions at home to be more important than economic opportunities or sociopolitical uncertainty elsewhere. Despite such constraints, migration does occur and has had a significant impact on the development of the human landscape:

1. Migration has accompanied and made possible the spread of settlement and the colonization of new lands (which, of course, may have already been settled by others, as was North America before European colonization).
2. Migration has fostered and been fostered by industrialization and urbanization. Frontier development within the United States was promoted by migration, and American industrialization and urbanization spurred migration to America between 1870 and 1920.
3. Migration has supplied the labor to develop new resources. Chinese and European immigrants, for example, provided the manpower for the construction of the Union-Pacific Railroad in the 1860s.
4. Migration has contributed to the diffusion of technology, language, religion, and customs. The spread of Christianity through the Roman Empire in the first two centuries after Christ was aided by migration, and new kinds of food, tools, weapons, and customs were introduced into centers of civilization by barbarians before and after the fall of the Roman Empire.
5. On a more local, but massive, scale, migration has supplied the population for rapid urbanization and for suburbanization and the expansion of urban settlements.

Reasons for Migration

People usually migrate to escape unsatisfactory conditions at home and to find better opportunities in another location. The excerpt from *The Grapes of Wrath* cited at the beginning of this chapter is an example: impoverished families from drought-stricken Oklahoma hoped to create a new life for themselves in California. These *push and pull factors* are usually economic or social; people who move do so because they want to improve their level of living or provide their children with better social and economic opportunities.

The vast rural-urban migration of the past two centuries was largely motivated by real or imagined economic incentives. Economic opportunities began to shift from the countryside to the city as the urban and industrial revolutions gathered momentum. Farming areas became overpopulated, and rural poverty grew severe. A job in the city, no matter how menial the labor or how filthy and miserable the living conditions, was preferable to underemployment in the countryside. At least 400 million persons moved from rural to urban areas in this century alone. Today the shift is most prevalent in the developing third world countries.

But not all rural-urban migration has been a response to potential jobs. In both developed and developing countries, some families will move for the sake of better educational opportunities for their children in or near the larger cities or to escape the social restrictions of rural life. Nor does rural-urban migration necessarily meet expectations: the anticipated jobs may not exist. Lack of jobs and poor housing may even result in a decline in the standard of living. Excessive rural-urban migration can be a great burden to cities, especially in developing countries, which are unable to absorb and house so many unskilled persons.

Brazil provides a good example of the nature of rapid rural-urban migration. The urban population grew from 19 million in 1950 (36 percent) to 66 million in 1976 (60 percent), about half of which came from in-migration. All rural areas in the populous coastal provinces were sources of migration, but the greatest exodus came from the poverty-stricken northeast, where urban and industrial development was less advanced. The local cities of the northeast, the preferred destinations, could not absorb this flood, which overflowed into the great metropolises of São Paulo and Rio de Janeiro, to the *favelas* (squatter settlements) where friends and relatives had migrated earlier.

During the process of urbanization, most migration waves flow from rural areas and small towns to larger cities (*rural-urban* migration). But after a high degree of urbanization is reached, *urban-urban* movements between cities and *intraurban* movements within cities predominate. Such urban movements occur as residents seek better jobs, respond to employment fluctuations in industries, take advantage of personnel transfers at the management level within a firm, or satisfy a desire to be near greater cultural or environmental amenities. *Suburban migration* flows result from population growth and inadequate city housing, from city residents' increasing affluence and the desire for more space, and from their desire to escape unpleasant conditions in the city. Figure 11.5 shows the rural-urban and urban-urban flows in a sample study area in the U.S. South.

Economic opportunities have attracted migrants to rural areas as well as to urban centers. One of the greatest population movements in history was the agricultural settlement of Canada and the United States during the

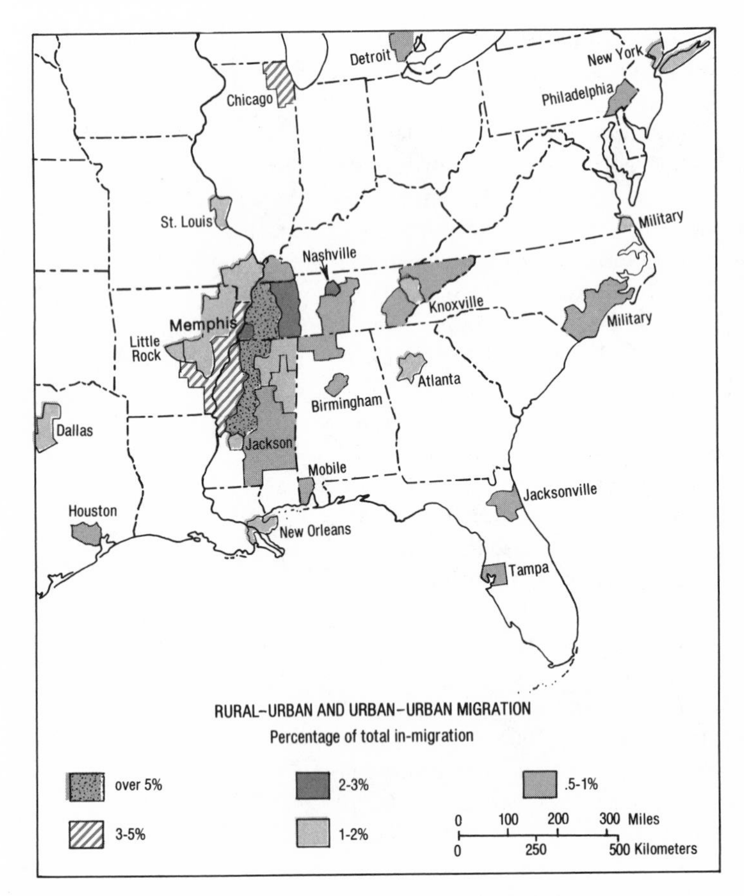

11.5. The migration field of Memphis. This migration pattern has a dual character. There is a continuing *rural-urban* flow within the Memphis region, indicated by the high proportion of in-migrants (over 3 percent) from surrounding nonmetropolitan areas. There is also a fairly substantial (0.5 to 3 percent) *urban-urban* flow into Memphis from the larger metropolitan areas of the South and Northeast. The latter flow partly represents military movements and the reverse migration of native Memphis people from cities like Chicago, New York, and Philadelphia; but it primarily reflects interaction between Memphis and other large cities.

eighteenth and nineteenth centuries. Some 30 million Europeans left their native villages and traveled thousands of miles overseas to a new and uncertain life in the open lands of North America. Given the natural conservatism of farmers, this movement attests to the sharp contrast between local nonagricultural opportunities in Europe and perceived agricultural opportunities in the New World. On a less dramatic scale, any opening or development of new farmland attracts many migrants. Recent examples include movements to the Columbia basin irrigation project in the United States, to interior areas in Brazil opened by new roads, and to irrigation developments in Egypt and the Sudan.

General international migration for political, economic, and social reasons remains significant, despite the reluctance of most countries to allow the emigration particularly of trained persons. The United States accepts 1 to 2 million immigrants each year from all parts of the world. Traditionally European immigrants were favored, but regional quotas have been replaced by preferences for skilled immigrants and for those with job expectations. This has led to the so-called brain drain, the loss of highly trained professionals from developing countries. Sometimes, as in the case of the Vietnamese, the United States is willing to accept political refugees. Still the largest migration flows into the United States are from Canada (balanced by a reverse flow) and Mexico (of which a fair proportion is illegal).

Both economic and social factors—poverty and racial discrimination—spurred the migration of blacks from the U.S. South to the North and West, particularly during World War I, when industrial jobs were first opened to them, and again in the decades following 1930, when mechanized agriculture in the South displaced sharecroppers and World War II created more industrial opportunities. Little migration occurred between 1880 and 1910, even though discrimination was at its most severe level, because the majority of blacks were legally tied to their home areas through indebtedness and, not legally, through intimidation.

Several million blacks moved directly from rural areas in the South, where many had been sharecroppers, to northern industrial centers. Lacking the skills needed for urban jobs, most found employment only at very low levels. Yet even the slum dwellers segregated in city ghettos preferred this life—which held the hope of a better education and better jobs for their children if not for them—to the desperate poverty and more pervasive discrimination they had experienced in the South. Improved economic and social conditions in the South, together with deteriorated conditions in large urban ghettos (including increased prejudice and segregation), have now apparently stemmed this flow.

Noneconomic motivations have become relatively more important in recent times. Scenic, recreational, and cultural attractions provide a significant pull for migrants today. Many are glad to escape the congestion, pollution, and visual blight of older metropolises for the slower pace, cleaner air and water, and friendlier neighborhoods of smaller cities and towns—even though they may earn lower monetary incomes in such places. Increasing numbers of people are moving for retirement purposes. Many people of retirement age now have higher incomes and greater freedom than in the past. In the United States, for example, many can afford to move to leisure communities in the sun belt, particularly in Florida, Arizona, and California, or to Mexico.

The Mechanism of Migration

Migration, like trade, depends on comparative advantage, complementarity, and transferability. Opportunities (comparative advantages) at the destination are expected to satisfy (complement) needs and desires unfulfilled at the origin. But if the economic or social costs of moving (the transferability factor) are greater than the potential increase in personal satisfaction or income, the migration probably will not take place. Transferability is often politically conditioned between or within countries, as in the Republic of South Africa, with passport requirements. At one extreme, people are expelled or displaced, particularly during and after wars; at the other, they are forbidden to emigrate (this is true of most centrally planned societies). All countries try to regulate immigration into their territory.

Human perception of opportunities is not strictly accurate; it is biased toward the earlier reputation of amenity areas and places of recent growth and the reports of friends and relatives. Some places—for example, industrial cities today—are perceived unfavorably, while others, such as the Mountain or Pacific West in the United States, are seen in a favorable light. Both images are exaggerated; nevertheless they still influence migration.

Distance decay may be observed in the flow of migration as well as in the flow of trade. The number of migrants declines with distance not so much because the cost of moving rises but because the level of information about opportunities falls. Moreover individuals usually select less profitable, closer opportunities over more distant ones for the sake of maintaining a stable cultural and physical environment for their families. For example, Appalachian migrants prefer nearby smaller cities like Roanoke, Virginia, or Asheville, North Carolina, to Detroit or Philadelphia, yet more actually move to the latter because of the greater number of job opportunities there.

One might expect the migration pattern of origins and destinations to be rather diffuse because of the subjective nature of migration decisions; but in fact migration is strongly concentrated along certain favored routes (figure 11.6). The persistence and selectivity of migration flows reflect the *feedback of information* that occurs when people migrate. Migrants inform friends and relatives back home about opportunities in the new location. Such feedback explains why people from particular communities, counties, and even countries consistently choose the same destination—why many Irish, for example, migrated to Boston during the past century and many Germans came to Milwaukee, or why a high proportion of migrants from a particular small town in Alabama have chosen a particular destination, among hundreds, in the North. In developing countries, where people clearly move to cities for expected economic gain, the choice of destination is mostly explained by the previous decisions of friends and relatives.

The Demographic Effects of Migration

Demography refers to the study of population—its size, characteristics, and distribution. The process of migration has important demographic implications. Most notably, it tends to redistribute population from economically weak, declining regions with aging populations to economically strong and growing ones with younger populations. Migration flows in most countries have tended to travel along the following channels:

1. From rural areas and small towns to large cities and metropolitan areas.
2. From economically stagnant regions to economically vital ones.
3. From regions where social or racial discrimination is greater to where it is less severe.
4. From environmentally harsh or culturally deprived regions to more attractive ones, as perceived at a given time. Air-conditioning, for example, has increased the attractiveness of much of the U.S. South and Southwest.

Flows along all four channels were prominent in the United States during the 1960s. Particularly in the South, Midwest, and Appalachia, substantial numbers of rural residents left their homes for urban and metropolitan centers in each region and beyond. Often this out-migration was selective, consisting of the younger and potentially more productive people. These metropolitan-directed flows reflected the migration from stagnant farming and mining areas to expanding opportunities in metropolitan centers. Flows of blacks from the South were relatively greater than those of whites, indicating the role of racial discrimination. By the late 1960s flows to environmentally more attractive areas such as Florida and the Rockies had become very prominent; large-scale out-migration had begun from the older, larger industrial metropolises, particularly in the North, now perceived to be poor environments in terms of quality of life. By the early 1970s the traditional metropolitan-directed flows had slowed as people and industries began to move to, or return to, small towns and rural areas in all parts of the country, but especially in areas with natural amenities.

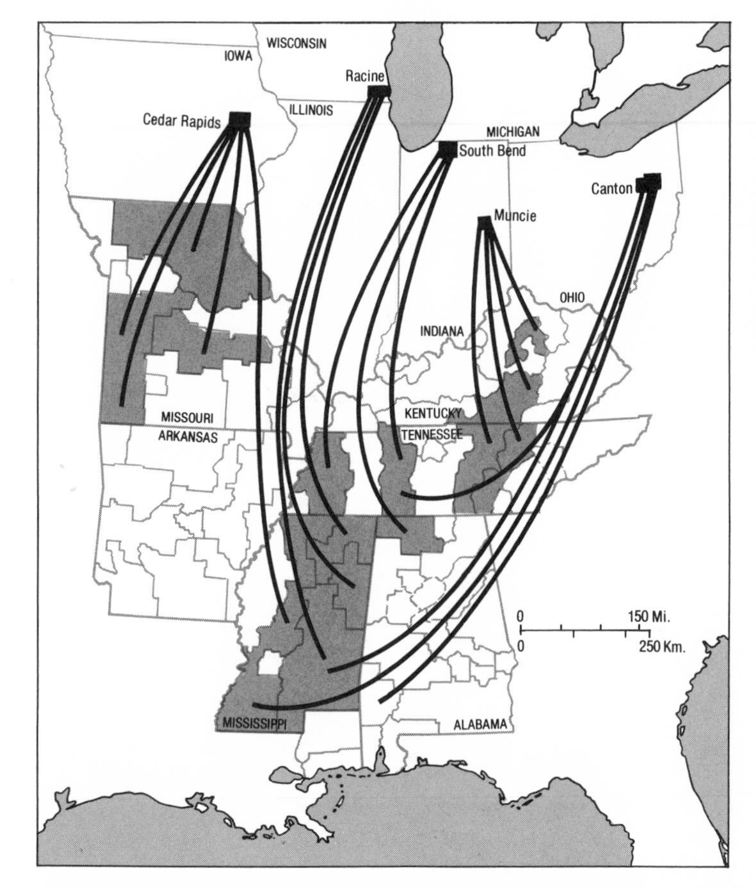

11.6. Persistence and selectivity in migration flows. This map shows major sources of migration flows to five cities in the Great Lakes region between 1955 and 1960. The pattern of origins is highly selective rather than diffuse or random, indicating the influence of friends and relatives who have moved north. Personal contacts play an important role in encouraging migration along traditional paths for long periods.

Source: Redrawn by permission from *Proceedings* of the Association of American Geographers, Vol. 3, 1971, C. Roseman.

Population change partly reflects rates of in- and out-migration and the resultant net migration to and from areas. But it is also determined by natural population increase or loss through the balance of births and deaths (figure 11.7). Population gains in the most rapidly growing areas in the 1960s—Florida, Nevada, and California—were due more to net in-migration than to natural increase. Gains in states growing just a little faster than the national average—among them, Texas, Georgia, Michigan, and Virginia—were due more to natural increase than to net in-migration. And gains in states growing less rapidly than the national average were due either to a natural increase that more than offset losses through net out-migration (as in twenty-two states) or to very low rates of natural increase or net out-migration (as in Wisconsin, Missouri, and Massachusetts). Some small areas, such as parts of Florida and the Ozarks, were growing rapidly through the in-migration of retired people, despite an excess of deaths over births, while a few other areas, such as Utah and parts of Idaho and Louisiana, were growing fairly rapidly, despite net out-migration, because of very high rates of

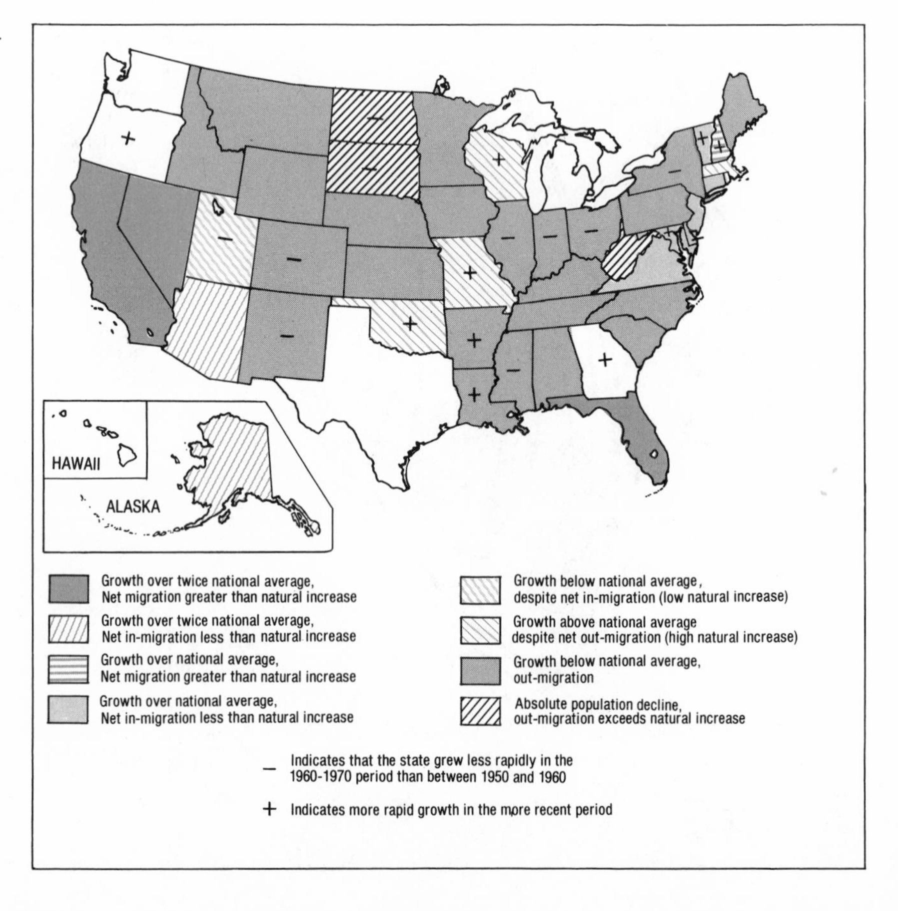

11.7. Components of population change: natural increase and migration, United States, 1960–1970. This map shows that there has been a significant redistribution of population even within a ten-year period and a marked variability in degree and rate of change within a single country. The fairly rapid growth of New England, the Ozark region, and Oregon and the relatively slow growth of the industrial lower Great Lakes region represent a major change from the previous decade.

natural increase. Finally a few areas—including West Virginia, North Dakota, and South Dakota—experienced population losses in the 1960s because net out-migration exceeded the natural increase.

The instability of migration patterns, and the difficulty of predicting them, is demonstrated by the great changes that have occurred in the pattern shown in figure 11.7 just since 1970. West Virginia, North Dakota, and South Dakota are growing again, while New York, Rhode Island, and Pennsylvania are now losing population. These changes indicate a general movement from the urban-industrial core to peripheral regions, fostered both by the decentralization of industry and by the lure of the environment.

Now consider the flow of migration and population change in a developing country like Brazil, which is in an aggressive stage of capitalist industrial growth. Brazil's development is characterized by rapid urbanization and centralization on the one hand, and by expansion of the rural settlement frontier, following new roads of penetration, on the other. Very high birthrates prevail in most areas, but economic opportunities are concentrated in urban centers, resulting in large-scale migration precisely along the channels noted earlier: up the urban ladder and from stagnant regions to areas with greater social and cultural opportunities.

Intraurban Migration

Although residential migration within the city (intraurban) is not migration in the sense that a person or family makes a total break with a place, including work as well as home, it is nonetheless a significant form of permanent movement. In America as much as 15 percent of the population moves in a given year. (The mobility rate is lower in Europe, perhaps 5 to 10 percent.) Sometimes a shift of work place triggers a residential move, as when a person is transferred to a different branch bank or when a factory is relocated to the suburbs, but more often the work place does not change.

Some of the reasons people move, often only a few blocks away, include changes in income, changes in family size, changes in local environment, and political and tax preferences. Improvement in income, in a society that venerates upward mobility, is often expressed through a move to a bigger or newer home in a better neighborhood. At the other end of the scale, many moves between apartments are made by those who cannot pay the month's rent. Marriage often entails a move for one or both partners; the appearance of children often leads to a shift from an apartment to a house, from renting to owning, and from city to suburb; and divorce, death of one partner, or old age may lead to abandonment of the too large home and yard for an apartment. Changes in the social environment—real or perceived—also cause many moves, including the flight of whites who fear racial integration and the flight of families who fear or dislike the entry of poorer families, of industry or business, or of too many other people. Finally some people move from one political jurisdiction (city school district) to another because of the level of taxes, the perceived quality of schools, or the perceived opportunity to wield more influence in the new area.

Some geographic consequences of intraurban migration have been the continuing physical spread of the built-up area as people demand new housing and more space, the filtering of older housing to somewhat poorer groups or to minorities, and a decline in the sense of community or neighborhood because of transience, although there is now increasing interest in the rehabilitation of older housing and in community organization.

Interaction Models

Just as trade models have been formulated to determine the most efficient, least costly paths for the flow of goods, *interaction models* have been devised to predict human movements such as recreation and shopping trips and migration and communication flows. The interaction models differ from the trade models in several important respects.

First, it is evident that economic factors are not necessarily the chief consideration in human interaction, and that the number of person movements is many times greater than the number of trade movements. Second, the personal satisfaction gained by taking a longer, more expensive route may outweigh the cost savings of the shorter routes. While business firms will avoid expensive routes and choose a limited number of least-cost routes to minimize costs, individuals will continue to use the more expensive, preferred routes and merely reduce the frequency of trips to cut costs. Thus for individuals the interaction approach shown in figure 11.8 is more realistic than the cost-minimization approach. And third, human perception of opportunities and the cost of distance is quite subjective. All the prospective migrants from a given town could, in theory, travel to the nearest city that offered attractive job opportunities. But in fact individuals interpret differently the opportunities that exist in other cities, as well as the cost of moving there. Thus some migrants will be willing to travel farther to take advantage of what they perceive to be a better opportunity. For these reasons, the trade models used to predict the movement of goods are inadequate for predicting the movement of people.

Most interaction models are modifications of the basic *gravity model* (named after Newton's Law of Gravitation), which states that there is a direct relationship between frequency of contact and size of the interacting masses, and an inverse relationship between frequency of contact and distance. Thus the more people there are in two towns, the more telephone calls and letters can be expected to flow between them, and the farther away the two towns, the fewer the telephone calls and letters. The gravity model describes fairly well a variety of movements, including migrations, vacation trips, shopping trips, trips to hospitals, mail and telephone flows, airline flows, and journeys to work. The gravity model is the most practical geographic one. It is commonly used in private and public planning—for example, in deciding upon highway improvements, in designing bus routing, and in locating hospitals, schools, or retail and service activities.

A variant of the gravity model, the *potential model,* is used to sum the expected level of interaction between one place and all others in a given region (figure 11.9). It measures potential interaction, or accessibility, and is widely used in studies of retail marketing, industrial location, trade, and communication. The pattern shown in figure 11.9 underscores the dominance of the Boston-Washington megalopolis and the Chicago–New York axis (compare with figures 10.10 and 10.11).

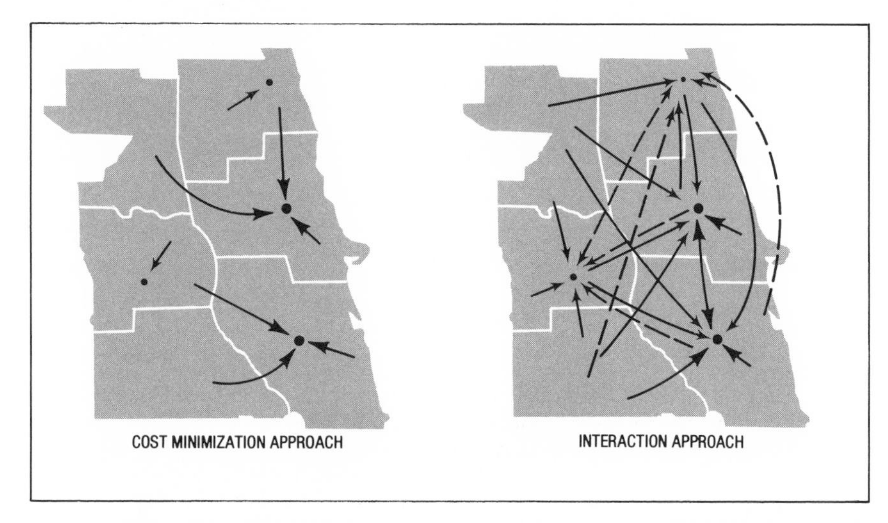

11.8. Two approaches to movement: cost minimization and interaction. This figure depicts the movement of people to hospitals in Chicago. Dots represent hospitals (larger dots are larger hospitals) and arrows symbolize flows of people. The cost (distance) minimization approach on the left, while useful for the movement of goods, is not so appropriate for the movement of people because not enough flows are permitted: patients must go to the closest available hospital. The interaction approach is more realistic: patients are as likely to go to larger, more distant hospitals for various reasons (such as prestige or religious affiliation) as to smaller, much closer ones.

Migration Models

The simple gravity model is mechanistic; it can describe a result—the pattern of flows—fairly well, but it does not indicate the processes involved. Thus geographers have formulated more complex models that can separate the decision to move from the decision on where to move. In the former, the individual or family evaluates both its present environment and some potential environments. This evaluation, of course, is colored by how much information the family has and by its perception of the various environments (figure 11.10). Even if comparison of present and potential environments or satisfactions suggests a move, a simple consideration of feasibility in terms of economic costs, social disruption, and the breaking of contracts may rule out the move. If the family does decide to move, then they evaluate a few of the best places more carefully, using additional information gained from visits, job interviews, and the like.

THE MOVEMENT OF IDEAS AND INNOVATIONS

The movement of ideas, innovations, rumors, and customs differs from the other movements we have studied in that they do not physically depart

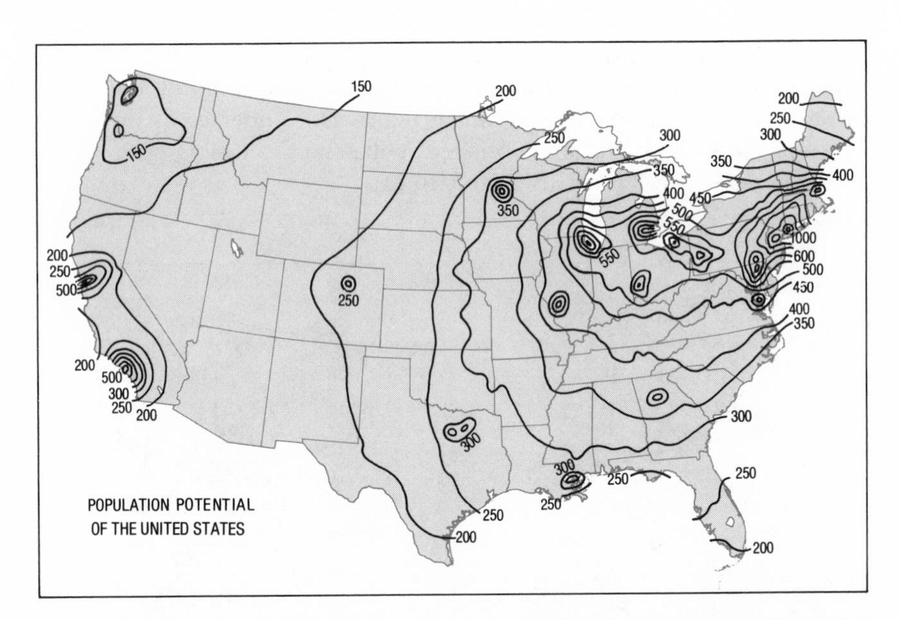

11.9. Potential model. This model measures accessibility of any given point to the total population. The total potential at any one point is found by summing the potential interaction (defined as the population of any other place or area divided by the distance to that point) between the given point and all other points. A variant of the gravity model, this model has proved valuable in studies of industrial location and airline and highway flows. According to the mathematical construct shown here, "300" can be interpreted as the line along which there is a potential volume of 300,000 interactions (contacts) per year between the population in a typical square mile and all the rest of the population. This population potential map of the United States, 1960, can also be regarded as a highly generalized or simplified description of the relative distribution of population.

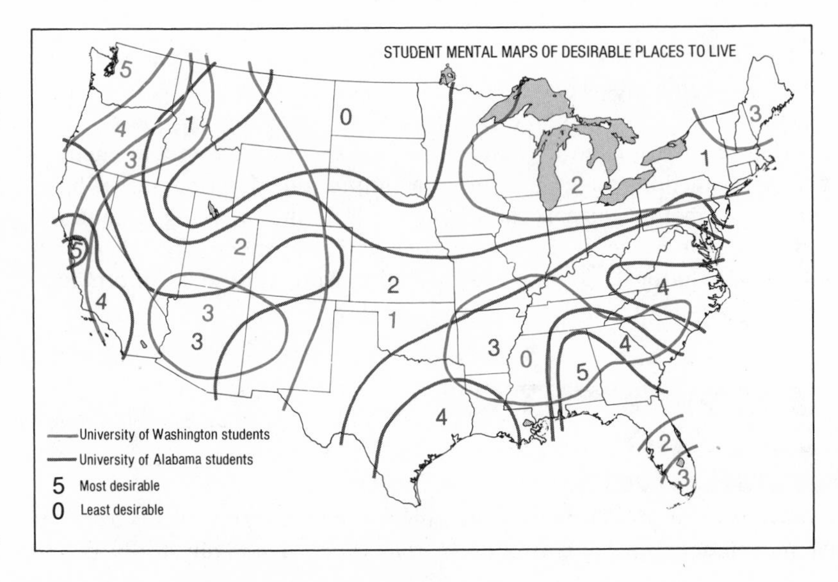

11.10. Mental maps: indexes of the desirability of places to live (five most desirable, five least desirable) from the perspective of Alabama and Washington State students. Note the maximum preference for the familiar home region, the tendency for northerners to discount the South and vice versa, and the generally favorable image of the Pacific Southwest. Such preferences have been shown to influence later migration of individuals and families.

Source: Data from University of Washington, Department of Geography, and Peter R. Gould.

from one place and arrive in another but instead spread from their place of origin. We touched on the process of *spatial diffusion* in chapter 3. Diffusion refers to the gradual spread of a phenomenon through a population and over space from limited origins.

Not only is the distribution of humankind a product of diffusion over hundreds of thousands of years, but the pattern of languages, the spread of agricultural techniques, the use of particular plants and animals, the technology of house construction, pottery making, weaving—in short, the whole array of techniques and customs that define human culture—are probably more a result of diffusion than of independent origin (although this is a matter of controversy). Just in recent centuries the physical and cultural landscape has been transformed through the diffusion of Western civilization via European colonization and the spread of European technology (including industrialization), religion, languages, ways of dress, and modes of behavior. Shorter-lived diffusions include the spread of influenza and other epidemics and the spread of particular styles of dress or forms of music. On a less global scale, major diffusions include the physical expansion of cities or of areas occupied by minority groups in a city, the spread of new store outlets or restaurants, the pattern of acceptance of new products, and the like.

How Diffusion Works

Some ideas and innovations spread directly, or contagiously, from one person or place to the next. Gossip and rumors are a prime example of *contagious diffusion*. Some innovations, such as rail transportation or hybrid corn, which foster economic development and human settlement, spread contagiously from the place or places of origin to neighboring areas (figures 11.11 and 11.12). Other examples are the spread of logging operations from more to less accessible areas (figure 5.18), the extension of urban land use into the rural fringe (figure 9.1) and the block-by-block expansion of the ghetto (figure 9.6).

Ideas and innovations that may be adopted by the entire population, such as conservation movements and color television, tend to spread not contagiously but in steps down the urban hierarchy, starting with the largest metropolises, then diffusing down to their satellite cities, and gradually reaching surrounding small towns and rural areas. Figure 11.13 shows the *hierarchical diffusion* of the Sierra Club, an environmental organization. Note that while the formation of new chapters rather clearly moved down the urban hierarchy, it also diffused contagiously on a local scale within regions, particularly in California, where the organization was founded. Another recent example is the adoption of container-ship facilities at major ports. Because of high capital cost and uncertainty, this innovation began at the large ports of New York, London, Rotterdam, and Tokyo twenty years ago and is gradually filtering down to smaller ports.

Some innovations, particularly products or services new to the market, may originate in smaller cities and peripheral regions and thus move up the

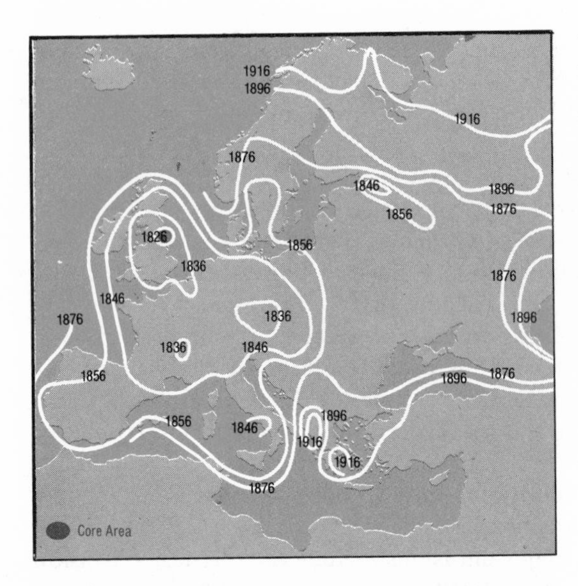

11.11. Diffusion of the railway in Europe, 1826–1916. Note the general contagious diffusion of the railway from the England-Belgium core area and the more limited diffusion from metropolitan capitals in other countries, such as Rome, Prague, and St. Petersburg.

Source: Data used with permission from S. Godlund, "Ein Innovations verlauf in Europa," *Lund Studies in Geography,* Series B, 6, University of Lund, Sweden, Department of Geography, 1952.

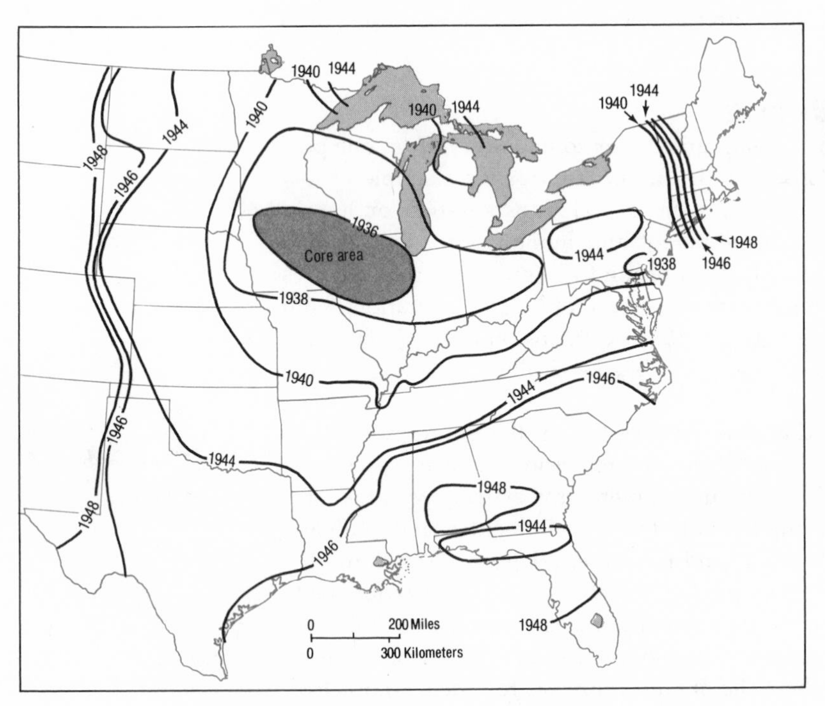

11.12. Diffusion of hybrid corn. This map shows how the use of hybrid corn spread from a core area in Iowa and Illinois from before 1936 until environmental barriers were met in 1948. The adoption of hybrid corn was largely a contagious diffusion.

Source: "Hybrid Corn and the Economics of Innovation," Griliches, Z., *Science,* Vol. 132, pp. 275–280, Fig. 29, July 1960. Copyright 1960 by the American Association for the Advancement of Science.

urban hierarchy because the firms want to develop their competitive position in a small area before branching out. Their subsequent diffusion is both contagious within regions and hierarchical throughout the nation. A chain of motel or restaurant franchises, for example, may begin in a small

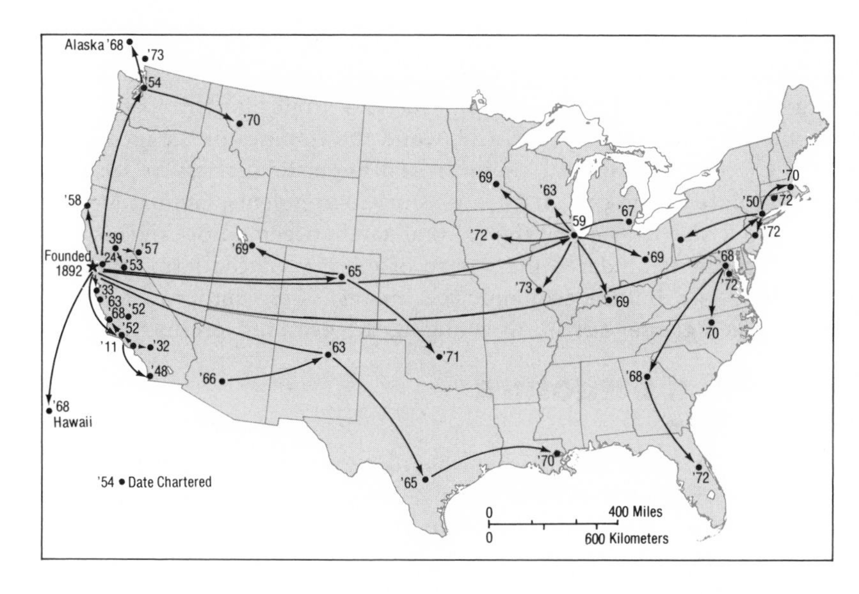

11.13. Diffusion of the Sierra Club. The spread of Sierra Club chapters is an example of hierarchical diffusion. After a long period of more local spread in California, regional offices were founded in New York and Chicago and then spread to lesser centers in those regions.

Source: Data used with permission from John F. Kolars and John D. Nystuen, *Geography: The Study of Location, Culture, and Environment* (New York: McGraw-Hill, 1974).

city and then spread both contagiously and hierarchically, with branches appearing early in larger, more distant cities. Indeed many new products are first tested in a regional market.

The Role of Interpersonal Relations

Ideas and innovations are often spread by transients and travelers—by college students, for example, or by out-of-area visitors. They may also be spread by newspapers, magazines, books, radio, and television. But the most effective vehicle for diffusion is *interpersonal relations.* One person tells a few friends about the advantages of an innovation such as citizens band radio. These people in turn tell their friends, and word gets around. Soon every other car sprouts a CB antenna, and drivers everywhere are conversing with each other on the road. To take another example, an important innovation like birth control illustrates the dual role of the media and professionals in spreading information and of the exchange of information between people prior to actual adoption. Even the adoption of highly technical and expensive changes, such as port facilities to handle containers, appears to have been influenced by interpersonal contacts—certain port commissioners know and visit certain other commissioners and exchange information. Fairs and exhibitions remain important sources of ideas and propagators of diffusion.

Diffusion through interpersonal relations is subject to at least four constraints. First, people vary in their ability to accept change. Persons with higher incomes and better education are usually better able to accept new ideas and thus may be more willing to try innovations than those less well off. Second, once having accepted the innovation, people vary in their willingness to tell others about it. A farmer who has found a better way to raise chickens may be reluctant to share the news with competitors. Third, an individual's network of contacts is usually quite limited spatially. It takes time for a new idea to spread beyond the confines of the local area. And fourth, physical, cultural, political, and linguistic barriers impede the progress of diffusion in certain directions or environments. Gunnar Myrdal points out in *Asian Drama* that the cultural gap between European colonists and Asian peasants hindered the spread of more advanced farming methods in South Asia. But the economic self-interest of the European colonists probably was a greater barrier to changing the peasants' productivity.

The Spatial Structure of Diffusion

The diffusion of innovations is somewhat like a wave rising from a point of origin and moving outward. The form of the wave reflects the pattern of adoption as it changes with distance and time from the origin of the innovation (figure 11.14). The crest of the wave is the area of most active adoption. It gradually moves outward from the origin as the population near that point becomes saturated and as new adopters of the innovation tell others beyond the local area. The final level of adoption typically declines slowly with distance. Over time, the cumulative adoption of the innovation at a given point in space will follow an S-shaped curve. The rate of adoption will be very low when the innovation is still far away and when few have adopted it; the rate will peak when the innovation reaches a given point and more adopters are telling others about the innovation; and the rate will then fall off because most people receptive to the innovation will already have adopted it. Perhaps the easiest example to grasp is the spread of a metropolis. Rural residents can literally see the crest of most active subdivision construction approaching over the years, preceded by a few early individual homes or small subdivisions; and years after the active building is over, they can see the filling in of remaining lots.

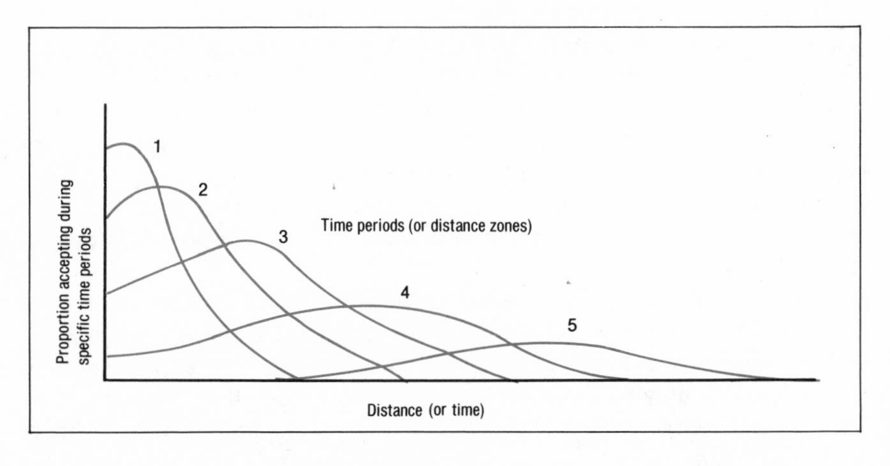

11.14. The pattern of innovation diffusion. The proportion of people accepting an innovation forms a wavelike pattern as it changes with distance and time from the place of origin. At any given distance, note how acceptance begins slowly, then speeds up, and later slows again. At greater distances acceptance begins later and never becomes as complete as in places closer to the origin.

The actual movement of the innovation wave is generated by those who first adopt the innovation and tell friends, neighbors, and relatives about it. Thus people in this **contact field** are the ones most likely to be told about the innovation and to adopt it. These people then tell others—some within and some beyond the range of the original contact field. (While it is the builder who physically extends the city, it is personal exploration and contacts with friends and relatives that fill those houses and provide the impetus for new construction.) The rate of diffusion depends on the adopters' persistence in spreading the news, on the degree of resistance to change, on the range of the contact field, and on the utility and value of the innovation.

CONCLUSION

The desire for stability—for identification with place—impels society to seek a unique, protected, and self-sufficient relationship to its local environment. This motivation has been a powerful force in human history, leading to the settlement and development of regions. Standing against this conservative goal is the equally compelling motivation of mobility, which drives society to explore, trade, interact, specialize, and ultimately to develop new patterns of location. In the next chapter we will show how these two geographic forces, existing always in creative tension, are integrated in the human landscape.

SUGGESTED READINGS

Beaujeu-Garnier, J. *Geography of Population.* London: Longmans, 1966.

Brown, Lawrence. *Diffusion Processes and Location.* Philadelphia: Regional Science Research Institute, 1968.

Clarke, J. I. *Population Geography.* Elmsford, N.Y.: Pergamon, 1972.

Cosgrove, I., and Jackson, R. *Geography of Recreation and Leisure.* London: Hutchinson University Library, 1972.

Gould, Peter. *Spatial Diffusion.* Resource Paper 4. Washington, D.C.: Association of American Geographers, 1969.

Hägerstrand, Torsten. *Innovation Diffusion as a Spatial Process.* Chicago: University of Chicago Press, 1967.

Hudson, John. *Geographic Diffusion Theory.* Studies in Geography 19. Evanston, Ill.: Northwestern University Press, 1972.

Morrill, R. L. *Migration and the Spread and Growth of Urban Settlement.* Lund Studies in Geography, Series B. Lund University, Sweden.

Moryadas, S., and Lowe, J. *The Geography of Movement.* Boston: Houghton Mifflin, 1975.

Olsson, G. *Distance and Human Interaction: A Review and Bibliography.* Philadelphia: Regional Science Research Institute, 1965.

Pred, Allan. *Systems of Cities and Information Flows.* 1973. Lund Studies in Geography, Series B, No. 38. Lund University, Sweden.

Rossi, Peter. *Why Families Move.* Glencoe, Ill.: Free Press, 1955.

Wilson, A. G. *Urban and Regional Models in Geography and Planning.* New York: Wiley, 1974.

Zelinsky, W. *A Bibliography of Population Geography.* Department of Geography Research Paper, No. 80, 1962.

———. *A Prologue to Population Geography.* Englewood Cliffs, N.J.: Prentice-Hall, 1966.

12

Spatial Organization

12 Spatial Organization

A major objective of this book has been to reveal the spatial order of society through location theory. Although our discussion has by no means been limited to location theory, we have devoted much attention to theoretical patterns of location and interaction. Before going on to the actual complexities of the human landscape, it would be useful to pause now and summarize the theories explaining society's organization of space. In this chapter we will review the basic patterns of location and interaction described and analyzed in previous chapters. We will also elaborate on the regional structure of the contemporary U.S. and world landscape. In the next chapter, we will discuss the discrepancy between the theoretical and actual human landscape.

SPATIAL STRUCTURE

A key to understanding spatial structure is the specialization of people and activities. A city or region establishes its identity in part by specializing in one or more activities and forming a set of spatial relationships with other places. Washington, D.C., for example, specializes in national government administration. The national capital has both formal linkages with its own regional branches, down to the local post office, and with state governments, and informal linkages through the mass media and the lobbying of special interest groups.

The structure of the human landscape is produced by a society's attempts to realize two goals: maximizing the utility of places at least cost or effort, and maximizing interaction at least cost or effort. Of course, other goals and values also affect the appearance of the landscape, but location theory is concerned primarily with the spatial order that results from the pursuit of these two important goals. To the extent that these goals are successfully followed, the resulting patterns can be considered spatially efficient.

People who decide on the location of economic, political, or social activities—farms, factories, stores, churches, schools, city halls, fire stations, homes, and so on—follow the same general goals of maximizing utility and interaction at least cost or effort. In selecting a location for a new factory or a new home, for example, one usually considers the cost of distance (transport costs). Often a substitution, or trade-off, may be made between transport costs and other costs.

A substitution may be made between land costs *and* transport costs *when seeking access to a central point, such as a market or work place.* More expensive sites closer to the center may be selected and then used for more intensive activities, or cheaper, more distant sites may be chosen and used for less intensive activities, resulting in the same total costs. For example, farmers use the more expensive land in New Jersey, near the largest concentrated market in the United States, the Northeast, for the intensive production of higher value, less transferable crops such as fruits and vegetables; and they use the cheaper land in Montana and Nebraska, far from that market, for the extensive production of lower value, more transferable crops such as wheat and livestock. As another example, some families may prefer more space and accept the higher transport costs incurred when living in the far suburbs, while others of the same income level may prefer centrally located, higher rent apartments, from which transport time and costs are lower.

A substitution may be made between production costs *at the selected site and* transport costs *when determining the optimal market size and scale of operations.* The benefits of a larger scale of production must be weighed against higher transport costs for procuring raw materials and distributing finished products. For example, American shoe manufacturers may decide to trade off higher transport costs for lower production costs by locating their plants in Italy, where labor is cheaper. Or a community may decide to build a single large library rather than several smaller ones. The costs savings and quality of the larger library may offset the poorer accessibility for much of the population.

A substitution may be made between possible production-cost savings from agglomeration *and* transport costs. Agglomeration economies in a complex of activities must be weighed against the risk of lacking even a partial monopoly over supplies or markets. By locating in a large city rather than near local paper suppliers, for example, book printers substitute higher transport costs for the agglomeration benefits of proximity to skilled labor and metropolitan services.

A substitution may be made between self-sufficiency *and* trade. Importing higher quality resources or goods from another region or nation may involve greater transport costs and incurs the risk of political or military interference. These factors must be weighed against the alternative of using lower quality local resources or producing goods locally at higher cost. A petroleum-importing nation must weigh the risk of another embargo from OPEC nations against the higher cost of developing alternative energy resources in its own territory. Similarly a very small state or province must balance the desirability of having its own medical school and teaching hospital against the very high costs of such facilities. Farmers, retailers, manufacturers, public officials, and other decision makers weigh these alternatives when making location decisions.

In trying to maximize the utility of places and interaction at least cost, activities compete for the limited space around a central point. On a homogeneous surface, such competition theoretically produces regular patterns of location and interaction, including spatial gradients of land use and a spatial hierarchy of regions. The surface of the earth, however, is not homogeneous but is highly differentiated. Variations in physical and cultural conditions over space result in less regular but still predictable patterns of location and interaction. If location decisions were based only on these economic goals, if technological, social, and economic conditions did not change with time, and if development did not diffuse gradually over space, then the spatial structure of the real world might conform more faithfully to the simple theoretical models we have introduced. But the modifying influences of environment, noneconomic values, cultural and technological change, and spatial diffusion result in a departure from theoretically ideal spatial structure.

Spatial Gradients

One of the fundamental patterns revealed by location theory is the spatial gradient, the increase in population density, land value, and intensity of land use as one approaches a central point, such as a large agricultural market. According to location theory, spatial gradients result from the competition of activities for land near the market. Theoretically the pattern of land use should maximize the value of land and minimize transport costs, while satisfying the demands of society for goods.

The theory holds that activities fall into a gradient pattern because of differences in inherent productivity and transferability (see figure 5.3). People engaged in activities with very high transport costs, such as the production of perishable vegetables and fluid milk, have limited ranges where their activities can survive. They must therefore use land intensively near the market. Conversely those engaged in activities entailing lower transport costs, such as wheat or cattle raising, are displaced to locations farther from market, substituting more land for greater intensity. Thus a wheat farmer will use more land but less fertilizer, irrigation, and other inputs than will the vegetable grower close to market.

The competitive bidding of potential producers for the more accessible land raises its value (purchase price or rent). This more expensive land near the market can be afforded only by producers who use land more intensively; cheaper land at the periphery will be left to those who use land less intensively but profitably. The gradient is continuous from the farthest edge of the agricultural hinterland to the most intensive urban core.

We cannot expect to find perfectly regular gradient patterns in the real world. But gradient tendencies do in fact exist, as is revealed, for example, by the value of crops per acre in the United States (figure 12.1). Values are highest in the areas in and around major metropolitan areas, but there is also a wide gradient tendency with respect to the major core areas of "Megalopolis" (Boston-Washington), Chicago, and Los Angeles.

Spatial gradients occur in all societies, but the extent of the gradient, or territory, related to a given market tends to be much greater in more developed countries than in less developed ones. For example, the areas serving London, New York, and Tokyo are larger than those serving Bangkok, Tombouctou, and New Delhi. The reason is partly because much more of

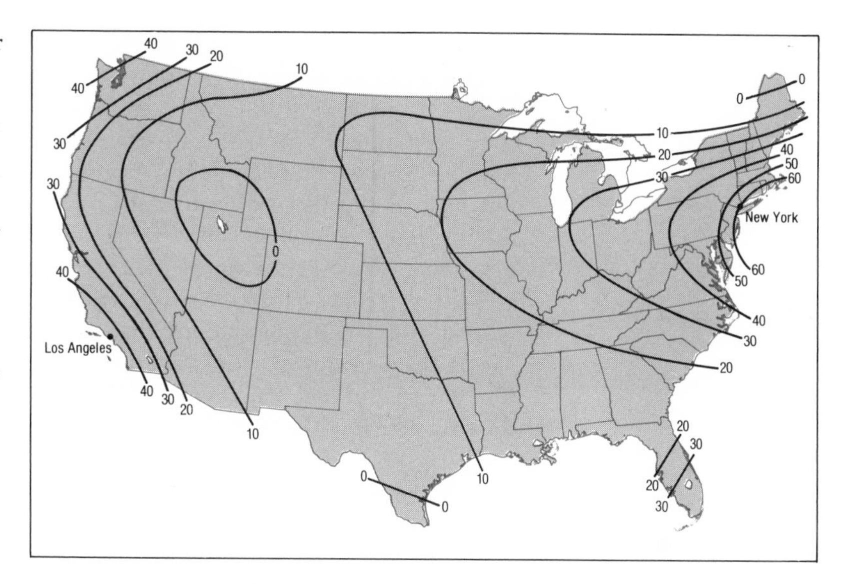

12.1. Net farm income per acre for the United States. This generalized trend surface (contour interval is ten dollars) demonstrates that the intensity of our agricultural system is oriented to the northeastern megalopolis, even though the greatest total output occurs in the Midwest.

Source: Data used with permission from P. O. Miller, "Trend Surface of American Agricultural Patterns," *Economic Geography*, Vol. 49, 1973, fig. 8.

the food in advanced countries is marketed and processed before reaching ultimate consumers.

Spatial Hierarchy

Location theory has enabled us to perceive not only the gradient structure of the landscape but also its hierarchical structure. The central points around which gradients form differ in variety of activities, population size, number, and spacing. Such variation constitutes a hierarchy of places, from hamlets to metropolises, and a spatial hierarchy of market areas, from local to regional and national ones. Each center in the hierarchy attracts producers and serves customers within its market area. Each one also maintains a comfortable distance between itself and competing centers of similar rank. Thus space is efficiently divided into regional markets.

Many activities, including administration, trade, and other service functions, require only a small area in which to operate. The mutual attraction of such activities is basic to the market concept. Activities with similar thresholds (sales volume needed to survive) and ranges (distances customers are willing to travel) have the same spatial needs and therefore agglomerate. The agglomeration of women's and children's clothing stores, bookstores, toy stores, and hairdressing salons in a shopping center is an example, as is the gathering together of farmers or peddlers in a periodic market in an Asian or African village. These arrangements are also efficient for customers, who can save time and effort by satisfying more than one need on a single trip.

While different activities with similar thresholds and ranges are mutually attractive and tend to agglomerate, many identical activities are mutually repellent and seek separation. As we noted in the example of the two ice cream vendors (p. 253), each seller tries to obtain a spatial monopoly—an area in which a seller has a competitive advantage by having all the customers. Competition thus induces sellers to locate just far enough away from each other to provide a profitable spatial monopoly but close enough together to prevent a competitor from springing up between them. Ideally this process results in a triangular arrangement of sellers, each one equidistant from the others. These triangular patterns, when combined, form a hexagonal network (see figure 7.14).

A hierarchical structure arises when certain activities require a larger market (threshold population) than in the basic system just described. Movie theaters, for example, need a larger market than do grocery stores. One rather optimal arrangement is for alternate central places to acquire the new activity. These alternate centers then become second-level centers. Similarly alternate second-level centers will acquire activities that call for an even larger market and thus become third-level centers. In this way, a territory develops a system of central places with as many hierarchical levels as the economy will support. The hierarchy represents an equilibrium between the desire to locate as many activities as possible in the fewest places for the sake of agglomeration economies, and the need to minimize the total, aggregate distance traveled by customers.

In the more developed countries, this hierarchy tends to be closely spaced and to consist of many levels, from hamlets to villages, towns, small cities, large cities, small metropolises, and perhaps large metropolises. In less developed countries, central places may be fewer, more widely spaced, and more limited in levels, usually consisting of market villages, market towns, and a few cities.

The Gradient-Hierarchical Structure of the Landscape

The theoretical spatial structure of the landscape comprises the two elements described above, gradients and hierarchies (see figure 7.12). Competition for access to a central point gives rise to the gradient structure of land use and leads to some dispersal of population and activities. If there is enough space around the main center for smaller competing centers to emerge, a central place hierarchy will develop. The larger the territory of an economy, the more elaborate the hierarchical structure, with less likelihood of the territory's being dominated by a single primary center. Modifying the overall gradient around the main center, or market, are local gradients around regional markets. The population potential map of the United States shown in figure 11.9 suggests this joint gradient-hierarchical pattern, with central places dispersed over the entire territory, an overall gradient of intensity surging outward from the Northeast (the major market), and smaller gradients radiating around regional centers.

The internal structure of the city also forms a joint gradient-hierarchical pattern. When a city is young, its center contains all distribution facilities and constitutes the focal point of the intensity gradient (see chapter

9). As the city ages and increases in area, the threshold for the emergence of smaller centers is reached. The original gradient thus becomes punctuated by areas of greater intensity around district shopping centers. When the city becomes a metropolis, it exhibits a complex central place hierarchy consisting of hundreds of nodes at many levels with overlapping gradients.

Environmental and Cultural Variation

The spatial structure projected by location theory assumes that the earth's surface is undifferentiated. In fact, however, the face of the earth is extremely varied in its environmental productivity, perceived amenities, and, most important, its distribution of human and physical resources. The uneven quality of the landscape modifies the theoretical gradient-hierarchical structure by causing irregularities in the intensity of land use and in the concentration of activities.

Both the agricultural gradient and the spacing of central places are affected by variations in land productivity over large areas. In the Soviet Union the intensity of agricultural production and the density of settlement fall much more rapidly in the colder and harsher areas north of Moscow. Somewhat predictable local irregularities are created by more localized differences in land productivity. A pocket of inferior land close to market may be used less intensively than one might expect from its favorable location, or the size of local farms may be larger than suggested by theory. If differences in land productivity are pronounced and transportation is relatively cheap, the agricultural gradient may be significantly modified. Lower transport costs reduce the friction of distance and permit farmers to take advantage of better environmental conditions that occur at greater distances from the market.

The location of industrial activities is strongly influenced by variations in the character of the landscape over space—for example, by differences in resource endowment, labor supply, or business services. Many firms procure raw materials from a limited number of specific sources and distribute semifinished and finished products to other firms in a few specific locations rather than to widespread final consumer markets. Such firms tend to concentrate at large markets or at material sources. Many manufacturing firms cluster at locations that have proved profitable, hoping to share most, if not all, of the regional market, rather than seek locations away from other firms in order to acquire a spatial monopoly. Manufacturers of printing machinery locate as a group in New York, and manufacturers of certain kinds of mining equipment are concentrated in Minneapolis. Variations in labor supply and cost induce certain more labor-intensive sectors, such as textiles, to cluster in regions of lower cost labor, as in the U.S. South.

The location of market-oriented industries with high transport costs for finished products and with diffuse markets, such as soft-drink bottlers, is similar to the location of central place activities. But if the cost of transporting raw materials is a significant variable, the industry will seek a spatial monopoly over a local supply area. Thus pulp mills tend to locate near lumber supplies, and meatpackers near livestock sources. Industries with high transport costs for both raw materials and finished products, such as steel, have three choices: they may locate at the material source, at the market, or at major transshipment points in between. The less important the transport costs and the more important other cost differences between sites, the more the industrial pattern will be clustered, with producers concentrated in several least-cost locations. The machine tool industry, for example, relies on highly skilled labor and thus is concentrated in a few major centers, such as Cincinnati, Ohio, and Providence, Rhode Island. Finally industries that share national or large regional markets and whose transport costs are relatively insignificant locate near their largest single market, usually a metropolitan area. The concentration of the photographic products and the optical goods industries in Rochester, New York, is an example. The net effect of variations in the landscape on industrial location is, first, to create clusters of urban settlements near large concentrations of resources or other favorable conditions, and, second, to concentrate population and production in fewer, larger places (figure 12.2).

Within a nation, cultural variation stemming from racial, religious, linguistic, or ethnic differences may also modify the ideal landscape through, for example, different crop preferences and fairly distinct subnational agricultural markets and central place hierarchies. As you recall from chapter 9, the additional goal of maintaining and enhancing social identity and status has exerted a major influence on the spatial organization of the city, as in zones of different class or race. This goal also affects the organization of the countryside, as demonstrated by the linguistic or religious divisions existing in many countries.

Uncertainty and Noneconomic Values

While varying conditions over space produce irregularities in the structure of the human landscape, uncertainty and noneconomic values also distort the ideal patterns suggested by location theory. Uncertainty refers to the fact that the decision maker is not likely to know what the optimal choice might be, especially over a period of time. An individual might decide to open a clothing store in a suburban location without knowing that an established competitor had the same intention. With better information, a more profitable site could have been chosen. But even with adequate information, error may be introduced when the data are interpreted. The store owner may know that a sufficient market exists but may select merchandise that does not meet the particular tastes of the local population.

People make decisions on the basis of many values other than economic optimality—for example, on the basis of acceptable or satisfactory economic performance, environmental preferences, maintenance of tradition, or preservation of cultural values and social status. Many are satisfied with locations that ensure only a moderate, or satisfactory, level of profits.

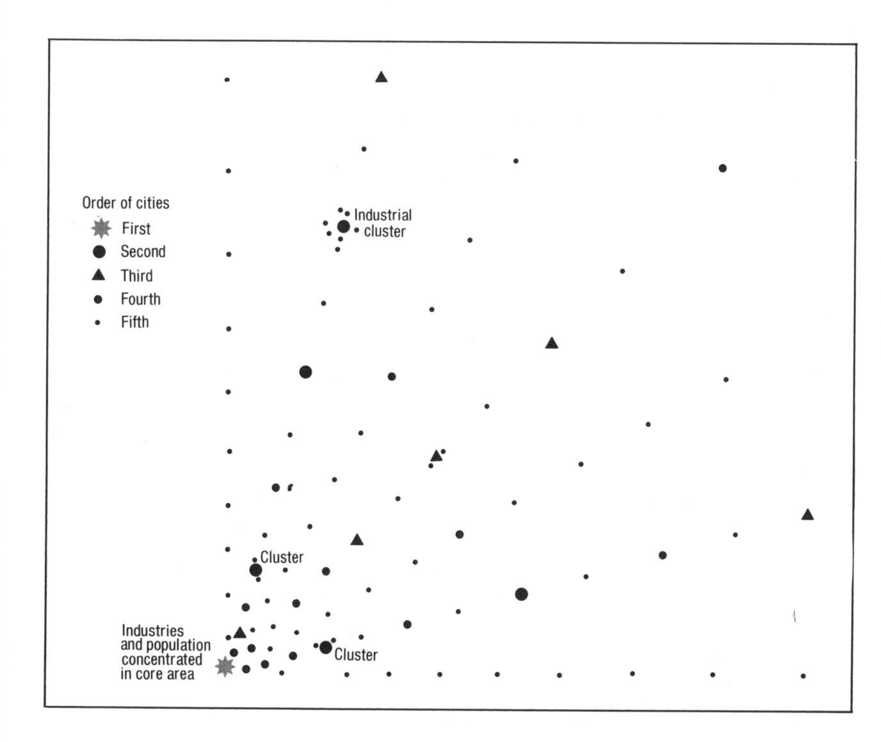

12.2. Impact of variations in spatial quality (such as resource location) on the gradient-hierarchical structure. Clusters of industrial centers emerge near large concentrations of resources or other favorable conditions, modifying the theoretical gradient-hierarchy landscape. (Only urban places are shown here.)

They may not wish to risk their investment in a potentially optimal but unfamiliar site. Such behavior is fairly common across the whole range of human activity, from the location of houses and churches to the choice of crops and the location of factories and office buildings. Many people prefer small town or rural life, even if they could do better in the city. Some like to change environments frequently, even if their income suffers. Others may want to maintain cultural separation from a dominant group, even if integration would lead to economic advancement.

Decisions based on noneconomic values can result in economic failure. But more commonly, profits and income are not maximized and businesses operate below capacity. Geographically the pursuit of noneconomic goals blurs the precise outlines of the gradient-hierarchical structure. The agricultural gradient becomes distorted by zones of interpenetration among crops of different intensity, and the central place hierarchy becomes weakened by zones of indifference in the range of customer shopping trips (see figure 9.8). The spatial effect of noneconomic values will be discussed further in chapter 13.

Cultural and Technological Change and Spatial Diffusion

Population, activities, and ideas spread outward from a small number of places endowed with superior power, access to markets and capital, and ability to organize space effectively. Industrialization and urbanization in the United States, for instance, gradually diffused westward and southward from early centers of economic power, such as Boston, New York, and Philadelphia. Such spatial diffusion creates a characteristic gradient in which the density of population and the fineness of the transport network decline outward from the hearth areas of settlement (the Atlantic seaboard in the United States) to the periphery (see figure 10.10).

Social, economic, and technological conditions that affect location change with time. The shift in land transportation from trains to trucks and automobiles caused the decline of many older places and forced firms and individuals to make spatial adjustments in order to survive. A change to a socialist economic structure may result in drastic changes, including the disappearance of many private firms and establishments and their replacement by fewer, larger ones. (Such changes and their effects will also be discussed in the next chapter.)

The classical theories of agricultural, central place, and industrial location were developed, for simplicity, on the basis of static and uniform conditions. An attempt to combine these theories was made in figure 12.2. The introduction of uneven landscape quality, uncertainty and noneconomic values, diffusion of development, and especially cultural and technological change should tell us that a search for ideal landscapes in the real world is likely to be futile. Nevertheless the theories help us find order in a seemingly random and chaotic landscape. The patterns, though imperfect, are evidence that principles of spatial efficiency operate in the making of many location decisions in the real world.

SPATIAL INTERACTION AND SPATIAL STRUCTURE

Spatial structure provides a framework for the dynamic element of geography—the movement of goods, people, and ideas among different places. The pattern of movements directly reflects the observed pattern of specialization. Without movement, the contemporary gradient-hierarchical structure of the landscape could not exist (figure 12.3). There would be only individual man-land relationships, as in a primitive society.

In a sense, location theory is a theory of movement or interaction. Minimizing the friction of distance is an essential part of the theory. The agricultural gradient, for instance, results from the desire to minimize the cost of transporting crops to market, and the urban gradient is tied to the cost of the journey to work and of shopping trips. The size of markets and cities is governed by transport costs as well as by the value of land.

The friction of distance and the role of transport costs are central to location theory. What, then, is the effect of modern trends toward better and cheaper transportation on the landscape and on the theory? Advances in

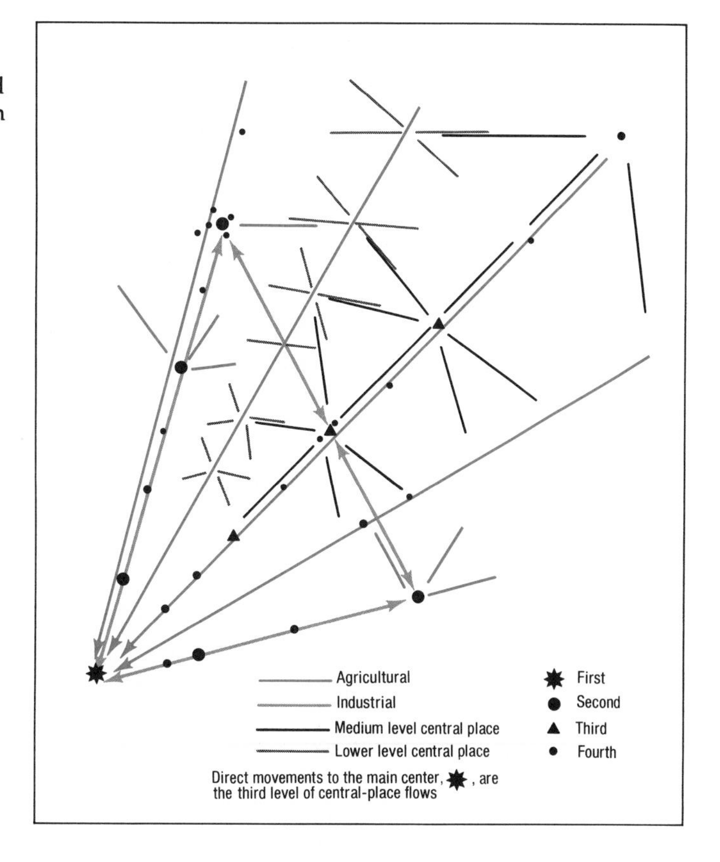

12.3. Partial movement pattern on a gradient-hierarchy-cluster surface. Movements represented by the dashed and solid lines serve the spatial pattern shown in figure 12.2. Agricultural and industrial flows tend to be longer and less regularly patterned than flows to central places.

technology have improved the quality and, at least until recently, have reduced the cost of transportation and communication. As a result, people are taking greater advantage of environmentally superior locations, and the landscape has become more highly differentiated. The influence of transport costs on location has weakened. Agricultural activities tend to be governed more by variations in soil, climate, and terrain and by the differential response of crops to inputs than by distance from market. Many services and industries are becoming more responsive to economies of scale and agglomeration. They are concentrating in metropolitan areas that offer greater market and labor stability and provide a higher level of amenities for employees. And families with the opportunity to buy tend to select homes on the basis of such criteria as view, open space, recreational resources, quality of services, security, and prestige rather than proximity to work.

One can speak of the new freedom of location, underplay the role of accessibility and centrality, question the viability of cities, the central business district, and the central place hierarchy—and indeed cast doubt on the relevance of location theory itself. The fact is, however, that distance remains a critical factor in location. Our ability to travel greater distances at lower cost has extended the range of our activities and relationships, but the desire to minimize the friction of distance still influences our location decisions: witness the continued concentration of activities near, if no longer in, the few locations that are most accessible—for example, the metropolitan areas of New York City, Chicago, and Los Angeles.

To be sure, the shift from public transportation to the private automobile has had a profound impact on the structure of cities (see figure 9.12). The entire urban area may now fall within a person's shopping or commuting range. The unprecedented personal mobility of the urban population has transformed the city center's greatest asset—its accessibility—into a liability—congestion. Consequently some activities have dispersed to more distant locations with less accessibility but more space. In such cases, a more complicated pattern of rent and land values has developed, displaying far greater emphasis on local differences in environment and amenities.

Yet the automobile has not really granted us true freedom of location. Despite the changes that have taken place, spatial order still exists. Businesses remain aggregated to facilitate the exchange of information and to reduce transfer costs. Even more aggregations become appropriate as cities increase in size and decrease in density. This trend is particularly evident in cities like Los Angeles, a prime example of urban sprawl (figure 12.4). The dispersal of shopping centers in metropolitan Los Angeles is due not to freedom of location but to the friction of distance. Residents would find it difficult to reach a single CBD from such a vast built-up area. Rather than being randomly scattered, stores and services are clustered in district shopping centers that are rationally located with respect to the distribution of a dispersed population.

Affluence, quality transportation, the availability of urban amenities in rural areas, and earlier retirement, along with metropolitan decay, are beginning to permit a resurgence of nonmetropolitan and small-city growth in the most advanced societies. Fewer people need to locate in metropolitan areas because of their jobs. Many employers, too, are decentralizing, as exemplified by the flight of large corporations from New York City. They are finding that lower land and labor costs and greater employee satisfaction in nonmetropolitan areas more than offset the loss of agglomeration benefits.

But in much of the world, transportation remains limited and expensive, and location decisions still clearly reflect the need to minimize transport costs. There is less freedom of choice, and investment tends to be concentrated in the few locations best served by transportation. Even in the richest countries, the rapidly rising costs of fuel are impelling us once again to consider more seriously the cost of overcoming distance and the possibility of substituting communication for actual movement whenever feasible.

REGIONAL STRUCTURE

The patterns of location and interaction described above form a complex spatial structure. A more practical way to view human spatial organization is to divide the landscape into regions, areas unified by one or more common characteristics *(homogeneous regions)* or by economic and social dependence on a particular center *(nodal regions)*.

The Utility and Meaning of Regions

Regionalizations (dividing the earth into regions) are made for two basic reasons: first, to explain or understand the landscape in a scientific manner, and second, to establish convenient units for the exercise of government and other human activities.

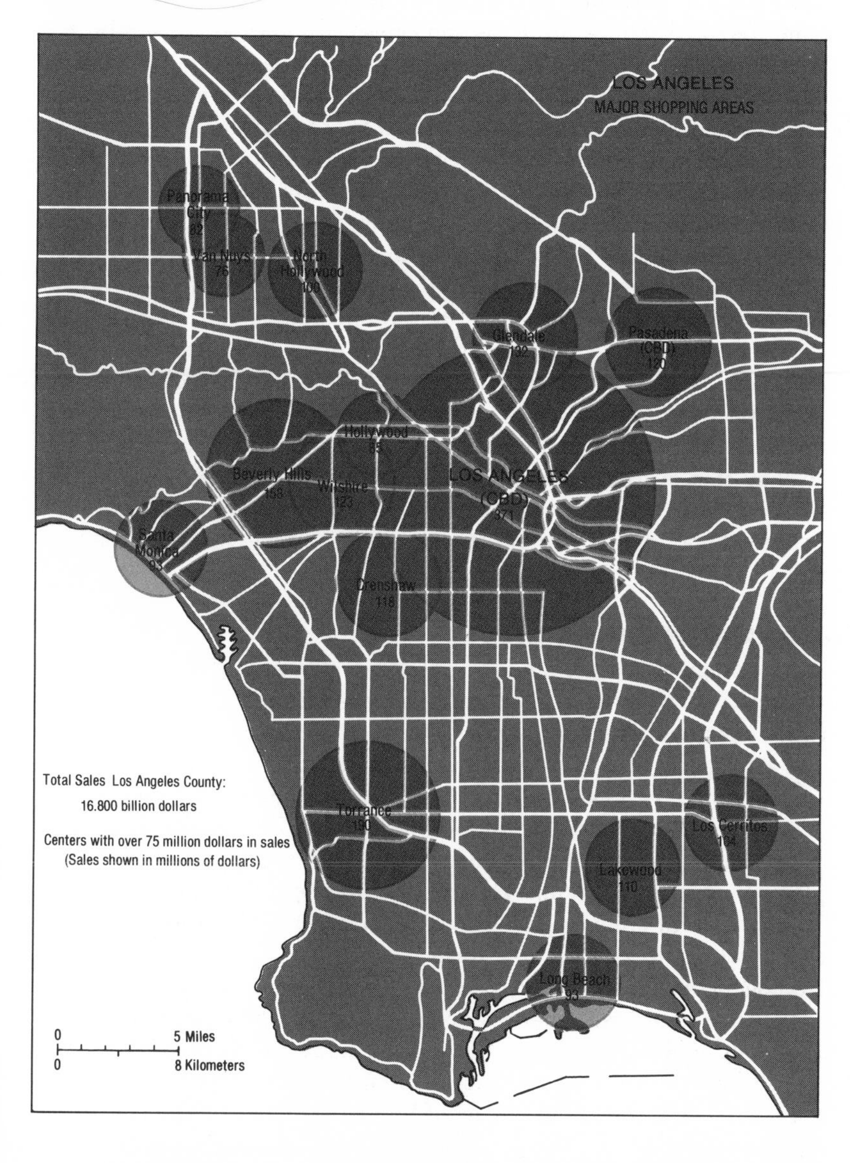

12.4. Major shopping centers in Los Angeles. Los Angeles is one example of a multicentered metropolis. Even these fifteen largest shopping centers contain only 12 percent of metropolitan retail sales.

A major task of any science is to classify the phenomena studied, dividing the whole field into a logical structure. Just as biologists divide plants and animals into phyla, genera, and species, geographers divide the earth into various regions. There are many kinds of regional classifications. The division of the earth into land and water regions or into a system of nations is an obvious one. Others are more subtle—the earth may be divided into regions of poverty or hunger, for example, or into regions of conservative and liberal political tendencies. All the location theories we have discussed should predict a particular regional structure based on certain combinations of activities (wheat-barley and livestock-grain regions, for example) or on the dominance of particular centers, as in the central place hierarchy.

Some regionalizations are based on physical processes, others on human decisions. Regardless of the approach, societies have found it essential to draw up regions for practical purposes of administration, defense, and service. Politically regions have been defined for several thousands of years as territories claimed or defended by particular groups or as areas in which particular laws or taxes apply, such as irrigation districts. Today nations legally define a large number of regions, not only for government administration at all levels, but for the provision of a wide range of services, such as schools and water supply, and even for the collection of data (census areas). Legally defined regions are necessary for administrative efficiency. It is socially and economically efficient to match territories to the performance of certain functions, such as sewer districts for waste disposal, and it is beyond human capabilities to administer as socially complex a system as a nation without decentralizing control into regional jurisdictions. Perhaps the main reason why this is true is again a matter of distance. One center cannot exercise continued control over a large territory but must delegate authority to areas small enough to maintain law and order, a basic function of all societies.

Homogeneous Regions

One way to regionalize the landscape is to identify sets of distinguishing, homogeneous characteristics. The features that separate *homogeneous* regions from their surroundings may be physical, such as climate, terrain, vegetation, or wildlife, or they may consist of visible human elements, such as various urban and rural land uses, or less visible human traits such as language, race, religion, income, or political persuasion. The same area may be claimed by many different homogeneous regions. For example, Iowa is part of a plains topographic region, a continental climatic region, a natural grassland region, a grain-animal agricultural region, an English-speaking, mainly Protestant cultural region, and so forth.

Homogeneous regions are seldom truly homogeneous, and their boundaries are imprecise. There is no consensus on the boundary of such broad regions as the Great Plains or the corn belt. Nonetheless the concept is useful. Homogeneous regions based on land use are particularly meaningful because they reveal the impact of human activities on the landscape.

The national land use map in figure 12.5, for example, not only indicates the broad outlines of the U.S. physical base (desert, swamp or marsh, alpine region, tundra, and forest), but it summarizes human modifications of the landscape: irrigated land, cropland, and the tiny proportion (less than 1 percent) of urbanized land. The local land use map of Chicago in figure 12.6 shows the more elaborate structuring of the urban landscape and illuminates several important features of that landscape: the close relationship between transport routes and industry; the concentration of commercial activities at and between highly accessible nodes, or centers; the patterns of residential density that have developed partly in response to those centers; and the degree of spatial segregation or specialization of land use.

Several location theories predict the emergence of homogeneous regions. Agricultural location theory predicts a zonation of homogeneous crop specialty regions based on intensity, modified by variations in physical conditions. Within the city, the intensity gradient, combined with the efforts of various groups to maintain their social identity, results in homogeneous zones of land use based on intensity and type, as well as homo-

12.5. Major land use regions in the United States.

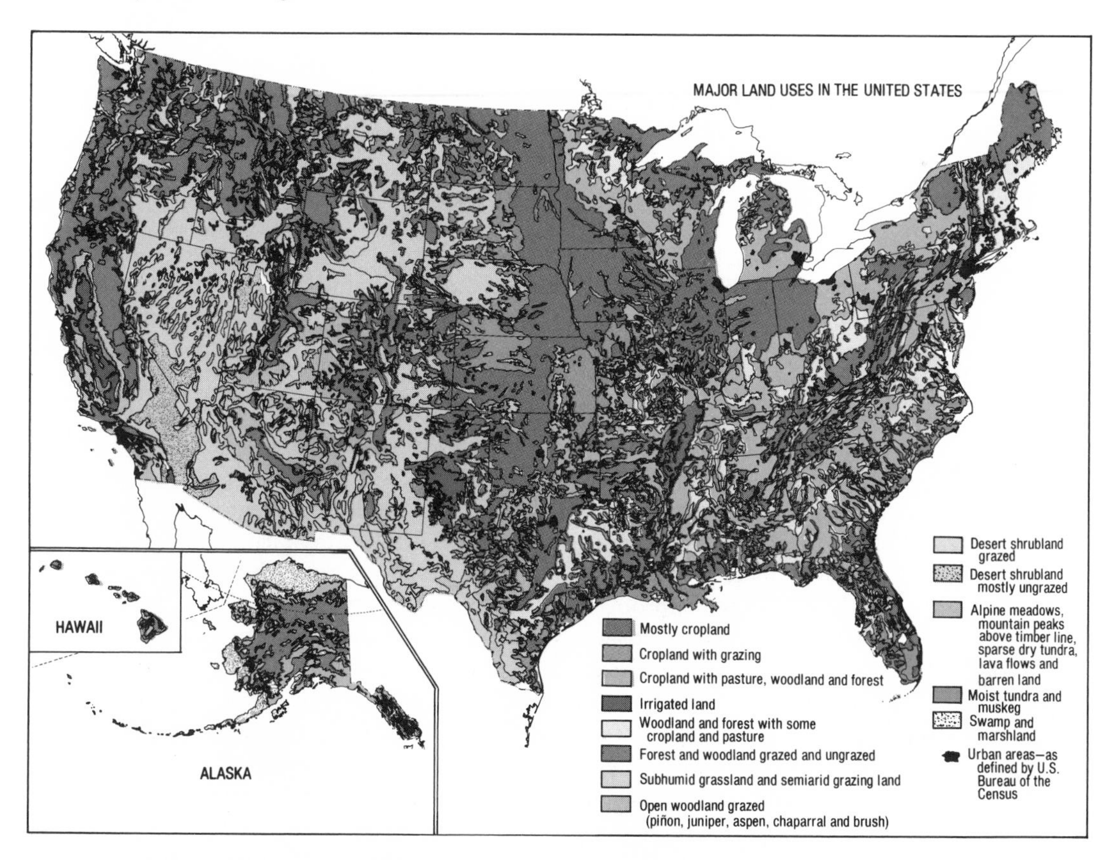

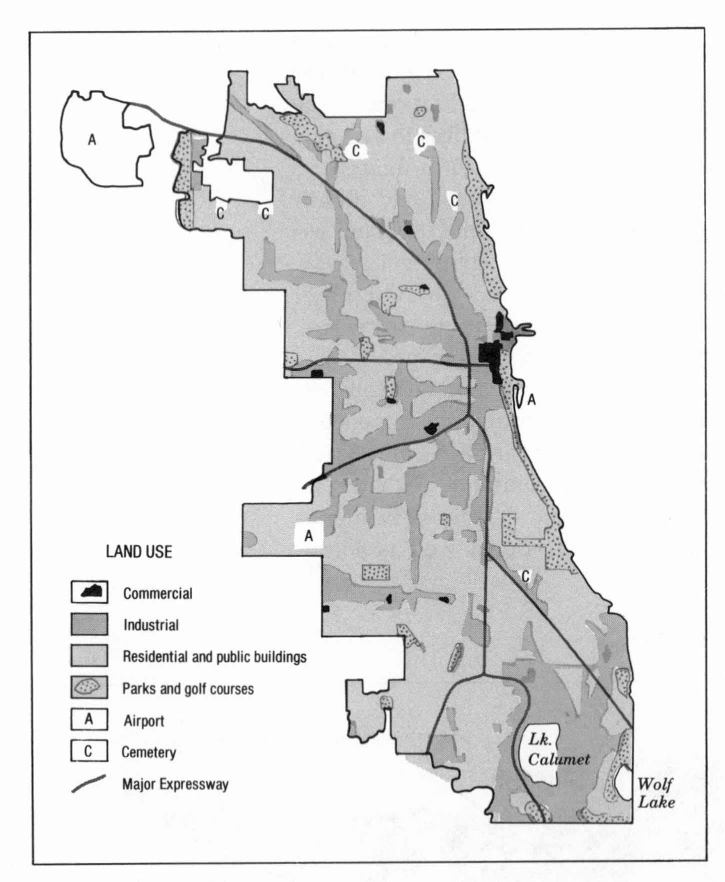

12.6. Major land use regions in Chicago. Note the roughly wedge-shaped pattern of separate industrial and residential zones. While distinct commercial areas stand out, particularly the commercial nucleus (the Loop) in the east central portion of Chicago, near Lake Michigan, the residential zones also contain commercial activities.
Source: Redrawn with permission from a map by Ying Cheng Kiang in Irving Cutler, *Chicago: Metropolis of the Mid-Continent* (Chicago: Geographic Survey of Chicago, 1973), p. 72.

geneous sectors based on cultural and economic class distinctions. Finally industrial location theory predicts the clustering of related industries, such as textiles or automobile parts suppliers, again resulting in homogeneous industrial regions.

In sum, regional homogeneity stems from the interplay of three factors: the capabilities of the local environment, the social and economic benefits of agglomeration, and the relative ability of an activity or group of people to compete for a particular territory. Thus a hard-wheat region exists in North Dakota because the area is physically suitable, the farmers find it valuable to agglomerate and share specialized markets and other services, and the wheat is of the appropriate intensity to compete at that distance from market. Metropolitan Boston contains industrial regions specializing in electronics because producers have agglomerated to take advantage of local qualified labor and research facilities, and they can afford to locate in such areas. Within cities, homogeneous land use regions develop because of local topography, agglomeration benefits for commerce and industry, differences in ability to pay rent or buy homes, and perceived social and economic incompatibility.

Many homogeneous regions are legally defined. Zoning regulations often reinforce the trend toward segregation of land use. To the extent that these regulations precede rather than follow land use decisions, they help create the regions, as in the development of industrial tracts or suburban residential zones. Other homogeneous regions serve the purposes of particular laws. Thus shorelines may be defined and protected, and wildlife habitats set aside. Areas with high unemployment may constitute regions eligible for special benefits, as may parts of a city with blighted housing or a large minority population.

Homogeneous regions may be defined at all scales, from within a building or park to the entire earth's surface. Regions are defined at a more local scale in detailed rural and urban land use maps, and at a national or world scale in atlas maps displaying variations in landforms, climate, vegetation, race, language, religion, and principal crops and industries.

The emergence of distinct homogeneous regions at the world scale, and to some extent at the national scale, may be attributed partly to the fact that spatial separation gives rise to cultural diversity. This important geographic principle underlies differences in race, language, religion, and, less obviously, in food and housing preferences. Societies that have little contact with others develop distinctive languages as well as other unique cultural traits. The map in figure 1.6 shows that dominant societies have imposed their own language, through conquest and colonization, on regions far from the cultural hearth. But many peoples have been able to preserve their native language despite conquest or colonization—the Chinese and Amerindians are examples.

Nodal Regions

Another way to regionalize the landscape is to delineate areas or fields of influence around central points, called nodes. Chicago and its hinterland constitute a *nodal* region.

Because zones of influence between competing regions overlap, the boundaries of nodal regions, like those of homogeneous regions, cannot be rigidly defined. Some communities and individuals in southeastern Indiana identify with the Cincinnati, Ohio, nodal region. They read Cincinnati newspapers and set their clocks by Cincinnati time. Other communities and people in the same area identify with their own state capital, Indianapolis. Thus the Cincinnati and Indianapolis nodal regions overlap. Generally the boundaries of nodal regions are based on patterns of economic and social strength, including patterns of commuting, newspaper circulation, dependence on wholesale suppliers, and retail shopping or service trips. Nodal regions, such as those of New York City and Boston, do not observe state boundaries but capture the actual fields of economic and cultural domination (see figure 7.6).

Nodal regions are predicted and defined by central place theory. Any small nodal region is part of a whole set, ranging from the tiniest hamlet to the national economy (figures 12.7 and 12.8). The smaller regions usually

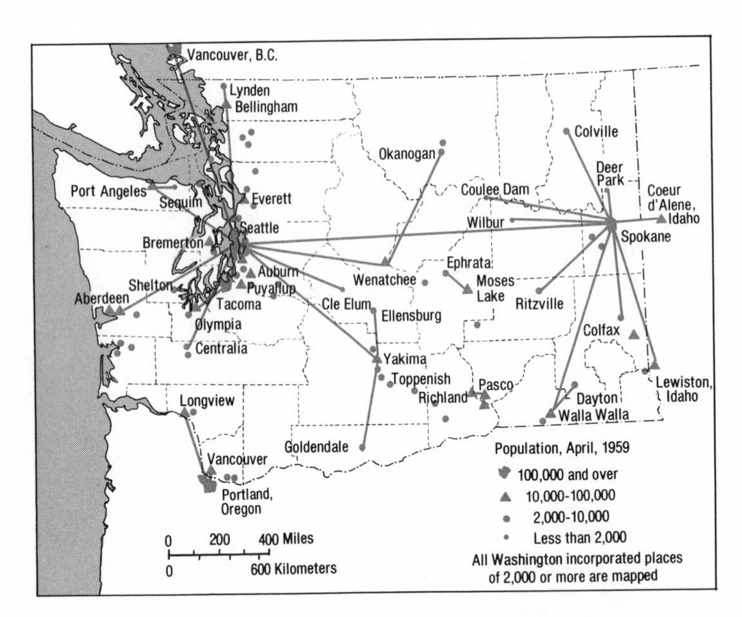

12.7. Nodal regions of Washington. One clue to the nesting of smaller nodal regions within larger ones is provided by an analysis of telephone data. In this example, it was assumed that one town depends on another if it places more calls to the latter than it receives. Seattle receives from all places more calls than it sends, including calls from Spokane, the next most important place. Thus Spokane and its local region nest within the Seattle region.

Source: Adapted with permission from J. Nystuen and M. Dacey, "A Graph Theory Interpretation of Nodal Regions," *Papers and Proceedings of the Regional Science Association*, Vol. 7, 1961.

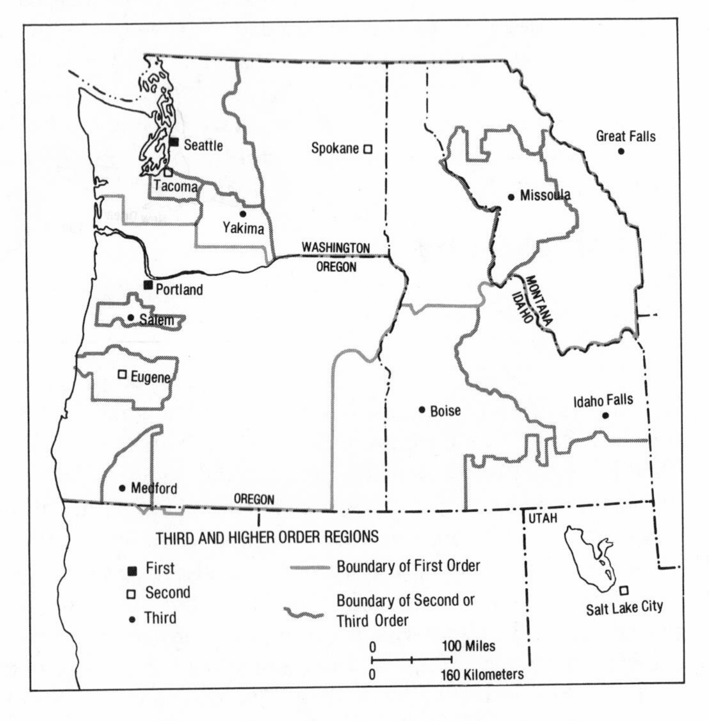

12.8. Nodal regions of the Pacific Northwest. This regionalization is based on numbers of branch offices whose headquarters are located in the larger centers. Note the dominance of Seattle and Portland. Areas beyond but closer to Eugene are still directly tributary to Portland.

Source: Adapted with permission from R. Preston, "The Structure of Central Place Systems," *Economic Geography*, Vol. 47, 1971.

nest within the larger ones, though there are many cases in which a smaller region is divided in its allegiance to larger ones. Regions farther up the hierarchical ladder, such as the Chicago metropolitan region, tend to be economically and culturally more self-contained.

The structure of metropolitan nodal regions has been intensively studied (figure 12.9). Metropolitan regions are based mainly on wholesale and financial patterns. Although boundaries are typically drawn between such regions, it should be clear that zones of influence overlap widely, particularly where metropolitan centers are close together and their hinterlands are large.

Nodal regions also may be legally defined. U.S. labor market regions, based mainly on commuting patterns, are used as a basis for measuring local welfare (unemployment rates and poverty, for instance) and eligibility for various federal programs.

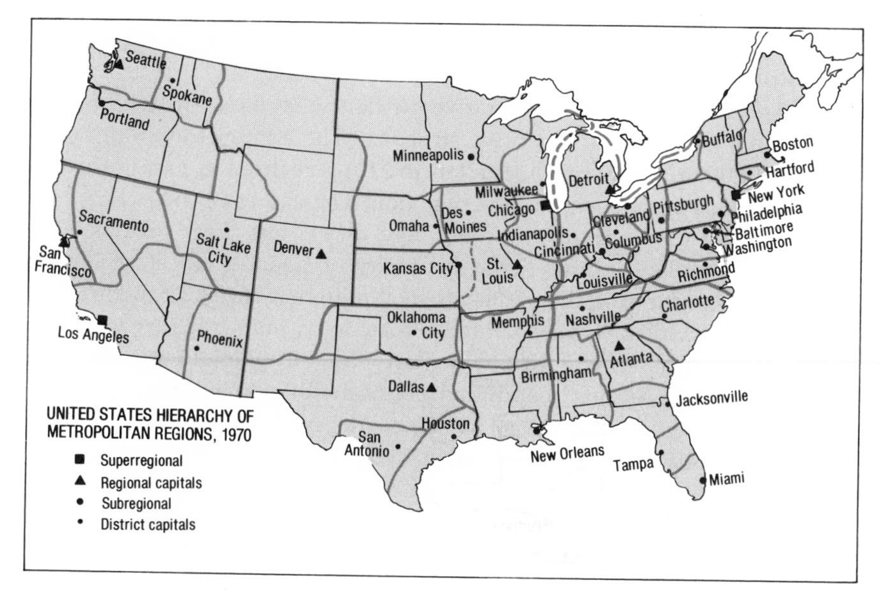

12.9. Hierarchy of metropolitan nodal regions, United States, 1970. The heavier lines define regions dominated by the largest centers—the superregional and regional capitals. The thinner lines define regions dominated by subregional and district capitals. Estimates (by R. L. Morrill) are based on wholesale trade data, airline patterns, commuting fields, and related data. The actual structure of metropolitan regions is rather unstable as a result of the rapid development of the United States during the past century. The growth of new centers such as Dallas, Houston, Kansas City, and Denver has led to the relative decline of older gateway cities such as New Orleans, St. Louis, and Cincinnati.

Functional and Political Regions

Some regions do not quite fit the homogeneous or nodal mold. They may be too complex to be considered homogeneous or they may have several nodes. Yet they have an identifiable economic or social role in the nation and can be defined as *functional* regions on the basis of interrelated activities and places. An example is Germany's Ruhr industrial region. It has too great a variety of land uses to be considered homogeneous, but it functions as an entity in the German and world economies. Functional economic regions, called *territorial production complexes,* are particularly well developed in the Soviet Union.

School districts, water districts, legislative districts, townships, and cities form a different kind of functional region. Known as *political* regions, they are created for a specific purpose. They are nodal in the sense that a center is exercising authority over a territory, and they are homogeneous in that the authority holds power uniformly over that territory. In contrast to most nodal or homogeneous regions, whose boundaries are often quite vague, political regions are defined by specific legal boundaries. Our lives are to some extent directly affected by these boundaries. Quality of public education, political representation, and tax structure, for instance, may vary from one political region to another. For economic and social purposes, of course, individuals often circumvent or ignore political boundaries, as demonstrated by flows of people crossing political boundaries to obtain quick marriages or divorces, abortions, or more available or cheaper alcoholic beverages or cigarettes. Figure 12.10 shows the inequities for black people that occurred as a result of the structuring of political regions in part of the South. Some other problems associated with the structuring of political regions will be discussed in the next chapter.

In large territories administrative efficiency requires the division of space into a hierarchical, repetitive structure. The administrative principle formulated theoretically by Christaller (p. 210) is realized in part in the territorial political organization of every nation. The names of the political regions and the distribution of power may differ, but most nations have found that a three- or four-tiered hierarchy of nodal political regions is most effective, reflecting the scales at which different societal functions are most readily handled. In the United States, local units such as towns or townships are parts of counties, which are basic divisions of states, which together form the nation. In addition to a nested hierarchy of general purpose governments, most nations have an overlapping system of special purpose administrative districts, usually created for the provision of some extra service, such as water supply, pollution control, maintenance of a port or airport, and schools (if not part of the local general purpose government).

The boundaries of political regions tend to become sacrosanct, despite the fact that changes in settlement patterns may have made them obsolete. Some countries, like the Soviet Union and India, have succeeded in changing local and regional jurisdictions. But this has proved extremely difficult in most individualist societies such as Germany, Great Britain, and the United States.

CULTURAL REGIONS

The regionalizations we have discussed are not the only possible ones; many variations have been proposed. In addition, geographers have long been trying to develop a composite regional structure. But there is yet no overall structure with which societies can identify, one that combines the different elements into a meaningful whole.

We may reasonably argue that a cultural basis comes closest to the desired structure, since "culture" implies a combination of the political, economic, and traditional ways in which individuals in a society interact with

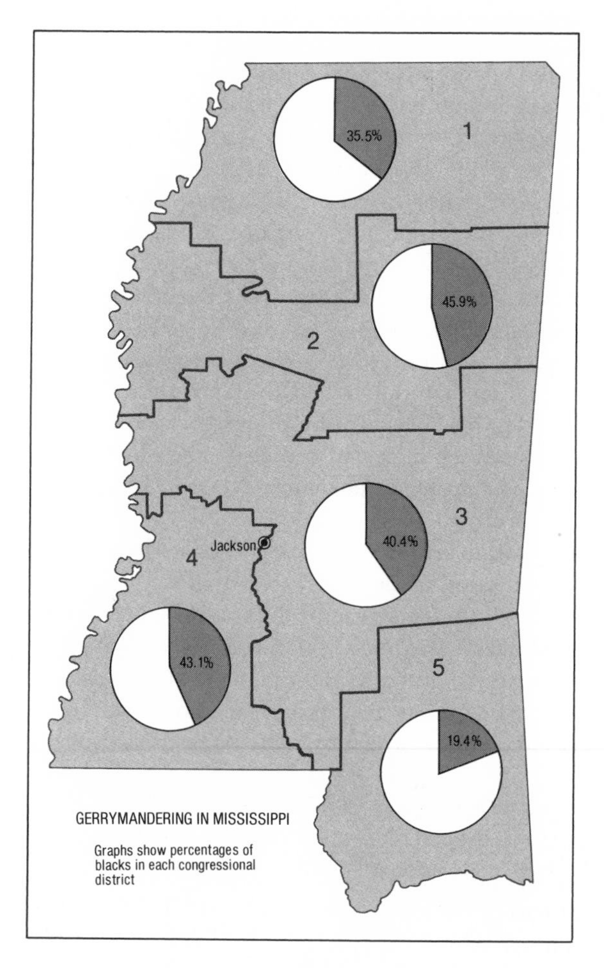

12.10. Gerrymandering in Mississippi. Gerrymandering refers to drawing political districts so as to minimize the representation of certain groups. Congressional districts 1–4 were laid out in an east-west fashion, against the grain of the distribution of races (blacks cluster along the Mississippi River, which defines the western boundary of the state), to prevent blacks from attaining a majority in any district.

each other and with their environment. Composite *cultural* regions form distinct landscapes, or *unique* regions; they possess a distinctive and nonduplicated combination of traits, which may themselves be systematically explained. Still the holistic concept, which emphasizes the relationship of parts to the whole, represents a departure from our general emphasis on the similarity and predictability of human landscapes. Our purpose here is to show not only that the study of the human landscape can be approached in more than one way but also that unique regions are themselves products of spatial forces—of spatial separation and interaction.

Cultural differentiation results from the interplay of internal development, diffusion, and interaction with other groups. Interaction in the form of intermarriage, shared settlement of a territory, and extensive trade generally tends to enrich a culture and make it more complex. But it also reduces differences among social groups. Therefore it is *spatial separation*—the fact that a people resided in a given territory for hundreds or even thousands of years, relatively free from outside influences—that permitted the development of differences in language, food preferences, housing styles, and many other traits. Racial differences evidently stemmed from separation over hundreds of thousands of years; major linguistic differences, over many thousands of years; and minor linguistic and religious differences, over hundreds of years. The differentiation of Romance and Germanic languages in Europe, for instance, evolved between the fifth and fifteenth centuries during a period of greatly reduced interaction.

Specific cultural differences partly reflect features of the physical environment, such as available building materials and native animals and plants (rice rather than wheat, for example, became the preferred food grain in areas where native rice developed). Until recent centuries, the physical environment also governed the basic means of economic survival, whether by hunting, herding, fishing, or farming.

A particular culture may rather suddenly break out of its hearth area and impose itself over vast territories, as demonstrated by the Russian settlement of Siberia and the American settlement of the Great Plains and the West. But once cultural differences arise, members of a distinct group place a very high value on maintaining the group's integrity, even at the cost of a lower standard of living or warfare, as has occurred in Lebanon. Even in the midst of the plains of Russia and America and in many large cities, isolated cultural groups survive.

Over a long period, contacts with other groups—desired or not—modify cultural differences or lead to their replacement through the diffusion of more effective traits. Whether in the quest for food or space or in an attempt to enlarge their territory through conquest, various peoples have spread their cultures—and their very genes—across the globe, bombarding the cultural fortresses of isolated societies. The Spanish conquest of Latin America is an example of large-scale cultural and genetic diffusion. In modern times, dominant societies have imposed an international technological way of life on many indigenous cultures, further blurring cultural distinctions. The Alaskan Eskimo (Inuit) culture and native Hawaiian culture, for example, have been modernized through the influence of American technology.

Yet identifiable cultural regions do exist and have meaning in the modern world. We will briefly consider the cultural regions of the United States and then of the world as a whole. A close examination of any given region will reveal far more internal variation, but certain traits stand out and give regions their distinctive character.

Cultural Regions of the United States

The mobility of the U.S. population has fostered the development of a national culture. This culture retains several elements from the early colonial period: a spirit of individualism, with an emphasis on property rights and personal responsibility and an inherent distrust of government intervention; acceptance of change and mobility and a virtual expectation of continual growth and improvement; and faith in technology and in our ability to succeed against all obstacles.

But it is clear that regional cultural variation also exists. Differences among regions can be perceived in the national origin of the population; in their religion, language, political philosophy, and attitudes toward violence; and in their food and housing preferences, income, socioeconomic status, and preference for urban or rural life. Variations are also apparent in patterns of speech, voting behavior, crime, consumption, and settlement. Mindful that regions at best reveal tendencies toward distinctive cultural landscapes and that many other plausible regionalizations may be drawn (figure 12.11), we suggest the following classification.

Northeast
1 Yankee (New England, New York)
2 Metropolitan (ethnic)

South
1 Border South
2 Deep South
3 Southwest
4 Black Subculture
5 Peninsular Florida

Midwest
1 Upper Midwest
2 Central Midwest

West
1 Rocky Mountain
2 Mormon
3 Southwest
4 Pacific Coast

The Northeast. The Northeast remains the core region of cultural and economic control. A metropolitan industrial and commercial culture overlies an original Protestant English culture. Enormous waves of non-British immigrants have been absorbed into this region—mostly Catholics and Jews from Central, Southern, and Eastern Europe, followed by blacks from the rural South. Culturally the Northeast is both a melting pot and a stronghold of ethnic variation. Within the region, two subregions can be distinguished: an upstate rural, relatively more Protestant hinterland, and a densely populated, predominantly Catholic (and Jewish) cosmopolitan megalopolis.

The Midwest. The Midwest is thought of as quintessential America. Its rich agricultural base and generally dispersed settlement pattern has been spangled with towns, small cities, and industrial centers. Settled by migrants from the Northeast and Europe, it is often considered somewhat more provincial, conservative, technology oriented, and individualist than the parent region. Two major subregions include a predominantly Scandinavian and German Lutheran upper Midwest, and a predominantly mainstream English Methodist-Presbyterian central Midwest.

The South. The South is readily perceived as a cultural region. Its population derives from early English-Scotch stock and from African slaves. The South was slower to industrialize and urbanize than were the

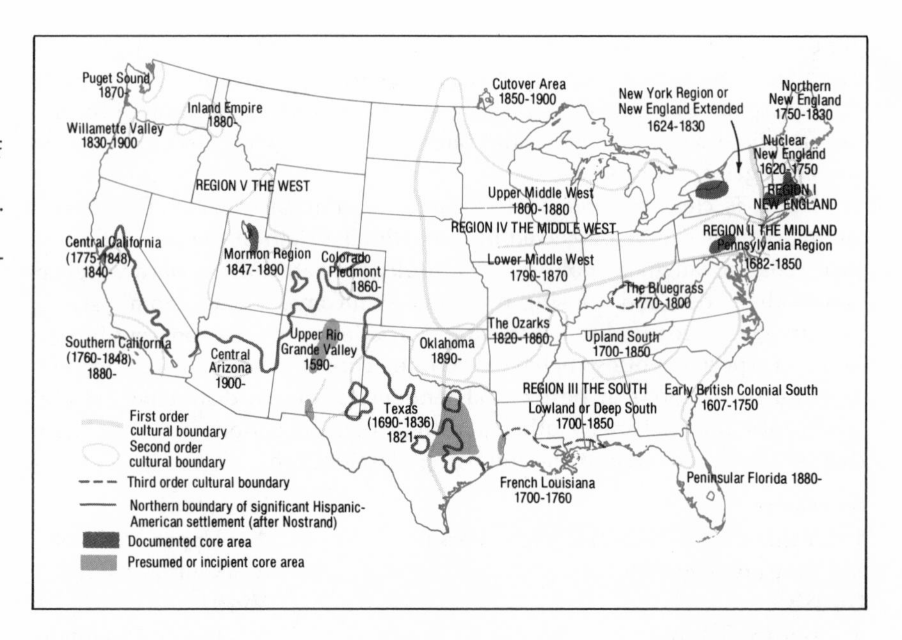

12.11. U.S. cultural regions. These cultural regions are based on Zelinsky's regionalization. Dates refer to the period of settlement and the emergence of regional character. Regional character derives from the interaction of particular population groups with a particular physical environment.

Source: Wilbur Zelinsky, *The Cultural Geography of the United States.* ©1973, pg. 118. Adapted by permission of Prentice-Hall, Inc., Englewood Cliffs, New Jersey.

other regions, but it is now perhaps the fastest-growing region of the nation. Its relative isolation after the Civil War fostered social and economic conservatism, reinforced by a strongly conservative Christian influence. Racial conflict and local rather than state governmental control are part of its heritage. The Deep South core reflects the legacy of its former plantation economy, while the Border South was a region of individual, and often isolated, farms. The Southwest, primarily Texas, shares much of the South's culture but has its own frontier tradition, strongly influenced by Mexican and Catholic elements. The Black Subculture, created partly by discrimination and modified by urban ghettoization, is an offshoot of the Deep South tradition. Peninsular Florida is a distinctive subregion of the South. Its early Spanish cultural heritage has been revitalized by the migration of large numbers of Cubans to Miami in the 1960s. The region's thriving tourist resorts and retirement communities provide a setting for many other cultural influences.

The West. Many people view the West as a frontier and mining region, but the reality is more difficult to pin down. Its population is a mixture of migrants from the East, Midwest, and South, supplemented by Asians and Mexicans. The Rocky Mountain subregion has received a southern influence through the diffusion of cattle ranching northwest from Texas. It maintains a strong tradition of frontier independence, perhaps because of the relatively harsh environment and sparse settlement typical of the area. The Mormon subregion in Utah–south Idaho stands out as a distinct enclave of Mormon traditions as expressed in religion, settlement form, and social behavior. The interior Southwest is a subregion in which

Spanish and native American influences still survive. The Pacific Coast, particularly California, is a microcosm of all the subcultures and the national culture as a whole, while Hawaii, with a majority population of Asian stock, has no mainland counterpart. Alaska is still a frontier economy, heavily dependent on resources (before this century, gold; now, petroleum).

These cultural regions are large and broadly defined, but such areas may be discovered at all scales. Within New England, for example, are subregions of predominantly French Catholic peoples as well as Yankee Protestants. And within cities throughout the world, the very concept of neighborhood is linked to culture; each neighborhood is a zone whose population shares certain characteristics, such as religion, ethnic background, and socioeconomic class, certain institutions (schools, scouting programs, and the like), and certain markets or shopping facilities.

Cultural Regions of the World

Variations in most cultural traits are far more pronounced across the world than within a single advanced nation such as the United States. A plausible composite regional structure may be based on race and language, religion, economic development, and political and economic system. The cultural regionalization briefly discussed below is just one possibility. At a broad level it rests on religious differences more than on other differences because of the close relationship between religion and other forms of development. Thus the Western cultural region is basically Christian, the Indian is mostly Hindu, and the Eastern primarily Buddhist or Buddhist influenced.

The Western Cultural Region

Western culture is essentially a product of European development—a combination of Caucasian peoples who speak related European languages, who are governed by political and legal structures derived from Egypt, who share a Judeo-Christian religious tradition, and whose recent history includes the development of capitalism and the emergence of the nation-state. During the past few centuries, Western societies tolerated enough individualism to permit significant change and to stimulate the growth of science, technology, and democratic institutions. The early technical and military superiority of these societies reinforced their cultural aggressiveness: they carved out colonies from large portions of the world—the Americas, South Africa, and Australia–New Zealand—and temporarily subjugated many other areas.

The West, so defined, contains about one-third of the world's population and more of its area, and controls about three-quarters of the world's present wealth. Within the general realm of Western culture are three major subregions: Western Europe, North America, and Australia–New Zealand; the Soviet Union and Eastern Europe; and Latin America. A smaller subregion, the Republic of South Africa, completes the list.

Western Europe, North America, and Australia–New Zealand. The most influential and economically advanced subregion comprises Western Europe and its offspring, the United States, Canada, and Australia–New Zealand. The Nordic (mainly Protestant) subculture is at present dominant in Great Britain, Germany, France, the Low Countries, Scandinavia, Austria, Switzerland, the United States, Canada, Australia, and New Zealand. The Mediterranean subculture (largely Roman Catholic or Eastern Orthodox) is represented in Italy, Spain, Portugal, Greece, and France, the last sharing in both subcultures. While aggressive in the post-Renaissance period of exploration and colonization, most of this subculture was slower to industrialize and urbanize than its Nordic counterpart. The urban-industrial economies of the Western European and North American cultural region are clearly the richest and most advanced of all regions. They are highly interconnected by exchange and are still predominantly capitalist, although the extent of government control is increasing.

The Soviet Union and Eastern Europe. The Soviet Union and Eastern Europe share most of the Western European traditions. However the region historically was somewhat less developed than Western Europe. Since World War II, its governments have been united through their commitment to a socialist economy and social order. Industrial and urban development has been rapid and centrally planned. Political and cultural isolation is gradually lessening under pressure of economic interdependence.

Latin America. Latin America, a cultural colony of Mediterranean Europe (Catholic; Portuguese, Spanish, Italian), retains some of its indigenous Amerindian culture, which survives both in pure form and intermixed with other cultures. While much of Latin America is rapidly urbanizing, it remains a dual society: peasant farming is still practiced in some regions, and vestiges of hunting-gathering societies still exist. Democratic institutions are weak or fragile, possibly because of the region's lower level of development or because the early colonizers were bent on exploiting rather than developing the region.

The Republic of South Africa. An uneasy outlying region of the Western cultural region, the Republic of South Africa is politically, economically, and culturally dominated by a white Nordic minority within a black African majority population. The future of this troubled area is unclear, but change is inevitable.

The Islamic Cultural Region

The Islamic character of the vast region extending across North Africa and the Sahara through Southwest Asia and into Central Asia is related rather consistently to a primarily nomadic economy and an oasis-irrigation agriculture. Both the religion and the economy unify at least three different racial groups (and many linguistic ones): Negroid (Bantu), Caucasian (Hamitic-Semitic, Iranian, Hindi), and Mongoloid (Turkic, Indonesian). The Islamic faith was as aggressive as Christianity. For a time, the Islamic core area led the world in scientific scholarship. In the twelfth century, however, feudal aristocratic forces prevailed, and development was brought to a virtual standstill. In recent years, petroleum wealth and Islamic reform movements have begun to provide a base for rapid commercialization and new development, although, as in Iran, conservative Islamic groups are opposing change.

About 10 to 15 percent of the world's people dwell in the Islamic cultural region. The core subregion, which is Arab in nationality and Hamitic-Semitic in language, is located at the crossroads of Europe and Asia, a position inviting cultural contact and inevitable conflict. Israel, a European cultural enclave within the Islamic realm, is the focus of particularly intense political and cultural conflict today.

To the south, a group of predominantly black Islamic states, including Nigeria and the Sudan, cross the southern rim of Saharan Africa. Today traditional and sub-Saharan African and Islamic cultures frequently clash in this zone, and imported Christianity is in conflict with both. To the northeast are Persian and Turkic Islamic peoples. Soviet Central Asia to the north is now an area in transition: its Islamic base is being supplemented by Russian European cultural elements and a socialist political and economic structure. Finally Islamic and Indian cultural elements are partially intermixed to the east in Pakistan and, just beyond, in Bangladesh, Malaysia, Indonesia, and the south Philippines.

The Indian Cultural Region

About one-seventh of the world's population inhabits the relatively small territory comprising India. This region has evolved a unique culture over many centuries. The majority of its peoples are racially Caucasian and linguistically Indo-European. They have been sufficiently isolated from other Caucasian peoples to have developed one of the two dominant Eastern civilizations—historically a feudal system based on intensive peasant agriculture, with religious constraints on individuality and this-worldliness, an acceptance of social (occupational) caste and cultural continuity, and an autocratic local or regional government structure. British imperialism modified the traditional culture in some ways, notably through the imposition of British law, language, and democratic institutions.

The Eastern Cultural Region

The Chinese culture dominates an East Asian region numerically equal to the economically stronger Western cultural region, each containing about one-third of the world's people. The ancient Chinese civilization emerged in the second millennium B.C. as a feudal aristocracy based on intensive peasant irrigation agriculture, and spread from northern China into Southeast Asia and across to Japan. China's imperial social and economic organization was more centralized than the comparable structures of India and Europe and was supported by the religious and philosophical systems of Buddhism, Taoism, and Confucianism, which stressed the subordination of the individual to the broader interests of the community and state. While the Chinese produced highly important innovations (paper and gunpowder, to name only two), they valued cultural refinement and continuity over science and technology and did not challenge the prevailing autocratic social structure until modern times.

For convenience, the Eastern cultural region may be divided into four subregions.

China. Despite its immense population (estimated at about 900 million), China is unified by a strong centralized tradition. The nation is evolving into the world's first large-scale agrarian-based communist society. While in the process of industrializing and urbanizing, China seems to be creating a social structure different from that of Japan: cooperative rather than competitive and far more egalitarian.

Japan. An outstanding example of the overlayering of different cultures, Japan is rooted in an ancient feudal tradition and centralized culture derived from China, but it has adopted many Western elements, such as capitalism, urbanism, industrialism, and democratic institutions. These acquired elements have modified but not fully supplanted traditional cultural forms, such as housing styles, food preferences and preparation, and social relations.

Southeast Asia. Southeast Asia is a complex subregion settled mostly by Mongoloid peoples from the north, including the Chinese. The area has also been influenced by both the Islamic and the Indian cultures. An interesting contrast is provided by the feudal aristocracy and intensive peasant agriculture of the rich lowlands and the tribal hunting-gathering societies and primitive agriculture of the hills and islands.

Siberia, Canada, and the Pacific Islands. The Mongoloid peoples of the Siberian and Canadian arctic tundra regions have developed a unique way of life based on animal husbandry, fishing, and hunting. It will be difficult for these small groups of Eskimo (Inuit) and related peoples to preserve their identity against the preponderantly urban-industrial society of the Soviet Union, Canada, and the United States.

The isolated peoples of the Pacific Islands, scattered over a large area, are gaining political independence. Their culture, too, may not survive the modernizing influence of tourism and exchange and the earlier culturally shattering impacts of World War II.

The African Cultural Region

Africa, though perhaps the cradle of humankind, was sparsely settled and weakly organized throughout most of its history, falling victim to Islamic invasions from the north and, more drastically, to European subjugation, slavery, colonization, and economic exploitation. Now that the colonial period has virtually drawn to a close, Africans are faced with the complex problem of rediscovering or reforming their indigenous cultural heritage and merging it with the cultural elements imposed by the Europeans, including religion, political and economic structure, and urban-industrial technology. The new nations of Africa (figure 12.12) are variously adopting capitalist or socialist strategies. The structure of the states themselves is still in a formative stage.

CONCLUSION

In the perspective of location theory, geography is a discipline in distance. It concerns the ways in which variations in spatial qualities, transport costs, and the demand of activities for space create a complex but orderly structure, exhibiting predictable spatial patterns at local, national, and world scales. We cannot observe purely theoretical patterns in the real world because they are modified by noneconomic human spatial behavior and because conditions affecting location continually change. In addition, development itself diffuses across space, so static theories of location and interaction are inevitably incomplete. The principal value of location theory for our purposes is that it gives us a useful framework for understanding the human landscape.

The alternative is to view each place or region as the unique outcome of the interaction of a local people with a local environment. Each place or region is unique, of course, but we know that its character is largely determined by its role in and relation to the world around it. These relations are illumined by theories of spatial organization. In sum, the spatial order of society can best be defined by characteristic patterns of land use and by the complex patterns of interaction each place has with the world around it.

To deal with the landscape in a more practical manner for planning or administrative purposes, it is necessary to simplify spatial structure by dividing space into broad general regions. These can be defined on the basis of certain common characteristics, by the relationship of an area to a central point, or by an area's economic or social role. National and world landscapes can also be divided into composite cultural regions. While the Western urban-industrial technological pattern is being adopted in nearly all parts of the world, helping to diffuse an international culture, all societies still identify with their traditional culture too. Culture differences give regions a distinctive character, but they are also a source of misunderstanding and conflict. These and other problems related to human spatial organization will be explored in the next chapter.

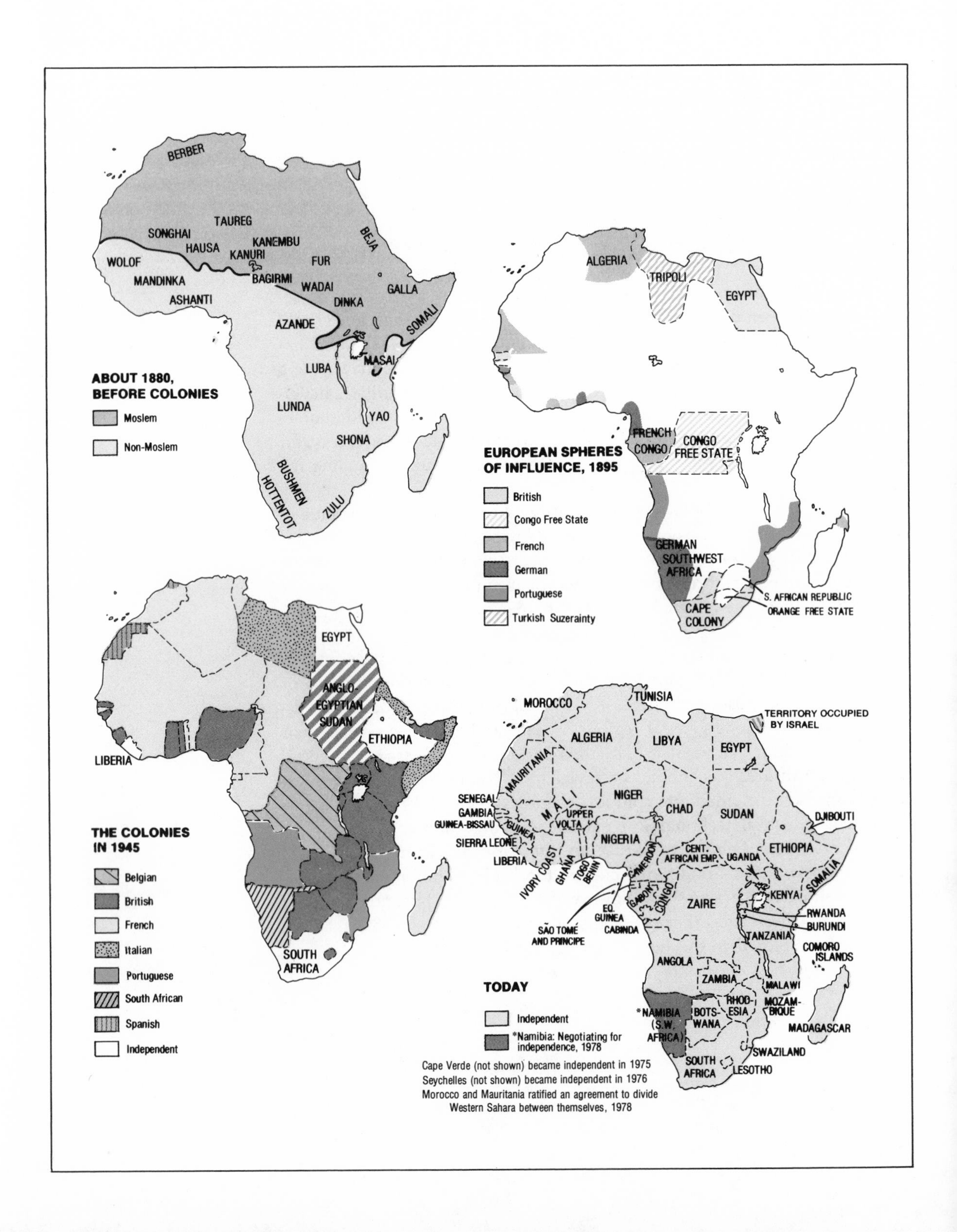
BERBER
TAUREG
SONGHAI
HAUSA
KANEMBU
KANURI
BEJA
WOLOF
FUR
MANDINKA
BAGIRMI
WADAI
GALLA
ASHANTI
DINKA
SOMALI
AZANDE
MASAI
LUBA
LUNDA
YAO
SHONA
BUSHMEN
HOTTENTOT
ZULU
ABOUT 1880, BEFORE COLONIES
Moslem
Non-Moslem
ALGERIA
TRIPOLI
EGYPT
FRENCH CONGO
CONGO FREE STATE
EUROPEAN SPHERES OF INFLUENCE, 1895
British
Congo Free State
French
German
Portuguese
Turkish Suzerainty
GERMAN SOUTHWEST AFRICA
S. AFRICAN REPUBLIC
ORANGE FREE STATE
CAPE COLONY
EGYPT
ANGLO-EGYPTIAN SUDAN
ETHIOPIA
LIBERIA
THE COLONIES IN 1945
Belgian
British
French
Italian
Portuguese
South African
Spanish
Independent
SOUTH AFRICA
MOROCCO
TUNISIA
TERRITORY OCCUPIED BY ISRAEL
ALGERIA
LIBYA
EGYPT
MAURITANIA
NIGER
SENEGAL
GAMBIA
GUINEA-BISSAU
MALI
UPPER VOLTA
CHAD
SUDAN
DJIBOUTI
GUINEA
SIERRA LEONE
NIGERIA
LIBERIA
IVORY COAST
GHANA
TOGO
BENIN
CAMEROON
CENT. AFRICAN EMP.
UGANDA
ETHIOPIA
SOMALIA
KENYA
GABON
CONGO
ZAIRE
EQ. GUINEA
SÃO TOMÉ AND PRINCIPE
CABINDA
RWANDA
BURUNDI
TANZANIA
COMORO ISLANDS
ANGOLA
ZAMBIA
MALAWI
*NAMIBIA (S.W. AFRICA)
BOTS-WANA
RHOD-ESIA
MOZAM-BIQUE
MADAGASCAR
SWAZILAND
SOUTH AFRICA
LESOTHO
TODAY
Independent
*Namibia: Negotiating for independence, 1978
Cape Verde (not shown) became independent in 1975
Seychelles (not shown) became independent in 1976
Morocco and Mauritania ratified an agreement to divide Western Sahara between themselves, 1978

12.12. Tribal, colonial, and national regions of Africa. Long before European colonization, Moslem merchants and traders extended the influence of Islam over tribal regions in east Africa and in much of the north and west. Although Europeans established trade and communication links in Africa between the sixteenth and nineteenth centuries, their interests were limited mainly to the exploitation of gold and slaves. In the 1880s the slave trade ended, and European powers began to compete for colonies in Africa. European spheres of influence were concentrated in coastal areas at first, but they soon spread to the interior of the continent, cutting across tribal and physically homogeneous regions. After about fifty years of colonial rule, African colonies began to achieve independence. Today nearly all of Africa is independent.

SUGGESTED READINGS

Berry, B.J.L. *Growth Centers in the American Urban System.* Boston: Ballinger, 1973.

Dickinson, R. E. *City and Region.* London: Routledge, 1964.

Henderson, J. M. *The Efficiency of the Coal Industry.* Cambridge, Mass.: Harvard University Press, 1958.

Hoover, E. *An Introduction to Regional Economics.* New York: Knopf, 1975.

Isard, Walter. *Location and Space Economy.* New York: Wiley, 1956.

Losch, August. *Economics of Location.* Translated by W. Woglom. New Haven, Conn.: Yale University Press, 1954.

Perloff, Harvey, et al. *Regions, Resources and Economic Growth.* Baltimore: Johns Hopkins Press, 1960.

Russett, Bruce. *International Regions and International Systems.* Chicago: Rand McNally, 1967.

Soja, Edward. *The Political Organization of Space.* Resource Paper 8. Washington, D.C.: Association of American Geographers, 1971.

Stewart, George R. *American Ways of Life.* New York: Doubleday, 1954.

Warntz, William. *Toward a Geography of Price.* Philadelphia: University of Pennsylvania Press, 1959.

Zelinsky, Wilbur. *The Cultural Geography of the United States.* Englewood Cliffs, N.J.: Prentice-Hall, 1973.

13

The Ideal Landscape and Problems of the Real World

Efficiency and Equity

Barriers to Efficiency and Equity

The Geographic Impact of Economic Inequality

Problems of Developing Countries

13 The Ideal Landscape and Problems of the Real World

In our discussion of the human landscape, we have often referred to principles of spatial behavior. Economic location theorists assume that human spatial behavior is governed by the principle or goal of maximizing the utility of places and interaction at least cost. But there is another less strictly economic goal of spatial behavior: to maximize human well-being. At least in some societies, this goal could well lead to decisions and patterns that serve societal needs and values but do not result in maximum output or the wisest use of resources.

We can reasonably say that in a free-market economy comprising individuals who wish to maximize their well-being, the rational person uses land as productively and efficiently as possible; movements of people and trade flows are optimal. Consistently and predictably (in theory), individuals make sensible decisions on where to locate activities, and they use those locations efficiently. In a centrally planned economy, the same should be true of the collective locations decisions of society.

How well does this principle actually operate in various societies? An owner of semiarid grassland in the American West will probably discover by experience that breeding and grazing beef cattle is the most productive way to use the land. An owner of land near a crowded shopping center in an American or West European city may find it most profitable to convert the parcel from residential to commercial or parking use (if permitted by zoning or other regulations). In a less developed agrarian country, a farmer—pursuing a similar goal—might find it best to grow a variety of crops, using land in a less specialized way than does the American farmer. (As you recall from chapter 4, poor transport conditions, small landholdings, and limited technology, among other things, discourage specialization.) In a centrally planned economy, where government owns the land, it might prove more efficient in terms of the long-range needs of society not to convert the residential land near a commercial center to other uses. A young person growing up in a rural village in any of these societies may find it most satisfying to move to a nearby city to obtain the best income. In all these cases, people are trying to maximize their well-being.

The spatial patterns formed by the rational individual and society are, in theory, both efficient and equitable: *efficient,* insofar as any change in land use or pattern of trade would result in a lower total level of production or consumption (that is, would upset the spatial equilibrium); *equitable,* to the degree that each person, presumed free to compete on an equal basis, is allocated what he or she is worth as an economically productive member of society.

In many respects, our sample landscape, the Chicago area, reflects the behavior of the optimizing individual, firm, and society. Entrepreneurs have taken advantage of Chicago's location in the midcontinent, its proximity to industrial raw materials, its access to the agricultural interior, and its central position on the transport network. Residents have taken advantage of Chicago's job opportunities, its beaches and other recreational resources, its museums, theaters, hospitals, and universities. The metropolis as a whole testifies to the human ability to create and maintain a complex, productive system. Productivity and personal satisfaction seem to be maximized.

Yet any resident or visitor knows that the Chicago landscape is far from ideal. Rush-hour traffic can be a nightmare; the air is often polluted with dust and fumes, and the lake streaked with industrial wastes; many stores and services are poorly located and barely surviving; and the slums are burdened with poverty, drugs, violence, and crime. Thus the Chicago landscape—and every other landscape in the real world—reveals substantial inefficiency, inequity, and insensitivity toward the environment.

In this chapter we will examine some of these problems and discuss their effect on the landscape. The first step is to establish criteria for evaluating efficiency and equity; then we can consider some barriers to their attainment and the resultant impact on the urban and regional landscape. Finally we will attempt to evaluate efficiency and equity at the world scale.

EFFICIENCY AND EQUITY

Location theorists might use the following criteria to assess the qualities of efficiency and equity in the human landscape.

1. Do enterprises successfully produce demanded goods and services at maximum, but not excessive, profit? That is, is location optimal and are economies of scale realized?
2. Is the allocation of value fair? Are wages, salaries, rent, royalties, interest, and profits allocated to people according to their level of productivity, and are they adequate to provide all with acceptable levels of food, shelter, and security? Are natural resources, including land, utilized as needed but not wasted?
3. Does the pattern of interaction reflect in a simple way the structure of economic specialization and social differentiation, avoiding unnecessary cross-movements? And does the level of interaction satisfy the needs and desires of the population?
4. Are regional and intraurban differences in income and wealth relatively insignificant?
5. Are all costs of production—including waste disposal, restoration of damaged land, and provision of adequate space—accounted for by the producer, thereby eliminating pollution, blight, and other social costs?

If the answers to these questions were yes, then a spatial equilibrium would exist in the location of people and activities. The spatial structure of the economy—private enterprise, centrally planned, or peasant—could be considered both efficient and equitable. Spatial equilibrium implies the greatest good for the greatest number of people. Yet the allocation of value might still seem unfair for those people and regions less able to compete. Immigrants, for example, may be handicapped by language and other cultural barriers. Hence government intervention may be necessary to ensure greater equity.

Although the criteria listed above appear to apply mainly to more developed countries, they are applicable, in altered form, to any society. Even in a primitive hunting society or an area of shifting cultivation, the questions are the same: (1) Do economic units (the family, hunting parties) produce the needed food and other essentials? (2) Are these goods allocated justly? (3) Is the resource base maintained? (4) Is the pattern of movement for hunting or for cultivation reasonably efficient? (5) Are the existing inequalities acceptable to the society?

Location theory stresses efficiency over equity. But in the real world the desire to make the most profitable use of space may be tempered by a responsible attitude toward those less able to compete, as well as toward the physical environment. In a primitive communal society, the more successful will, to a varying degree, share with the less successful. In early capitalist societies, efficiency was particularly stressed, whatever the damage to the environment or to people. The shortcomings of such an approach led to the socialist alternative, which modifies the efficiency/survival-of-the-fittest criterion of location theory by stressing the objectives of participation and equality. Modern capitalist economies have also come to accept increasing government intervention in the interests of greater equity. To achieve greater equity, society may be willing to sacrifice some efficiency—to locate health services, for example, close to the poor, even if it means an increase in the average cost of medical care (figure 13.1). To protect the physical environment, society may be willing to limit short-term profits—to resist clear-cutting a forest or strip mining a mountain slope—even if it means less income for the entrepreneur and higher costs for the consumer.

BARRIERS TO EFFICIENCY AND EQUITY

In the real world, truly efficient and equitable patterns are the exception rather than the rule. Some barriers preventing society from achieving ideal spatial order are technological change and obsolescence; individual, corporate, and societal error; the precedence of noneconomic values; monopoly control and the misallocation of resources; and external, unallocated social costs (misuse of common resources).

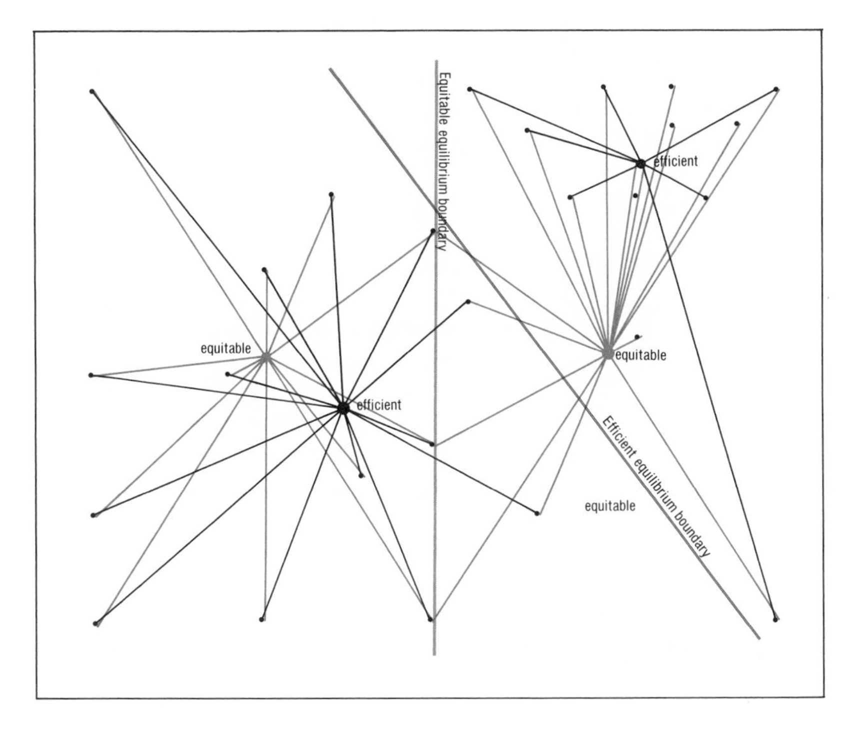

13.1. Equity versus efficiency in the location of services. This diagram shows alternative solutions to the problem of locating service facilities, such as hospitals. The efficient solution, in black, minimizes the aggregate distance individuals (small dots) must travel to reach two such facilities (large dots); however, a few people have to travel very far (see long black travel lines connecting individuals in the upper left-hand and lower right-hand corners to the two efficiently located facilities). The equitable solution, on the other hand, results in greater aggregate distance traveled by all individuals but minimizes as much as possible the distance the farthest people must travel. The equilibrium boundaries define the service areas of each set of facilities. If you look only at the equitable solution, for example, all people on the left side of the equitable equilibrium boundary travel to the left-hand facility (everyone travels to the closest facility). Planners must decide whether to locate facilities in the most efficient sites, which will maximize interaction at least cost but create hardships for some individuals, or to sacrifice some efficiency for the sake of those who live far away.

Technological Change and Obsolescence

Many inefficiently located activities are casualties of technological change and shifts in patterns of land use. Stores and services now faltering may once have been optimally located. The corner grocery store, for instance, was within walking distance of local shoppers; but today most people would rather drive to the wider range of shopping opportunities and lower prices at the supermarket than walk to the more limited corner store. Similarly nineteenth-century New England textile mills were well located. They enjoyed a plentiful source of energy from waterfalls, and they hired surplus rural or immigrant labor at low wages. But electric power has replaced water power, unionized workers have demanded higher wages, and new methods of production have supplanted the old. Today many old New England mill towns are obsolete, their abandoned buildings standing as a bleak reminder that times have changed, and lower cost opportunities exist elsewhere. It is more common, however, to find stores, factories, and farms that are still successful and worth maintaining but are no longer optimally located to serve their customers or process materials.

While technological advances have greatly benefited some people and places, they have brought ruin to many others. Machines have displaced countless farm workers and miners, their skills and entire way of life made obsolete. The shift from trains to trucks and automobiles has immobilized thousands of miles of local rail lines—at least for a time (Los Angeles and other cities now wish they had not abandoned their streetcar and interurban networks)—and has weakened thousands of small trade centers.

In rural areas, entire towns may be abandoned once local mineral resources are depleted. Inevitable shifts in patterns of land use also affect growing metropolitan areas. The flight of middle-class residents and businesses from the city to the suburbs has reduced demand for some older residential and commercial areas and fostered the decay of the inner city. And while the shift from streetcars to private cars and freeways has created new shopping districts, it has also destroyed some older ones.

Technological change is ultimately desirable if it raises the well-being of individuals and societies. But it does mean that an ideal equilibrium is never reached; the landscape is always a mix of the outcomes of decisions made in different eras under different technologies and preferences. In the short to medium run, technological changes—even planned obsolescence—are used in capitalist societies to maximize annual output and to maintain a labor reservoir. And in the long run most changes of this kind have probably been beneficial.

In developing countries, new technologies, new ways of life, and new patterns of land use conflict with the older ones and gradually replace them. Under conditions of surplus labor, characteristic of many less developed countries, advanced capital-intensive technology may be inappropriate and damaging; intermediate technologies are more suitable. Even so, both modern and traditional modes may exist side by side, as exemplified by supermarkets and itinerant vendors, and modern textile mills and handcrafted cloth.

Individual, Corporate, and Societal Error

The landscape in many places reflects the outcome of human error. A planner's underestimation of population growth results in overcrowded schools and hospitals, and overestimation results in empty and underutilized facilities. A manager's misguided instinct about the perfect location for the firm adds another failure to the heap of marginal activities.

Decision makers often do not research all the facts necessary for an optimal decision, nor do they weigh the alternatives. A public official or agency may construct a road along a particular route without even considering other routes, let alone other uses of the investment capital. Families—perhaps in their haste to escape integration—may move to a suburb without inquiring about tax rates, services, distance from work, and other conditions. Unemployed individuals may not move to available job opportunities because they lack information.

Some failures occur because information is deliberately withheld. A firm may acquire and monopolize new information about a potential government contract, using it to seize short-term advantages that will damage basically more efficient or better located competitors. But more often, poor decision making can be blamed on excessive haste, competitive pressure, lack of entrepreneurial skills, or, quite commonly, the failure to follow directions. Producers may misinterpret marketing information, for instance, and locate their firms in unpromising sites, or farmers may adopt useful innovations but implement them ineptly.

Finally people often choose nonoptimal locations because they fear making an error. Reluctant to risk everything, most people will decide to locate in well-established areas that guarantee satisfactory profits. And who can blame them? Ideal conditions are notoriously unstable, and the value of even the most carefully selected location may erode with time.

The Precedence of Noneconomic Values

In real life the perfectly rational individual is rare. Most of us do not make purely rational economic location decisions; we are influenced by inertia, impulse, fear, prejudice—and even by altruism. Technologically unemployed farm workers and miners may linger at home because the social cost of moving to better opportunities is too high. Variety store owners may cling to their unprofitable, small street-corner shops because they are unable to shift to higher-paying jobs or to a better location, or because they enjoy their way of life. College graduates may forgo immediate job opportunities to serve in the Peace Corps. People may change occupations or move simply because they want new experiences.

At a local scale, shoppers may cross the city because another center seems more attractive. At a regional or national scale, society may decide to forgo economic gains in favor of preserving greater aesthetic, biological, or recreational value over a wider region and a longer time. For example, a tall-grass prairie evolving near Chicago may be set aside as a wilderness area rather than continued as corn or soybean fields, or a virgin forest near Seattle may be preserved rather than harvested.

All these decisions may be satisfying to individuals and to society, but they are noneconomic in that maximizing productivity and profitability was not the primary goal. When noneconomic behavior prevails, the productive potential of people, activities, and locations is not realized. We do not mean to imply that following noneconomic values is necessarily undesirable; it is simply that the landscape will be different and somewhat more difficult to understand.

Noneconomic decisions prevent the achievement of theoretically optimal patterns of interaction as well as location. When executives choose to locate their firms in prestigious office buildings downtown rather than near most of their employees, the journey to work is prolonged. Similarly when consumers bypass good local stores that appear slightly outdated, patterns of interaction acquire excessive cross-movements.

Problems of the urban landscape stem in part from noneconomic decision making. The fragmentation of services into separate political units, typical of most metropolitan areas, spawns inefficient spatial patterns (figure 13.2). Many small police and fire departments exist, when fewer, larger ones might save the taxpayers money; many overlapping service districts exist, when fewer, larger ones would serve the public more efficiently—and perhaps more responsively. Inequities in tax rates and in quality of services occur when political factors influence district boundaries (figure 13.3). The political separation of city and suburb, which geographically and economically form a unit, has produced many inequities: much of the income of

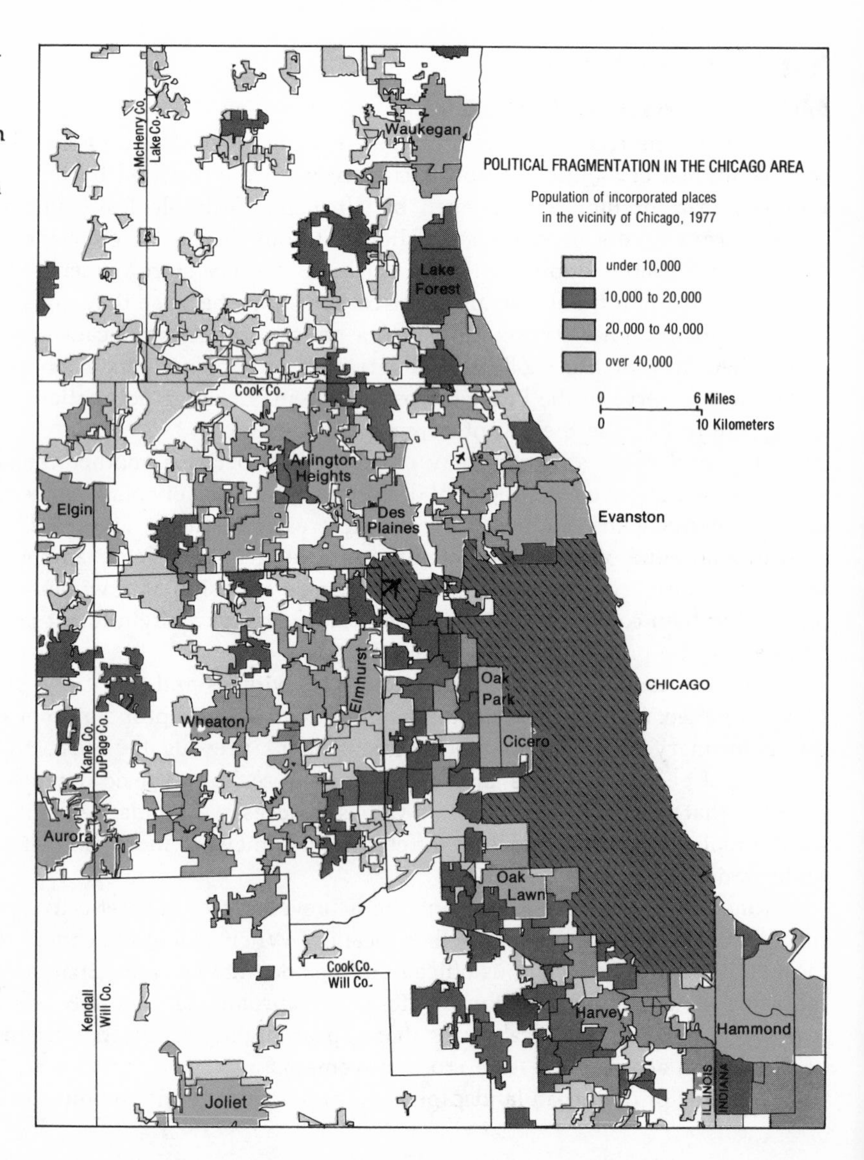

13.2. Political fragmentation in the Chicago area. Like pieces of a jigsaw puzzle, the many suburban jurisdictions around Chicago illustrate the problem of political fragmentation.
Source: Base map redrawn with permission from Rand McNally *Road Atlas*, © Rand McNally & Company, R.L. 79-Y-3.

suburban commuters is generated in the city but is spent in the suburbs, and much of the pollution, congestion, and police- and fire-protection costs are borne by city residents.

Although the separate political units of city and suburb may be inefficient and inequitable, many people place great value on local government, which they perceive to be important in maintaining some degree of control over their own lives. Few would blame the contented residents of a small Westchester suburban village for wishing to avoid New York City's seemingly overwhelming problems. Similarly many Americans and, increasingly, Europeans are willing to bear higher real costs and to deplete energy stocks by driving their cars long distances to work because they value the privacy, convenience, and security of the automobile. And they are willing to maintain large lots and sprawling homes because they value their space

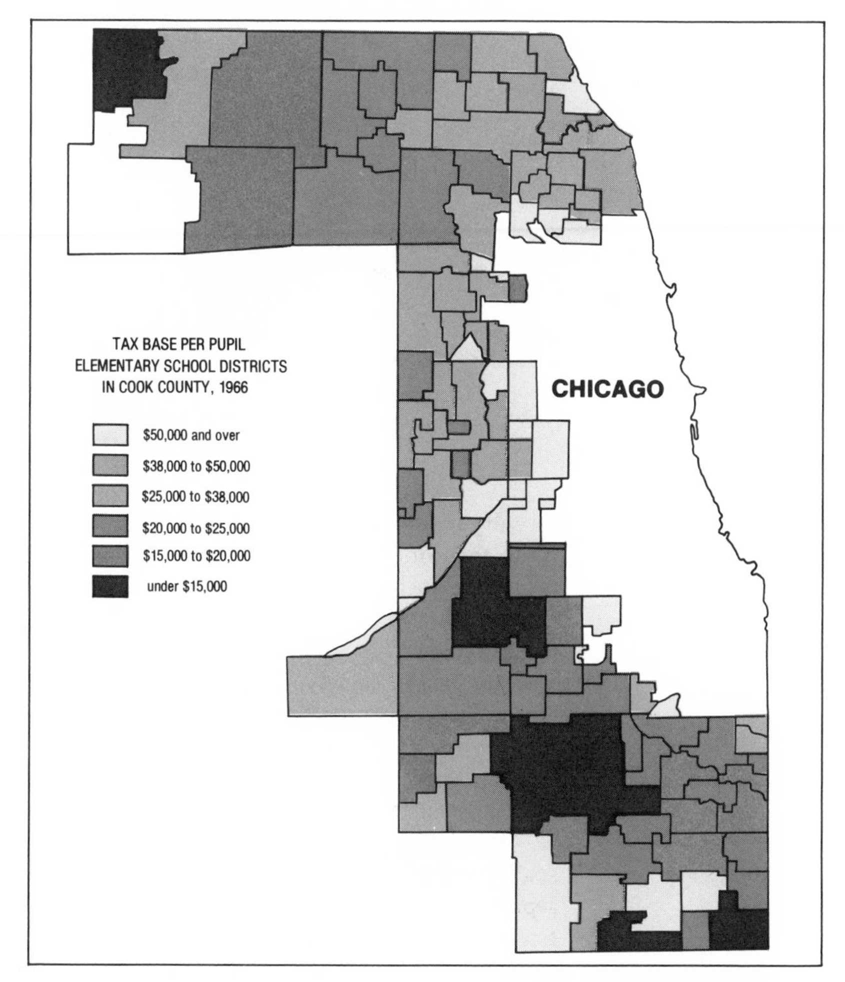

13.3. Political fragmentation and tax discrimination, Chicago, 1966. Suburban Chicago is divided into scores of school districts. This map shows elementary school districts in Cook County. The tax base (assessed value of taxable property) ranges from less than $15,000 per pupil in some districts to $50,000 and more in other districts. Some areas have formed separate school districts around industrial zones to permit good support of schools at a very low property tax rate (taxes paid by the industries make up a high proportion of revenues). Areas with many children but little industry must endure very high tax rates or poorer schools, or both.

Source: Data from Cook County Superintendent of Schools, Assessed Valuation of Taxable Property per Resident Pupil in Average Daily Attendance.

and enjoy their gardens, even if such a life-style consumes more land and fuel.

Location theory does not deal directly with the effects of racial discrimination—another noneconomic factor—on the landscape. But since spatial order is created by the competition of people and activities for desirable locations, the outcome for racial and other minorities can be understood in the light of their competitive weakness. At the national scale, the defeated and powerless native Americans were pushed outward virtually beyond the reach of national markets and opportunities (figure 13.4). At the regional scale, much of the South—with its large proportion of blacks—until fairly

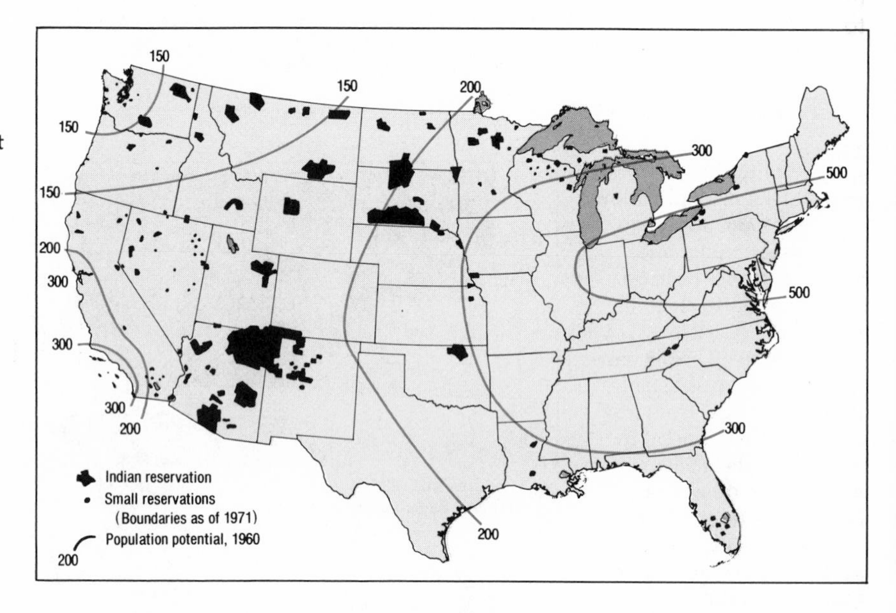

13.4. Indian reservations and population potential, United States, 1960. Note that most Indian reservations are in zones of lowest accessibility to the total U.S. population. (See the population potential surface in figure 11.9.)

recently had lower levels of agricultural productivity, urbanization, and income than one might expect from its favorable competitive position with respect to such major markets as the Boston-Washington megalopolis and Chicago. At the local scale, urban racial ghettos exist partly because racial discrimination has prevented minorities from competing on an equal basis with the rest of society. Higher-status social groups have the economic and political power to locate in desired areas, relegating lower status groups to inferior ones. In many North American cities, inferior portions of the inner city are where the poor and minorities must live, while in many developing countries the poor are forced to live in suburban squatter settlements.

The most visible geographic evidence of racial discrimination is the spatial separation of minorities in urban ghettos or rural reservations. The separation of blacks and whites in the United States clearly illustrates the tension between the geographic principles of cultural separation and social interaction. During the pre–Civil War era, when the inferior status of blacks was legally defined, the two races often lived in close proximity. After legal equality was affirmed, the white majority created a complex structure of sanctions aimed at maintaining white supremacy. As legal social distance was gradually reduced, greater spatial separation was seen as a way to avoid the presumed stigma of associating with blacks. When increasing numbers of blacks left the South for what they believed to be better opportunities in the North and West, whites felt threatened by the competition for jobs and housing. The unwelcome migrants were thus relegated to neighborhoods from which many whites fled. Racial prejudice has been such a barrier to mobility that even middle-class blacks have found it difficult to escape the ghetto. They have instead often been accommodated by its radial extension outward into middle-class housing vacated by whites (see figure 9.6). The spatial separation of blacks has been enforced by several factors: legal constraints until 1949, the discriminatory practices of real estate and financial agents, the threat and practice of terrorism, and the need to band together for self-protection. But the black (and in many cities the Puerto Rican, Chicano, and Chinese) ghetto does not persist just because of external forces. To many residents, ethnic concentration can be a source of some social, economic, and political strength—although far less than they might achieve in the absence of discrimination.

Social and economic constraints prevent ghetto residents from making the best use of space. Many activities are not optimally located, and the access of ghetto residents to the labor market and services of the larger community is often poor. Social barriers between blacks and whites have resulted in the wasteful duplication of facilities, such as recreation and health services, and have reinforced interracial fears and prejudices. On the other hand, the existence of an ethnic community offers some protection from outside threats and helps cushion the shock of cultural and economic integration.

Overshadowing the inefficiencies of ghetto life are the inequities. Public services may be inferior or lacking altogether, and employment opportunities are inadequate (much of the blue-collar employment has shifted to the suburbs). Spatial isolation and lack of opportunities have often produced a landscape of poverty, decay, and despair, further undermining the ability of ghetto residents to succeed.

Although we have concentrated on familiar American problems, the effects of social or racial discrimination can be observed in most, if not all, cultures and nations. Discrimination against Algerians in France and French-speaking people in parts of Canada, for example, has created similar problems.

External, Unallocated Social Costs

Location theory suggests that productivity and interaction are maximized at least cost when related activities agglomerate, most notably in cities. Producers benefit from concentrating a great many activities in the CBD, but residents of the city have to pay the costs of the pollution, congestion, loss of open space, and visual blight that may result from agglomeration. For example, while skyscrapers may create a magnificent skyline in our metropolises and generate property taxes benefiting the cities, the population has to bear the extra costs of high-rise construction, including congestion and even lack of sunlight (a serious problem in Tokyo).

In the short run at least, location theory ignores these external, **unallocated** social **costs,** which are not directly chargeable as production costs or taxes to those responsible for them. The city tends, in effect, to undercharge CBD activities through relatively low property taxes, encouraging excessive concentration.

Another problem is the abuse of common, or "free," resources, such as air (figure 13.5). Industries and motorists have carelessly polluted this resource, causing damage to human health and physical structures and generating unallocated costs to society.

The low-density patchwork of suburban homes and subdivisions also generates many unallocated social and economic burdens. Because distances are greater in the sprawling suburbs, the cost of utilities and fire and police protection is higher, expensive school busing may be required, more roads and highways are needed, and the journey to work may be longer. As more commuters take to the road, the level of air pollution and congestion also rises. To the extent that these costs are subsidized by the more efficiently located residences and businesses in the city, the distant suburban residents are unfairly exploiting the common resources of city revenues. Another unallocated cost is the loss of productive farmland to suburban development. Many farmers have found it more profitable to sell their land to speculators, particularly if the local government taxes the land according to its fair market value (the price that could be obtained for it if sold on the open market) rather than its value as agricultural land. In this case, the speculator is exploiting value created by the growth of the city, not by his entrepreneurship.

Other costs arise from primary activities, such as mining, forestry, and ranching, as well as from urban and industrial activities. According to location theory, primary activities can be considered efficient as long as the direct costs of exploitation and transport do not exceed revenues at the market. Thus short-term profits may be gained from strip mining, excessive clear-cutting, and overgrazing; but society has to shoulder much of the visual blight and long-term hidden costs of environmental damage from erosion, deforestation, flooding, and the removal of land from productive use.

Monopoly Control and the Misallocation of Resources

Monopoly control (in the economic, not spatial, sense) blocks the play of free competition among the equals assumed by location theory. Inefficient and inequitable spatial patterns may result when individuals, businesses,

13.5. Air pollution: an unallocated social cost in the Ruhr industrial district of West Germany. (Photo from Keystone Press Agency.)

government agencies, or unions wield more than their fair share of power. More fundamentally, monopoly control may create imbalances in the use of resources and unnecessary inequality in wealth and income across a city or a nation. An interesting example of collusion among public and private institutions is the routing of the Alaska oil pipeline, which may not be the least-cost solution but which concentrates benefits in Alaska. Another example of public-private or private restraint on potential trade is the very limited numbers of businesses located on turnpikes. In many countries, single enterprises in some regions may be strong enough to keep out potential competitors for resources and labor.

When production is concentrated in fewer than the optimal number of locations, longer movements are required, prices may be higher, and consumption lower, thereby reducing the efficiency of the industry. For example, if a particular firm gained monopolistic control over a regional meatpacking industry, it could close several plants. Products would have to be shipped farther to market; prices would rise; and consumption, but not profits, would probably decline. The result would be lower efficiency and further deviation from a theoretically ideal spatial equilibrium.

Inequities may also result from the concentration of production in fewer places than would arise under free competition. For example, if a firm producing small appliances in Massachusetts gained monopolistic control over the industry, it might be able to prevent others from producing small appliances in California. Northeastern customers could be made to subsidize transport costs to California by paying a slightly higher price, while California customers would pay a relatively low price (low enough to prevent local competition from arising in California).

Monopoly control over investment, whether by industry or government, may result in the misallocation of resources and lead to greater disparity in wealth and income. Investment capital may become concentrated in select locations or economic sectors. Private investors may show preference for certain regions, such as the Southwest and other growing areas, ignoring regions where opportunities are perceived to be less promising, such as Appalachia or the inner city.

In the public sector, favoritism and earmarked revenues may foster inefficient investments. Excessive road construction in the United States may be promoted by the exclusive earmarking of federal and state gasoline taxes for new roads. Benefit-cost comparisons for public investment in irrigation systems, dam construction, river-diversion schemes, and the like are often exaggerated, and alternate investments are rarely considered. The Aswan High Dam, for example, has brought many benefits to Egypt, such as hydroelectric power and additional irrigation, but the costs have been far greater than expected: much more fertilizer must be used now that silt is no longer deposited by annual flooding; the salinity of the soil has risen because salts are no longer washed away by the floodwaters; and a debilitating disease, schistosomiasis, has spread more rapidly because disease-bearing snails breed in the new irrigation channels.

THE GEOGRAPHIC IMPACT OF ECONOMIC INEQUALITY

Perhaps the most important effect of unequal power is that those with some degree of monopoly control over prices, labor, and resources are able to secure higher incomes and maintain greater wealth (through tax laws) than would be true in a purely competitive social order. Indeed the clearest sign of spatial inefficiency and inequity in urban and regional landscapes is extreme divergence in levels of income. If the economy were unfettered by monopoly control, if decision makers had equal access to information, and if individual and regional power were balanced, any existing economic inequalities would be the limited and inevitable outcome of relative location and individual differences. But the fact is, income levels vary greatly not only among different individuals but also among different sectors of the city and different regions, not to mention different countries (figures 13.6 and

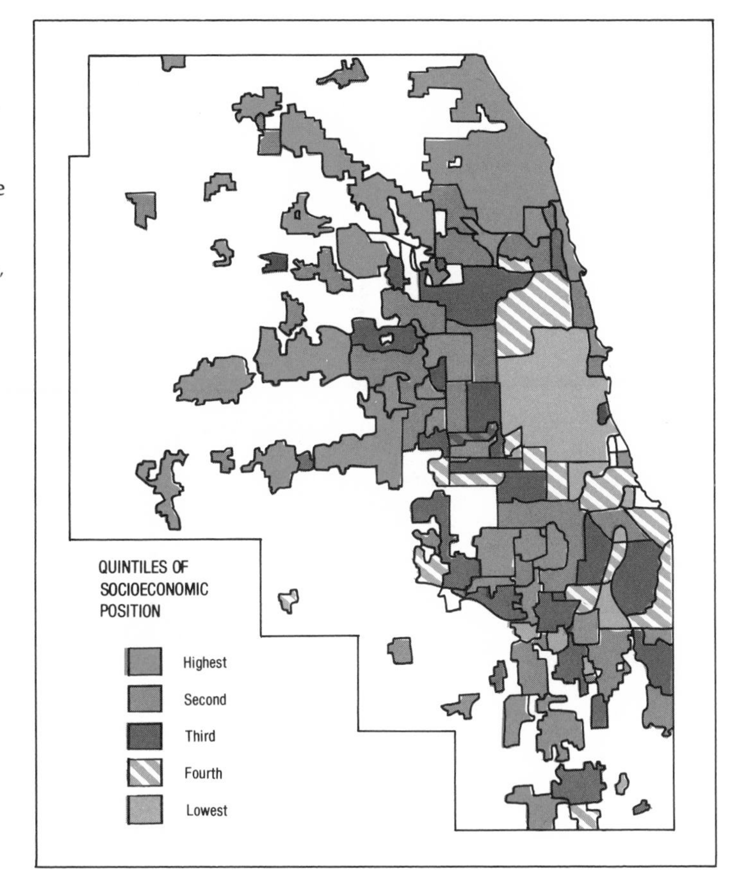

13.6. The socioeconomic position of Chicago-area communities. *Quintiles* is a statistical term referring to the division of a population into five equal proportions. Most high-income communities are located on the periphery, particularly along the north shore, while low-income communities are concentrated in the inner city.
Source: Redrawn with permission from J. Simmons, *The Changing Pattern of Retail Location,* Research Paper No. 94, University of Chicago, Department of Geography, 1964.

13.7; see also figure 1.5). Because the economic disparity is greater than one would expect from inherent differences in productivity, we can only conclude that the spatial structure of the economy is inefficient and inequitable. We will examine some of the causes and effects first of economic inequality in the city and then of economic inequality among regions. The discussion will emphasize the United States but is applicable to other societies.

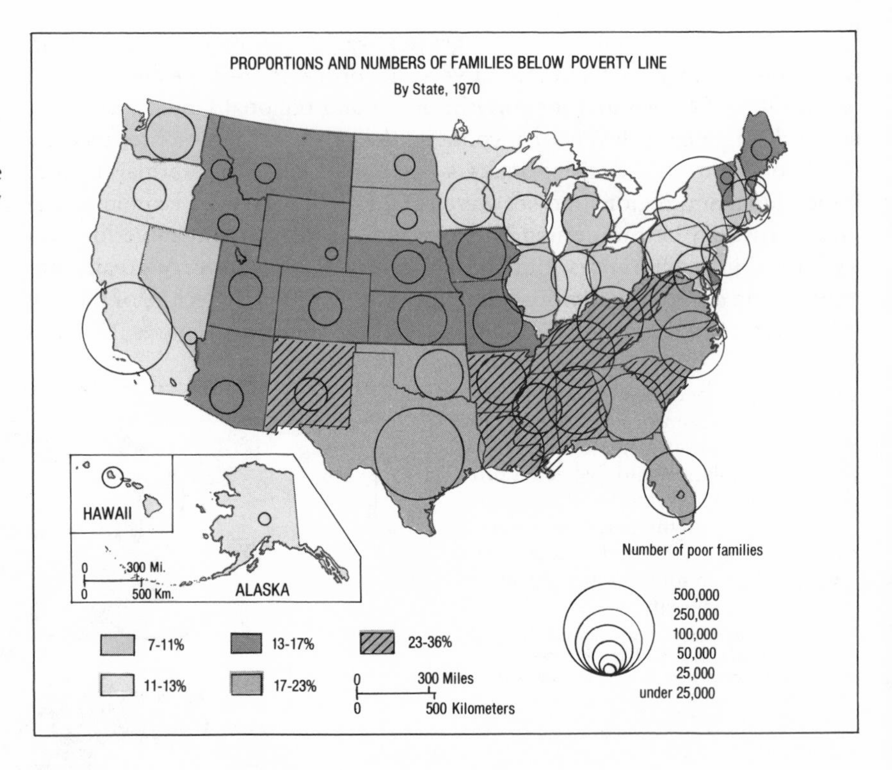

13.7. Distribution of poverty, United States, 1970. The shading indicates the proportion of families classified as poor by state. The circles indicate the absolute number of poor families by state. Note the large numbers of poor in the generally more affluent northeastern states and California.

The Urban Landscape

Ideally the urban landscape should reflect the spatial patterns described in chapter 9—patterns formed by competition for land and for access to central locations (see figures 9.3, 9.4, and 9.9). But not all members of society can compete as equals; racial discrimination, for example, has hindered the achievement of theoretically ideal patterns.

Many spatial problems stem from extreme economic inequality in the city, which in turn is the result of racial and social discrimination, the competitive weakness of recent rural migrants, and differences in the political and economic power of individuals and groups. Such problems include incomplete business center development, segregation by socioeconomic class, declining accessibility to services for the poor, excessive cross-movements and travel, and inner-city decay.

To understand why many central cities have deteriorated, we must look to the past. The poor from the countryside or from abroad usually moved into industrial and commercial areas shunned by those better off. As the poor increased in number, part of the middle class moved to newer housing in the outer city or suburbs, and their homes were left for the poor. The latter often could not afford the high rents needed to cover high taxes for these homes, yet they required access to inner-city jobs. Thus landlords found it both necessary and profitable to subdivide homes into low-rent apartments. The density of such neighborhoods increased as people crowded in. Eventually the subdivided homes degenerated into slum tenements, where a transient and poor population no longer produced enough income to cover taxes, maintenance, and profits.

The suburbanization of the middle class, which became pronounced in the 1950s in U.S. cities, was accompanied by a shift in the location of many retail stores and services from the city to the suburbs. Access to goods and services thus worsened for the urban poor. The erosion of downtown trade, services, and blue-collar employment hastened the social, economic, and physical decay of the inner city. Many inner-city neighborhoods have suffered decades of neglect and poverty; formerly occupied homes now lie vacant and derelict (figure 13.8).

Middle-class flight from the central city has led to underutilization of urban land, diminished the productivity of existing investments, and raised transport costs both for the poor, many of whom must commute to industrial jobs in the suburbs, and for middle-class and white-collar workers,

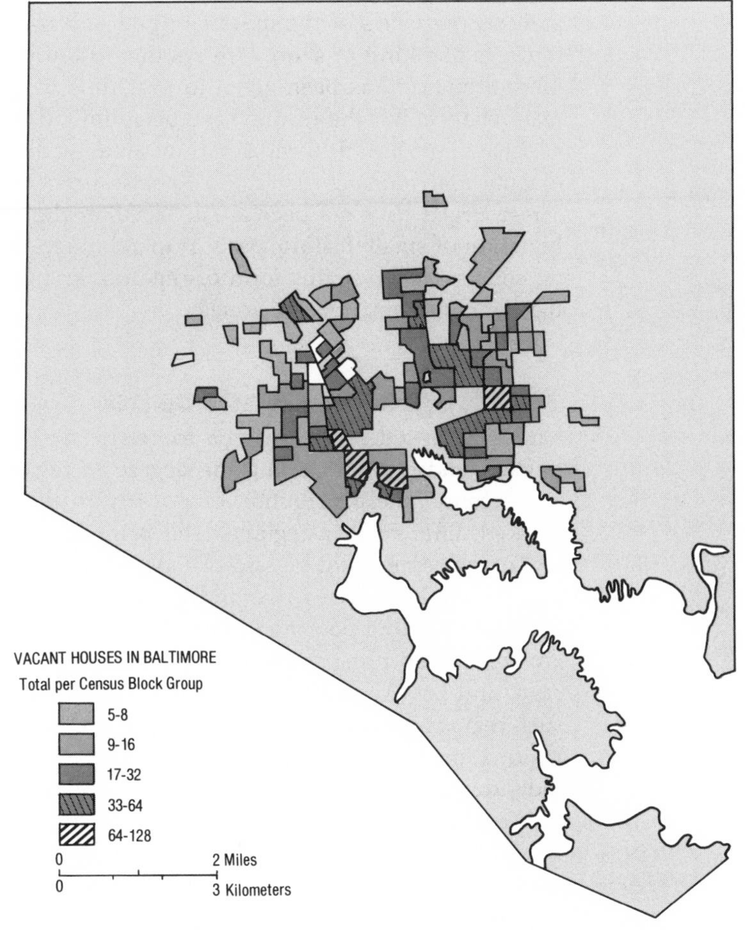

13.8. Vacant houses, Baltimore, 1968. The shading indicates the range of the numbers of houses vacated per census block. Most of this housing is abandoned because of long-term neglect by owners, often absentee, or the inability and unwillingness of the poor to pay sufficient rent or cover property taxes. Since 1974 some areas have been the focus of private renewal, or "gentrification," as middle-class professionals rehabilitate and restore older housing in favored areas.

Source: D. Harvey, *Society, the City, and the Space-Economy of Urbanism,* (Washington, D.C.: Association of American Geographers, Resource Papers for College Geography #18, 1972). Redrawn by permission.

who commute across the inner city to downtown jobs. In addition, the exodus of the middle class has reduced the availability of health services for the urban poor (figure 13.9). High death rates tend to coincide with the inner-city distribution of poverty, substandard housing, and racial segregation (figure 13.10). The income of the city has declined because of the loss of middle-class consumers and retail stores, the shift of profits and rents to the suburbs (many store owners and landlords live in the suburbs), the loss of population and housing, and the subsidy of urban commuters and other forms of income transfer (figure 13.11). In a vicious cycle, the city is forced to increase taxes on the declining tax base of a poorer population to pay for the higher costs of services and of crime, further increasing the tax burden on the remaining poor.

Thus income inequalities have played a major role in the deterioration of inner city neighborhoods. The planning response to these problems in American cities has typically been architectural: massive slum clearance measures in urban renewal projects that usually replace tenements with high-rise towers. Belatedly it has been recognized that unless employment and income levels of the inner-city poor are raised, urban renewal merely replaces one kind of slum with another, often less desirable one. Recently more attention has been given to programs of housing rehabilitation and maintenance, but the results have been limited. Many urban renewal projects have attempted to replace slum areas with offices and higher income tenants and housing—Detroit's Renaissance Center is a recent example. These efforts have not been as successful as the private, smaller-scale rehabilitation of small, historic parts of many inner cities by upper-income professionals. However this form of renewal, known as *gentrification,* displaces the poor into neighboring zones.

The Regional Landscape

Earlier we listed several criteria for assessing efficiency and equity in the human landscape, including the degree of regional differences in income and wealth. The distribution of poverty in the United States shows that marked differences in regional well-being exist in this country (see figure 13.7). The same could be said of most other nations, advanced or developing, private enterprise or centrally planned.

Fully half the poor and unemployed in the United States live in what are considered prosperous metropolitan regions, such as New York, Chicago, and Los Angeles. The fact that large numbers of poor people live in such regions suggests that economic inequality may stem not so much from regional problems as from the basic imbalance of economic power among individuals and groups. Yet striking differences in income and wealth appear in the regional landscape. Relatively prosperous regions include the large metropolitan areas, California, the Boston-Washington megalopolis, the lower Great Lakes, and some rural areas in the Northeast and Far West. Relatively poor regions include most nonmetropolitan areas, the rural

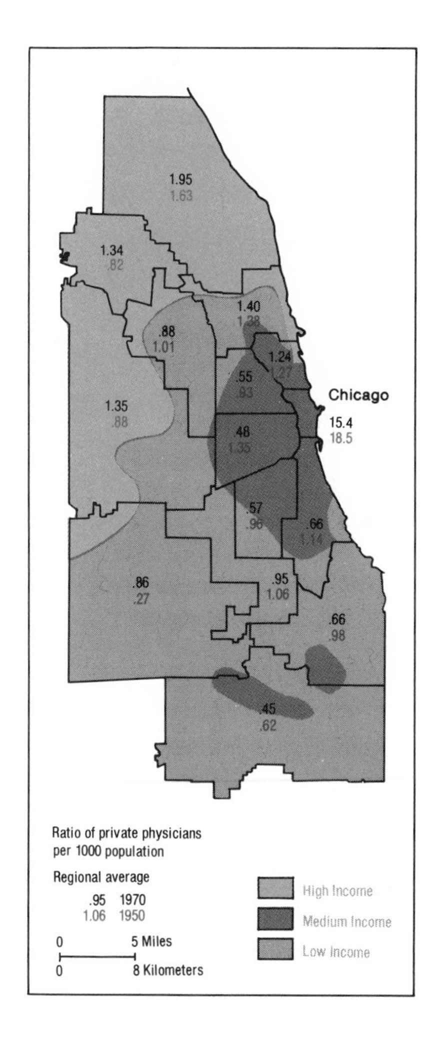

13.9. Changes in the availability of physicians' services, Chicago, 1950–1970. Except for the high concentration of specialists in downtown Chicago (not shown), poorer inner-city neighborhoods have suffered sharp declines in number of physicians. The level of accessibility to physicians' services is less than half that of affluent inner suburban areas.

Source: Data from D. Dewey, "Where the Doctors Have Gone," Research Paper, Illinois Regional Medical Program, 1973.

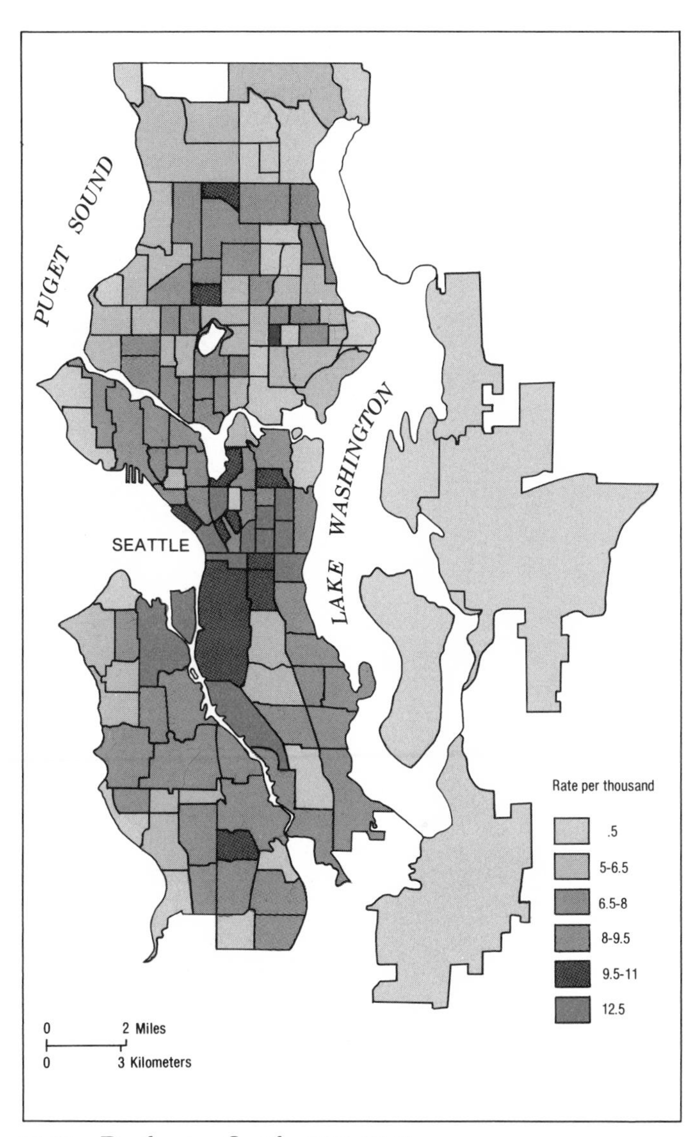

13.10. Death rates, Seattle, 1968–1969. These age- and sex-adjusted death rates show a concentration of higher death rates in the poorest, most industrial parts of the city with the worst housing.

Source: Adapted with permission from D. Johnson, "Differential Mortality Within Cities," M.A. Thesis, University of Washington, Department of Geography, 1972.

South (including smaller urban areas), Appalachia, the Ozarks, peripheral areas in New England and the Great Lakes region, the Great Plains, and some parts of the West, particularly Indian reservations.

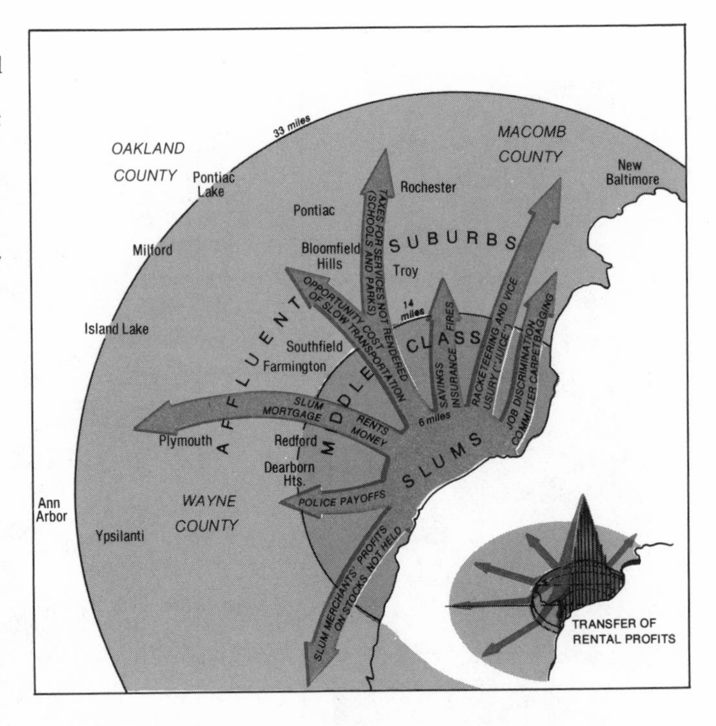

13.11 Income transfers, Detroit. This figure suggests the pattern of estimated income transfer from the poorer inner city to the more affluent suburbs. Most important are rents, profits from outside-owned businesses, and local regressive taxation. Actual figures, however, are very difficult to determine.
Source: Redrawn with permission from W. Bunge, *Fitzgerald* (Cambridge, Mass.: Schenkman, 1971).

The role of the physical environment and the uneven distribution of resources explains some of these disparities in regional well-being. But poverty in some of the richer natural settings, like the Mississippi delta, and prosperity in arid or semiarid regions, like Wyoming and Nevada, suggest that the answer lies elsewhere. Because poverty tends to increase with distance from large metropolitan areas, and the most severe poverty is found in areas with the poorest accessibility, such as the Ozarks and Appalachia, a more plausible factor would be access to major metropolitan markets. Perhaps the most convincing explanation of regional inequality has been advanced in a theory that postulates a lag in regional development. As the economy develops, according to this theory, some rather large areas cannot adjust to changes in transportation, technology, and economic demand. Consequently these regions have a much higher level of unemployment and lower per capita income than the regions that can adjust. For example, mining and lumbering areas declined in the upper Great Lakes region because labor requirements fell, iron ore deposits became depleted, and timber resources thinned out. Technological unemployment in such areas spread rapidly. In addition, many agricultural areas in the South and West have declined because they contained small, inefficient farms that could not yield sufficient income per family. Of course, increasing efficiency and productivity in mining, forestry, and agriculture continually displace labor. But most rural areas have not been able to attract enough industry and services to absorb the surplus workers. Hence many people in such regions are either unemployed, underemployed, or can be hired at relatively low wages.

In the adjustment process, surplus workers from declining regions migrate to more prosperous ones—usually large cities. The rate of migration is often too slow to rescue the declining regions from unemployment and poverty, and too fast for the prosperous regions to absorb, given the migrants' lack of skills and education. In theory, an equilibrium will eventually be reached in the backward areas: the surplus population will leave, and the remaining farms, mines, and forestry operations will be large and efficient. Labor-intensive industries will move out from the higher cost metropolises to take advantage of the surplus rural labor, and the rural population will be more fully employed. Regional economic convergence (equality in wealth and income), theoretically at least, will have supplanted economic inequality. That this process is occurring is demonstrated by the significant slowdown of rural-urban migration and by the decentralization of manufacturing out of metropolitan regions in both Europe and the United States. Economic development has been rapid particularly in white-settled parts of the U.S. South since 1960. Other poor regions, like the Ozarks, are experiencing economic and population growth due to the return of retired persons as well as industrialization.

Although relative incomes have already begun to converge, regional unemployment and levels of poverty are still high. The out-migration from rural areas often consists of the better educated and more motivated members of the population, leaving behind a large proportion of elderly people (or middle-aged people who feel it is too late to start over again) and a high degree of economic dependency, as in Appalachia. Consequently the level and variety of services decline. But despite the poor return on much of the farming and the lack of alternative opportunities in rural areas, many people prefer to remain in the countryside. Perhaps they realize that moving to the city will not guarantee any real improvement in the quality of their lives.

Even though advanced societies are becoming increasingly dominated by metropolitan areas, some question remains as to whether such a pattern is indeed the most efficient. Productivity may be lost by underutilizing human and natural resources in areas relegated to extensive activities such as farming and forestry. Rising pollution, congestion, crowding, and visual blight in the city may offset the benefits of agglomeration. Moreover even the most prosperous metropolitan regions have been unable to achieve full employment and to eliminate poverty. The United States has considered the alternative of planning economic development more evenly across the nation, a course of action preferred by many advanced nations. There is recent evidence, however, that employers and individuals are acting on their own to restore growth to many of these lagging regions. This may or may not lead to greater equality among regions. But it is clearly purposeful behavior in response to the desire to maximize well-being, even if achievement of this goal cannot be measured in purely economic terms.

The developed countries, despite their wealth and technology, have not succeeded in fully meeting and maintaining the most basic societal objectives: adequate food, shelter, and security. Perhaps 5 percent of the people in the United States have severely inadequate diets and substandard housing. In the richest countries such failure cannot be for lack of resources; it stems from unequal access to resources and from the traditional Western emphasis on individualism rather than social responsibility. Economic inequality in the United States also reflects higher levels of unemployment for many minority groups. Given sufficient income redistribution and full employment, families and individuals would readily be able to obtain adequate food and shelter. Some developed countries, like Sweden and Austria, have been able to reduce regional and class inequality and eliminate outright poverty—in part because of a rather homogeneous, well-educated population. Others, like France and Italy, have greater regional and class inequality than is found in the United States.

PROBLEMS OF DEVELOPING COUNTRIES

Now let us turn from the urban and regional problems of advanced countries to world problems related to human spatial organization, particularly the problems of less developed countries and their relations to the developed world. It is not strictly possible, of course, to evaluate the equity and efficiency of the world, for there exists not just one single world economy or society but many national and even local economies with only tenuous connections to each other. These societies vary greatly in their organizational complexity, spatial interdependence, and cultural characteristics. It would be impossible to unravel these complexities in a few pages, but we will attempt a brief introduction to the problems of world social and economic development, which reflects the ability of societies to provide for the well-being of their people in terms of food, shelter, and quality of life.

We observed earlier that all societies may be evaluated by whether the economic units provide the necessary food, shelter, and other goods; the allocation of goods and services is just and the resource base is maintained; the pattern of movements is efficient; the level of inequality is acceptable; and real costs are borne by those responsible. We also noted that the performance of all societies is affected by technological change and obsolescence; individual and societal error; the precedence of noneconomic, or nonoptimizing, values; inequality in power and the misallocation of resources; and the presence of unallocated social costs. We may reasonably state that the importance of noneconomic values is even greater in most developing countries than in the more advanced ones and that such values are as basic to understanding the world landscape as the economic forces we have been stressing.

The World Economy: Inequality and Inefficiency

International trade may be considered a first step in establishing a world economy. Most international trade is rational in the sense that it is profitable to both sides, but in many ways it is nonoptimal. First, the theoretically ideal balance of trade is distorted by ideological conflicts and political loyalties. There may be too much trade within a political bloc, as within the EEC or COMECON (Soviet Union and Eastern Europe), while good opportunities to trade outside the bloc are ignored.

Second, the goal of self-sufficiency and the fear of dependence on others may result in more costly internal production, even though greater exchange would be more profitable. The United States, for example, intends to work toward energy self-sufficiency for national security reasons, despite the fact that it would probably be much cheaper to continue to rely on energy imports (conservation might be cheaper still).

Third, the more powerful nations may be able to control prices to their own advantage, thus aggravating international differences in development. In particular, this takes the form of low prices for the raw materials of less developed nations and higher prices for the manufactured goods of richer nations. (The OPEC petroleum cartel, however, has placed a special set of countries in the position of great power with respect to other developed and less developed nations.)

Fourth, most nations impose barriers against trade from their foreign competitors. It can be argued that at times during the developmental process, barriers to protect infant industries may be required. But too often even the most developed nations use trade barriers to protect inefficient industries from healthy competition.

And fifth, much more trade would undoubtedly exist if the poorer countries were able to pay for more imports. Canada, the United States, and other countries, for example, could in theory produce much more food and relieve world hunger. But the neediest countries are the least able to pay for more imported food.

The principal evidence of economic imbalance and inefficiency is the extreme inequality of the world distribution of income and consumption of energy (see table 1.1). Variations in level of development and income among nations are, of course, far greater than within a single nation (figure 13.12). Technology, organizational complexity, kinds of production, and patterns of consumption vary greatly among nations.

Development in an earlier era could be defined by the achievement of a nation's few rich people. But in the modern era, when an individual's expectations are no longer tied to social position, development is better measured by the average income per consuming unit—usually the family. All other measures of development—energy consumption, literacy, health, quality of transport—are closely related to the typical family's income. Measures of income and energy use generally reflect only the organized,

A

B

13.12. Economic inequality. These street scenes of Calcutta, India, in photo A, and Stockholm, Sweden, in photo B, show wide disparities in standard of living, reflecting differences in average per capita income ($150 in India, $8,670 in Sweden). (Photo A from Keystone Press Agency; photo B by Staffan Wennberg from Black Star.)

commercial sectors (the goods and services sold to others), so that differences in actual well-being are markedly less than the figures suggest. Still they are indicative of a nation's ability to provide for the well-being of its population.

The wealthy, advanced countries include those of northwestern Europe and their direct offshoots, the United States, Canada, Australia, and New Zealand, as well as Japan, all with per capita incomes exceeding $3,000. The economy and culture of most of these countries have been closely linked for centuries. Except for Japan, they were the first to abandon feudalism, adopt capitalism, and develop open political and social institutions. They are highly industrialized and urbanized. Few people remain in agriculture; all are involved in the commercial economy. The countries are highly interconnected, internally and externally, and they dominate world trade. Although containing but 12 percent of the world's population, they control perhaps 50 percent of the world's wealth. They have reached a level of affluence that allows the population to be concerned over internal economic inequality and environmental deterioration.

A somewhat larger set of countries is moderately to highly developed, with per capita incomes of $1,000 to $3,000. Included are such diverse areas as southern Europe, eastern Europe and the Soviet Union, a few countries of Latin America (Venezuela and Argentina), Africa (the Republic of South Africa), and Asia (Israel and Singapore). These countries are fairly industrial and urban, but they retain a large, less productive rural sector. Some display great economic inequalities (most noncommunist countries), underdevelopment of service activities (the communist countries), or severe racial discrimination (the Republic of South Africa). Although further internal specialization and interconnection may be required for development, these countries have established significant industrial and export bases and may soon be in a position to assure their citizens reasonable material affluence (figure 13.13).

Saudi Arabia, Kuwait, Qatar, and other OPEC countries have very high per capita incomes, but this prosperity has been achieved only recently. Hence while modernizing fairly rapidly, they still have many of the characteristics of much less developed countries—little industry, high infant mortality, low levels of literacy, and low status of women. They have, however, adopted the Western technological model of development, emphasizing specialization, urbanization, and trade, and are in a favorble position for further economic and social development, depending perhaps on international security and energy costs.

The remaining countries comprise the third world. Containing about two-thirds of the world's population but controlling only one-fourth of the world's wealth, they may be considered developing nations, with per capita income levels generally under $750 per year. Included are much of Latin America, most of Africa, and much of Asia. In this group are such major nations as Nigeria, China, Indonesia, India, and Pakistan. While all are in the process of industrialization, urbanization, and infrastructure development,

13.13. Successful industrial development in Taiwan: the Kao Siung oil refinery. (Photo from Keystone Press Agency.)

they may still be considered *dual economies:* a minority of the population is moderately affluent and fully involved in the commercial economy, and the majority remains in a traditional, near-subsistence economy. Consequently income inequality and regional disparities are unusually great (except, perhaps, in China). A few countries, such as Burma, Cambodia, and Tanzania, are trying to avoid a dual economy by following a slower, more dispersed and egalitarian model of development.

To account fully for variations in national develoment would be beyond the scope of this book. Some of the reasons, however, will be suggested below.

Barriers to Development

Economic and social development are based on the interdependent processes of industrialization, urbanization, commercialization, mechanization of agriculture, and improved education and health. Normally a rather long period of heavy savings and capital investment is required to provide an efficient system of education and communication—the foundation of development—as well as to raise the productivity of labor and land.

Such development is a major goal of most countries. Those who suggest that not all people want to be or are meant to be developed simply do not understand history or human nature. There are *no* people who do not want greater economic security or material well-being. It is true, however, that most societies are also concerned with military security, political ideology, and protection of cultural integrity. Rich and poor nations alike spend an often unnecessarily high proportion of their limited capital on nonproductive armaments, thereby slowing development. Other factors that impede development include population imbalances, environmental and cultural constraints, competition with wealthy nations, colonial trade patterns, lack of capital, and international conflict.

Population Imbalances

Population imbalances create a basic developmental problem. Nations with very small populations or areas, or both, such as many of the new nations of Africa and Oceania, lack the market and labor forces necessary for even modest scale and agglomeration economies, let alone the accumulation of enough capital to finance a variety of industries. The creation of larger nations, or at least the formation of cooperating economic blocs, may be necessary to development, but this is difficult to accomplish in the face of the militant nationalism of most new nations.

Consider the problems confronting the new nation of Maldives, off the southwest tip of India. Its small population of 120,000 persons, scattered across 220 inhabited islands over an area of 298 square kilometers, subsists on fishing. Perhaps the only form of development possible in such a tiny and fragmented nation is through the promotion of tourism. But the Maldivians themselves recognize that success in this endeavor might destroy the special quality of their lives and are therefore reluctant to embark upon such a course.

Too large a population can also be a serious problem. Often the ratio of population to arable land is too high, few economic opportunities other than farming exist, the level of technology is low, and the rate of population increase is excessive. India, for example, is one-third the size of the United States but has three times as many people (620 million). Its population grows by 35,000 people each day—more than 13 million a year. In countries where land already is used intensively, most of the society's investment and labor is commonly required just to increase food production enough to feed the additional population (figure 13.14). Despite a large potential market and labor force for other kinds of production, not enough surplus food can be generated to provide capital for allocation to alternative activities. In these countries birth control programs are needed to reduce the rate of population growth below that of economic expansion and capital formation. (India, Indonesia, and several other countries have succeeded in raising food production slightly above the rate of population growth and, through

13.14. World distribution of hunger. The U.N. Food and Agriculture Organization has established that people need an average of 2,350 calories a day to maintain health and strength, although there are individual variations depending on climate, level of physical activity, age, sex, and body build. The map shows that nutritional requirements are not being met in much of Asia, Africa, and Latin America.
Source: Data from 1976 FAO Production Yearbook.

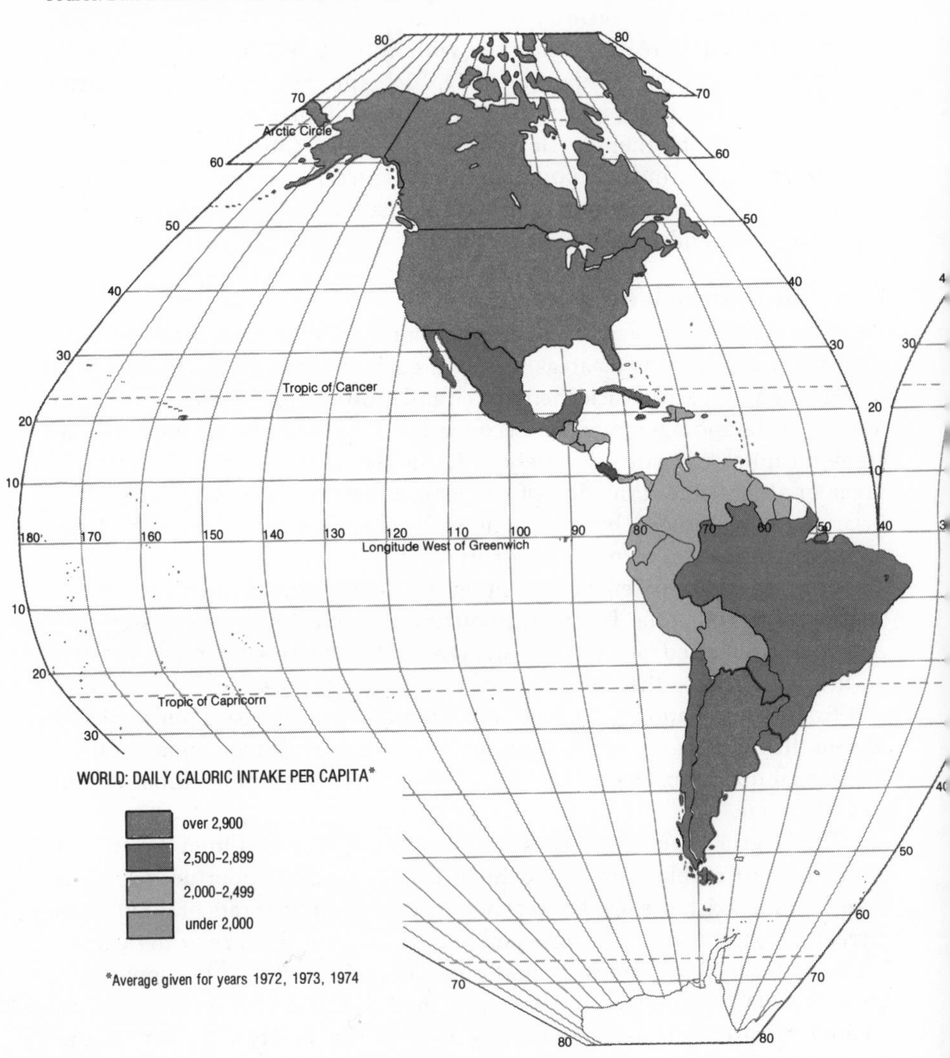

strenuous efforts, are beginning to reduce birthrates.) Birth control programs are often difficult to institute, however, under conditions of little education, traditionally large families (for a number of reasons, one being that parents want to be guaranteed care in their old age), inadequate nutrition, and, in some cases, religious and other cultural barriers.

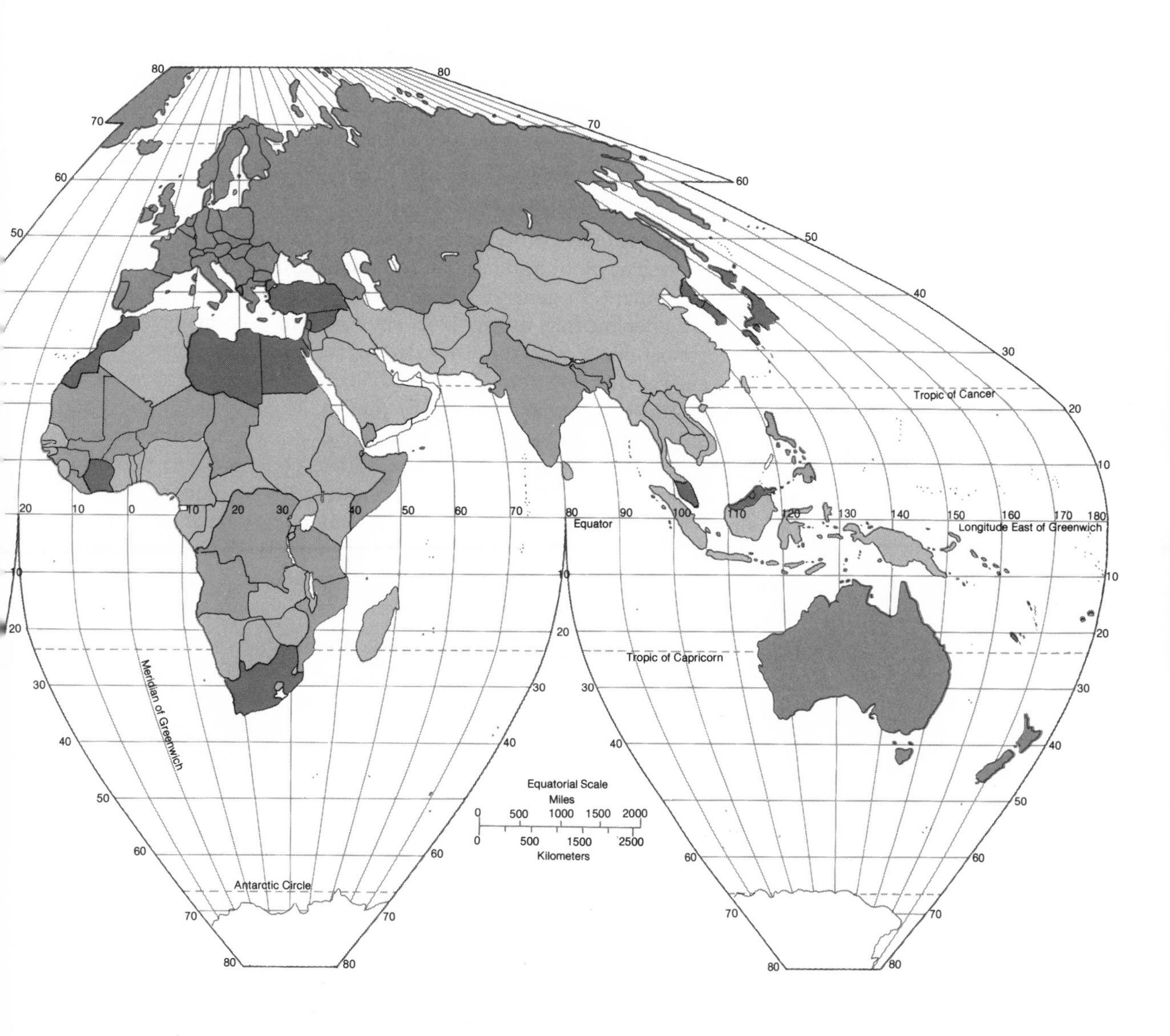

In many developing nations, such as India, Egypt, Indonesia, and Brazil, part of the surplus population has chosen or been forced to leave the overcrowded countryside and move to the cities, even when there are no jobs. Incredibly overcrowded housing results. The capital needed to create jobs may thus have to be diverted into minimal housing and services. Much of the labor force is engaged in retail and service activities, as peddlers and servants, for example, generating only marginal incomes. (A few, however, earn fairly high incomes in the modern industrial sector or in the bureaucracy.) Current capital-intensive Western technology is not always appropriate to such an economy, given the large number of surplus workers. Most successful, perhaps, has been the luring of labor-intensive industries, such as textiles and electric components, out of Europe and the United States to Singapore, Taiwan, Korea, the Philippines, and other less developed countries. But this is at the risk of dependence on multinational corporations and uncertainty as to their long-term behavior and permanence.

Environmental Constraints

The kinds and qualities of resources available to a nation, including land and water as well as biotic and mineral resources, strongly condition the nature of development. In general, the less technically advanced the society, the greater its dependence on the local environment. The major source of income for most developing countries derives from the exportation of local minerals, such as iron ore, petroleum, and bauxite, and of foods appropriate to the land and climate, notably sugar, pineapple, bananas, rubber, and cacao. Lack of significant natural resources makes development particularly difficult. Bangladesh, for example, is so resource poor that it is unable to import the technology it needs for the maintenance of food production. Lack of fresh water may also constrain development, as it does even in such rich Middle Eastern countries as Saudi Arabia and Kuwait. Cycles of inadequate moisture, as was experienced by the West African Sahel in the mid-1970s, can severely set back both economic and social progress.

Cultural Constraints

The measures necessary for development often conflict with existing attitudes and traditions. Religious practices and beliefs or traditional diets may militate against economic efficiency, as demonstrated by the underutilization of cattle in India. More important, social structures often show a lack of enthusiasm for development. If a ruling group has achieved great wealth through the small surpluses of many laborers or by controlling resources, it will gain little from development in the modern sense, which is based on the concept of a mass market. The upper class can gain more security and a greater return by investing funds in more developed countries. A successful transition from such a structure requires land reform. But unless there are simultaneous urban-industrial opportunities for the surplus farm labor, land reform will in itself be futile. For example, Iran, with the help of great oil revenues, is struggling to overcome centuries of feudal tradition, including a pervasive discrimination against the equality of women. Despite Iran's petroleum wealth, its development is hindered by the lack of skilled workers, by the desire to become an important military power, and especially by conservative Islamic opposition to modernization.

In many countries—Egypt and India are two—the entrenched bureaucracy makes it very difficult to begin new enterprises or to be innovative. Many less developed countries with largely illiterate populations tend to remain conservative peasant societies, distrusting rapid change and intolerant of individual innovation. (The failure to adopt innovations is also due to lack of capital.)

Competition with Wealthy Nations

Another barrier to development is the superior position of already developed nations. In the short run, a new industry in a poor country will rarely, even with extreme savings on labor costs, be able to compete with a highly efficient industry in a rich country. Until the local labor force increases its productivity and until local markets develop, the new industry will usually require financial protection. It is difficult to determine whether a given industry will eventually be competitive or whether purchasing the products from another country will always be cheaper.

Colonial Trade Patterns

A developing nation's trade patterns, often inherited from its colonial past, typically depend on one or a few exported goods subject to wide price fluctuations. The variations in capital formation make developmental schedules and payments difficult to meet. In addition, the colonial pattern of exporting primary goods and importing manufactured products leads to excessive emphasis on the importation of consumer goods. Thus there is a persistent conflict between importing capital goods for future gain and importing consumer goods to satisfy present needs and desires. The effect of this competitive imbalance and the colonial trade pattern is to maintain a pattern of industry and exchange based more on historical relationships than on the inherent comparative advantage of resources or location. But the formation of producers' cartels, such as OPEC, has turned the dependency of advanced nations on imported raw materials into an advantage—at least for a while.

Lack of Capital

The principal barrier to development is lack of capital. Industrial nations may be reluctant to invest capital in developing nations, because they fear investments will encourage direct competition with their own industries or will be nationalized by the developing nations. Internal capital is difficult to accumulate if much of the economy remains near-subsistent and if population and consumption pressures are high. A farmer, for example, may want to buy a small tractor or hybrid seed, but his holdings may be too small to justify the investment (at the prices he could receive for any surplus). Democratic but uneducated societies are unlikely to restrict consumption enough to permit the high investment necessary for sustained growth. Dictatorial societies are often unwilling to invest in needed public works (transport, roads, schools, water systems), which create the basis for later productivity; they also tend to discourage individual initiative.

International and Internal Conflicts

The international and internal conflicts spawned by nationalism have dismantled trade and communication ties and destroyed settlements and major cities, as demonstrated during the past decade by hostilities in Northern Ireland, Cyprus, Rhodesia, Angola, Ethiopia, Iran, Israel, and Lebanon. The Israeli and Rhodesian conflicts are to a large extent classically economic; they involve groups struggling for land, livelihood, and survival. These goals have characterized countless wars for thousands of years. But all the current conflicts are ideological as well: they are motivated by attempts to maintain the status of a religious group, race, tribe, or other unit, whatever the damage to the environment or the destruction of life and property.

Nonviolent conflict or rivalry among countries also affects the nature of development, but in a less damaging manner. Most countries are concerned with maintaining their cultural identity. This kind of nationalism is expressed in the careful control over immigration and emigration, the content of radio and television, the availability of literature, and the controls over foreign investment and land ownership. Such policies hinder the emergence of an international culture (or the too rapid acceptance of an aggressive culture, like the American culture today), and they preserve national cultural differences. These nationalistic policies are likely to impede the economic and social development of less advanced countries. But with some thought, those of us who live in an aggressive, technological, highly individualistic, and rapidly changing society can understand why other societies may wish to screen out the less desirable of our cultural attributes.

Forces Promoting Development

Despite all these barriers, development does occur. Most developing nations are highly motivated and determined to modernize. They are aided in this process by the availability of a large body of technology already created by the developed nations, although much of this technology may be too capital intensive.

The most successful developing nations are those with strong governments dedicated to growth and capable of enforcing an unpopular emphasis on investment to the neglect of present living standards—examples are Korea and Brazil. It is clear that austerity and very high reinvestment levels were common during the industrialization of all the now wealthy nations.

Totally independent development seems almost impossible, but growth financed by exports is possible, even without external aid. However loans and gifts facilitate the development of less advanced nations. Some forms of assistance, such as technical aid for health, agriculture, and water supplies, are channeled through the United Nations and related agencies. Larger in scope are private investments (primarily to develop the exportation of desired commodities, such as petroleum and other minerals) and government loans and gifts.

There is some hope that international ideological conflict may be diminishing, that trade and investment can occur fairly normally, even between capitalist and communist economies, and that more capital will become available to developing nations through international sources. Raw material shortages in the richest nations may also continue to result in higher and more stable prices for some exports from developing countries. Thus per capita output, both in food and in other goods, continues to rise slowly, despite serious international conflicts, population growth, natural disasters, poor weather, and recession.

Problems of Internal Efficiency and Capital Allocation

The richer nations have national economies in which virtually all the people participate commercially. In contrast, most poorer nations have dual economies: a minority of the population is involved in the commercial sector, which is often closely tied to exports, and the majority remains in the semi-subsistence sector, mainly engaged in producing food crops. The distinction between the two groups is profound. Thus a major problem of less developed nations is to commercialize the subsistence sector of their economies.

Typically the principal sources of capital for such nations are fairly small exports of foods or minerals (see tables 10.1 and 10.2). There is also a temptation to spend heavily on consumer goods and to invest in secure overseas markets or, at best, to invest in urban services, ignoring the subsistence sector of the economy. This behavior is spatially inefficient because resources are underutilized, and the distribution of income remains inequitable.

Given a desire for widespread development and less dependence on a few overseas markets, developing nations are faced with the difficult task of finding the most effective allocation of limited capital resources within the country, both geographically and among sectors of industry, agriculture, and services. Spatially their alternatives are to disperse investment evenly over the entire territory (a strategy espoused by Tanzania, for example), to concentrate investment in the single best location, or perhaps to distribute it to a set of growth points. Dispersal is a popular choice when political control is diffuse and all parts of the country want to develop at once. But such a pattern is generally inefficient: the investment in each area will be on an unprofitably small scale and will probably do no more than enable food production to keep up with population growth.

In the short run, the most profitable use of capital in a small country is to concentrate industrial development in the single best location, usually the capital city, where threshold markets can be found and adequate labor is available. Part of the population will migrate to the city from the countryside to supply additional labor, and the surrounding agriculture should then commercialize in order to serve the urban market. A primary center is not inefficient as long as most of the population is within a few hours of the center and an adequate structure of local service centers is developed.

In a fairly large and populous country, however, extreme concentration of capital in one place will not generate the most widespread development. Talented people will refuse to leave the economic capital, and the capital's economy will be undermined by the continued subsistence level of the economy in the periphery. Investment is better allocated to a set of regional centers, which may in turn help commercialize their hinterlands.

Total concentration of investment in strategic growth centers, whether one or a few, is unrealistic. More diffuse investment in agriculture, including local transportation, is essential to raise rural efficiency and productivity and to avoid costly food importation. Poor areas not selected for development cannot be ignored without risking social and economic unrest that may drag down the entire economy.

The best allocation of capital among sectors of the economy is also a matter of controversy. The traditional emphasis on industry has been criticized as unbalanced; many planners favor higher allocations to overhead services, such as roads and education, and to agriculture. No simple formula can be applied to all countries. It seems clear that there is no real substitute for industrialization; but industrialization can be most effective only if accompanied by adequate investment in agriculture, services, and the acceptance of improved standards of consumption.

If the chosen strategy is successful, development will spread from the growth centers to their hinterlands, until all regions and peoples of the nation are brought into a unified economy. Such orderly development should bring about a more balanced distribution of population, production, and income—and perhaps even a closer approximation to the theoretical landscapes outlined in the preceding chapters than is now true of most advanced nations.

CONCLUSION

Despite the severity of the problems we have described, the human landscape is an incredible achievement—a functioning, interdependent system of people and activities, working and interacting. No one can define, and no society can achieve, the perfect landscape. While the efficient patterns projected by location theory often disregard those people less able to compete, governmental attempts to introduce more equity into the landscape are often inefficient. Thus human societies continue to strive for spatial equilibrium, and the landscapes they create continue to evolve. In the process of developing new landscapes, we must attempt not only to balance efficiency with equity but also to maintain the integrity of the physical environment. The mutual impact of human society and the physical environment will be explored further in the next chapter.

SUGGESTED READINGS

Albaum, M., ed. *Geography and Contemporary Issues.* New York: Wiley, 1973.

Bunge, William. *Fitzgerald: Geography of a Revolution.* Cambridge, Mass.: Schenkman, 1971.

Coates, B. E.; Johnston, R. J.; and Knox, P. L. *Geography and Inequality.* New York: Oxford University Press, 1977.

Coppock, J. T., and Sewell, W.R.D. *Spatial Dimensions of Public Policy.* New York: Pergamon, 1976.

Cox, Kevin. *Conflict, Power and Politics in the City.* New York: McGraw-Hill, 1973.

Friedmann, John. *Regional Development and Planning.* Cambridge, Mass.: MIT Press, 1964.

Gilbert, Alan, ed. *Development Planning and Spatial Structure.* New York: Wiley, 1976.

Hansen, Niles. *Rural Poverty and the Urban Crisis.* Bloomington, Ind.: Indiana University Press, 1970.

Harvey, David. *Social Justice and the City.* Baltimore: Johns Hopkins Press, 1973.

Hunter, J., ed. *Geography of Health and Disease.* University of North Carolina, Studies in Geography, 1974.

Klaasen, L. *Areal Economic and Social Redevelopment.* Paris: OECD, 1965.

Morrill, R. L., and Wohlenberg, E. *Geography of Poverty.* New York: McGraw-Hill, 1971.

Myrdal, Gunnar. *Rich Lands and Poor.* New York: Harper & Row, 1957.

Peet, Richard. *Radical Geography.* Chicago: Maaroufa Press, 1977.

———. *Geographic Perspective on American Poverty.* Worcester, Mass.: Antipode, 1975.

Rose, Harold. *The Black Ghetto.* New York: McGraw-Hill, 1971.

Smith, David. *Geography of Social Well-being in the U.S.* New York: McGraw-Hill, 1973.

14

The Interaction of Society, Space, and Environment

14 The Interaction of Society, Space, and Environment

We have emphasized the role of space as an influence on human behavior and location. Spatial considerations helped direct the evolution of the landscape. The space needed to produce food for a growing population forced the dispersal of human beings over vast areas. In the process, many groups became isolated by barrier mountains, rivers, and oceans or by distance. They developed distinctive languages, physical traits, beliefs, customs, and other cultural variations. At the same time, however, the human desire for exchange, social contact, and greater living space encouraged trade and migration, fostered the diffusion of ideas, and, despite frequent conflict and warfare, may ultimately bring about cultural integration throughout the world.

Our primary aim has been to show how a few principles governing location decisions tend to result in predictable spatial patterns. We have argued that the desire to use space efficiently tends to create repetitive patterns in which people and activities are clustered in a hierarchy of nodes, from hamlet to metropolis. These nodes, in turn, are surrounded by space-consuming activities, most notably agriculture, which are arranged in some predictable manner.

We have used location theory to help explain the human landscape. Rather than taking a strictly theoretical approach to geography, we have tried to integrate the spatial-economic factors of location theory with the environmental and cultural factors that also influence location in the real world. Location theory casts environment and culture into secondary roles, viewing them as modifiers or conditioners of the ideal patterns that might be observed on a featureless plain. This bias is intentional. Scientific inquiry requires limiting the number of variables in order to isolate the contribution of one or more factors. We could have organized our discussion around theories dealing with the interaction of society and the physical environment. In that case, spatial factors would have been viewed as modifiers. But we have focused on spatial factors because, though they are less obvious than physical ones, they nonetheless appear to contribute the most to regularity in the landscape.

But of course we do not wish to deny the importance of the physical environment. In this chapter we remind ourselves of our ultimate and total dependence upon nature. After reviewing the relationship between society and the physical environment, we will evaluate the relative importance of spatial and environmental influences on the distribution of population and the location of human activities.

THE PHYSICAL ENVIRONMENT AND HUMAN LOCATION

In chapter 2 we discussed various aspects of the physical environment in terms of their influence on location decisions. Now we will briefly consider both the net effect the environment has had on human spatial organization and the impact humans have had on the environment.

We all know that food and water are basic necessities of life. Even the most sophisticated urban-industrial society relies on the support of an agricultural base. Despite human efforts to remold the physical environment and to increase nature's productivity, and despite greenhouses, hydroponics (raising plants in nutrient solutions), and technology, virtually all food production and most human habitation is restricted to areas inherently favorable for agriculture.

The constraints imposed by landforms, climate, and soils limit significant human settlement to a surprisingly small portion of the earth. It has been calculated that of the 36.6 billion acres comprising the earth's land surface, 40 percent is too cold or too dry for significant human habitation, 20 percent is too steep, and 10 percent is too barren. That leaves only 30 percent, or about 11 billion acres—fewer than three acres per person—for the ready use of human society.

Landforms

In search of food, in flight from enemies, or through expulsion by dominant societies, human beings have migrated to areas of incredibly rugged terrain. The Hatia of Nepal dwell in small villages nestled in the cloud-covered valleys and ridges of the Himalayas, the world's highest mountains, at altitudes averaging 4,500 meters (15,000 feet) (photo A in figure 14.1). Only limited forms of agriculture are possible at such heights, and population is therefore sparse. The large expanses of forested or grassy hills and mountains covering the earth support very few people, except in a few areas where local population pressure has induced the terracing of slopes for farming (photo B) or where valuable mineral resources can be exchanged for food (photo C). The totally uninhabited parts of the earth's surface consist of steep and barren mountain slopes and deserts of shifting sands, where not even pasturage for sheep and goats can be wrested from the land. By far the greatest number of people—90 percent of the world's population—live in gentle plains and valleys, and nearly all food is produced in such areas.

Not surprisingly, we can best discern the patterns predicted by location theory on the vast plains of North America, northern Europe, the Soviet Union, and northern China and India. But areas of rugged topography—the Appalachian and Rocky mountains are examples—themselves often display patterns that can be predicted from a knowledge of physical processes, such

14.1. Human habitation in harsh physical environments. Photo A: the Hatia of East Nepal have built settlements high in the cloud-covered Himalayas. Photo B: peasants in Sanchiang Tung Autonomous County, People's Republic of China, have transformed barren hills into fertile terraced farmlands. Photo C: copper from the Andes Mountains supports La Paz, the capital of Bolivia, and its population of over 525,000 people. Photo D: nomads of Ethiopia's Danakil Depression, an area of volcanic debris and dried salt, live in tents made of straw mats and goatskins. The tribesmen often fight over the limited water supplies of this barren land. (Photo A by Martin Etter from Anthro-Photo; photo B from China Photo Service, Sovfoto/Eastfoto; photo C by Don Briscoe from De Wys; photo D by Victor Englebert from De Wys.)

A

B

C

D

as mountain building and erosion. And these topographical patterns tend to modify human location patterns in predictable ways.

Climate

People have settled in all climates but perpetual snow and ice. A few Danakil nomads of Ethiopia manage to survive in lava-strewn deserts where temperatures soar to 50°C (120°F) (photo D in figure 14.1). At the opposite extreme, Eskimos have adapted to the bitter cold of North American and Siberian polar regions. The small population of Greenland depends on the surrounding ocean, not the land, to support itself.

Population is sparse in these rigorous climates: deserts are too dry, and subarctic tundra has too brief a growing season to support large populations. Significant population growth can occur only when economic development justifies bringing in food from the outside. Access to nickel ores helped create the city of Norilsk on the Siberian Arctic coast, and access to oil is building settlements at Prudhoe Bay in the Alaskan Arctic. Oil wealth readily permits Kuwait to import needed foods and to desalinate ocean water. But more than 90 percent of the world's population lives in more temperate zones, where moisture levels and growing seasons allow the production of at least a reasonable grain crop or where irrigation systems can be developed fairly easily.

Patterns predicted by location theory are best observed in regions of adequate moisture and temperature, where agriculture is most productive. Harsher climates, in which moisture or temperature levels, or both, are low, modify spatial patterns. Settlements and central places are more widely spaced in frigid northern Canada and Siberia; and they are highly clustered in most desert regions, where population concentrates around limited sources of water. Since water—a critical resource—is sporadically located, industrial location theory (that is, resource orientation) is the most useful in helping to explain settlement patterns in such climates.

Soils

The productivity of soils depends on climate, terrain, underlying rock, and vegetation cover. Richer soils tend to develop in semiarid and subhumid climates; less productive soils in very humid climates. Only a rather small portion of the earth's land surface has both rich soils and plentiful moisture. Most agriculture must be carried on in regions where moisture is marginal but soils are rich—such as the Russian steppes and the U.S. and Canadian plains—or in regions where soils are poorer but moisture is abundant—much of the tropics and other areas that were once heavily forested, such as Europe, southern China, and the eastern United States. Inadequate soils and drainage remain a major barrier to the development of the well-watered Amazon basin.

Soil utilization also depends on human perception and technology. In a classic study on the Great Plains, W. P. Webb describes how pioneers from the humid, wooded Northeast were at first confounded by the semiarid, treeless plains that stretched infinitely across the West. So stark was the contrast between the two regions that for many years the plains were perceived as harsh, unfruitful, and unarable. It was not until settlers learned new techniques, such as dry farming, that Kansas, Colorado, Nebraska, the Dakotas, and parts of Montana and Oklahoma became useful and were eventually transformed into a productive granary.

Resources

Water and soils have been key resources throughout most of history. Disputes over water and land rights led to the development of perhaps the earliest laws—and to much warfare. Water still exercises significant control over the location and cost of human settlement. Limited water, for example, is a source of conflict between mining and ranching interests in the semiarid U.S. West. (The United States hopes to shift from dependence on petroleum to the greater use of coal, much of which lies in semiarid areas of the West; both the strip mining of coal and power production limit water left for irrigation and urban use.)

In Egypt controversy also rages over water limitations caused by the Aswan High Dam. Constructed at great expense to control the Nile's seasonal flow and to open new areas to farming through irrigation, the dam has unfortunately lowered productivity in parts of the Nile delta that benefited from annual flooding.

Grandiose proposals have been advanced to counteract regional imbalances of water supply. Examples include drawing Columbia River water southward to California and reversing north Russian or Siberian rivers to supplement the Volga River and Caspian Sea and to increase water supplies in central Asia. In desert countries like Saudi Arabia and its neighbors, water shortages are spurring research into desalinization and have even raised interest in such schemes as towing icebergs from the Antarctic.

Urban-industrial societies depend heavily on mineral resources—building materials, metals, and especially fuels—most of which are sporadically located (figure 14.2). Some countries, often because of size, are well endowed with rich deposits, and others are deficient. The resource endowment of a region or nation has two major effects on the landscape: it conditions the kind of industrial specialization that will eventually occur (for example, oil-rich areas tend to specialize in petroleum refining and petrochemicals), and it concentrates or distorts patterns of urban settlement.

Nations lacking critical minerals can or must import them. Some, like Japan, have been very successful at exporting a wide range of technical goods that enable them to import the needed minerals. But for many developing countries, the high cost of importing essential fuels like petroleum and coal severely limits economic growth. Dependence on other countries for petroleum and other minerals and the fluctuating or rising prices of such commodities not only strains internal development but affects international relations as well. Most nations are careful not to offend OPEC countries. Even large, rich nations like the United States are vulnerable to the disrup-

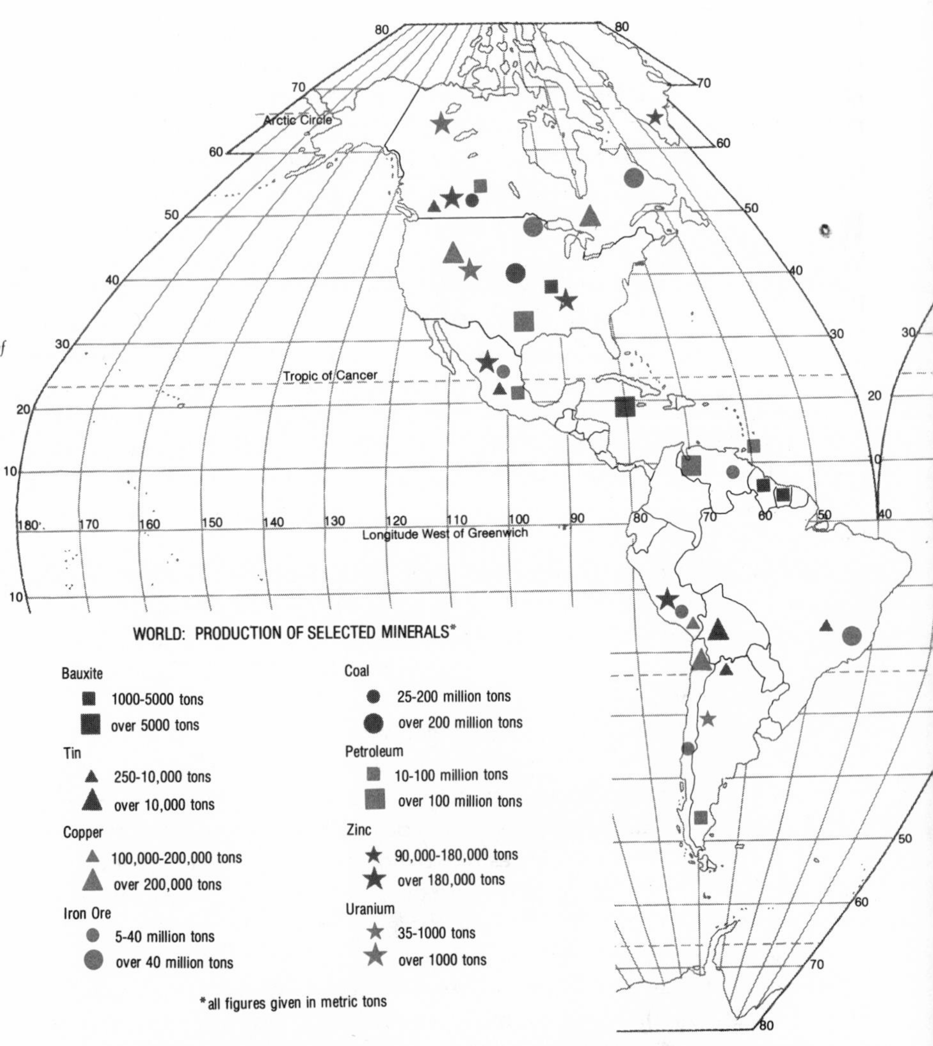

14.2. World distribution of mineral production. Many of the minerals needed for industrial production in North America and Europe are concentrated in Asia (most notably oil in the Middle East—see bar graph), Africa, and Latin America.
Source: Data from *Minerals Yearbook*, 1975, U.S. Department of the Interior, and *B.P. Statistical Review of the World Oil Industry*, 1977, British Petroleum Co., London, England.

tion of imported resource supplies. The 1974 OPEC oil embargo and price rise caused serious economic dislocations in North America and Europe. And U.S. dependence on countries like the Soviet Union or the Republic of South Africa for supplies of gold, chromium, and manganese cannot help but influence its political relations with those countries.

Disease and Natural Hazards

Over time human societies have evolved a mutually beneficial relationship with the physical environment, occasionally improving on nature's productivity. Until recently, however, humanity has had little control over the spread of disease and the threat of natural calamities. Throughout history diseases such as plague, smallpox, malaria, and sleeping sickness have deci-

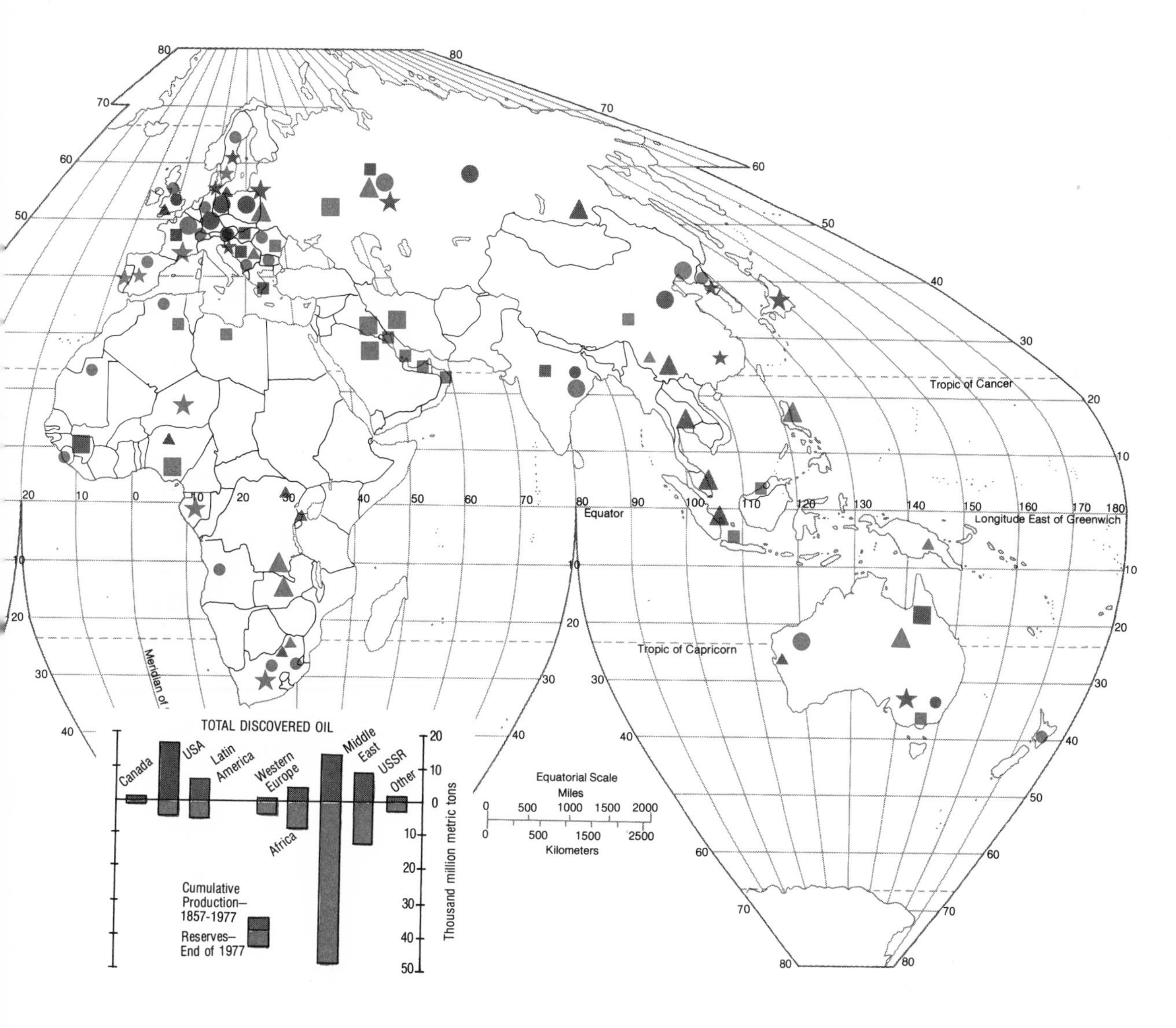

mated the population of the most developed areas (Europe, China, India) and have impeded the progress of less advanced ones, particularly in the tropics (figure 14.3). Great strides have been made in this century in the control of some of these diseases, notably plague, smallpox, and malaria. While these gains must be applauded, it is unfortunately also true that consequent population growth has often introduced additional problems, such as malnutrition.

We are continually reminded of the power of nature through devastating earthquakes, tidal waves, volcanic eruptions, hurricanes, typhoons, tornadoes, floods, fires, and droughts. Although society has developed ways to deal with natural catastrophes through warning systems, improved construction, storage, and insurance, the cost to human life and property is still

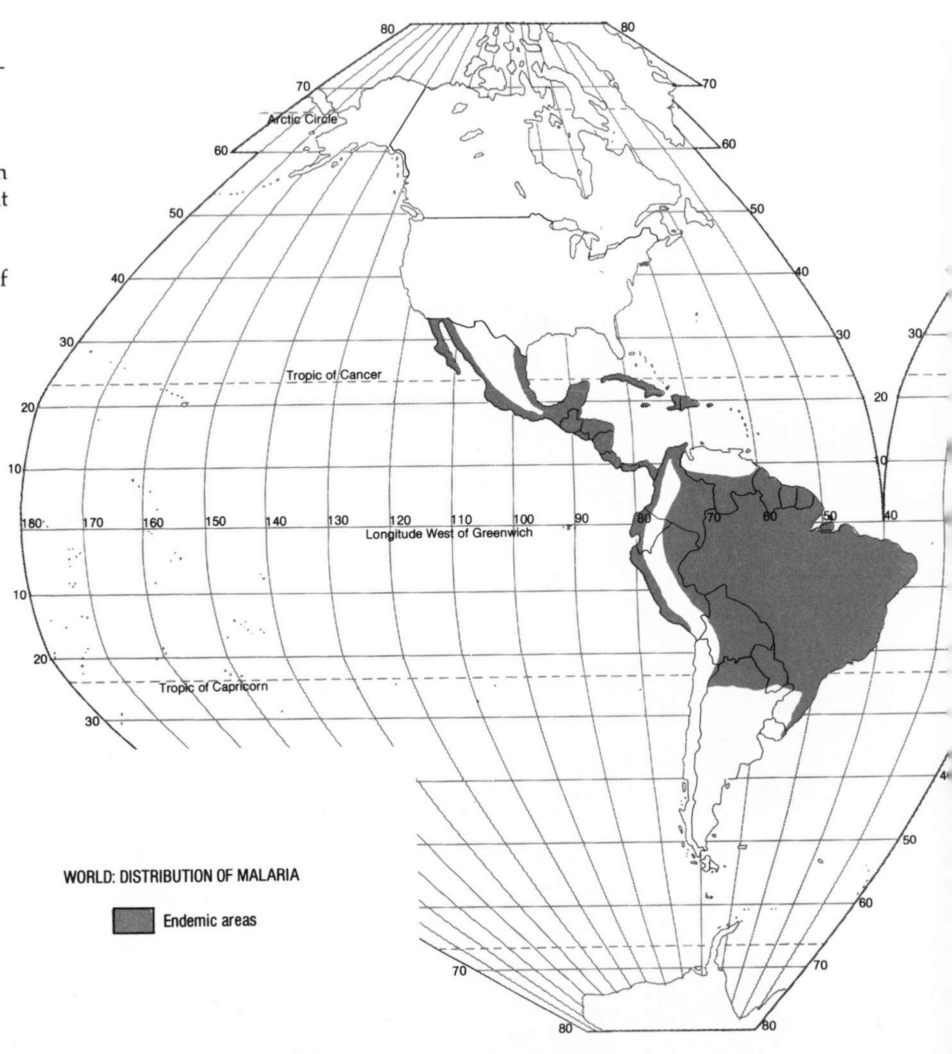

14.3. World distribution of malaria. Spread by mosquitoes, malaria is most prevalent in tropical and subtropical areas. Insecticides have been effective in combatting the disease, but after long-term exposure, mosquitoes may become resistant to certain kinds of poison.

extremely high. Measures to counteract the effects of natural hazards tend to be adopted only in areas that are frequently attacked.

Most people seem to be overly optimistic about their chances of avoiding catastrophe. Residents of Anchorage, Alaska, live over an earthquake-prone fault zone. A severe earthquake struck the city in 1964, destroying many homes. But rather than move to safer ground, people rebuilt their homes in exactly the same site as before. Most believed that such a disaster would not recur during their lifetimes.

Similarly people who live in places periodically devastated by floods often refuse to remove activities from the most vulnerable sites (commonly areas of intensive industrial capacity). Instead they rebuild their farms and factories and attempt to control the floods by constructing dams, reservoirs,

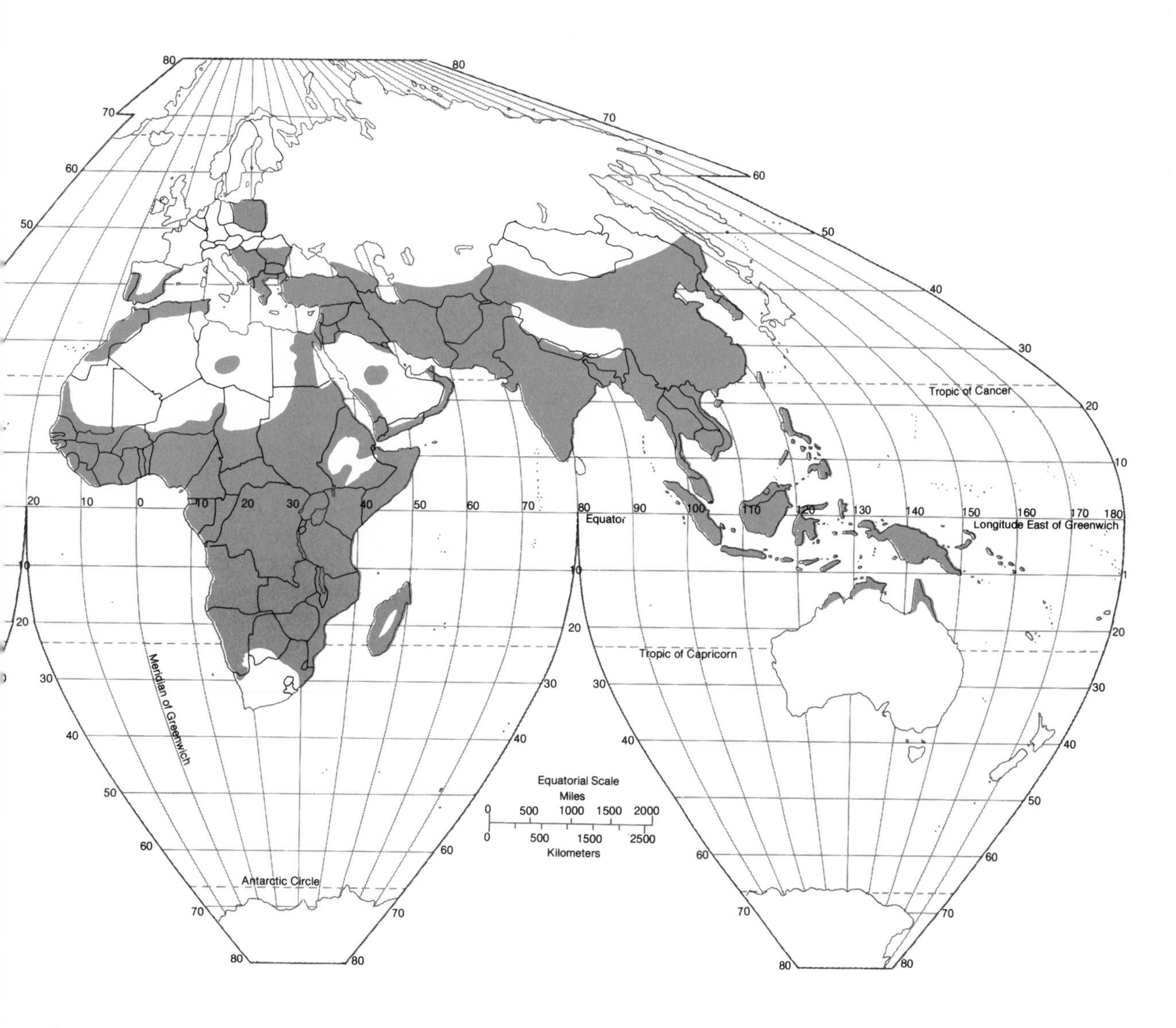

levees, channels, and the like. Other examples include settlements crowded around soil-rich but active volcanoes and homes built on scenic but steep, wooded slopes subject to landslides and fire.

ENVIRONMENTAL MODIFICATION

In the continuing effort to survive and increase well-being, human populations have tried to manipulate the physical environment to serve their own needs. From prehistoric times to the present, these efforts have often disrupted the current balance of natural systems. Only more recently has society concerned itself with preserving the integrity of nature, in the sense of diversity of species or variety of natural habitats, and maintaining the earth's productive capacity.

The Impact of Early Humans

Though few in numbers, primitive tribes exercised some control over nature through effective hunting techniques and the use of fire. Migrating in search of food, hunters killed an extraordinary number of animals, killing off entire species. Their most efficient and destructive hunting technique was to flush out game by fire, a practice believed to have deforested vast areas in East Asia and parts of North America.

As humans progressed from hunting and gathering to the domestication of animals, human ability to modify the environment also advanced, though unwittingly. It appears almost certain that some four thousand to five thousand years of overgrazing have severely damaged natural grasslands over much of the world, especially in North Africa and in southwest and inner Asia (figure 14.4). Rich grasses have been replaced by inferior species; fertile topsoil, stripped of its protective grass cover, has been blown or washed away; and even the climate may have changed for the worse. In addition, primitive agriculture (slash-and-burn), though involving relatively few people, is believed to have entailed large-scale deforestation and subsequent erosion, most notably in China, India, Latin America, and Mediterranean Europe. Such practices undoubtedly aggravated the impact of floods and may have contributed to drought and climatic change. But it is also true that the damage is often reversible. Superior grasses have reappeared when grazing has been limited, and other areas have been reforested.

Agriculture

The development of intensive peasant agriculture marked a new stage in society's relationship with the environment. Rather than exploiting an area and then abandoning it, people for the first time began to stay in one place. They improved upon nature in terms of increased food production, but perhaps at the expense of plant and animal diversity. Irrigation greatly increased the productivity of the soil by enabling farmers to schedule and control the amount of water applied to crops and by depositing organic fertilizer. In Mediterranean Europe, China, India, and other areas, the agricultural system gradually evolved into a set of practices—irrigation, fertilization, and crop rotation—that maintained the soil's productive capacity indefinitely and at levels far higher than before.

Intensive peasant agriculture was practiced without irrigation in more humid zones. The deforestation necessary to clear space for fields led to erosion and flooding in many hilly areas. But in extensive plains areas, such as northern Europe, farming practices generated a higher level of productivity, and of course utility, than was provided by the original forest.

The advantages of level land for sedentary agriculture shifted population from protective hills and forests to open plains and valleys. For defense

14.4. The effects of overgrazing and changes in climate on land in Algeria. Only patches of vegetation survive outside this north African settlement. (Photo by Georg Gerster from Rapho Photo Researchers.)

against brigands and invaders, as well as for social contact, farmers in most societies settled in villages rather than in isolated homesteads. A new landscape emerged, composed of fields and villages, fences, walls, and roads serving the rural population.

When population growth or demand outpaced a society's agricultural production on its arable land, farming was pushed into marginal areas that were drier, steeper, or wetter than desirable for farming. Arid or sloping lands were often eroded by wind or water. Wetlands were useful only when drained, at considerable cost. At times space needs become so pressing that arable land was reclaimed from the sea. Examples may be seen along the coast of Japan and the Netherlands (figure 14.5). Over time, the spread of agriculture, with its limited variety of useful plants and animals and its fertilizers, herbicides, and pesticides, has inevitably reduced wildlife habitats and has tended to limit or destroy many kinds of plants and animals.

14.5. Polder development in the Netherlands. Since the year 1200, the Dutch have reclaimed close to 1.7 million acres of arable land from the sea. They have significantly altered their landscape through the construction of dams, dikes, and canals.
Source: Excerpted from *Of Dikes and Windmills* by Peter Spier. Copyright © 1969 by Peter Spier. Reprinted by permission of Doubleday & Company, Inc.

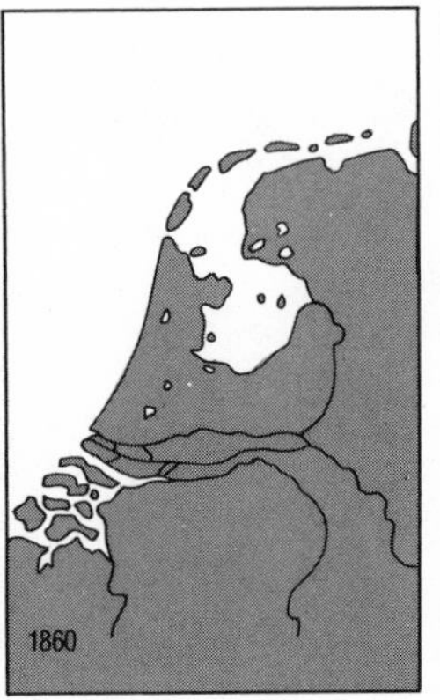

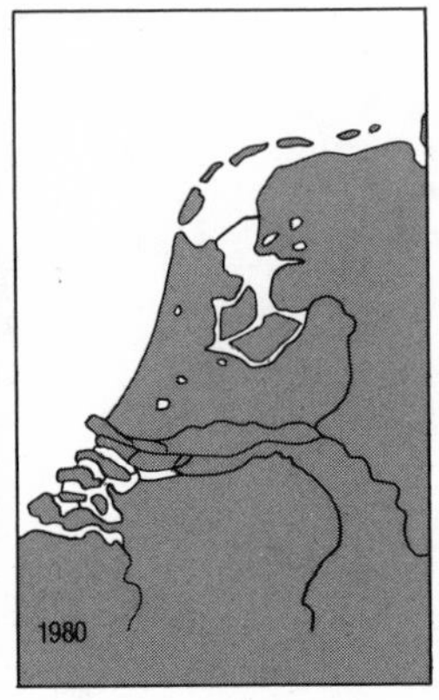

Urbanization and Industrialization

While humanity has modified more of the earth's surface through agriculture than by any other means, the most dramatic transformation has occurred in limited areas through urbanization and industrialization. The built environment has obliterated the physical environment in many urban areas. Roads, houses, buildings, and factories have changed drainage patterns, reduced vegetation and wildlife, and altered the microclimate. The unusually frequent thunderstorms over London, for example, have been attributed partly to the city's higher level of heat from buildings and streets and to air pollution from coal-burning.

Mighty though human works may appear, almost none are visible in satellite photographs. No more than one-tenth of 1 percent of the earth's land surfaces has been paved—at this scale, a seemingly modest human imprint on the face of the earth. If all the paved areas of the world, including cities, towns, roads, and highways, were gathered together, they would fill only a space roughly equivalent to the area of Illinois. Nonetheless urban and industrial air and water pollution does show up in satellite photographs, and the impact of human activity spreads far beyond the immediate site. The tremendous mass of industrial and urban waste has seriously polluted many bodies of water, such as Lake Erie and parts of Japan's Inland Sea as well as countless rivers. Carbon dioxide generated by the burning of fossil fuels and ozone produced by high-flying aircraft may be capable of modifying the climate adversely over the long term.

Costs and Benefits of Environmental Modification

The **ecosystem** (relationship between organisms and their environment) is so complex that we can only guess at the net costs and benefits of human intervention. From society's standpoint, environmental modification has been necessary. To produce enough food for expanding populations, forested areas had to be cleared. To provide reliable moisture for crops, river water had to be diverted. To move goods easily between distant places, roads had to be carved into the terrain. All these modifications are justified by human need.

Preindustrial societies were perhaps as destructive as modern ones, at least with respect to large areas. We are now able to reclaim land previously damaged through overgrazing, overcutting, and poor farming methods. For the first time in history, programs advocating reforestation, grassland upgrading, and wildlife and scenic protection can realistically be implemented on a large scale. Improved knowledge of physical systems, together with better technology, also is permitting the reclamation of pollution-damaged lakes, rivers, and shorelines. Of course, urban and industrial pollution is by no means disappearing. Indeed some ecologists predict global ecological catastrophe (see chapter 15).

The benefits of environmental modification, then, include gains in society's well-being, particularly with respect to adequate food and shelter. The costs are of two kinds: the depletion of energy and other resources needed to create and maintain the modifications that sustain our civilization, and the strains imposed on the environment itself, as measured by the decline in natural diversity and in the ability to absorb and recycle wastes.

In less developed societies, the struggle to survive prevails over environmental concern. Farmers in some parts of Indonesia, for example, deforest hillsides and farm the slopes in order to feed their families. Illiterate and impoverished, they do not know how or cannot afford to prevent erosion by contour farming. The diffusion of such innovations, or the money to pay for them, trickles very slowly down to the poor, especially in remote areas. Moreover many farmers are unwilling to risk losing the little food they have by trying new methods they fear may fail. Cultural as well as physical barriers prevent the rapid spread of techniques that might both raise human levels of living and protect the environment.

ENVIRONMENT VERSUS SPACE

It would be impossible to evaluate precisely the roles played by spatial and environmental factors in shaping the human landscape. But our emphasis on spatial factors requires some attempt to justify our position. We will therefore try to weigh the relative importance of space and the physical environment in the distribution of population and in the location of human activities.

The Distribution of Population

The migration and dispersal of humans over the earth was primarily a response to the space requirements of food gathering and food production. The larger human populations became, the more space they needed and the farther they migrated. To a considerable extent, however, the density of population in a given area depends on the productive quality of the local physical environment. The fact that Australia has an average of only 2 persons per square kilometer and that the Netherlands has an average of 375 persons per square kilometer may be attributed in part to differences in climate and soil. Much of Australia is covered by deserts or steppes where average rainfall is less than twenty-five centimeters (ten inches) a year, while the Netherlands is fertile and relatively humid. To some degree, technical and cultural innovations have permitted a greater concentration of people in some areas than environmental potential alone would suggest. Thus the Netherlands' position in the heart of a highly developed Western Europe has helped its citizens prosper and its industries succeed, permitting the importation of needed food. In contrast, Australia's remote location relative to other highly developed parts of the world has slowed its potential growth.

Continuous intensification of farm practices over thousands of years has made possible exceptionally dense populations in Europe, China, and India (97, 87, and 175 persons per square kilometer, respectively; the world average is 29 persons per square kilometer). Later settlement or development helps explain the relatively sparse populations of North and South America (11 and 15 persons per square kilometer, respectively). (Recall from chapter 3 that the population explosion in Europe resulting from the industrial revolution encouraged migration to the New World.)

While no really unfavorable environments are populous and no really favorable ones are empty, it is also evident that some very hospitable environments, as found in parts of East Africa from the Sudan to Tanzania, have relatively few people, and some rather difficult environments, as in Switzerland, have relatively many. In sum, the environment appears to set broad limits for settlement. But within these limits, differences in technology and economic development are responsible for marked variations in the distribution of population.

Agricultural Production

Throughout most of history, agricultural production has been influenced by both physical environment and technology. The growth of metropolitan markets, particularly in the nineteenth century, led to a partial reordering of agriculture somewhat independently of the physical environment. But in recent decades physical quality has again become a more dominant influence. This is because improvements in transportation have substantially reduced the friction and cost of distance. Strawberries and asparagus, for example, can be flown to distant markets within a few hours of harvesting. Farmers can now raise perishable crops in peripheral locations if the climate and soil favor such production. On the other hand, it was the existence of affluent metropolitan markets and quality transport services that enabled people to take advantage of superior local environments. As with population, then, the environment sets broad limits to what may be grown, but other factors tend to determine the precise arrangements of different crops.

The Location of Cities, Industries, and Transport Routes

The technological advances that fostered urbanization and industrialization permitted the more efficient organization of human activities. Production became concentrated in and around cities. Food and other resources were brought into these centers. Spatial more than environmental considerations began to dictate the location of cities and surrounding intensive farm production (truck and dairy farms). Cities like Dallas and Fort Worth, Texas, sprang up at the junction of rail lines and became prosperous trade centers. Yet if you consider the distribution of major cities, you will find that most are located at physically advantageous sites: natural harbors (New York, Canton), river junctions or fords (St. Louis, Lyon, Frankfurt), or the foot of mountain passes (Denver; Torino, Italy; Peshawar, Pakistan).

It is difficult to determine whether environmental or spatial forces are most important in industrial location. It depends largely on whether the industry is resource oriented. Some industries are tied to concentrations of material or energy resources. Aluminum processing, for example, is often located near large hydroelectric resources, such as the Columbia River basin in Oregon-Washington. Nonresource-oriented industries locate in or near population centers to take advantage of agglomeration benefits, economies of scale, and access to markets and labor pools; examples are the publishing and fashion industries in New York, London, and Paris.

Industrialization tended to shift population from productive farmlands to mineral-rich areas. Similarly postindustrial affluence and leisure are now redistributing people to amenity environments and recreational resources. Growth centers in the United States today include such sun belt cities as Sarasota and Fort Lauderdale, Florida, and Phoenix, Arizona.

Transport links are created to facilitate trade and travel between centers of human activity. Physical barriers are overcome if necessary, but major transport routes often follow natural corridors, as along the Hudson-Mohawk Valley in New York. (Note, however, that new construction technologies permitted Interstate Highway 80 to cross directly the ridges and valleys of northern Pennsylvania at relatively low grade using extensive cuts, fills, and bridges. The road has shortened the trip between New York and Chicago several hours in comparison to the longer Mohawk Corridor or the southern Pennsylvania route.)

Flows and activities still concentrate at the few bridges that span major barrier rivers, particularly in developing countries. A notable example is the great importance of Wuhan, China, a large industrial city that possesses the only rail and highway bridge crossing the main stem of the Yangtze River.

In sum, the specific location of larger cities appears to be determined more by local environmental factors than by spatial ones. Most major cities are located at physically advantageous sites. But the overall number and spacing of urban places depends on population density and level of economic development rather than on environmental factors, for reasons given in chapters 6 and 7. The more undifferentiated the physical environment, the more the principle of spatial efficiency appears to govern the location of smaller centers.

The Layout of the City

The city's built environment reflects the greatest degree of spatial as opposed to environmental influence. The principle of spatial efficiency generally determines the distribution of people and activities in cities built on a relatively uniform surface, such as Chicago, Des Moines, and Indianapolis. When physical features are varied, however, the location of the central business district, industrial sites, transport routes, and residential areas will be affected. The layout of some cities, such as Boston, San Francisco, Hong Kong, and Stockholm, is constrained by their placement on peninsulas or islands. But in the more typical city, industries tend to be situated on waterways for ease of transportation (and waste disposal), the city center is probably located at the most convenient crossing of the main river, and the homes of the affluent are often built on hills with a pleasant view or are located favorably with respect to prevailing winds.

CONCLUSION

Technologically advanced societies can locate people and activities wherever they wish, regardless of the local physical environment's ability to support human life. We can create cities like Los Angeles in arid rather than well-watered regions, and we can build homes on wooded hillsides with spectacular views rather than on level ground. But such freedom of location is expensive. Water must be brought into the city at high cost, and homes must be protected against fire and sinking land. Increased costs and disastrous environmental side effects call into question the wisdom of ignoring physical constraints.

Indeed our very perception of nature seems to be shifting. We no longer regard the physical environment primarily as a force or obstacle to be overcome and dominated. More and more people are adopting an ecological perspective and are beginning to plan future development in greater harmony with the inherent quality of local environments. If the future landscape is to be worth living in, such planning is essential. In the next chapter we take a look at several possible futures.

SUGGESTED READINGS

Bernarde, Melvin. *Our Precarious Habitat.* New York: Norton, 1970.

Detwyler, T. R. *Man's Impact on Environment.* New York: McGraw-Hill, 1971.

———, and Marcus, M. *Urbanization and Environment.* North Scituate, Mass.: Duxbury Press, 1972.

Kates, Robert W. *Natural Hazards in Human Ecological Perspective.* Toronto: Department of Geography, University of Toronto, 1971.

Lowenthal, David, ed. *Environmental Perception and Behavior.* Department of Geography Research Paper 109. Chicago: University of Chicago, 1967.

Mikesell, M., and Manners, I. A. *Perspectives on Environment.* Washington, D.C.: Association of American Geographers, 1974.

Saarinen, T. *Environmental Planning, Perception and Behavior.* Boston: Houghton Mifflin, 1976.

Strahler, A. *Environmental Geoscience.* Santa Barbara, Calif.: Hamilton, 1973.

Ward, Barbara. *Spaceship Earth.* New York: Columbia University Press, 1966.

15

The Future Landscape

15 The Future Landscape

On January 1, 1881, the *Boston Globe* published "a prophetic look ahead" to 1981. The following is an excerpt from that article, supposedly written a hundred years later, at the close of 1980:

> The year 1980 will be forever memorable in the annals of the American Republic, and therefore of the world, of which it now forms, in point of population, wealth, commerce, and all that makes up the sum total of power, the greater half.
>
> We are now [in 1980] an indissoluble union of 139 indestructible states embracing, according to the latest monthly census, a little over 8,00,000,000 [*sic*] of people, speaking five distinct languages, besides 111 marked varieties of United States (or English, as the old writers used to say). If the application of Brazil, Chile, and Peru to be admitted to the Union, which comes before Congress at the January session, shall be granted, as seems probable, we shall speedily be a nation of over 1,000,000,000 inhabitants. There is, indeed, no reason why they should not be admitted, and the admission of Canada in 1903, Cuba in 1910, and Australia in 1924, make a formidable array of precedents in their favor.

Although the article went on to make some surprisingly accurate predictions—trans-Atlantic air travel and closed-circuit television—those cited above have obviously not come to pass nor are likely to by 1981. Even much shorter-range predictions tend to fall wide of the mark. The fact is, no one can predict the future with unfailing certainty; fortunately we still have the ability, at least in part, to design our own future. The best anyone can do is to sketch patterns that might evolve from present ways of life and suggest possible alternative futures.

Scenarios of the future written today range from glorious visions of a new world order to the most appalling nightmares of ecological catastrophe. While neither extreme is likely to materialize, the reasoning behind such views must be assessed if we are to gain some insight into the possible consequences of our present actions. By examining forces that threaten to disrupt natural and human landscapes, we can determine what courses of action might help us approach a markedly better state.

UNCERTAINTY

No one can see into the future; there are too many uncertainties and unknowns. The forces influencing the future landscape are either unpredictable, subject to change, or beyond our understanding. Volatile birth, marriage, and death rates play havoc with population predictions. Migration patterns are virtually impossible to predict. Weather patterns change capriciously, and natural disasters take us by surprise, frustrating our efforts to generate food surpluses. Life-style patterns, the location choices and investment preferences of individuals, corporations, and governments, the balance between private and public enterprise and between consumption and investment, the willingness of richer nations to invest in poorer ones, and the balance of power within and between societies are all bound to change. We cannot foretell when the next embargo, economic depression, or war is likely to occur. We can only speculate on the nature and cost of technological improvements and on the volume, quality, and nature of resources yet to be discovered. What are the long-term effects of present-day pollution? Will a breakthrough in technology end our search for a limitless source of energy? What are our chances of avoiding nuclear warfare?

The future is veiled by so many uncertainties that we cannot predict whether national and world populations will grow or decline, or whether national products will rise or fall faster or slower than the rate of population. If, as is probable, population continues to increase, will there be mass starvation and global warfare, or will society find some way to accommodate the multitudes? Where will people choose to live? Concentrated in mammoth megalopolises or dispersed in small cities and villages?

These are just some of the questions without answers. But we do not have to buckle under the uncertainties and resign ourselves to whatever fate may bring. By measuring and evaluating current trends, we can estimate probable consequences. Then, as individuals and collectively, we can direct change toward a desirable state.

THE FUTURE LANDSCAPE OF THE UNITED STATES

The most reasonable predictions of the future are those inferred from current trends. Current trends, in turn, are determined by examining present tendencies, recent patterns of change, and the long-term evolution of such patterns. Though extrapolation of this sort is a common and indeed necessary step in planning, it is by no means error free. Until the 1974 energy crisis, for example, energy consumption and automobile ownership in the United States had been projected at a continually increasing rate. Despite warnings from ecologists that such exponential increases as a doubling in energy use every twenty years could not last, most people were quite unprepared for the sharply rising prices and slowing of consumption caused by diminishing internal supplies of oil, a situation aggravated by the external shock of the oil embargo. The problem is that extrapolation often ignores the feedback processes (such as the rise in petroleum prices) that alter trends and change human behavior.

Forewarned of the limitations of prediction, let us outline some of the current trends affecting the landscape of technically advanced nations and consider four scenarios extrapolated from the U.S. experience. Later we will expand our discussion to the future landscape of the world.

Current Trends in the United States

1. The rate of population growth is slowing down. Families today are planning to have an average of two children or fewer (statistically, 1.7). If this trend continues, the U.S. population could be stabilized at 250 million or less by the year 2000. Suburban development might then taper off, and the pressure on adjacent farmland might diminish; any necessary urban planning could be achieved more easily. Less pressure might be placed on water, forest, and mineral resources, owing to a stabilization in the numbers of consumers. But if people choose to have larger families, the birthrate could rise again. If immigration quotas are raised or if incentives to enter the United States illegally are sufficiently strong, legal and illegal immigration could increase. The strain on land and resources would thus begin anew. And even if internal demand for American resources and goods stabilizes, demand from the rest of the world may rise substantially.

2. The population is becoming older as a result of the declining birthrate. Hence fewer schools and other youth-oriented facilities are likely to be built, and those already existing may be converted to other uses. Labor surpluses and unemployment should disappear as a diminished labor force supports a much larger retired population. With more older persons retiring earlier (this may change), living longer, and perhaps enjoying higher income and greater mobility, there should be even more pressure on amenity areas than at present, particularly in warm-winter zones (figure 15.1). If many retirees were to build second homes in these favored environments, unoccupied areas would gradually be filled in. Pressure would be imposed on environments incapable of supporting large populations, such as the desert of Arizona, and the landscape would be visibly altered.

3. Affluence has been rising, though this trend may be halted or even reversed by inflation or recession and resource constraints. More people are enjoying greater leisure time, mobility, and income that can be spent on nonessentials. These trends may also decline as a result of inflation and recession. But if the per capita consumption of goods and level of mobility continue at the rate of increase that prevailed before the 1974 energy crisis and recession, changes in the landscape will be evident. The demand for new cars and second homes, for example, will place more stress on energy resources, raw materials, and amenity environments. If pre-1974 trends are not regained, people will spend more on local services than on goods and mobility, and the impact on the landscape will be less conspicuous.

4. Although the population has become more mobile over the years, the energy crisis of the 1970s, the reduction in military forces (resulting in

15.1. Sun City, Arizona. This community, encircled by a golf course and centering around a shopping nucleus, contains 40,000 people and is expected to grow to 64,000. (Photo from Wide World.)

fewer temporary migrations to army bases), high levels of unemployment, and the lessening of regional differences in racial discrimination have served to reduce internal movement, at least temporarily. If this trend holds, greater stability is likely to occur in the distribution of population and in the landscape itself. But ease of mobility is highly valued in our society, so the decline may not last. Then, too, possible programs of guaranteed annual income with no work requirement may increase migration to amenity areas in the future.

5. The rising cost of gasoline has slowed the growth of automobile travel, and since 1973 there has been a slight recovery of the use of public transportation. If mobility via public carriers is promoted and the use of

private automobiles restricted (a rather unlikely prospect), the effect on the landscape could be dramatic. For example, suburban sprawl would probably give way to more concentrated settlement, especially in areas well served by public transit; highway-oriented businesses would be weakened (perhaps making the highway landscape even less attractive); and pressure would diminish on diffuse recreation areas while increasing on closer, more accessible ones. Spatial patterns in the United States might come to resemble those of the 1920s, when most people relied on public transportation. This trend, however, is too weak at present to be really effective.

6. Population and employment have become increasingly megalopolitan. In the South and the plains, urbanization is progressing on the strength of ongoing industrialization, rising farm income, and surplus rural population. New suburbs and shopping centers, improved rural roads, and other signs of development continue to modify the landscape. In the more urbanized northeastern states and in California, however, urbanization appears to have peaked; people and jobs are leaving the metropolis for small cities and towns and even rural areas. The revival of smaller centers is visible at a very local scale, and exurbanization is spreading a low-density veneer of industry, urbanization, construction, traffic, and pollution over wider and wider areas.

7. Americans are showing a greater concern for preserving environmental integrity, not only for health and aesthetic reasons, but also because more people realize that development may damage natural environments worth maintaining in their own right. Such attitudes may promote more careful planning of development, decreasing the rate of environmental deterioration.

8. People are becoming more conscious of and attracted to amenity environments. Many are now willing to accept a lower income for the sake of living in pleasanter surroundings, especially in scenic, low-density, sunny, coastal, or mountainous regions. However the movement of large numbers of people to such environments may destroy the very values that were sought by filling open spaces with homes, roads, schools, shopping centers, and the like (figure 15.2).

9. Despite heightened environmental concern, some of the most pressing issues in the United States center around problems of unemployment and inflation. Extremes of wealth and poverty still exist; we have so far not been able to organize a secure, stable economy assuring reasonable equality of opportunity and reward. Poverty and discrimination remain highly visible in slums and ghettos, and blight continues to spread through the heart of many cities. There is little indication that this trend will be reversed in the near future, except that the aging of the population may indirectly reduce unemployment.

10. Americans live in a state of perpetual tension between the tradition of individualism in their treatment of land, resources, and other people, and their growing willingness to accept social responsibility for the welfare of others and greater restrictions on property rights through land-use controls

15.2. Residential development in southern California. As increasing numbers of people are attracted to scenic areas, the natural scenery disappears. This photo shows a Los Angeles housing development encroaching on the range of mountains bordering the city. (Photo from Wide World.)

and other measures. The latter trend, however, appears confined to a strictly local level. Regional self-interest and even dreams of regional self-sufficiency are on the rise. Thus regional landscapes may continue to show marked differences in levels of human well-being.

Various scenarios of the future have been extrapolated from these trends. The four that we will describe deal with the medium-range future, about the year 2000.

Scenario I: A Continuation of Present Trends

The most probable future, for lack of contrary evidence, is derived from a cautious extrapolation of current trends. But since several of these trends are contradictory—such as growing affluence and mobility versus greater environmental concern and serious energy shortages—even such a conservative scenario is not easy to design.

According to this scenario, the U.S. population will slowly increase to about 250 million by 2040 (300 million if the rate of immigration remains fairly high) and will gradually age. Real incomes are likely to rise despite possible energy shortages, mainly because population density will be low enough to ensure an adequate supply of food. (Under conditions of scarcity, food prices would rise and absorb a much greater proportion of personal income.) Though transportation will become more expensive, mobility is not likely to decrease; people will depend more on public transportation than they do at present. Urbanization will continue in the sun belt, while crime, pollution, and obsolescence may reduce metropolitan populations in the Northeast. The high cost of materials and energy will prevent all but the rich from enjoying the luxury of suburban sprawl. Larger numbers of people will live in apartments, perhaps in smaller cities and towns.

The general aging of the population will help solve employment problems, but severe social and economic inequality and discrimination are likely to persist. Older central cities will be further abandoned by the middle class and left to the poor and minority populations. The more affluent will move to amenity areas, perhaps on a seasonal basis. Although people will argue for changes in the social and economic system, its essential character will remain much the same: economic power will still be concentrated in the hands of a few, and tensions will persist between private and public interests and between individualism and social responsibility.

This scenario stops short of major technical and social breakthroughs and does not imply a dramatic modification of the landscape. Fewer acres of farmland will be lost through urbanization. Strip mining, clear-cutting, and other forms of resource exploitation, however, may become more extensive in the effort to maintain high levels of mobility and consumption. Racial discrimination and economic inequality will continue to be expressed in the landscape, particularly in urban ghettos and slums. Most land use decisions will still be made by individuals, corporations, and local governments rather than by central planners. The greatest change would stem from the institution of guaranteed annual incomes, which would probably encourage large-scale migrations to amenity areas.

The landscape envisioned in this scenario is not very different in character from that of today. Indeed it is perhaps a playing out of the social, technical, and economic forces already present at the time of World War I: by then the automobile and airplane had been invented, the mass market and suburbanization had begun, people had started to move to amenity areas, and a trend toward greater social responsibility could be discerned.

Scenario II: A Postindustrial Society

The most popular scenario among futurologists depicts a highly affluent **postindustrial society** characterized by a stage of development in which most kinds of factory work have been automated and the large majority of the population is engaged in service activities. This scenario proclaims the triumph of technology over want and, to a great extent, over the friction of distance itself. New technologies and sources of energy will eliminate the need for a large industrial labor force and will make possible faster, more efficient modes of transportation. Most people will be employed in service activities and will have far more leisure time than at present. Production will be efficient and pollution free, and the environment will at last be cleansed. The problems of race and poverty will be resolved, and the tensions between private and public interest and between individualism and social responsibility will be overcome.

Futurologists have projected two very different postindustrial landscapes. Some believe that most people will be concentrated in enormous supercities (figure 15.3). Only a small permanent rural population will be needed to supply agricultural, mineral, and forest products from the vast open spaces surrounding the population centers. City dwellers will probably build a network of second homes in the countryside. Older cities, revitalized and reconstructed, will be served by superior transport and communication systems.

In this particular portrait of the postindustrial landscape, there is a clear dichotomy between the relative emptiness of rural areas and the density of metropolitan areas. The older, less child-oriented population will come to

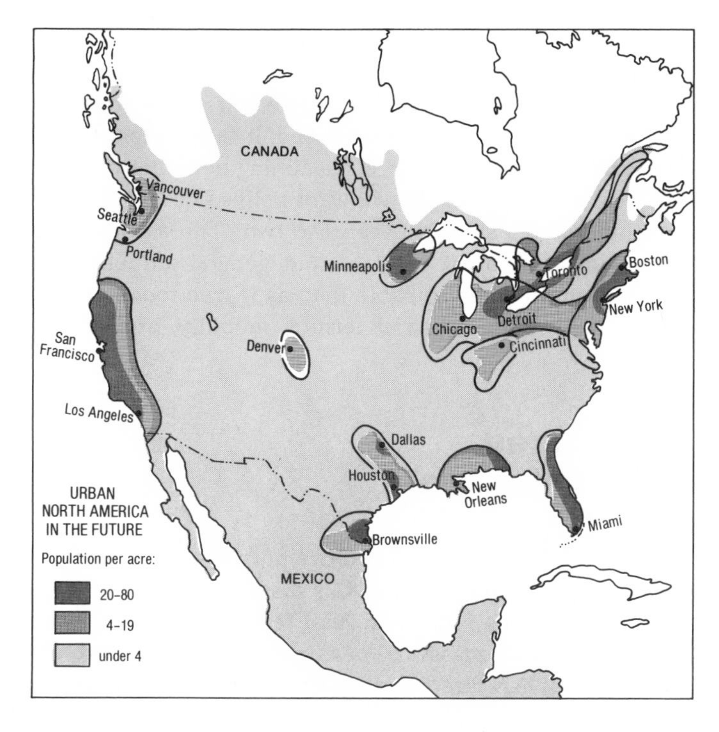

15.3. Urban North America in the future. If metropolitan regions continue to expand, it is believed that several great megalopolises will emerge by the middle of the next century.

prefer high-rise central living. Note that in this version people agglomerate in metropolitan centers not because they have to but because they want to.

Interestingly enough, other futurologists take the opposite view. They assume that most people will abandon the increasingly obsolete cities and move to rural or village settings in a variety of amenity environments. Families might own several homes: a ski lodge in the mountains, a cottage on the seashore, and a farmhouse in the countryside. Since production will be mostly automated, work will typically involve organization, management, analysis, or handcrafts and could be done at home rather than at an office or factory. Telecommunications will be so advanced and universal that physical presence will become less and less necessary. Traditional identification with place will give way to identification with enjoyable people and environments. Residential location will be governed not by job opportunities but by recreational opportunities, and travel for social or recreational purposes will supplant the journey to work.

This portrait prescribes a true landscape revolution. In contrast to the eighteenth-century industrial revolution, which forced people into cities (the "centers of human progress"), the postindustrial revolution could permit a return to the rural-village pattern of settlement. People will no longer have to agglomerate to earn their living. Telecommunications and superior transport will substitute for physical proximity. Dispersed into what has been called the global village, the population will enjoy a more intimate contact with nature, and the problem of social and economic incompatibility will be solved by means of sheer physical separation.

This postindustrial scenario is based on humanity's age-old dream of overcoming the friction of distance through superior modes of transport and communication. One version assumes that people will choose to live in cities; the other, that rural residence will be preferred. It seems more likely, however, that a dual landscape would emerge, containing both metropolises and rural-village dispersion. Some people will want to live in the city, some in the country, and others will shift between the two. The assumptions made about new technological breakthroughs and general affluence are much less likely to be realized. It is probable that, as is true today, only a wealthy minority will enjoy postindustrial leisure, mobility, and spatial freedom.

Scenario III: A Mixed Industrial-Postindustrial Society

A somewhat more realistic scenario than the above portrays a new form of dual society. We examined the old form in our discussion on low technology rural landscapes (chapter 4), in which rural feudalism, consisting of a small upper class dependent on masses of peasants, exists along with urban capitalism. The new dualism predicted by this scenario consists of a less affluent urban-industrial society and a more affluent, freer-moving postindustrial society. The poor, elderly, and minority populations may still be tied to urban-industrial centers, with the prospect of earning enough money to move into suburban homes abandoned by the postindustrial rich, who will tend to disperse into exurban rural or village locations, leaving the obsolescent cities to the industrial poorer classes. This dual landscape reflects the realities of human experience in that inequality of power, income, and mobility are retained. (It does not seem likely that the entire population will achieve material prosperity; individual and social differences will continue to produce differences in levels of ambition, achievement, and opportunity.) The physical separation between the less and more affluent societies projected by this scenario offers a spatial solution—for some—to the social problems of the recent past. Such a scenario is almost as cautious an extrapolation as scenario I, differing from it mainly in the assumption that many people will attain postindustrial affluence, mobility, and freedom from the journey to work, and that abundant energy will be available to make this possible. (See table 15.1 for a point-by-point comparison of the various scenarios.)

Scenario IV: A Less Affluent, More Planned Society

In striking contrast to postindustrial scenario II, based on limitless energy and rising affluence, scenario IV assumes that scarcity, not abundance, will prevail. Futurologists holding this more pessimistic view proclaim that drastic changes will result from environmental constraints—insufficient water, land, energy, and mineral resources—and environmental deterioration stemming from pollution. The mushrooming growth of world population, coupled with the demands of poorer nations not only for more food but for a share of the wealth, will push us to the limits of our resources. Our land will no longer provide us with sufficient food and minerals, and our technology will be unable to create further sources of energy or dispose of the mounting volumes of waste. Some ecologists warn that unless we learn to live more simply within the constraints of nature, ecological disaster will befall us.

This scenario, then, predicts lower levels of energy and resource consumption and a steep decline in at least private mobility. Shortages of energy and raw materials and the high cost of maintaining the environment will force us to reduce our annual production and consumption of goods. Achieving this reversal will probably require much greater economic planning and control than at present and a fundamental shift from private to public interest and from individual to societal control.

Less affluence and more planning would result in a landscape similar to Europe's today. Cities would be more densely populated and compact; more housing would be government controlled; industrial and commercial activities would utilize space more efficiently and be better served by public

Table 15.1. Scenarios of the Future Landscape of the United States

Features	*Scenario I: Continuation of Present Trends*	*Scenario II: Postindustrial Society*
Population growth	Slow increase	Slow increase
Income	Rising, but threatened by inflation/recession	General affluence
Transportation and mobility	Rise in cost of private transport; more emphasis on public transport; continuing high mobility	Fast, efficient transport; high mobility
Urbanization	Continues in sunbelt; metropolitan decline in Northeast; smaller cities and towns favored	Population concentrated in supercities or dispersed in global village; older cities revitalized
Socioeconomic conditions	Continuation of inequality and discrimination; economic power concentrated in hands of a few; tension between private and public interests	Problems of race and poverty resolved; tensions between private and public interests overcome
Landscape	Central cities inhabited by poor and minorities; more extensive resource exploitation; large-scale migration to amenity areas	Pollution free; dense metropolitan settlement; relatively empty rural areas; or rural-village settlement

transportation; the private automobile would become a luxury; and urban and interurban public transit would be widely used. Economic growth would probably be focused in smaller cities, if they are found to be more efficient than larger ones. The tremendous latent value of now-declining central business districts would once again be realized, inducing revitalization and reconstruction.

In many ways this landscape recalls that of the 1920s—particularly the greater dependence on public transit, the less dispersed settlement patterns, and the greater importance of the CBD—with the exception of far more advanced production and communication technologies. (A materially less affluent society is thus compatible with some aspects of the postindustrial future.) Some futurologists even argue for a widespread return to the land—perhaps after the Chinese model of social and economic organization—on the theory that large cities will be too costly to maintain. It is doubtful, however, that our society, which has for centuries emphasized individualism and the acquisition of material wealth, would be willing or able to undergo such a radical transformation. But this is one of the options.

Scenario III: Mixed Industrial-Postindustrial	*Scenario IV: Less Affluent, More Planned*
Slow increase	**Slow increase**
Less affluent industrial society/affluent post-industrial society	**General decline**
Greater mobility for rich; less mobility for poor	**Decline in private mobility; increased use of public transport**
Obsolescent cities left to poor; exurban settlement for rich	**Cities more densely populated and compact; revitalization of CBD**
Inequality of power and income	**Greater economic planning; shift from private to public interest and societal control**
Dual landscape: concentrated urban and dispersed rural	**Similar to European landscape today; efficient use of urban space; shortages of energy and raw materials**

These different scenarios show that the future is open, not preordained, and that humans have a limited role as agents of change. Our efforts are unlikely to transform the landscape as drastically as we might hope or fear.

THE FUTURE LANDSCAPE OF THE WORLD

The scenarios discussed above refer specifically to the United States. We cannot apply them to the world as a whole because some of the current trends around the world differ significantly from those in the United States. Moreover far more variation occurs on the world scale than in a single nation like the United States. Indeed when contemplating the world landscape, one is struck by the incredible variety not only of physical environments but also built environments. Variations in the human landscape result from differences in level of technological and economic development, including degrees of industrialization and urbanization, and differences in cultural development, including norms and values, political, economic, and religious systems, and attitudes toward the physical environment. These variables influence how and to what extent societies civilize the landscape.

The most significant question that can be asked about the world landscape is whether portions of it are converging toward a single model, and if so, which one. If patterns of highly unequal development persist, how will these differences be expressed in the future landscape? More specifically, will most of the world pursue the Western model of industrialization, urbanization, and energy dependency, or are there alternative ways to develop? Does the nature of the economic system (capitalist or socialist) really make a difference? Will the current emphasis on national and even subnational self-interest and cultural identity prevail in the future, or will a world government and world culture develop? Will the increase in world population, accompanied by rising levels of energy consumption and environmental pollution, lead to disastrous human and environmental crises? If so, how will these crises be resolved? Will all societies come to enjoy reasonable security, comfort, and opportunity, or will levels of living remain highly unequal? The scenarios presented below attempt to answer these questions. As before, let us first review current trends that point the way to the future.

Current Trends around the World

1. World population continues to grow rapidly, with a population equivalent to that of the entire United States (220 million) being added every four years. The developed nations are growing at a fairly slow rate—less than 1 percent a year—and are gradually aging. The developing nations, particularly in Africa, South America, and South Asia, are increasing by at least 2 percent a year and have a rather young population.

2. World food supplies have kept pace with population growth over the years, but uneven weather and problems of distribution are increasing the frequency of malnutrition and famine. Particularly in parts of Africa and South Asia (see figure 13.14). The investment required to produce food leaves very little for other forms of development in these areas.

3. Energy consumption has been rising even faster than population as less developed nations industrialize and advanced nations strive for even greater mobility and productivity.

4. Rising demand for food, energy, and raw materials has resulted in higher prices and inflation, thereby slowing economic growth and development.

5. Despite greater environmental awareness and concern and better methods of pollution control, the assault on air, water, and soil seems to continue.

6. The gap between rich and poor nations remains very wide, with the ten richest enjoying levels of income perhaps fifty times as great as the ten

poorest. Over the world, a dual or even triple socioeconomic division exists, ranging from peasant agricultural to urban-industrial to affluent and mobile societies.

7. Despite rhetoric to the contrary, virtually all societies, no matter how different their socioeconomic systems, are adopting the Western technological model of development. Industrialization and especially urbanization are progressing fairly rapidly in most less developed nations (figure 15.4).

8. The competition between diverse socioeconomic systems is becoming more and more complex. Besides the ideological competition between capitalism and socialism, there is also much conflict between giant multinational enterprises and smaller businesses, particularly in less developed nations.

9. Though some degree of international cooperation and interdependence is developing under the uneasy superpower hegemony of the United States and the Soviet Union, the overall trend shows a preponderance of national self-interest, dependence on military force, and economic and cultural protectiveness. Wars and disputes over territory continue to erupt around the world, impeding economic development as well as social interchange.

These trends have provided material for the extrapolation of several possible futures, six of which will be described below. They deal with the next thirty years (anything beyond that is but science fiction) and even within this limited span, the uncertainties are so great that it would be dishonest to express much confidence in any one scenario.

Scenario I: Relative Stability

A scenario predicting little change does not rate highly with **futurologists,** most of whom prefer visions either of grandeur or of disaster. And in some ways stability is not a particularly valid projection, if past experience is any indication of the future. Yet several trends in today's world support the projection of relative stability. For example, while the poorer nations endeavor to raise their living standards through agricultural, industrial, urban, and cultural development, their efforts are continually frustrated by shortages of food, energy, and other essential resources, by political and military instability, and, most important, by explosive population growth.

Rich nations have had little incentive in the past to share with the poor. Thus it seems reasonable to predict that the former will maintain at least present standards through existing technology and future innovations and that the latter will be prevented from achieving high levels of development for the reasons given above. Periodic warfare over access to land and resources will probably continue within and between poorer nations, as well as between rich and poor nations. Urbanization and industrialization may continue to progress in the less advanced nations, though perhaps more slowly than in the recent past because of the overriding need to produce sufficient food.

15.4. Millet-processing plant, Zinder, Niger. Machines now do work that traditionally required many hours of human labor. The processing of millet is becoming energy intensive rather than labor intensive in this developing region of Africa. (Photo from United Nations.)

In essence, this scenario implies that the **Malthusian crisis**—in which population growth exceeds agricultural productivity, resulting in mass starvation and warfare—can be postponed, if not prevented. The landscape it envisions differs little from the one we know today, though environmental deterioration will probably increase because of the necessary expansion of both agricultural production and resource exploitation.

Scenario II: Worldwide Economic Security with Continued Inequality and Conflict

More optimistic futures have also been extrapolated from current trends. In recent years advances in technology and the development of both old and new sources of energy have permitted production and income to rise faster than population. Scenario II assumes that this favorable ratio will hold. Not only will there be enough food (through an extension of the green revolution and the development of new plant varieties) for the over 6 billion mouths demanding to be fed by the year 2000, but the food will be of superior quality and will be distributed faster and more efficiently than before.

In addition, either the rich nations will find it profitable to assist the poor into the mass-consumption stage through industrialization, urbanization, and agricultural modernization, or the poor will be able to achieve this on their own through their control of the raw materials needed by the rich nations (see figure 14.2). Most of the world will adopt the urban-industrial model of development, despite differing social and political systems. Nations that are already affluent will pass into the postindustrial stage. The highly visible disparity between industrial and postindustrial societies will result in a perception of relative deprivation (no one will really be suffering from want; some societies will simply be less well off than others). Economic, social, and political systems will still vary, and international conflict, especially over resources, will continue to flare up.

The resultant world landscape will be dual in character: postindustrial and industrial. The landscape in advanced nations will resemble one of the two postindustrial models described earlier, or perhaps a combination of the two. With larger populations (lower per capita income) and a greater lag in development, the less advanced nations will come to look somewhat like the present-day Soviet Union or Western Europe, emphasizing traditional industry. The particular model adopted—Soviet or West European—would depend on whether a centrally planned or private enterprise socioeconomic system prevails in a given nation.

Scenario III: A New World Order

A spectacularly radical vision of the future portrays the establishment of a new world order, with economic security for all, the more nearly equal distribution of wealth, and the achievement of world community. To realize this dream, great technological advances would first have to be made not only in the production of food (perhaps synthetic food will be produced by chemical processes), but in the development of energy supplies (solar energy; nuclear fusion). If most of the world were to achieve by the year 2000 the level of living enjoyed by the richer nations today, incredibly rapid progress would be necessary. In particular, the pace of industrialization and urbanization would have to accelerate. Ideally if resources were recycled and social and economic organization improved, income could rise much faster than the consumption of resources. Some kind of world government would have to be created, economic systems would have to converge, and enough harmony would have to prevail to permit the more nearly equal distribution of wealth. Individual, corporate, and national self-interest would have to give way to international cooperation.

The landscape associated with the new world order would probably look like the postindustrial metropolitan model, with a high degree of automation, advanced modes of transportation (primarily mass transit), and a high level of mobility, with the addition of an internationalization of culture and a widespread intermingling of the races.

The realities of human experience cast doubt on the credibility of such a future. There is not much evidence to support the notion of the world's changing, in one generation, from its present state to the ideal conditions described above. Many nations have barely begun to industrialize, and the overwhelming needs of the less developed nations may well prevent the already developed from achieving higher levels—or even from maintaining present levels—of mobility and wealth.

Scenario IV: Greater Disparity between Societies

More pessimistically, some futurologists feel that there will be a greater polarization between the more and the less advanced societies. They predict that advanced nations with stable populations but high resource needs will find it necessary to maintain, by whatever force or violence may be required, the burgeoning poorer nations in a quasi-colonial state of semidevelopment. While the advanced nations will probably try to ensure minimal standards of living in the less-developed ones, they would, if it seemed necessary to them, resort to preemptive strikes or allow a Malthusian course of famine and internal warfare to reduce excessive populations.

The advanced nations probably could get away with such inhumanity. Pessimists hold that the more successful competitors in the fight for survival will be forced to use brutal methods as limits are reached in the earth's ability to produce food and supply energy.

The widening gap between rich and poor will be expressed in the world landscape by obvious contrasts between the gleaming postindustrial communities of the most advanced nations and the squalid rural and urban hovels of the least developed, most overcrowded subsistence societies.

Scenario V: Eco-Catastrophe

The grimmest scenario of all, as fervently predicted as the new world order, portrays a Malthusian apocalypse. In this most hopeless future, technology will finally have met its match. Nature will no longer furnish the food required by the teeming populations or yield the resources demanded by insatiable producers and consumers. Pollution and waste will irreversibly degrade the quality of water, air, and soil, and the earth will lose its capacity to sustain our way of life. In a futile attempt to maintain the standards of living they are accustomed to, the rich nations will continue to exploit the poor, only to postpone the inevitable decline in living standards everywhere. The natural landscape will deteriorate through massive pollution (or nuclear warfare) and through the desperate exploitation of forests, mineral resources, and marginal farmland, followed by abandonment and erosion. The human landscape will decay as a result of declining income, increasing inability to maintain the built environment (factories, housing, transport systems), and relentless warfare. In its extreme version this scenario projects mass death through nuclear holocaust, starvation, or poisoning and concludes with humanity's reversion to preindustrial hunting or agriculture.

Scenario VI: Ecological Adjustment to Simpler Ways of Life

Those who predict ecological catastrophe offer a way out: worldwide acceptance of a less mobile, less resource-oriented and materialistic way of life. If technology proves incapable of making the energy and production breakthroughs necessary to ensure universal affluence and mobility, then perhaps a better alternative to scenarios I, IV, and V would be one in which the richer nations voluntarily reduce their annual levels of production, consumption, and mobility. Theoretically if the small number of rich nations did not consume such a disproportionately large share of resources, most of the world's people would live in moderate security.

Many futurologists consider it essential to achieve this goal, the "ecological imperative." To do so would require a drastic transformation on the part of many societies. Settlement would become more concentrated. Cities, businesses, and institutions would tend to become smaller. Greater emphasis would be placed on agriculture. In a sense, this would mean a return to an earlier landscape and way of life.

Capitalist economies emphasizing annual output would have to strive instead to improve the general welfare. Individual and national self-interest would have to give way to world-regulated allocation of resources and rewards. Even one's right to have children might have to be controlled. In some ways, then, society would shift toward the Chinese model, in which self-interest is submerged for the common good. Such a scenario has special appeal to those who think humanity is too materialistic and has estranged itself from nature.

There is evidence that portions of society are beginning to heed the ecological imperative. For example, some newly formed nations, such as Tanzania, are consciously pursuing an alternative, less materialistic model than the Western technological one. And even in the most affluent nations, individuals and families are advocating and living a simpler life. Still it seems unlikely that many nations, developed or developing, will choose this path unless forced to do so.

CONCLUSION

We have no way of knowing whether specific features of the various scenarios (table 15.2) will become reality. The two scenarios of greatest interest to futurologists—the postindustrial vision of a frictionless world and the

Table 15.2. Scenario of the Future Landscape of the World

Features	*Scenario I: Relative Stability*	*Scenario II: Economic Security with Inequality and Conflict*	*Scenario III: New World Order*
Population growth	Slow increase in advanced nations; rapid rise in less advanced nations	Rising	Rising
Living standards	Relatively high in advanced nations; relatively low in less advanced nations	Improved for less advanced nations; continued high for advanced nations	Generally high
Mobility	Relatively high for advanced nations; relatively low for less advanced nations	Improved for less advanced nations; continued high for advanced nations	Generally high
Urbanization and industrialization	Continuing or "peaked out" in advanced nations; continuing, but perhaps more slowly in less advanced nations	General urbanization and industrialization	General urbanization and industrialization
International relations	Periodic warfare over access to land and resources	Economic, social, and political divergence; conflict over resources	Achievement of world community and harmony
Landscape	Differs little from today, but some increase in environmental deterioration	Megalopolitan-rural dichotomy and/or dispersed global village	Postindustrial metropolitan

portrait of ecological disaster—depend essentially on unknowns. To be achieved, the postindustrial scenario would require the availability of enormous quantities of energy and the achievement of prodigious improvements in the technology of transportation and communication. In addition, the world would have to become united politically and economically if all societies were to benefit from these technological marvels. The other scenario assumes that human ingenuity and technological potential will suddenly collapse, that birthrates cannot be controlled, and that new resources will not be discovered—in short, that what we now know is all that we will ever know. None of these conditions is likely.

The weakness of these forecasts is that they are based on one-dimensional extrapolations of observable trends, mathematical extensions of curves beyond human or scientific experience. The postindustrial vision of time-space convergence through supertransport and communications ignores the tremendous economic and human costs of even beginning to achieve freedom from the tyranny of space. The scenario predicting irrever-

Scenario IV: Greater Disparity	Scenario V: Eco-Catastrophe	Scenario VI: Ecological Adjustment
Stable in advanced nations; rapid rise in less advanced nations	**Explosive**	**Controlled**
High in advanced nations; low in less advanced nations	**General decline**	**Generally moderate**
High in advanced nations; low in less advanced nations	**General decline**	**Voluntary decline in advanced nations**
No further development in less advanced	**General decline**	**General decline**
Neocolonialism or suppression of less advanced nations by more advanced nations	**Global conflict**	**Cooperative; globally regulated**
Great disparity between postindustrial advanced and subsistence-level less advanced nations	**Deterioration through pollution and overexploitation; reversion to preindustrial agriculture and hunting**	**Ecologically stable; emphasis on general welfare**

sible environmental deterioration fails to take into account our ability to learn and our capacity to grow.

No one knows what the future will bring, but we may safely predict that it will contain elements of most of the scenarios presented above. It is certainly premature to foresee, within the next generation, an end to the so-called tyranny of space or to the identification with place associated with the postindustrial future. We repudiate this view not because we fear that geography will become obsolete—its very essence being that of understanding spatial order— but because we feel that identification with place is an instinctive human trait. Even more important, the desire to escape the constraints of space and environment is but one more manifestation of the human desire to emulate the gods. No matter how impressive our technological achievements, we will always sense physical separation, respond to personal and environmental presence, and pay higher and higher prices for marginally smaller and smaller gains in mobility and locational freedom. Perhaps our major task in the future will be to plan development in accordance with the principles discussed in this book: to create spatial structures that temper efficiency with equity, to find that balance between technological potential and human well-being that will keep us in control of technology, and to preserve that vital link to nature from which ultimately no living thing can escape.

SUGGESTED READINGS

Abler, R., et al. *Human Geography in a Shrinking World.* North Scituate, Mass.: Duxbury Press, 1975.

Bell, Daniel. *The Coming Post-Industrial Society.* New York: Basic Books, 1973.

Darling, F., and Milton, J. P. *Future Environments of North America.* New York: Natural History Press, 1964.

Forrester, Jay. *World Dynamics.* Cambridge, Mass.: Wright-Allen Press, 1971.

Kahn, H., et al. *The Next Two Hundred Years: Scenarios for America and the World.* New York: Morrow, 1976.

Landsberg, H., et al. *Resources in America's Future.* Baltimore: Johns Hopkins, 1963.

Meadows, D. *Limits to Growth.* New York Universe Books, 1972.

Toffler, A., ed. *The Futurists.* New York: Random House, 1972.

Webber, M. *Explorations into Urban Structure.* Philadelphia: University of Pennsylvania Press, 1964.

Afterword: But What Do Geographers Do?

We hope it is now clear to you that geographers seek to understand how societies organize themselves in space and adapt to and modify the landscape. For a long time most geographers were content to teach these facts as a service to general education. Others became cartographers (and drew maps) or area specialists, working in foreign trade, development, or intelligence. These remain important occupations: accurate maps not only depict facts but communicate ideas and plans; and precise knowledge of foreign areas and customs promotes effective foreign policy making. But still more important today are positions in regional and environmental planning. To a large degree, geography is, or should be, the science behind the art or profession of planning the landscape. Geographers are trained to understand how private and public decisions can change patterns of employment, housing, or transport and to evaluate the effects of proposed changes—in short, to fill analytical research positions in a wide range of private firms (where geographers may be called location analysts) and local to international public agencies. The many issues raised in chapter 13 are among the kinds of problems geographers deal with. To be successful, geographers must acquire broad knowledge from many other disciplines and master a variety of regional-analysis methods, for example, techniques to find the best location for stores, shopping centers, schools, factories, and the like, or to extract the geographic distribution of the effects (benefits and costs) of public and private decisions (such as new laws, regulations, programs, stores, factories). The work of a professional geographer is not only diversified, but it is useful, creative, and challenging.

Glossary

Accessibility The relative degree of ease with which a location can be reached from other locations.

Administrative principle In central place theory, the principle governing a spatial arrangement that maximizes spatial efficiency for administrative purposes: a higher level center wholly dominates the areas of the lower level central places that surround it.

Agglomeration The spatial grouping of activities or people for mutual benefit. Particular savings, or economies, are realized by such groupings of retailers or industries.

Agricultural gradient The progressive decline in intensity per acre of crop production with distance from market.

Basic activities Economic activities whose products or services are exported from the region.

Break of bulk The division of a large shipment into smaller parts, particularly upon transfer from a higher volume carrier to a lower volume one (as from ship to rail or from rail to truck).

Central place functions Services offered from a central place to the surrounding area.

Centrality High accessibility (such as location at the center of a transport network).

Colonial Used here in the limited sense of dependence of a less-developed area upon a more-developed one.

Comparative advantage The relative superiority of a location for certain activities.

Contact field The distribution of an individual's acquaintances over space; the geographic range of personal communication.

Complementarity A state that exists if the varying advantages of two or more locations or areas permit a mutually beneficial linkage, usually through trade.

Cultural group A group of people unified by a common language, religion, customs, technological level, or similar element.

Culture The way of life characterizing a particular group of people. Includes traditions, values, beliefs, political and social behavior, use of material resources, and ways of perceiving the physical environment.

Deforestation The clearing of forests, usually for cultivation.

Development Changes resulting in the improvement of human well-being.

Diffusion The spread over space and time of people, technology, and ideas, usually from very limited origins.

Dispersed cities Clusters of small urban places that display some functional specialization, such as an emphasis on industry or commerce; or built-up urban areas, together with their surrounding exurban nonfarming population.

Economic margin The point where revenues equal costs. In agricultural location theory, this term refers to the farthest location from which produce can be shipped profitably to commercial markets.

Economies of scale Cost savings brought about by the tendency for marginal costs (those of producing the next unit) and average costs (those of producing all units) to decrease with increasing volume of output. Diseconomies, however, may set in at excessively high volumes (called the *point of diminishing returns*).

Ecosystem A stable system of relationships among organisms and between the organisms and their environment. Any disturbance of one element in the ecosystem will have an effect on the others. Changes in water temperature, for example, may result in the decline or disappearance of certain plants in a marine ecosystem.

Efficiency The most productive use of space at least cost or effort.

Environment The physical landscape, including climate, vegetation, wildlife, bodies of water, and landforms.

Equilibrium A theoretical state of stability. Any change from this state would decrease efficiency or profitability. Equilibrium prices are values for goods, labor, capital, or land corresponding to these conditions.

Extensive Characterized by a relatively low level of inputs and/or outputs per unit area.

External diseconomies The rise in assembly and distribution costs that occurs when more distant sources and markets must be tapped as the scale of production increases.

Futurologists Scholars whose interest is focused on the future.

Gradient (urban intensity) The gradual increase in intensity of land use and density of population from the rural fringe to the city center.

Hearth area A region where an innovation first appears. Middle America, for example, is thought to be the hearth area for the cultivation of maize.

Human landscape The earth as inhabited, used, and modified by people.

Industrial complex A set of closely related firms, such as steel, fabricated metals, and chemicals. Each firm typically makes significant purchases from or sales to the others.

Information Knowledge about the physical environment, technology, and other conditions that would be necessary for making the best decisions.

Innovations Ideas that lead to change, typically increasing productivity in individual or corporate behavior.

Intensive Characterized by a relatively high level of inputs and/or outputs per unit area.

Interaction Relationships, such as trade and movement, between locations or people.

Interdependence Relationships resulting from specialization and trade among locations; what happens in one location affects many other locations.

Interindustry linkages The spatial and economic bonds that occur when a firm supplies parts for a larger firm or utilizes the by-products of another firm.

Internal diseconomies The rise in in-plant costs that may occur as production increases.

Laborshed Labor supply area of an employment center.

Location A place where human activities occur.

Location theory The body of principles that govern patterns of location and interaction.

Malthusian crisis A condition in which population growth exceeds agricultural productivity, resulting in mass starvation and warfare.

Marginal farmer The farm operator whose net return per man-hour is very low.

Market A place where goods and services are demanded and exchanged.

Market-oriented industries Industries that locate in or near their markets in order to save on distribution costs.

Market penetration A firm's share of the market.

Marketing principle In central place theory, the principle governing a spatial arrangement that minimizes distance traveled to centers. According to this principle, the service area of a larger, higher level center includes one-third of the service areas of each of the six neighboring lower-level centers.

Microclimate Local climatic conditions (temperature, precipitation, cloud cover, and so forth).

Migration The movement of population to a new area.

Milkshed The zone of an urban market's fluid milk supply.

Monoculture The long-term cultivation of a single kind of crop.

Near-subsistence farming The production of barely more than the minimal amount of crops needed to sustain life.

Nesting In central place theory, the tendency of service areas of lower level centers to be wholly included in the service areas of larger, higher level centers.

Nodes Intersection points or junctions on a transport network.

Nonbasic activities Activities whose products or services are generated and consumed in the local area.

Nonoptimal behavior Any decision-making behavior that results either intentionally or by default in less than maximum profitability.

Opportunity cost The opportunity that one passes up when making an alternative choice.

Postindustrial society A society characterized by automated industrial technology and a population largely engaged in service activities.

Primacy The concentration of cultural, political, and economic activity in one urban center of a nation or region.

Processing activities Activities that use a technological process to transform raw materials to finished goods.

Range In central place theory, the maximum distance over which a seller will offer a good or service or from which a customer will travel to make a purchase.

Rent gradient The range in the value of land, as a result of competition, from the highest value for the most accessible (central) location to the lowest value for the least accessible location.

Residential farmer A person who lives on a farm and works in the city.

Resource-oriented industries Industries that locate near their resources in order to save on procurement or assembly costs.

Resources In an economic sense, any valued aspect of the environment; here it refers to demanded natural materials, such as water, minerals, and soil.

Risk A condition in which the probability of success or failure is known.

Self-sufficiency The attempt of a local economy to provide by itself for all its own needs.

Shifting cultivation An agricultural economy in which fields must be abandoned and villages must migrate often because of the exhaustion of soil fertility.

Social space The area within which a social group carries on most of its interrelations.

Space The surface area of the earth—sometimes viewed abstractly as the separation of people and activities, sometimes concretely, as the varying character of different areas.

Spatial adjustment Changes in the location or behavior of a firm or its suppliers or markets in response to external change, such as an increase in the cost of materials.

Spatial behavior Actions resulting from the decisions people make regarding their use of space.

Spatial equilibrium *See* Equilibrium.

Spatial experience The range and intensity of a person's knowledge of his own and other areas, including awareness gained through travel.

Spatial monopoly In central place theory, the competitive advantage that a seller has in providing goods and services to the customers in his or her trade areas.

Spatial oligopoly A fairly stable, shared regional market controlled by a few industries.

Spatial organization The patterns of location and interaction that structure the space used by human society.

Spatial structure *See* Spatial organization.

Specialization The commitment of a particular location to producing one or a very few products.

Technology The means (tools, procedures) by which human society meets its needs for food, shelter, and knowledge.

Tenant farmer A farmer who does not own land but rents fields belonging to a landlord.

Threshold In central place theory, the minimum level of sales needed to attain marginal profitability.

Trade area The area from which most of a seller's customers originate.

Transferability The potential of shipping a product without incurring unprofitably high transport costs.

Transport network The physical system of links (railroads, roads, waterways) and nodes (junctions) on which movement can take place.

Transportation principle In central place theory, the principle governing a spatial arrangement that results in the most efficient transport linkage among centers.

Unallocated costs Costs *not directly chargeable* as part of the cost of production.

Uncertainty A condition in which the probability of success or failure is unknown.

Wholesale trade The handling and distribution of commodities in large quantities. Bulk goods from producers are divided into smaller lots for resale at the retail level.

Index

ARCTIC OCEAN
GREENLAND
Arctic Circle
ALASKA
Anchorage
PRIBILOF IS.
Juneau
CANADA
Vancouver
Seattle
NEWFOUNDLAND
Montreal
St. John's
Chicago
Boston
UNITED STATES
New York
Washington D.C.
St. Louis
San Francisco
Los Angeles
ATLANTIC
BERMUDA
MIDWAY IS. (U.S.A.)
GUADALUPE IS. (MEX.)
GULF OF MEXICO
MEXICO
Havana
BAHAMAS
Tropic of Cancer
Honolulu
HAWAII (U.S.A.)
CUBA
DOM. REP.
PUERTO RICO
JOHNSTON (U.S.A.)
IS. DE REVILLAGIGEDO (MEX.)
Mexico City
BELIZE
JAMAICA
HAITI
GUADELOUPE
MARTINIQUE
HOND.
BARBADOS
GUAT.
EL SAL.
NIC.
CARIBBEAN SEA
PACIFIC
CLIPPERTON (FR.)
TRINIDAD-TOBAGO
PALMYRA (U.S.A.)
FANNING (BR.)
COSTA RICA
PANAMA
VENEZUELA
GUYANA
SURINAM
WASHINGTON (BR.)
COLOMBIA
FR. GUIANA
HOWLAND (U.S.A.)
CHRISTMAS (BR.-U.S.A.)
BAKER (U.S.A.)
JARVIS (U.S.A.)
ECUADOR
Longitude West of Greenwich
PHOENIX IS. (BR.)
MALDEN (BR.-U.S.A.)
ARCH. DE COLÓN (GALAPAGOS I.) (ECUADOR)
TOKELAU (N.Z.)
STARBUCK (BR.-U.S.A.)
MARQUESAS IS. (FR.)
BRAZIL
Recife
WESTERN SAMOA
MANIHIKI IS. (N.Z.)
PERU
Lima
TUAMOTU (LOW) ARCHIPELAGO (FR.)
TUTUILA (U.S.A.)
IS. DE LA SOCIÉTÉ (FR.)
OCEAN
COOK IS. (N.Z.)
BOLIVIA
FIJI
TONGA
RAROTONGA (N.Z.)
IS. TUBUAI
PARAGUAY
Rio de Janeiro
São Paulo
Tropic of Capricorn
RAPA NUI (EASTER) (CHILE)
PITCAIRN (BR.)
DUCIE (BR.)
SALA-Y-GÓMEZ (CHILE)
I. DE SAN FELIX (CHILE)
I. DE SAN AMBROSIO
RAPA
KERMADEC IS. (N.Z.)
CHILE
Santiago
ARGENTINA
IS. DE JUAN FERNANDEZ (CHILE)
URUGUAY
Buenos Aires
CHATHAM IS. (N.Z.)
FALKLAND IS.
S. GEORGIA IS. (FALKLAND IS.)
S. SANDWICH IS. (FALKLAND IS.)
S. SHETLAND IS.
SOUTH ORKNEY (B.A.T.)
WEDDELL SEA
COATS LAND
ANTARCTICA
BYRD LAND
AÇORES (PORT.)
ISLAS CANARIAS (SP.)
CAPE VERDE
Dakar
GAMBIA
GUINEA-BISSAU
SIERRA LEONE
LIBERIA
FERANDO DE NORONHA (BRAZIL)
Interrupted Sinusoidal Equal-Area Projection